Hepatitis C

METHODS IN MOLECULAR BIOLOGY™

John M. Walker, SERIES EDITOR

METHODS IN MOLECULAR BIOLOGY™

Hepatitis C

Methods and Protocols

Second Edition

Edited by

Hengli Tang, PhD

Florida State University, Tallahassee, FL, USA

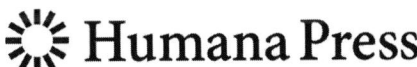 Humana Press

Editor
Hengli Tang
Florida State University
Tallahassee
FL, USA
tang@bio.fsu.edu

Series Editor
John M. Walker
University of Hertfordshire
Hatfield, Herts
UK

ISBN: 978-1-58829-970-3   e-ISBN: 978-1-59745-394-3
ISSN: 1064-3745   e-ISSN: 1940-6029
DOI 10.1007/978-1-59745-394-3

Library of Congress Control Number: 2008938170

Printed on acid-free paper

springer.com

Foreword I

When I was first approached by Dr. Hengli Tang about writing the foreword to this book, the second edition of *Hepatitis C Virus*, I was a little puzzled: Book forewords are penned by persons far more famous than I am. Surely there were better people to write this, with a handful of the true visionaries in the field immediately coming to my mind. However, after some thought and self-reflection, I realized that I might just be the right person for the job.

I remember the publication of the first edition of *Hepatitis C Virus* in 1998, the year I began reading about hepatitis C virus with great interest. I purchased a copy of the first edition from my meager graduate student stipend and read it from cover to cover, along with a variety of hepatitis C publications, to try to get a feel if this hepatitis C virus field was something I might want to work on in the future. At the time, the first edition was among the most comprehensive collection of hepatitis C protocols and methods available in one volume. I was amazed at the progress that had been made in hepatitis C research, and even more amazed as I learned more about the experimental challenges facing the field at that time. Here was a system with only a single infectious molecular clone, no cell culture replication system, no infectious virus culture system, no small animal model, and only very limited structural information on the viral NS3 protein. Despite these significant limitations, a considerable amount of information had been gathered on this virus using patient data, bioinformatics, surrogate expression systems, the chimpanzee model, and a vast array of in vitro methods. Even at that time, it was clear to me (and, I am sure, to many others) that HCV was an unusual and fascinating virus. The previous edition of this book helped me understand the major limitations facing the field, and highlighted the lengths and efforts required to make progress on what was then one of the most challenging areas of virology ... both of which were invaluable in focusing my later research directions. I joined the hepatitis C field in early 2000, at what might have been the perfect time, with cell culture replicon systems making the study of RNA replication tractable, and I have watched the field explode with amazing discoveries since that time. After this little bit of retrospection, I thought I might actually be a good person to put the second edition of *Hepatitis C Virus* in perspective. I have been actively working at the bench for the last eight years, witnessing many of the new discoveries and technologies outlined in this second edition. I have been lucky enough to have been involved in a few of these projects, and observed many others at conferences and in the scientific literature with great respect for the skill and insight of my colleagues. At every turn, hepatitis C virus and the dedicated people who study it continue to amaze me.

So, here we are with a brand new second edition of *Hepatitis C Virus*, 10 years after the publication of the first edition. It is clearly time for a new edition of this book, as so much of hepatitis C research has changed in the past decade. I have already mentioned some of the shortcomings of hepatitis C research ca. 1998, mostly focusing on the significant issue of studying a virus that was clearly infectious in humans and chimpanzees, yet that did not replicate or produce infectious virions in the cell culture environment. Perhaps the biggest breakthrough in hepatitis C virus research in the last decade is the development of efficient cell culture replicon–based replication systems, and more recently, the development of a cell culture infectious virus system. These two developments, separated by more than half a decade, have opened hepatitis C research to many labs and allowed for detailed genetic and functional analysis of the hepatitis C life cycle. The replicon has evolved to an efficient system allowing for forward and reverse genetic screens, and the information garnered from these types of experiments has fueled incredible discoveries. This volume includes chapters on the use of replicons, identification of revertant replicons by RT-PCR sequencing, and inhibition of replication by siRNAs. There are also chapters on applying biochemistry to the replicon system, including the visualization and purification of the active RNA replication complex from cells, the analysis of NS5A phosphorylation in replication experiments, and the development of in vitro replication systems. Mathematical modeling of the kinetics of RNA replication is also described. The replicon has also served as a valuable screening tool for small molecule anti-viral development. At the time the last edition of this book was published, interferon-ribavirin was state of the art, with pegylated interferon and ribavirin therapy arriving three years later. These therapies have increase sustained virologic responses in patients dramatically, but more work is clearly needed. This volume describes the development of drug screening replicon tools used and in use to identify more effective specific small molecule anti-virals. The development of anti-viral resistance upon exposure of replicons to these new anti-virals, as described herein, has allowed rapid identification of potential compound binding sites and aided in drug development, as well as highlighted the need for more anti-virals. The virus system has also allowed for significant advancements in our understanding of hepatitis C biology. This book features chapters on the use of the infectious virus system, measuring virus infectivity, generating chimeric viruses, and adaptation of virus to cell culture to increase the utility of the system. The virus system has led to interesting discoveries in the area of the functions of p7 and NS2 to virus production, and more recently, the roles of NS5A and NS3 in this process. It also features sections on the application of the infectious cell culture virus system and the pseudoparticle systems to the poorly understood areas of virus receptors and entry. The vast majority of replicon and virus research has been conducted on a variety of adapted subclones of a human hepatoma cell line. Although these lines have been incredibly useful in characterizing many aspects of hepatitis C biology, they are clearly quite different than authentic human hepatocytes. A chapter in this book describes recent advances in the infection of isolated primary hepatocytes in cell culture allowing for a more direct view of the virus lifecycle in the presumed authentic target cell.

The structural biology of hepatitis C has seen great advances in the last decade, progressing from isolated structures of the NS3 protease and helicase domain at the time the first edition was published to where we are today with crystal structures of NS2, the complete NS3/NS4A complex, a portion of NS5A, and NS5B. It is worth noting that structures of NS3 and NS5B with ligands ranging from nucleic acids to small molecule inhibitors have greatly enhanced our knowledge of these proteins. NMR structures have also been determined for the membrane anchors of NS5A and NS5B and provided a view of a portion of the core protein. Cryo-electron microscopy and structural biology of the viral RNA in the IRES region has greatly enhanced our understanding of how HCV RNA functionally engages the ribosome. The text in this edition focuses on the preparation of NS3, NS5A, and NS5B for structural studies, as well as some exciting new work on the structural characterization of the p7 protein.

Immunology has seen great progress as well. The investigation of the complex interplay between the virus and the host immune response has produced some amazing and unexpected results. It is clear that hepatitis C plays an active role in immune evasion by a variety of methods including mutation-based evasion of immune recognition, cleavage of a number of innate immune response pathway adaptor proteins, manipulation of host cell signaling, modulation of cytokine levels, masking of virions with lipids, and many other fascinating mechanisms that remain to be discovered. It is increasingly clear that the maintenance of long-term persistent infections requires a complex, multi-faceted approach by the virus … one we are only beginning to decipher. This text features chapters on the interplay of hepatitis C with IRF3 signaling and the characterization of the immune response to infection, as well as a chapter dealing with the exciting idea of using of mass spectrometry-based analysis of hypervariable regions of the hepatitis C proteins.

Another area that has seen significant development in the past decade has been the technology used to detect and quantitate HCV RNA and proteins in both clinical and research-based environments. The sensitivity and reproducibility of these assays have made great strides in the past decade, allowing very accurate measures of viral load in a clinical setting. It has become commonplace to monitor RNA and protein levels in the patient, genotype the patient's infection, and apply this information to optimize therapy. A chapter in this text compares these quantitation methods. Our understanding of hepatitis C genotype distribution, in light of the large numbers of HCV isolates sequenced to date, has continued to evolve, and the current nomenclature and classifications are also presented. Instead of trying to be exhaustive and provide methods for all the work that has been performed since the publication of the first edition, which would be impossible, this book focuses on significant methods that can be applied to a wide range of projects. Looking back at the last decade, the discovery and progress toward understanding hepatitis C has occurred at an incredible pace. This is largely due to the incredibly bright, dedicated, and creative people working on this important pathogen. I can hardly wait to see what the next edition of this book will contain. Where do I reserve my copy of the 3rd edition?

Timothy Tellinghuisen
Jupiter, Florida
June 2008

Foreword II

Our understanding of hepatitis C virus (HCV) has been largely restricted by the availability of tools to dissect the virus and its relationship with the host. Thus, new aspects of HCV biology have been revealed through a number of technological breakthroughs and clever experimental designs. As we close in on the third decade of HCV research, we can be sure that these advances have carried us a long way. The study of HCV began with its identification through cDNA cloning of the viral genome. From there, the viral genes and their gene products were studied, the first chimp-infectious cDNA clones were constructed, and subgenomic replicons were built. The importance of this last point cannot be understated, as it ushered in a new era of studying viral replication, aided in the discovery of HCV-specific inhibitors, and most recently, culminated in the development of systems to grow infectious HCV in cell culture. At the host-pathogen interface, we have identified some of the cellular machinery necessary for infection and genome replication, gained a basic understanding of innate and acquired immune responses to the virus, and learned of some tricks the virus uses to counteract these defenses. Of no less importance, we have developed small animal models of HCV infection and improved our understanding of HCV dynamics in vivo.

In many ways, the book you hold in your hands reflects the progress of discovery in HCV research. In the first part we learn how to detect the virus (Chapters 1–3), characterize the viral genome (Chapters 4–6), and study its gene products (Chapters 7–10, 16). The introduction of HCV replicons (Chapter 11) provides us with the basic tools to examine functional HCV RNA replication complexes (Chapters 12–15) and develop inhibitors of their function (Chapters 17–19). Expanding our view of the virus life cycle, we learn how to study HCV entry by using surrogate systems (Chapters 20–22) and to tame the virus in cell culture (Chapters 23–27). We next turn to understanding the host-pathogen interaction, with protocols to study HCV infection in culture primary hepatocytes (Chapter 28) and mice bearing human liver grafts (Chapter 29), as well as a series of methods to study the adaptive immune response to the virus (Chapters 30–32). Tying all of this together, Dahari and colleagues demonstrate how mathematical modeling can be used to provide the big picture of HCV dynamics in vivo.

Still, our understanding of HCV remains incomplete, and there is much work to be done. This book is not so much a chronicle of the past as it is a guide to the future. The protocols here represent the state of the art in HCV research, and importantly, were all written by researchers directly involved in developing these techniques.

Congratulations to Hengli for editing this volume, and a sincere thanks to my fellow contributors. To readers of the book, I say "Good luck," but must confess a certain kind of envy ... I only wish I had been able to find such a great book of protocols years ago!

Brett Lindenbach
New Haven, Connecticut
September 2008

Preface to the Second Edition

The first edition of this book, published in 1998, presented a large collection of authoritative protocols from the first decade of HCV research, outlining clinical diagnostics, genotyping, and molecular biology of the virus. One of the largest obstacles to productive research on the HCV life cycle had been the lack of a robust cell-culturing system capable of HCV replication and infection. We now possess such systems, and as a result, the past decade has witnessed an explosion of innovative research directed at understanding HCV biology and, more specifically, RNA replication in cell-culture replicon models. More recent advances in the production of infectious particles and infection of cultured hepatoma cells as well as chimpanzees will surely open a new chapter of studies examining the full life cycle of the virus by means of the new infection model. This advance is especially important for traditional virologists and immunologists, as they are now able to study viral assembly, entry, and pathogenesis as well as the innate and adapted immunity of the host. In other words, we have finally reached a point where virtually all the important questions regarding this human pathogen and its interaction with the host can be addressed in a laboratory setting.

This new edition of *Hepatitis C* strives to reflect this major thrust in the basic research of HCV and to provide a compilation of cutting-edge research techniques that are currently used by HCV labs worldwide to study HCV infection and replication in vitro. In addition, in keeping with the style of the first edition, the book opens with updated chapters discussing important methods of clinical detection and quantification of HCV infection. These are followed by several chapters detailing current systems for classifying HCV genotypes, numbering HCV sequences, and determining quasispecies. Finally, the volume concludes with protocols for the characterization of immune responses to HCV and a unique chapter describing mathematical modeling of HCV RNA kinetics. We hope that both basic scientists and clinical researchers will find this book useful as a reference guide to lab research using HCV as a model.

I would like to express my gratitude to Dr. Anne B. Thistle of the Department of Biological Science at Florida State University for her tremendous efforts in providing excellent language editing assistance, to all the authors for their understanding and support, and to my mentors, Profs. Z. G. Li and Flossie Wong-Staal for their guidance. Last but certainly not least, I would like to thank my wife Jian, my daughter Ashley (Ting-Ting), and my son Nathan for always being there for me.

Hengli Tang
London, England
August 2008

Contents

Contributors

GRAEME J. M. ALEXANDER • *Cambridge Hepatobiliary Service, Addenbrooke's Hospital, Cambridge, UK*

FILIPPO ANSALDI • *Department of Health Sciences, University of Genoa, Genoa, Italy*

MELISSA AYERS • *Hospital for Sick Children, Toronto, Ontario, Canada*

RALF BARTENSCHLAGER • *Department of Molecular Virology, University of Heidelberg, Heidelberg, Germany*

BIRKE BARTOSCH • *ENS de Lyon, Lyon, France*

MATTHEW BOTTOMLEY • *Istituto di Ricerche di Biologia Molecolare "P. Angeletti," Pomezia (Rome), Italy*

VOLKER BRASS • *Department of Medicine II, University of Freiburg, Freiburg, Germany*

ANDREA CARFÍ • *Istituto di Ricerche di Biologia Molecolare "P. Angeletti," Pomezia (Rome), Italy*

KYUNGSOO CHANG • *Department of Microbiology, Immunology and Molecular Genetics, University of Kentucky, Lexington, KY, USA*

ISABELLE CHEMIN • *INSERM U871, Lyon, France*

PHILIPPE CHEVALLIER • *Laboratoire de virologie, Hospices Civils, Lyon, France*

CHRISTOPHE CHIPOT • *Equipe de Dynamique des Assemblages Membranaires, Université Henri Poincaré, Vandoeuvre-lès-Nancy Cedex, France*

FRANÇOIS-LOÏC COSSET • *ENS de Lyon, Lyon, France*

MARTIN D. CURRAN • *Health Protection Agency, Cambridge Microbiology and Public Health Laboratory, Addenbrooke's Hospital, Cambridge, UK*

HAREL DAHARI • *Theoretical Biology and Biophysics, Los Alamos National Laboratory, Los Alamos, NM, USA*

STEFANIA DI MARCO • *Istituto di Ricerche di Biologia Molecolare "P. Angeletti," Pomezia (Rome), Italy*

MARLÈNE DREUX • *Université de Lyon, (UCB-Lyon1), IFR128, Lyon, France; INSERM, U758, Lyon, France; Ecole Normale Supérieure de Lyon, Lyon, France*

ROHIT DUGGAL • *Department of Virology, Pfizer Global Research and Development, Pfizer Inc., San Diego, CA, USA*

JOHN ELYAR • *Department of Pathology, Immunology, and Laboratory Medicine, University of Florida, Gainesville, FL, USA*

MIKE FLINT • *Wyeth Antiviral Research, Pearl River, NY, USA*

ROBIN FOSS • *Department of Pathology, Immunology, and Laboratory Medicine, University of Florida, Gainesville, FL, USA*

OIVIER GALY • *INSERM U871, Lyon, France*

RAINER GOSERT • *Transplantation Virology, Institute for Medical Microbiology, University of Basel, Basel, Switzerland*

ERIC HALL • *CellzDirect, Inc., Pittsboro, NC, USA*

WEIDONG HAO • *Pfizer Global Research and Development, Pfizer Inc., San Diego, CA, USA*

ALAN HEMMING • *Department of Surgery, University of Florida, Gainesville, FL, USA*

MINGJUN HUANG • *Achillion Pharmaceuticals, 300 George Street, New Haven, CT, USA*

GIANCARLO ICARDI • *Department of Health Sciences, University of Genoa, Genoa, Italy*

TATSUO KANDA • *Division of Infectious Diseases & Immunology, Department of Internal Medicine, Saint Louis University, St. Louis, MO, USA*

ARTUR KAUL • *Department of Molecular Virology, University of Heidelberg, Heidelberg, Germany*

NORMAN M. KNETEMAN • *Department of Surgery, Faculty of Medicine and Oral Health Sciences, University of Alberta, Edmonton, Canada*

CARLA KUIKEN • *Los Alamos National Laboratory, Los Alamos, NM, USA*

EDWARD L. LECLUYSE • *CellzDirect, Inc., Pittsboro, NC 27312, USA*

HELENE LE GUILLOU-GUILLEMETTE • *Laboratory of Virology, University Hospital of Angers and Laboratory HIFI, University of Angers, France*

STANLEY M. LEMON • *Center for Hepatitis Research, Institute for Human Infections and Immunity, and the Department of Microbiology & Immunology, University of Texas Medical Branch, Galveston, TX, USA*

KUI LI • *Department of Microbiology and Immunology, University of Texas Medical Branch, Galveston, TX, USA*

BRETT D. LINDENBACH • *Section of Microbial Pathogenesis, Yale University School of Medicine, New Haven, CT, USA*

CHEN LIU • *Department of Pathology, Immunology, and Laboratory Medicine, University of Florida, Gainesville, FL, USA*

VOLKER LOHMANN • *University of Heidelberg, Heidelberg, Germany*

FRANÇOISE LUNEL-FABIANI • *Laboratory of Virology, University Hospital of Angers and Laboratory HIFI, University of Angers, France*

GUANGXIANG LUO • *Department of Microbiology, Immunology and Molecular Genetics, University of Kentucky, Lexington, KY, USA*

NICOLE LYANDRAT • *IARC, Lyon, France*

JOSEPH MARCOTRIGIANO • *Department of Chemistry and Chemical Biology, Rutgers University, Piscataway, NJ, USA*

ROLAND MONTSERRET • *Institut de Biologie et Chimie des Protéines, IFR128 BioSciences Gerland-Lyon Sud, Université de Lyon, Lyon, France*

DARIUS MORADPOUR • *Division of Gastroenterology and Hepatology, Centre Hospitalier Universitaire Vaudois, University of Lausanne, Lausanne, Switzerland*

PETRA NEDDERMANN • *Istituto di Ricerche di Biologia Molecolare "P. Angeletti," Pomezia (Rome), Italy*

HEATHER B. NELSON • *Department of Biological Science, Florida State University, Tallahassee, FL, USA*

FRANÇOIS PENIN • *Institut de Biologie et Chimie des Protéines, IFR128 BioSciences Gerland-Lyon Sud, Université de Lyon, Lyon, France*

ALAN S. PERELSON • *Theoretical Biology and Biophysics, MS-K710, Los Alamos National Laboratory, Los Alamos, NM, USA*

THOMAS PIETSCHMANN • *TWINCORE , Center for Experimental and Clinical Infection Research, Department of Experimental Virology, Hannover, Germany*

RATNA B. RAY • *Department of Pathology, Saint Louis University, St. Louis, MO, USA*

BARBARA REHERMANN • *Immunology Section, Liver Diseases Branch, National Institute of Diabetes and Digestive and Kidney diseases, National Institutes of Health, Bethesda, MD, USA*

RUY M. RIBEIRO • *Theoretical Biology and Biophysics, Los Alamos National Laboratory, Los Alamos, NM, USA*

JASON M ROBOTHAM • *Department of Biological Science, Florida State University, Tallahassee, FL, USA*

KATHRYN J. ROLFE • *Health Protection Agency, Cambridge Microbiology and Public Health Laboratory, Addenbrooke's Hospital, Cambridge, UK*

EDWIN RYDBERG • *Istituto di Ricerche di Biologia Molecolare "P. Angeletti," Pomezia (Rome), Italy*

NATHALIE SAINT • *Centre de Biochimie Structurale, Université de Montpellier I et II, Montpellier, France*

MASAAKI SHIINA • *Immunology Section, Liver Diseases Branch, National Institute of Diabetes and Digestive and Kidney diseases, National Institutes of Health, Bethesda, MD, USA*

EMI SHUDO • *Theoretical Biology and Biophysics, MS-K710, Los Alamos National Laboratory, Los Alamos, NM, USA*

PETER SIMMONDS • *Centre for Infectious Diseases, University of Edinburgh, Summerhall, Edinburgh, UK*

HENGLI TANG • *Department of Biological Science, Florida State University, Tallahassee, FL, USA*

RAYMOND TELLIER • *Hospital for Sick Children, Toronto, Ontario, Canada*

TIMOTHY TELLINGHUISEN • *Department of Infectology, The Scripps Research Institute, Jupiter, FL, USA*

CHRISTIAN TOSO • *Department of Surgery, Faculty of Medicine and Oral Health Sciences, University of Alberta, Edmonton, Canada*

DONNA M. TSCHERNE • *Department of Microbiology, Mount Sinai School of Medicine, New York, NY, USA*

ISABELLE VINCENT • *INSERM U871, Lyon, France*

TAKAJI WAKITA • *Second Department of Virology, National Institute of Infectious Diseases, Tokyo, Japan*

TIANYI WANG • *Department of Infectious Diseases and Microbiology, University of Pittsburgh, Pittsburgh, PA, USA*

FLOSSIE WONG-STAAL • *ItheX Pharmaceuticals, Inc., San Diego, CA, USA*

ILKA WÖRZ • *Department of Molecular Virology, University of Heidelberg, Heidelberg, Germany*

TIM G. WREGHITT • *Health Protection Agency, Cambridge Microbiology and Public Health Laboratory, Addenbrooke's Hospital, Cambridge, UK*

JIAN-PING YANG • *Immusol, Inc., San Diego, CA, USA*

WENGANG YANG • *Achillion Pharmaceuticals, New Haven, CT, USA*

CARMEN YEA • *Hospital for Sick Children, Toronto, Ontario, Canada*

MINKYUNG YI • *Center for Hepatitis Research, Institute for Human Infections and Immunity, and the Department of Microbiology & Immunology, University of Texas Medical Branch, Galveston, TX, USA*

JOO CHUN YOON • *Immunology Section, Liver Diseases Branch, National Institute of Diabetes and Digestive and Kidney diseases, National Institutes of Health, Bethesda, MD, USA*

DEMIN ZHOU • *Immusol, Inc., San Diego, CA, USA*

HAIZHEN ZHU • *Department of Pathology, Immunology, and Laboratory Medicine, University of Florida, Gainesville, FL, USA*

Color Plate

Color Plate 1.   Sequential construction of the monomeric p7 protein. (**a**) Cartoon representation of the protein sequence and secondary-structure elements. (**b**) Protein–protein molecular docking of the TM1 and TM2 α-helices. (**c**) Inclusion of the loop domain, yielding the hairpin motif. (**d**) The N-terminal α-helix is added to the hairpin. (**e**) The model is completed by insertion of the remaining unstructured C-terminal segment. (**f**) Incorporation of the protein in a fully hydrated POPC phospholipid bilayer shown in semitransparent van der Waals spheres. The N-terminal α-helix and the unstructured C-terminal part protrude into the aqueous medium of the ER lumen, whereas the loop region interacts with the head groups of the membrane. (*see* discussion on p. 138)

Part I

Detection and Quantification of HCV

Chapter 1

Detection and Quantification of Serum or Plasma HCV RNA: Mini Review of Commercially Available Assays

Helene Le Guillou-Guillemette and Francoise Lunel-Fabiani

Abstract

The treatment schedule (combination of compounds, doses, and duration) and the virological follow-up for management of antiviral treatment in patients chronically infected by HCV is now well standardized, but to ensure good monitoring of the treated patients, physicians need rapid, reproducible, and sensitive molecular virological tools with a wide range of detection and quantification of HCV RNA in blood samples. Several assays for detection and/or quantification of HCV RNA are currently commercially available. Here, all these assays are detailed, and a brief description of each step of the assay is provided. They are divided into two categories by method: those based on signal amplification and those based on target amplification. These two categories are then divided into qualitative, quantitative, and quantitative detection assays. The real-time reverse-transcription polymerase chain reaction (RT-PCR)–based assays are the most promising strategy in the HCV virological area.

Key words: HCV RNA, detection and quantification, commercialized assays.

1. Introduction

HCV is one of the major causes of chronic liver diseases; 170 million people are estimated to be infected by this virus. A prominent characteristic of HCV is its genetic variability, due to the lack of 5′ exonuclease activity and a high replication level (1). These two features have consequences for therapeutic strategy and molecular diagnosis tests, and they form the basis of HCV's classification into six clades (comprising 11 genotypes and more than 70 subtypes) (2). More than 80% of HCV infections lead to chronic liver disease that may evolve into cirrhosis and hepatocellular carcinoma. Interferon-α was used first to treat HCV infection.

Hengli Tang (ed.), *Hepatitis C: Methods and Protocols, Second Edition, vol. 510*
© 2009 Humana Press, a part of Springer Science+Business Media
DOI 10.1007/978-1-59745-394-3_1 Springerprotocols.com

Current antiviral therapy combines pegylated interferon-α and ribavirin and leads to a level of sustained virological response of around 60%. The chance of success depends strongly on HCV genotype and the pretreatment viral load. Numerous studies have reported a correlation between HCV viral load and the progression of the liver disease, and the level of sustained response ranges from 52% for genotypes 1 and 4 to more than 85% for genotypes 2 and 3 *(3–5)*. Moreover, treatments and virological follow-up during the therapeutic course are designed according to the genotype; detection of HCV RNA and quantification of the viral load under treatment are essential to adequate management or adaptation of the therapeutic schedule: These virological parameters provide to the clinician the possibility of adjusting dose and duration of treatment and managing therapeutic choices in nonresponders and relapsing patients. Algorithms are now well established, and they clearly define how to treat an active HCV infection depending on the HCV genotype, the initial viral load level, and the degree of liver disease. Moreover, quantitative detection of HCV RNA appears to be the gold standard for management of patients under and after treatment. The positive predictive value of a decline of 2 log IU/mL or an undetectable viral load at month 1 and month 3 under therapy is particularly relevant in the therapy response in genotypes 1 and 4 HCV-infected patients. Conversely, discontinuation of the antiviral therapy is recommended when the HCV RNA at month 3 shows no decrease from baseline *(6)*. Recently, the predictive value of viral load at week 4 has been studied in patients coinfected with HCV and HIV, or not coinfected, and it seems relevant to the chance of therapeutic success *(7, 8)*. Detection of HCV RNA 6 months after the end of treatment permits assessment of the virological response; an undetectable viral load defines a sustained virological responder.

Detection and quantification of HCV RNA have become useful tools for management of patients chronically infected with HCV. These molecular tests are particularly relevant (1) for assessment of an active HCV infection, (2) for the decision to initiate treatment, (3) for predicting the efficiency of the antiviral therapy, and (4) for determining the virological response 6 months after the end of therapy. Rapid, sensitive, and standardized automated assays for follow-up on patients treated for HCV seem very important. For more than 10 years, standardized assays have been available for detection or quantification of HCV RNA. We are now at an important turn in the area of virological diagnosis, because the recent development of real-time nucleic acid amplification assays allows qualitative and quantitative detection of HCV RNA in the same run.

Here, we review the different commercially available assays that detect and quantify HCV viral load. We detail main characteristics of each assay and describe their technological features and performances.

2. HCV RNA Detection and Quantification

2.1. General Performance Characteristics

All HCV RNA detection and quantification tests must have certain characteristics, whatever their technical features.

To standardize all molecular tests, the First International Standard for HCV RNA has been established by the World Health Organization (WHO) Collaborative Study Group and the WHO Expert Committee on Biological Standardization (9). This standard has established international units per milliliter as the only unit of measure to be used in the assays. This standardization simplifies the follow-up on infected patients, providing an easy comparison of the HCV viremia level, independent of the assay. All commercially available assays include an internal standard or an external quantification panel quantified against the international standard. Furthermore, every assay must quantify HCV RNA viral load equally, regardless of the HCV genotype.

The standard also specifies minimum sensitivity. The minimal threshold limit required is 50 IU/mL for quantitative assays and tends to be around 15 IU/mL for new real-time quantitative assays. This low limit of detection/quantification allows confidence in defining sustained virological response 6 months after the end of treatment in patients maintaining an undetectable HCV RNA load. Qualitative commercially available assays do not have the same sensitivity, and the threshold must be considered in the implementation of an HCV RNA detection or quantification test in a lab.

Strong specificity is also required. Despite HCV's great genetic diversity, the assay must detect all genotypes, whatever the level of viral load. This requirement always presents a challenge for the development of a new molecular assay in the HCV diagnosis domain. In addition, the linear range of HCV RNA detection should be large, so that samples with high viral loads need not be diluted. The minimal linear range currently required for HCV RNA detection ranges from 50 IU/mL to 6 or 7 log IU/mL. It must be applicable in all virological situations (untreated patient, treated patient, or relapser).

Finally, the assay's repeatability and reproducibility must be known; even treated patients are followed in the same laboratory with the same assay. We have evaluated reproducibility by comparing replicates of HCV RNA-positive samples carried out by several operators in separate runs at different times and repeatability by measuring replicates of the same HCV RNA-positive sample in the same run by the same operator; runs were generally repeated three times. These features are essential because differences between replicates must be minimal if the true variations of HCV RNA level during the virological monitoring of HCV-infected patients are to be detected; a variation of 0.5 log UI/mL is a significant value in therapeutic follow-up.

2.2. Description of the Commercially Available Tests

2.2.1. Materials

Detection and quantification of HCV RNA can be carried on serum or plasma. After venous punctures, samples must be sent to the laboratory within 2 h, under refrigeration. After centrifugation, plasma or serum must be frozen rapidly and stored at −80°C unless the molecular test is performed immediately.

2.2.2. Commercially Available Assays for HCV RNA Detection and/or Quantification

All methods currently commercially available belong to one of the two general methods of molecular assay: amplification of the signal (branched DNA, bDNA) or amplification of the target (transcription-mediated amplification, TMA; classical reverse-transcription polymerase chain reaction, RT-PCR; and real-time RT-PCR), and can therefore be classified into four categories, depending on their features. Another possible classification is that methods are either qualitative (HCV RNA detection) or quantitative (HCV viral load determination). The classifications and main characteristics of all the commercialized assays are shown in **Table 1.1**

2.2.2.1. Method Based on Signal Amplification: Branched DNA (bDNA)

The Versant® HCV RNA 3.0 Assay is a bDNA assay commercialized by Siemens Medical Solutions Diagnostics; it involves a signal-amplification nucleic acid probe assay strategy and could be compared to an enzyme immunosorbent assay (ELISA) in a schematic view. It is quantitative *(10)*.

Extraction: HCV RNA is released by chemical lysis of the virions and directly hybridized in capture wells coated with specific oligonucleotidic capture probes (8 strips of 12 wells per kit); 15 h are required for this lysis-hybridization phase.

Amplification: The signal is amplified by a succession of probe hybridizations that are sequentially added to each well. Captured HCV RNA is first hybridized in a sandwich manner by a set of target probes. The viral genomic domains 5′ untranslated region (5′UTR) and Core are the targets of the capture and target probes. Target probes then bind to preamplifier probes, which in turn hybridize to amplifier probes, and finally, a branched DNA complex is built.

Detection: Probes labeled by alkaline phosphatase are added and hybridize to the DNA complex, and incubation with a chemiluminescent substrate can achieve the detection of RNA in the initial sample. At this stage, a strong signal for each HCV RNA molecule is detectable. A calibration curve is established at each run with five standards and permits the quantification of HCV RNA in each patient's sample.

Table 1.1

Main characteristics of each commercially available assay for HCV RNA detection and/or quantification. Siemens, Medical Solutions Diagnostics; Roche, Roche Molecular Systems; Abbott, Abbott Molecular Diagnostics. Versant 3.0, Versant® HCV RNA 3.0 Assay; Amplicor v2.0, Amplicor® HCV v2.0; COBAS v2.0, COBAS® Amplicor® HCV v2.0; Versant, Versant® HCV Qualitative Assay; LCx, LCx® HCV RNA Quantitative Assay; Amplicor, Amplicor® HCV Monitor v2.0; COBAS, COBAS® Amplicor® HCV Monitor v2.0; Abbott, Abbott RealTime™ HCV Assay; COBAS Taq, COBAS® TaqMan HCV Test

Assay principle	Assay results	Assay method	Supplier	Commercial name	Automation level	Lower limit of detection (qualitative, IU/mL)	Lower limit of quantification (IU/mL)	Range of quantification (IU/mL)
Signal amplification	Quantification	bDNA	Siemens	Versant 3.0	Average[a]		615	615–7,700,000
Target amplification	Detection	RT-PCR	Roche	Amplicor v2.0	Manual	50		
		RT-PCR	Roche	COBAS v2.0	Semiautomated[b]	50		
		TMA	Siemens	Versant	Manual	10		
	Quantification	RT-PCR	Abbott	LCx	Average[c]		25	25–2,630,000
		RT-PCR	Roche	Amplicor	Manual		600	600–500,000
		RT-PCR	Roche	COBAS	Semiautomated[b]		600	600–500,000
	Quantitative detection	Real-time RT-PCR	Abbott	Abbott	High[d]	12 (or 30)[f]	12 (or 30)[f]	12–100,000,000
		Real-time RT-PCR	Roche	COBAS Taq	High[e]	15	43	43–69,000,000

[a] On the Versant 440 Molecular System. [b] On the COBAS®. [c] On the LCx Analyzer. [d] On the Abbott m2000sp and Abbott m2000rt. [e] On the COBAS Ampliprep and COBAS TaqMan 48. [f] Depending on the sample volume: 0.2 mL or 0.5 mL.

Technical characteristics: The limit of quantification is 615 IU/mL and that of detection is 1 IU/mL. All genotypes are quantified with good accuracy. Advantages of the Versant® HCV RNA 3.0 Assay are its high specificity, good reproducibility, and standardization. Major drawbacks are its low sensitivity, the use of external control (and not an internal control, which would provide better traceability), and the systematic need for five standards per run *(11, 12)*.

2.2.2.2. Methods Based on Target Amplification

Depending on the stage of the HCV disease and the time of treatment, HCV RNA level may be too low for detection by classical hybridization assays. A method that can increase sensitivity is based on target amplification, that is, production of a large number of viral genome copies during a cyclic enzymatic reaction. Three such methods are available: conventional RT-PCR, TMA, and real-time RT-PCR.

The genomic target for amplification is generally the 5′UTR. Commercially available tests are well standardized and have low limits of detection. All genotypes are detected. An advantage of these tests is relatively high sensitivity, but they share with all PCR-based tests the disadvantages of lack of reproducibility with samples of low HCV viral load and the risk of false-positive reactions due to contamination during sample preparation or hybridization of amplified products. Internal controls must therefore be used to detect false-negative results due to amplification inhibitors in the sample.

2.2.2.2.1. Transcription-Mediated Amplification (TMA)

TMA, a qualitative assay developed by Siemens Medical Solutions Diagnostics, belongs to the second class of methods based on target amplification.

Extraction: The Versant® HCV RNA Qualitative Assay starts with lysis of the sample, which releases HCV RNA from virions. The internal control is also added during this phase.

Amplification: Viral genomes are then captured by magnetic particles coated with oligonucleotides hybridizing with the 5′UTR HCV region. HCV RNA is copied in anti-sense single-stranded RNA during an isothermal TMA process using a T7 RNA polymerase and adequate primers.

Detection: Finally, RNA amplification products are detected by chemiluminescent signal, determined by a ratio of signal-to-cutoff signal.

Technical characteristics: The linear range of detection is narrow and ranges from 10 to 615 IU/mL. This assay is currently the most widely available sensitive test among qualitative assays, particularly useful for determining sustained virological response to antiviral therapy *(13, 14)*.

2.2.2.2.2. Classical Reverse-Transcription Polymerase Chain Reaction (RT-PCR)

Several commercially available assays based on target amplification and using the PCR method are available for either detection or quantification of HCV RNA. All the assays achieve a reverse transcription of the HCV RNA into cDNA, followed by an amplification step (in which many copies of double-stranded DNA are synthesized from the cDNA matrix) and an end-point detection of the amplification products by colorimetry, which differentiates HCV and internal control amplifications. The detection step is the major point differentiating classical PCR from the real-time PCR–based assays (developed below).

Roche Diagnostics proposes the manual Amplicor® HCV v2.0 and the semiautomated COBAS® Amplicor® HCV v2.0 for HCV RNA detection and the Amplicor® HCV Monitor v2.0 and the semiautomated COBAS® Amplicor® HCV Monitor v2.0 for HCV RNA quantification. All the assays commercialized by Roche are based on a competitive RT-PCR *(15)*.

Extraction: This first step consists of lysis with a chaotropic buffer, which releases HCV RNA from virions, followed by a precipitation of nucleic acid by alcoholic solution. Standard of detection or quantification is added with the lysis buffer, which further validates the phases of extraction, amplification, and detection.

Amplification: A one-step RT-PCR is then performed with a DNA polymerase combining a reverse transcriptase and DNA polymerase activities; a selective amplification is done with uracile-*N*-glycosylase and dUTP. Primer targets include both the 5′UTR of the HCV RNA of the patient and the synthetic RNA used as a standard. The reverse primer is labeled with biotin.

Detection/quantification: Magnetic beads coated with oligonucleotidic probes, which are specific for HCV or synthetic standard amplicon PCR products, are added to aliquots of the amplification mix, leading to a selective hybridization. Aliquots are then separately distributed into detection/quantification wells. Finally, a colorimetric reaction is performed by addition of the conjugate avidine–horseradish peroxydase and the substrate containing 3, 3′, 5, 5′-tetramethylbenzidine and H_2O_2.

Technical characteristics: Addition of an internal control before the assay begins ensures control of the entire process. These assays are sensitive, but a risk of contamination is always present *(16)*.

The LCx® HCV RNA Quantitative Assay from Abbott Molecular Diagnostics is the second RT-PCR–based test. It uses competitive RT-PCR followed by microparticle enzyme immunoassay (MEIA) detection to quantify HCV RNA *(17, 18)*.

Extraction: HCV RNA is extracted from a plasma or serum sample with a modified Qiagen® sample-preparation method. The process consists of a lysis step in a chaotropic buffer still containing a quantified internal standard RNA transcript, followed by multiple cleanings of the nucleic acid, capture of all the nucleic acids (RNA and DNA) in silica particles fixed in a minicolumn system, and finally an elution step with RNase-free water.

Amplification: Competitive RT-PCR is performed with a reaction mix containing DNA polymerase, HCV-specific unlabeled forward primer, HCV-specific carbazole-labeled reverse primer, HCV-specific adamantine-labeled probe, and internal standard–specific dansyl-labeled probe (whose target is the 5′UTR HCV region).

Detection: The last step occurs in the LCx® Analyzer, where aliquots of amplification products are captured by anti-carbazole-coated microparticles in reaction wells. These complexes are then detected by addition of conjugates (anti-adamantine alkaline phosphatase and anti-dansyl beta-galactosidase) and successive addition–detection of specific substrates (7-beta galactopyranosyl coumarin-4-acetic acid-2-hydroxyethylamide to detect the internal standard and 4-methylumbelliferyl phosphate to detect HCV amplification products); the resulting fluorescence is measured at each addition of substrate by LCx® Analyzer.

Technical characteristics: This assay provides an ultrasensitive quantification with a limit at 25 IU/mL. It also has a wide range, from 25 to 2,630,000 IU/mL, but this assay is slightly automated, and risks of contamination during the process are also present.

2.2.2.2.3. Real-Time RT-PCR

The real-time RT-PCR assays are the most recent technology developed and commercialized; assays designed for HCV RNA detection and quantification have only recently been approved for virological diagnosis. These methods are available for both qualitative and quantitative detection. Unlike that of classical RT-PCR, the amplification can be detected during the enzymatic reaction at each PCR cycle. The limit of detection is lower than those of classical RT-PCR and TMA. Quantitative measure occurs during the exponential phase of the amplification step; the quantity of the amplification products is proportional to the quantity of the HCV or internal standard initially present in the sample. Two tests are now available; both are developed on automated platforms combining extraction and amplification.

The COBAS® Taqman HCV test offered by Roche combines an extraction on the COBAS AmpliPrep™ followed by an amplification-quantitative detection on the COBAS Taqman 48 Analyer™ *(19)*.

Extraction: The HCV RNA is automatically extracted from controls and clinical samples; the HCV quantification standard is first added to each specimen to control the whole process. After a phase of incubation with protease, the addition of lysis buffer and magnetic glass particles permits the capture of RNA in the particles. The absorbed RNAs are then washed several times, and elution is finally performed with aqueous buffer. RNA extracts are transferred into 24-well microplates, RT-PCR master mix is distributed into each sample, and the plate is transferred into the COBAS® Taqman 48 Analyzer^{TM}.

Amplification real-time detection: An RT-PCR is carried out with primers hybridizing with the 5′UTR HCV region and Z05 polymerase combining reverse transcriptase and DNA polymerase activities. The quantification standard and HCV RNA are both amplified and lead to amplification products with the same length and base composition. Detection of amplification products is performed at each PCR cycle; the process relies on the use of two different dual-labeled fluorescent oligonucleotide probes, one of which binds to HCV RNA and the other to HCV standard quantification amplicons at the regions spanned by the primers. At each annealing-elongation step, the hybridized dual-labeled probe is cleaved by the Z05 polymerase, and the reporter is separated from the quencher, leading to fluorescence emission. Levels of fluorescence light for HCV RNA and HCV standard quantification are measured at each PCR cycle at different wavelengths. For each sample, the critical threshold values, that is, the earlier emissions of fluorescence detected above a certain level, are collected for HCV RNA and HCV standard quantification. The initial quantity of HCV RNA in the sample is then calculated by comparison of the two critical threshold values.

The Abbott RealTime^{TM} HCV assay is performed on the m2000sp^{TM} for sample extraction and m2000rt^{TM} for amplification-quantitative detection *(20)*.

Extraction: The automated extraction process starts with the addition of lysis buffer to patient samples, external controls, and quantification standards. The RNA internal control is also introduced at this step to control for recovery and inhibition during the process. HCV RNA is then captured by magnetic microparticles, washed to remove unbound components, and finally eluted. Aliquots of extracted nucleic acid samples are distributed into the wells of a 96-well optical reaction plate, and a master mix containing oligonucleotide primers and probes and reverse transcriptase polymerase enzyme is then added. The microplate is then covered and transferred to the thermocycler.

Amplification real-time detection: RT-PCR is performed on the m2000rt. The target of one specific set of primers and probes is the HCV RNA (5′UTR region) and that of another is the internal control. Probes are labeled with a fluorophore in the 5′ position and a quencher in the 3′ position; hybridization to the sequence target allows fluorescence emission and detection. As previously explained, a real-time detection occurs at each PCR cycle. Measurement of fluorescence at different wavelengths allows a simultaneous detection of the HCV and the internal control-amplified products. Quantification of the HCV viral load in the sample is calculated from a comparison between the critical threshold value of the sample and critical thresholds of quantification standards.

Technical characteristics: Roche and Abbott assay standardization is based on the Second WHO International Standard for HCV RNA. Several evaluations are currently in progress that will assess performances of these new assays. Studies generally show that the two commercially available real-time HCV RNA RT-PCR tests are sensitive, specific, accurate, and reproducible and have linear amplification over a broad dynamic range. Dilutions of samples with HCV RNA levels outside the quantification range in other technical processes will probably not be necessary with this novel technology. These assays combine all the quality at a higher performance level than other traditional assays so far available. Moreover, automation of the entire process leads to a high throughput of HCV RNA screening with a high safety level.

Recent publications report two critical points for the COBAS AmpliPrep/COBAS TaqMan HCV RNA Assay, however. An overestimation of HCV RNA levels has been identified in undiluted samples (at around 0.6 log IU/mL) and seems to increase with higher viral load. The assay may also underestimate loads of genotypes 2 and 4 (15 and 30% of the samples tested), probably because of a mismatch between probe or primers and HCV RNA target sequence *(21)*, but a good correlation between the two real-time RT-PCR assays has been demonstrated independently of the HCV genotype. They are therefore suitable for adequate clinical and therapeutic management of chronically infected patients *(20)*.

3. Conclusion

Highly sensitive HCV RNA quantitative-detection assays, all based on real-time RT-PCR methods, will be soon recommended for management of patients under anti-HCV therapy. Other similar types of tests will soon be commercialized by other firms and may include probes using other processes of fluorescence emission.

References

1. Steinhauer, D. A., Domingo, E., and Holland, J. J. (1992) Lack of evidence for proofreading mechanisms associated with an RNA virus polymerase. *Gene* **122**, 281–218.
2. Simmonds, P., Bukh, J., Combet, C., Deleage, G., Enomoto, N., Feinstone, S., Halfon, P., Inchauspe, G., Kuiken, C., Maertens, G., Mizokami, M., Murphy, D. G., Okamoto, H., Pawlotsky, J. M., Penin, F., Sablon, E., Shin, I. T., Stuyver, L. J., Thiel, H. J., Viazov, S., Weiner, A. J., and Widell, A. (2005) Consensus proposals for a unified system of

nomenclature of hepatitis C virus genotypes. *Hepatology* **42**, 962–973.

3. Hadziyannis, S. J., Sette, H., Jr., Morgan, T. R., Balan, V., Diago, M., Marcellin, P., Ramadori, G., Bodenheimer, H., Jr., Bernstein, D., Rizzetto, M., Zeuzem, S., Pockros, P. J., Lin, A., and Ackrill, A. M. (2004) Peginterferon-α2a and ribavirin combination therapy in chronic hepatitis C: a randomized study of treatment duration and ribavirin dose. *Ann. Intern. Med.* **140**, 346–355.

4. Fried, M. W., Shiffman, M. L., Reddy, K. R., Smith, C., Marinos, G., Goncales, F. L., Jr., Haussinger, D., Diago, M., Carosi, G., Dhumeaux, D., Craxi, A., Lin, A., Hoffman, J., Yu, J. (2002) Peginterferon alfa-2a plus ribavirin for chronic hepatitis C virus infection. *N. Engl. J. Med.* **347**, 975–982.

5. Manns, M. P., McHutchison, J. G., Gordon, S. C., Rustgi, V. K., Shiffman, M., Reindollar, R., Goodman, Z. D., Koury, K., Ling, M., and Albrecht, J. K. (2001) Peginterferon alfa-2b plus ribavirin compared with interferon alfa-2b plus ribavirin for initial treatment of chronic hepatitis C: a randomised trial. *Lancet* **358**, 958–965.

6. Berg, T., Sarrazin, C., Herrmann, E., Hinrichsen, H., Gerlach, T., Zachoval, R., Wiedenmann, B., Hopf, U., and Zeuzem, S. (2003) Prediction of treatment outcome in patients with chronic hepatitis C: significance of baseline parameters and viral dynamics during therapy. *Hepatology* **37**, 600–609.

7. Payan, C., Pivert, A., Morand, P., Fafi-Kremer, S., Carrat, F., Pol, S., Cacoub, P., Perronne, C., and Lunel, F. (2007) Rapid and early virological response to chronic hepatitis C treatment with IFN α-2b or PEG- IFN α-2b plus ribavirin in HIV/HCV co-infected patients. *Gut* **56**, 1111–1116.

8. Poordad, F., Reddy, K.R., Martin, P. (2008) Rapid virologic response: a new milestone in the management of chronic hepatitis C. *Clin. Infect. Dis.* **46**, 78–84.

9. Jorgensen, P. A., and Neuwald, P. D. (2001) Standardized hepatitis C virus RNA panels for nucleic acid testing assays. *J. Clin. Virol.* **20**, 35–40.

10. Trimoulet, P., Halfon, P., Pohier, E., Khiri, H., Chene, G., and Fleury, H. (2002) Evaluation of the VERSANT HCV RNA 3.0 assay for quantification of hepatitis C virus RNA in serum. *J. Clin. Microbiol.* **40**, 2031–2036.

11. Lunel, F., Cresta, P., Vitour, D., Payan, C., Dumont, B., Frangeul, L., Reboul, D., Brault, C., Piette, J. C., and Huraux, J. M. (1999) Comparative evaluation of hepatitis C virus RNA quantitation by branched DNA, NASBA, and monitor assays. *Hepatology* **29**, 528–535.

12. Veillon, P., Payan, C., Picchio, G., Maniez-Montreuil, M., Guntz, P., and Lunel, F. (2003) Comparative evaluation of the total hepatitis C virus core antigen, branched-DNA, and amplicor monitor assays in determining viremia for patients with chronic hepatitis C during interferon plus ribavirin combination therapy. *J. Clin. Microbiol.* **41**, 3212–3120.

13. Krajden, M., Ziermann, R., Khan, A., Mak, A., Leung, K., Hendricks, D., and Comanor, L. (2002) Qualitative detection of hepatitis C virus RNA: comparison of analytical sensitivity, clinical performance, and workflow of the Cobas Amplicor HCV test version 2.0 and the HCV RNA transcription-mediated amplification qualitative assay. *J. Clin. Microbiol.* **40**, 2903–2907.

14. Morishima, C., Morgan, T. R., Everhart, J. E., Wright, E. C., Shiffman, M. L., Everson, G. T., Lindsay, K. L., Lok, A. S., Bonkovsky, H. L., Di Bisceglie, A. M., Lee, W. M., Dienstag, J. L., Ghany, M. G., and Gretch, D. R. (2006) HCV RNA detection by TMA during the hepatitis C antiviral long-term treatment against cirrhosis (Halt-C) trial. *Hepatology* **44**, 360–367.

15. Morishima, C., Chung, M., Ng, K. W., Brambilla, D. J., and Gretch, D. R. (2004) Strengths and limitations of commercial tests for hepatitis C virus RNA quantification. *J. Clin. Microbiol.* **42**, 421–425.

16. Lunel, F., Mariotti, M., Cresta, P., De la Croix, I., Huraux, J. M., and Lefrere, J. J. (1995) Comparative study of conventional and novel strategies for the detection of hepatitis C virus RNA in serum: amplicor, branched-DNA, NASBA and in-house PCR. *J. Virol. Methods* **54**, 159–171.

17. Leckie, G., Schneider, G., Abravaya, K., Hoenle, R., Johanson, J., Lampinen, J., Ofsaiof, R., Rundle, L., Shah, S., Frank, A., Toolsie, D., Vijesurier, R., Wang, H., and Robinson, J. (2004) Performance attributes of the LCx HCV RNA quantitative assay. *J. Virol. Methods* **115**, 207–215.

18. Bertuzis, R., Hardie, A., Hottentraeger, B., Izopet, J., Jilg, W., Kaesdorf, B., Leckie, G., Leete, J., Perrin, L., Qiu, C., Ran, I., Schneider, G., Simmonds, P., and Robinson, J. (2005) Clinical performance of the LCx HCV RNA quantitative assay. *J. Virol. Methods* **123**, 171–178.

19. Sarrazin, C., Gartner, B. C., Sizmann, D., Babiel, R., Mihm, U., Hofmann, W. P., von Wagner, M., and Zeuzem, S. (2006) Comparison of conventional PCR with real-time PCR

and branched DNA-based assays for hepatitis C virus RNA quantification and clinical significance for genotypes 1 to 5. *J. Clin. Microbiol.* **44,** 729–737.

20. Halfon, P., Bourliere, M., Penaranda, G., Khiri, H., and Ouzan, D. (2006) Real-time PCR assays for hepatitis C virus (HCV) RNA quantitation are adequate for clinical management of patients with chronic HCV infection. *J. Clin. Microbiol.* **44,** 2507–2511.

21. Chevaliez, S., Bouvier-Alias, M., Brillet, R., Pawlotsky, J. M. (2007) Overestimation and underestimation of hepatitis C virus RNA levels in a widely used real-time polymerase chain reaction-based method. *Hepatology* **46,** 22–31.

Chapter 2

Simultaneous Detection of Anti-HCV Antibody and HCV Core Antigen

Filippo Ansaldi and Giancarlo Icardi

Abstract

HCV infection is usually diagnosed by means of an enzyme immune assay for the detection of antibody against HCV. The window period between infection and seroconversion remains a dramatic problem in the transfusional and diagnostic setting. In this chapter, we report (i) procedures for assays using two different approaches designed to reduce the window period and (ii) performance in terms of specificity and sensitivity in the detection of both antibody and antigen, and we compare their efficacy with that of commercial assays.

Key words: HCV, diagnosis, window period, antigen, antibody, combination assay, immunoassay, chemiluminescence.

1. Introduction

The method currently recommended for diagnosis of HCV infection is an enzyme immune assay (EIA) for the detection of anti-HCV antibodies. Although second- and third-generation HCV antibody tests provide better sensitivity and specificity than did the first-generation assays, the screening method is still not fully effective, because it is based on detection of antibodies that appear in the blood only approximately 8–10 weeks after infection (1,2). Two approaches currently used to reduce this so-called window period and to detect the virus replicating in blood are based on the detection (i) of HCV RNA by nucleic acid technology (NAT)-based tests and (ii) of the HCV core antigen by EIA or chemiluminescence assays. The first approach has been applied in screening of small pools of multiple samples from potential blood donors in the United States since mid-1999 and in other

Hengli Tang (ed.), *Hepatitis C: Methods and Protocols, Second Edition, vol. 510*
© 2009 Humana Press, a part of Springer Science+Business Media
DOI 10.1007/978-1-59745-394-3_2 Springerprotocols.com

developed countries since a few months later *(3–7)*. The detection of the HCV core antigen by EIA, in the absence and in the presence of antibodies, allowed the design and the development of two commercial kits (the Antibody to Hepatitis C cAg ELISA Test System and the Ortho Trak-C, Ortho Clinical Diagnostics, Raritan, NJ) *(8–11)*, whereas the simultaneous detection of antibody and HCV core antigen led to standardized assays, both in-house and commercial (the Monolisa HCV Ag-Ab Ultra Assay, Biorad Laboratories Limited, Marnes la Coquette, France) *(12–17)*.

Here, we report (i) procedures for the two assays for simultaneous detection of HCV core antigen and anti-HCV antibody and (ii) their performance in terms of specificity, clinical, and analytical sensitivity in the detection of both antibody and antigen, and we compare their efficacy with that of commercial assays.

2. Materials

2.1. Monolisa HCV Ag-Ab Ultra

1. Microplate from the kit, coated with (i) monoclonal antibodies against the capsid protein of HCV, (ii) two recombinant proteins produced by *E. coli* from the NS3 region, (iii) one recombinant antigen from the nonstructural NS4 region, and (iv) a peptide from the capsid area of the viral genome.
2. (i) Conjugate 1, containing mouse biotinilated monoclonal antibodies against HCV capsid, which does not react with the capsid-mutated peptide with which the microplate is coated, and (ii) conjugate 2, containing mouse peroxidase antibodies to human IgG and peroxidase-labeled streptavidin.
3. Washing solution containing 10× concentrate Tris NaCl buffer, 1% Tween-20.
4. Chromogen solution containing tetramethylbenzidine (TMB).
5. 1 N sulfuric acid stop solution.
6. Negative (Tris HCl buffer: 10 mM Tris–HCl, 0.5 mM MgCl2, and 150 mM NaCl, pH 7.4), positive (human serum containing antibodies to HCV diluted in Tris HCl buffer, photochemically inactivated), and antigen-positive controls (lyophilized capsid synthetic peptide; to resuspend, add water with ProClin 300, 0.5%)

2.2. Chemiluminescence Immunoassay (Prototype Presented by Abbott Laboratories, Abbott Park, IL)

1. Slid-phase from the kit constituted of microparticles coated with (i) monoclonal antibodies (MoAb) against HCV core antigen and (ii) recombinant antigens representing amino acid sequences from core, NS3, and NS4 gene products. Antigens were constructed so as to avoid the epitope recognition of the HCV core antigen by MoAbs.

2. Specimen diluent buffer.
3. Two conjugates: (i) acridium-labeled anti-HCV core MoAb and (ii) acridium-labeled MoAb anti-human IgG.
4. Alkaline hydrogen peroxide solution.
5. Negative (recalcified human plasma negative for HBsAg, anti-HBc, anti-HCV, anti-HIV, anti-HTLVI/II), anti-HCV-positive (HCV-positive calibrator of PRISM HCV antibody assay), and core-antigen-positive (HCV core-antigen-control plasma, JV01720, Raleigh, NC) controls. The core-antigen control has high viral load (14,000,000 copies/mL by HCV RNA quantitative assay, LCx, Abbott Laboratories), is highly positive with an HCV core-antigen EIA, and is negative for the anti-HCV antibody assay.

3. Methods

3.1. Monolisa HCV Ag-Ab Ultra

1. Add 100 μL of conjugate 1, 50 μL of serum or plasma for each sample, and 50 μL of each control to the well and incubate for 90 min at 37°C.
2. Wash the plates four times with 200 μL wash buffer (1 × PBS, pH 7.4, 0.1% Tween-20).
3. After the washing step, add 100 μL of peroxidase-labeled antibodies against human IgG and streptavidine-peroxidase (conjugate 2).
4. After 30 min of incubation at 37°C, remove the unbound enzymatic conjugate by repeating the washing step and reveal the antigen–antibody complex by addition of the substrate (80 μL of TMB solution).
5. Stop the reaction by adding 100 μL of sulfuric acid stop solution.
6. Measure absorbance intensity with a photometer and read at 450 nm with a 620–700 nm reference. Color intensity is proportional to the quantity of bound conjugate and therefore to the concentration of HCV Ag and/or anti-HCV antibodies.

3.2. Chemiluminescence Immunoassay

The combined chemiluminescence assay was designed as a two-step protocol similar to the PRISM HCV antibody assay and was performed on a single-channel fully automated serologic instrument (PRISM, Abbot Laboratories).

1. In the first step, add 100 μL of samples, 50 μL of specimen diluent buffer, and 50 μL of blended recombinant antigen-coated microparticles and MoAb anti-HCV core-coated microparticles to the well of a reaction tray and incubate for 20 min at 37°C.
2. Move the tray to the transfer station and flush the reaction mixture from the sample well into the reaction well, capturing the microparticles on a filter.

3. In the second step, add 50 μL of conjugate solution containing acridium-labeled anti-HCV core MoAb and acridium-labeled MoAb anti-human IgG.

4. After 25 min incubation at 37°C, remove the unbound conjugate by conducting the washing step.

5. Move the reaction well to the reading station and add an alkaline hydrogen peroxide solution to trigger the chemiluminescence reaction. The intensity of the chemiluminescence signal is proportional to the quantity of bound conjugate and therefore depends on the amount of HCV core antigen and/or anti-HCV antibody in the sample.

3.3. Performance

3.3.1. Specificity

Specificity evaluation of the assays was performed by different authors testing fresh or frozen anti-HCV-negative sera from blood donors or open population *(12–16)*. Furthermore, 100 "difficult" or "sticky" sera, negative for anti-HCV and HCV RNA, taken from hemodialysis patients, patients with autoimmune disorders or positive for anti-HAV IgM, anti-HBc IgM, or anti-HIV, were tested with Monolisa HCV Ag-Ab Ultra, for evaluation of the performance of the assay under difficult conditions *(15)*. The results of the more representative studies, summarized in **Table 2.1**, showed excellent discriminatory ability for both assays when both open-population and blood-donor samples and "sticky" sera were tested. Indeed, virtually all HCV RNA and anti-HCV-negative samples among about 7000 tested gave chemiluminescence or absorbance values <80% of the cut-off

Table 2.1
Specificity evaluation of chemiluminescence immunoassay (Chemi.) and Monolisa HCV Ag-Ab Ultra assay (Monolisa)

| | | | | | (S or OD/CO ratio) | | |
| | | | | Negative | Gray zone | | Positive |
Assay	Reference	Sample from	n	<0.79 (%)	0.8–0.99 (%)	1–1.2 (%)	>1.2 (%)
Chemi.	*(12)*	Blood donors	3017	3011 (99.8)	0		6 (0.2)
Monolisa	*(14)*	Blood donors	2503	2498 (99.8)	2 (0.1)	0	3 (0.1)
	(15)	Open population	400	400 (100)	0	0	0
		Difficult and "sticky" patients	100	99 (99)	1 (1)	0	0
	Ansaldi, pers. obs.	Blood donors	900	899 (99.9)	0	1 (0.1)	0

(CO) value; only 3 (0.04%) had an OD/CO in the gray zone, and 10 (0.14%) were positive. Reaching specificity higher than 99.8% with open-population and blood-donor samples and 99% with "sticky" sera, the tests performed comparably to kits routinely used in screening for blood-transmitted infections, such as hepatitis B and human immunodeficiency virus *(12–16)*.

3.3.2. Clinical Sensitivity During the Window Period

The sensitivity of the Monolisa HCV Ag-Ab Ultra Assay and chemiluminescence immunoassay in detecting HCV antigen was tested in several studies, and its efficacy in reducing the window period relative to commercial assays for HCV RNA detection, natural and commercial seroconversion panels, and HCV RNA-positive anti-HCV-negative samples was determined *(12–16)*.

Different results should be compared cautiously, because the clinical sensitivity estimation and the difference in detection of HCV infection is greatly influenced by the number of panels, the number and the timing of bleeds included in the panels, the immunological status of patients, the methods and strategy (pool or single specimen) used for HCV RNA detection, among others.

Of the 89 HCV RNA-positive and anti-HCV-negative sera from 23 seroconversion panels used by Shah et al. *(12)* for the evaluation of chemiluminescence immunoassay, 80 (89.9%) tested positive (**Table 2.2**). In 18 out of the 23 commercial panels tested, the sampling dates on which the chemiluminescence assay and the PCR yielded the first positive result coincided exactly. The use of the combined assay in the eight panels in which the first bleed was HCV RNA negative resulted in a reduction of the window period by an average of 34.4 days, whereas when panels were tested in which the first bleed was positive, the reduction was 14.4 days. The corresponding mean delays in the detection of HCV infection by chemiluminescence combined assay, with respect to PCR, were 3.6 and 1.9 days (**Table 2.3**).

To assess the sensitivity of the Monolisa HCV Ag-Ab Ultra assay in detecting HCV antigen, investigators *(13–16)* tested 149 HCV RNA-positive and antibody-negative sera taken from seroconversion panels or immunocompetent subjects. As summarized in **Table 2.2**, 117 (78.5%) of samples tested positive, and 4 (2.7%) fell in the gray zone. Furthermore, for evaluation of the performance of the combination assay in high-risk settings, 79 HCV RNA-positive and anti-HCV-negative sera, collected from immunocompromised patients, such as hemodialyzed ($n = 59$) and anti-HIV-positive ($n = 20$) subjects, were tested; 33.3% of the specimens collected from hemodialyzed patients and 65% of those from anti-HIV-positive patients showed OD/CO > 1 (**Table 2.2**).

The Monolisa HCV Ag-Ab Ultra assay was able to detect HCV infection a mean of 21.6–50 days earlier than anti-HCV assay, as reported by various studies, whereas the mean delay by

Table 2.2
Sensitivity evaluation of chemiluminescence immunoassay (Chemi.) and Monolisa HCV Ag-Ab Ultra assay (Monolisa). All samples were anti-HCV negative

Assay	Ref.	Sample from	HCV RNA	n	(S or OD/C0 ratio)			
					Negative	Gray zone		Positive
					<0.79 (%)	0.8–0.99 (%)	1–1.2 (%)	>1.2 (%)
Chemi.	(12)	Seroconv. panels (first sample RNA neg.)	Positive	39	3 (7.7)	0		36 (92.3)
		Seroconv. panels (first sample RNA pos.)	Positive	50	6 (12)	0		44 (88)
Monolisa	(13)	Blood donor	Positive	12	6 (50)	2 (16.7)	1 (8.3)	3 (25)
		Hemodialyzed pts.	Positive	59	36 (61)	4 (6.8)		19 (32.2)
		Seroconv. panels	Negative	37	37 (100)			
			Positive	44	14 (31.8)	1 (2.3)		29 (65.9)
	(15)	Seroconver. pts.	Positive	76	4 (5.2)	1 (1.3)	4 (5.3)	67 (88.4)
		Serconv. panels	Negative	5	5 (100)			
			Positive	17	3 (17.7)	0	0	14 (82.4)
	(16)	Seroconv.HIV-pos. pts.	Positive	20	7 (35)	0		13 (65) 17 (85)[1]
	(17)	Pts. with occult infection	Negative[2] Positive[3]	115	113 (98.2)	1 (0.9)		1 (0.9)

[1] With the optimized threshold of 0.5, as recommended by Ansaldi et al. (15).
[2] In serum.
[3] In liver.

Table 2.3
Estimated reduction of window period by antigen and antibody combination assays.
Chemi., Chemiluminescence immunoassay; Monolisa, Monolisa HCV Ag-Ab Ultra assay

Assay	Ref.	Sample from	Combination-anti-HCV differential Mean (range, days)	HCV RNA-combination differential Mean (range, days)
Chemi.	(12)	Seroconv. panels (first sample RNA neg.)	34.3 (8–73)	3.6 (0–22)
		Seroconv. panels (first sample RNA pos.)	14.4 (n.a.)	1.9 (n.a.)
Monolisa	(13)	Hemodialyzed pts.	21.6 (–7–287)[1] 31.1 (0–287)[2]	30.3 (0–354)
	(14)	Seroconv. panels	26.8 (0–72)	5.1 (0–24)
	(15)	Seroconv. panels	24 (0–37)	10.4 (0–32)
	(16)	Seroconv. HIV-pos. pts.	50 (0–143) 68 (0–148)[3]	27 (0–131) 9 (0–126)[3]

n.a., data not available
[1] Delay between Monolisa HCV Ag-Ab Ultra and the most sensitive anti-HCV assay used in the study.
[2] Delay between Monolisa HCV Ag-Ab Ultra and Monolisa HCV.
[3] With the optimized threshold of 0.5, as recommended by Schnuriger et al. (16).

the combined assay ranged between 5.1 and 30.3 days after the first positive result for HCV RNA detection (**Table 2.3**).

3.3.3. Clinical Sensitivity After Seroconversion

All anti-HCV-positive sera from the seroconversion panels tested positive with both the chemiluminescence immunoassay and the Monolisa HCV Ag-Ab Ultra assay. The unique exception was a specimen collected from a patient who, 18 mo after seroconversion and treatment with interferon, tested negative by the anti-HCV ELISA assay (Monolisa HCV Ab, Biorad), RNA (Amplicor HCV, Roche Diagnostic), and core (HCV core Ag trak-C assay, Ortho Clinical Diagnostic) and resulted anti-core and anti-NS3 weak positive by the RIBA assay (Chiron RIBA HCV 3.0 SIA, Chiron Corporation) (13).

Furthermore, Ansaldi et al. tested 20 moderately anti-HCV-positive sera from a weak-positive panel, 38 RIBA-single-band positive specimens, and samples collected from patients infected with different HCV subtypes and coinfected with HBV or HIV, using the Monolisa HCV Ag-Ab Ultra assay. The assay was clearly reactive in the viremic and aviremic samples from the weak-positive panel both when the serological pattern was characterized by antibody against nonstructural proteins and when antibody against core predominated *(15)*. Of 38 RIBA-single-band positive specimens, 30 (79%) tested positive and 1 (2.6%) fell in the gray zone. The seven Monolisa-negative samples were considered anti-HCV false reactivities because the following specimens collected from those patients tested seronegative and aviremic (Ansaldi, pers. obs.). The analysis of samples collected from patients infected with different HCV subtypes and coinfected with HBV or HIV showed that genotype and coinfection did not affect the sensitivity of the assay *(15)*.

3.3.4. Clinical Sensitivity in HCV Occult Infection

HCV occult infection is defined as a clinical picture characterized by persistently abnormal results of unknown etiology in liver function tests in an anti-HCV– and serum HCV RNA–negative patient with a liver biopsy that demonstrates the presence of hepatic viral RNA *(17)*.

To determine whether the Monolisa HCV Ag-Ab Ultra assay could detect core antigen in HCV occult infection, Quiroga et al. *(18)* tested 115 HCV RNA- and anti-HCV-negative sera from patients with HCV RNA-positive liver biopsies. Only one serum (0.9%) resulted in a positive Monolisa HCV Ag-Ab Ultra assay, and another sample fell in the gray zone (**Table 2.2**). Both sera, despite the permanent lack of antibodies against HCV detectable by commercial EIA, showed weak NS3 reaction when tested by immunoblot assay (Dediscan HCV Plus, Biorad).

3.3.5. Analytical Sensitivity in Antigen Detection

Estimation of analytical sensitivity was not the main objective of the studies evaluating the Monolisa HCV Ag-Ab Ultra assay and chemiluminescence immunoassay, but results of the limited number of samples around the CO value (12 specimens showed a OD/CO ranging between 0.8 and 1.99) showed that the chemiluminescence immunoassay could detect antigen in samples with viremia as low as about 107,000 copies per mL, whereas on the basis of regression curve, the Monolisa HCV Ag-Ab Ultra assay meets the CO value at an HCV RNA concentration of about 250,000 IU/mL.

The viral load of RNA-positive Monolisa-negative "discordant" sera ranged between 930 and 1.8×10^7 IU/mL, although antigen-positive sera were found with viremia of 6.4×10^4 IU/mL (data not shown). The finding that the analytical sensitivity of antigen assays does not correspond constantly to the HCV RNA titer in different subjects has also emerged from other analyses, in which Antibody to Hepatitis C cAg ELISA Test System and Ortho Trac-C assays (Ortho Clinical Diagnostics) were used. This lack of correspondence does not seem to depend on HCV genotype *(19, 20)*.

References

1. Barrera, J. M., Francis, B., Ercilla, G., Nelles, M., Archord, D., Darner, J., et al. (1995) Improved detection of anti-HCV in post-transfusion hepatitis by a third-generation ELISA. *Vox Sang.* **68**, 15–18.

2. Brooks, G. F., Butel, J. S., and Morse S. A. (2001) *Jawetz, Melnick and Adelberg's Medical Microbiology,* 22nd ed., Lange Medical Books, New York, pp. 406–410.

3. Busch, M. P., and Kleinman, S. H. (2000) Committee report: nucleic acid amplification testing of blood donors for transfusion-transmitted infectious diseases. *Transfusion* **40**, 143–159.

4. Food and Drug Administration (2002) *Guidance for industry: use of nucleic acid tests on pooled and individual samples from blood donors to adequately and appropriately reduce the risk of transmission of HIV-1 and HCV.* Federal Register 67, 17077.

5. Velati, C., Baruffi, L., Romanò, L., Fomiatti, L., Carreri, V., and Zanetti A. (2003) Introduction of NAT method of screening blood donors in Italy: reports of the two years of experience and a re-evaluation of residual risk. *Blood Transfus.* **4**, 368–378

6. Anonymous (2002) Implementation of donor screening for infectious agents transmitted by blood by nucleic acid technology. *Vox Sang.* **82**, 87–111.

7. European Agency for the Evaluation of Medicinal Products (2001) *Committee for proprietary medicinal products.* Note for guidance on plasma0deried medicinal products. EMEA/CPMP/BWP 269/95 rev.3 London, 25 January 2001.

8. Courouce, A. M., Le Marrec, N., Bouchardeau, F., Razer, M., Maniez, W. P., Laperche S., et al. (2000) Efficacy of HCV core antigen detection during preseroconversion period. *Transfusion* **40**, 1198–1202.

9. Icardi, G., Ansaldi, F., Bruzzone, B. M., Durando, P., Lee, S., De Luigi, C., et al. (2001) Novel approach to reduce the hepatitis C virus (HCV) window period: clinical evaluation of a new enzyme-linked immunosorbent assay for HCV core antigen. *J. Clin. Microbiol.* **39**, 3110–3114.

10. Gaudy. C., Thevenas, C., Tichet, J., Mariotte, N., Goudeau, A., and Dubois, F. (2005) Usefulness of the hepatitis C virus core antigen assay for screening of a population undergoing routine medical check-up. *J. Clin. Microbiol.* **43**, 1722–1726.

11. Pivert, A., Payan, C., Morand, P., Fafi-Kremer, S., Deshayes, J., Carrat, F., et al. (2006) Comparison of serum hepatitis C virus (HCV) RNA and core antigen levels in patients coinfected with human immunodeficiency virus and HCV and treated with interferon plus ribavirin. *J. Clin. Microbiol.* **44**, 417–422.

12. Shah, D. O., Chang, C. D., Jiang, L. X., Cheng, K. Y., Muerhoff, A., Gutierrez, R. A., et al. (2003) Combination HCV core antigen and antibody assay on a fully automated chemiluminescence analyser. *Transfusion* **43**, 1067–1074.

13. Laperche, S., La Marrec, N., Girault, A., Bouchardeau, F., Servant-Dalmas, A., Maniez-Montreuil, M., et al. (2005) Simultaneous detection of hepatitis C virus (HCV) core antigen and anti-HCV antibodies improves the early detection of HCV infection. *J. Clin. Microbiol.* **43**, 3877–3883.

14. Laperche, S., Elghouzzi, M. H., Pascal, M., Asso-Bonnet, M., Le Marrec, N., Girault, A., et al. (2005) Is an assay for simultaneous detection of hepatitis C virus core antigen and antibody a valuable alternative to nucleic acid testing? *Transfusion* **45**, 1965–1972.

15. Ansaldi, F., Bruzzone, B., Testino, G., Bassetti M., Gasparini R., Crovari P., et al. (2006) Combination hepatitis C antigen and antibody assay as a new tool for early diagnosis of infection. *J. Viral Hepat.* **13**, 5–10.

16. Schnuriger, A., Dominguez, S., Valantin, M. A., Tubiana, R., Duvier, C., Ghosn, J., et al. (2006) Early detection of hepatitis C virus infection by use of a new combined antigen-antibody detection assay: potential use for high risk individuals. *J. Clin. Microbiol.* **44**, 1561–1563.

17. Castillo, I., Pardo, M., Bartholomé, J., Ortiz-Movilla, N., de Lucas, S., Salas, C., et al. (2004) Occult hepatitis C virus infection in patients in whom the etiology of persistently abnormal results of liver function tests is unknown. *J. Infect. Dis.* **189**, 7–14.

18. Quiroga, J. A., Castillo, I., Pardo, M., Rodriguez-Inigo, E., and Careno, V. (2006) Combined hepatitis antigen-antibody detection assay does not improve diagnosis for seronegative individuals with occult infection. *J. Clin. Microbio.* **44**, 4559–4560.

19. Nubling, C. M., Unger, G., Chudy, M., Raia, S., and Lower, J. (2002) Sensitivity of HCV core antigen and HCV RNA detection in the early infection phase. *Transfusion* **42**, 1037–1045.

20. Maynard, M., Pradat, P., Berthillon, P., Picchio, G., Voirin, N., Martinot, M., et al. (2003) Clinical relevance of total HCV core antigen testing for hepatitis C monitoring and for predicting patients response to therapy. *J Viral Hepat.* **10**, 318–323.

Chapter 3

Immunohistochemical Detection of HCV Proteins in Liver Tissue

Oivier Galy, Isabelle Vincent, Philippe Chevallier, Nicole Lyandrat, and Isabelle Chemin

Abstract

Detection and localization of HCV in liver tissue are vital for diagnostic purposes and clinical management of HCV-infected patients, as well as for the elucidation of viropathological mechanisms. The fragility of HCV RNA and the low levels of viral expression in infected tissues are a constant limitation in molecular assays for HCV characterization. HCV antigen detection, by immunochemistry, in liver biopsies is an attractive option for precise localization and quantification of viral proteins with direct access to histological patterns. We describe here a study using a novel immunohistochemical method effective on fixed, archived specimens, including liver biopsies and surgical resection samples. The initial protocol uses a biotin-detection system but can also be used in a polymer-detection system. This protocol offers easy, precise, and strong staining resolution with distinct patterns consistent with the liver pathology, irrespective of the viral HCV genotype examined. This approach provides applications for diagnosis as well as for exploratory pathological studies.

Key words: HCV, immunohistochemistry, envelope protein, chronic infection, hepatocellular carcinoma.

1. Introduction

The HCV genome is a single-stranded RNA molecule of about 9600 bp. It encodes for a large polyprotein that is processed by host and virus proteases into several structural and nonstructural viral proteins (Envelope 1/2, Capsid, p7and NS2, NS3, NS4A/B, and NS5A/B).

The detection of HCV replicative intermediates or virus antigen may be helpful for diagnosis or clinical management of

Hengli Tang (ed.), *Hepatitis C: Methods and Protocols, Second Edition, vol. 510*
© 2009 Humana Press, a part of Springer Science+Business Media
DOI 10.1007/978-1-59745-394-3_3 Springerprotocols.com

patients with HCV infection, and it is of crucial importance for monitoring patients before and after HCV-related liver transplantation. HCV replication level generally seems to be relatively low in infected liver, hampering the detection of HCV particles directly in the liver *(1, 2)*. Detection methods based on HCV RNA amplification, like *in situ* PCR or *in situ* hybridization, do not always detect HCV and can lead to conflicting results concerning localization of viral particles *(3)*, but they remain an interesting tool when used to complement classical methods *(4, 5)*. These methods are probably limited by rapid RNA degradation in tissues and the difficulty of designing efficient probes that overcome the high variability of HCV genomes.

Detection of HCV antigens by immunochemistry in liver biopsies is therefore an interesting option that allows both localization and quantification of viral proteins.

Quite a few antibodies have been raised against hepatitis C antigens and fulfill the conditions for use in immunochemistry (IHC). So far, except for the promising commercial TORDJI 22 and TORDJI 32, no commercial antibody allows specific, reproducible, and efficient staining *(6–8)*, one reason explaining why IHC has failed to become established in routine experiment in diagnostic labs. Among those tested for IHC are antibodies raised against nonstructural proteins (NS3, NS4, and NS5) *(9, 10)* and core or envelope protein *(10, 11)*.

Antibodies raised against envelope protein E2 provided the best efficiency for IHC techniques *(11, 12)*. Even though the region encoding envelope protein encompasses the hypervariable region I (HVR 1) of HCV, the overall conformation of E2 seems to be quite conserved *(13)*. In our experience *(12)*, anti-E2 antibody $D_{4.12.9}$ detected the E2 protein from all genotypes tested, including genotypes 1 to 5, indicating that this antibody is suitable for HCV detection irrespective of its precise genotype. In our personal experience *(12)*, the staining location for anti-E2 immunostaining is mainly cytoplasmic, although occasional perimembrane staining (including cytoplasmic and nuclear membrane) occurs. Staining pattern is mainly coarse granular with a microvesicular pattern (*see* **Fig. 3.1**).

HCV staining patterns appear to differ slightly according to the pathological status of the liver tissue. We observed a very strong staining of hepatocyte membrane, cytoplasm, and perinuclear regions in liver from patients with active HCV-related cirrhosis (Intense plasma and nuclear membrane staining was observed in cases with high inflammatory activity.) In noncirrhotic and nontumoral tissues, anti-E2 staining intensity increased with hepatitis fibrosis state. In HCV-related tumors, staining was exclusively detected within regeneration nodules and confined to hepatocytes whose morphology remained unchanged. Staining appeared in one of two distinct patterns: trabecular throughout the hepatic parenchyma or only in isolated cells.

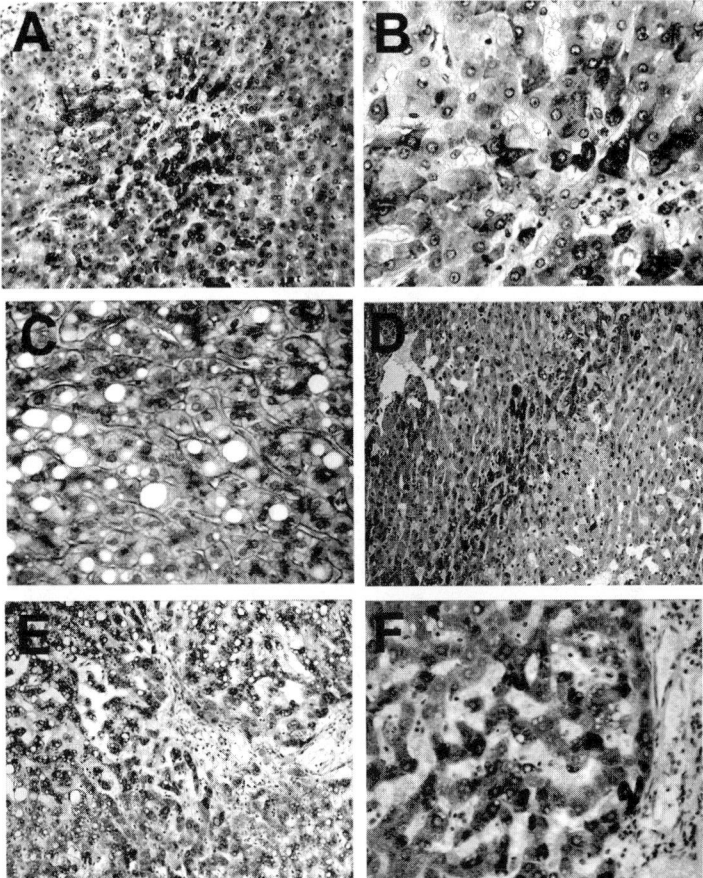

Fig. 3.1. D$_{4.12.9}$ immunostaining of liver sections from HCV patients. **A, B, C.** Severe HCV hepatitis cases. A, B: Patient before hepatectomy, immunostaining with polymer-based assay. C: Immunostaining with classic biotin-based assay. **D, E, F**: HCV-related hepatocellular carcinoma cases. D: Band of hepatocytes with high E2 accumulation among low-intensity stained hepatocytes in nontumorous area, suggesting clonal expansion of infected hepatocytes. E: Cirrhotic tissue surrounding area exhibiting important steatosis and strong E2 expression in cytoplasm and nucleocytoplasmic region. F: Accumulation of E2 in precancerous cirrhotic nodule with more important expression in isolated cells groups.

2. Materials

1. Normal source of samples: Transparietal biopsies or surgical biopsies.
2. Nature of samples: Frozen or formalin-fixed human liver.
3. Xylene dilutions in water.
4. Absolute ethanol dilutions in water: 95%, 70%.
5. Methanol solution (0.3% hydrogen peroxide).
6. Vector antigen unmasking solution (Vector Laboratories, Burlingame, CA).

7. Phosphate-buffered saline (PBS) solution (Sigma).
8. Skimmed milk.
9. Bovine serum albumin (BSA, Sigma).
10. Primary mAb, $D_{4.12.9}$ at 0.2 μg/mL.
11. Secondary antibody, Vectastain® ELITE ABC PEROXY-DASE KIT Rabbit IGG (Vector).
12. DAB Substrate kit (Vector).
13. Mayer's hematoxylin.
14. Mounting medium (DAKO Mounting medium, Dako, North America).
15. Slide racks and trays (Fisher Scientific).
16. Microwave oven.
17. Microscope cover glass.
18. Light microscope.

3. Method

1. Deparaffinize tissue (see **Note 1**) sections in xylene for 10 min (twice).
2. Rehydrate the tissue in graded ethanol concentrations (100%, 95%, and 70% for 5 min each) and proceed immediately to step 3.
3. Block endogenous peroxidase activities by incubation in methanol solution for 30 min at room temperature (see **Note 2**).
4. For unmasking, place the slides in a microwave oven for 15 min in antigen unmasking solution. Set the microwave power high enough to bring the solution to a boil, and then reduce power so that the solution continues boiling for the required time. Allow the slides to cool down for 30 min in the same unmasking solution (see **Note 3**).
5. Incubate with PBS solution (5% skimmed milk and 0.1% BSA) for 1 h at room temperature. Do not rinse.
6. Incubate sections overnight with the primary mAb at 4°C. Sections without primary antibody can be used as controls.
7. Wash in PBS three times, 5 min each.
8. Incubate for 30 min with secondary antibody at 5 μl/mL in PBS solution (0.1% BSA).
9. Wash in PBS three times, 5 min each.
10. Amplify signal at 37°C for 45 min (Vectastain peroxydase kit); use PBS solution (0.1% BSA).
11. Wash in PBS three times, 5 min each.
12. Incubate tissue sections with the DAB substrate at room temperature for 5 min,
13. Wash in PBS three times, 5 min each.
14. Counterstain the sections with Mayer's hematoxylin.

15. Dehydrate the section through successive ethanol baths (70% ethanol, 95% ethanol, absolute ethanol, and xylene, 5 min in each solution).

16. Mount slides using standard microscope cover glass and mounting medium.

4. Notes

1. Sensitivity of detection is significantly increased by use of fresh-frozen tissue.

2. Alternatively, a method based on polymer detection can be used with $D_{4.12.9}$ in IHC. It generally offers same sensitivity and a slightly better resolution because it eliminates endogenous biotin interference, the main source of nonspecific background staining. The method differs from the classic protocol in step 3, where we incubate 10 min in DAKO peroxydase block solution (DakoCytomation EnVision + Dual Link System Peroxidase, DAKO, France) and rinse gently with distilled water. For step 8, apply peroxidase-labeled polymer (DakoCytomation EnVision + Dual Link System Peroxidase) to cover specimen. Incubate for 30 min and rinse the slides with the buffer solution provided. Eventually, apply substrate-chromogen solution (DakoCytomation EnVision + Dual Link System Peroxidase) to cover the specimen, add DAB, and incubate for 2–10 min (optimal incubation time can be determined by verifying signal intensity under the microscope). Rinse gently with distilled water from a wash bottle (do not focus flow directly on tissue). Proceed with step 13.

3. During microwave treatment, ensure that solution level is sufficient to cover the tissue section throughout the treatment. Check solution level every 3–5 min.

Acknowledgments

We thank M. A. Petit for providing the anti-E2 Abs and C. Trépo for his constant interest and support.

References

1. Shimizu, Y. K., Feinstone, S. M., Kohara, M., Purcell, R. H., and Yoshikura, H. (1996) Hepatitis C virus: detection of intracellular virus particles by electron microscopy. *Hepatology* **23**, 205–209.

2. Negro, F. (1998) Detection of hepatitis C virus RNA in liver tissue: an overview. *Ital. J. Gastroenterol. Hepatol.* **30**, 205–210.

3. Lau, J. Y., Krawczynski, K., Negro, F., Gonzalez Peralta. R. P. (1996) In situ detection of hepatitis C virus—a critical appraisal. *J. Hepatol.* **24**(Suppl. 2), 43–51.

4. Biagini, P., Benkoel, L., Dodero, F., de Lamballerie, X., Chamlian, V., Nouhou, H., et al. (2001) Hepatitis C virus RNA detection by in situ RT-PCR in formalin-fixed

paraffin-embedded liver tissue. Comparison with serum and tissue results. *Cell. Mol. Biol.* (Noisy-le-Grand), **47** Online Pub OL167-71.

5. Comar, M., Dal Molin, G., D'Agaro, P., Croce, S. L., Tiribelli, C., and Campello, C. (2006) HBV, HCV, and TTV detection by *in situ* polymerase chain reaction could reveal occult infection in hepatocellular carcinoma: comparison with blood markers. *J. Clin. Pathol.* **59**, 526–529.

6. Komminoth, P., Adams, V., Long, A. A., Roth, J., Saremaslani, P., Flury, R., et al. (1994) Evaluation of methods for hepatitis C virus detection in archival liver biopsies. Comparison of histology, immunohistochemistry, *in situ* hybridization, reverse transcriptase polymerase chain reaction (RT-PCR) and *in situ* RT-PCR. *Pathol. Res. Pract.* **190**, 1017–1025.

7. Walker, F. M., Dazza, M. C., Dauge, M. C., Boucher, O., Bedel, C., Henin, D., et al. (1998) Detection and localization by *in situ* molecular biology techniques and immunohistochemistry of hepatitis C virus in livers of chronically infected patients. *J. Histochem. Cytochem.* **46**, 653–660.

8. Brody, R. I., Eng, S., Melamed, J., Mizrachi, H., Schneider, R. J., Tobias, H., et al. (1998) Immunohistochemical detection of hepatitis C antigen by monoclonal antibody TORDJI-22 compared with PCR viral detection. *Am. J. Clin. Pathol.* **110**, 32–37.

9. Blight, K., Lesniewski, R., Labrooy, J., Trowbridge, R., and Gowans, E. (1993) Localisation of hepatitis C virus proteins in infected liver tissue by immunofluorescence. *Gastroenterol. Jpn.* **28**(Suppl 5), 55–58.

10. Gonzalez-Peralta, R. P., Fang, J. W. S., Davis, G. L., Gish, R., Tsukiyama-Kohara, K., Kohara, M, et al. (1994) Optimization for the detection of hepatitis C virus antigens in the liver. *J. Hepatol.* **20**, 143–147.

11. Verslype, C., Nevens, F., Sinelli, N., Clarysse, C., Pirenne, J., Depla, E., et al. (2003) Hepatic immunohistochemical staining with a monoclonal antibody against HCV-E2 to evaluate antiviral therapy and reinfection of liver grafts in hepatitis C viral infection. *J. Hepatol.* **38**, 208–214.

12. Galy, O., Petit, M. A., Benjelloun, S., Chevallier, P., Chevallier, M., Srivatanakul, P., et al. (2007) Efficient hepatitis C antigen immunohistological staining in sections of normal, cirrhotic and tumoral liver using a new monoclonal antibody directed against serum-derived HCV E2 glycoproteins. *Cancer Lett.* **248**, 81–88.

13. Penin, F., Combet, C., Germanidis, G., Frainais, P. O., Deleage, G., and Pawlotsky, J. M. (2001) Conservation of the conformation and positive charges of hepatitis C virus E2 envelope glycoprotein hypervariable region 1 points to a role in cell attachment. *J. Virol.* **75**, 5703–5710.

Part II

HCV Nomenclature, Genotyping, and Quasispecies

Chapter 4

Nomenclature and Numbering of the Hepatitis C Virus

Carla Kuiken and Peter Simmonds

Abstract

International standardization and coordination of the nomenclature of variants of hepatitis C virus (HCV) is increasingly needed as more is discovered about the scale of HCV-related liver disease and important biological and antigenic differences that exist between variants. Consistency in numbering is also increasingly required for functional and clinical studies of HCV. For example, an unambiguous method for referring to amino acid substitutions at specific positions in NS3 and NS5B coding sequences associated with resistance to specific HCV inhibitors is essential in the investigation of antiviral treatment. Inconsistent and inaccurate numbering of locations in DNA and protein sequences is becoming a problem in the HCV scientific literature.

A group of experts in the field of HCV genetic variability, and those involved in development of HCV sequence databases, the Hepatitis Virus Database (Japan), euHCVdb (France), and the Los Alamos National Laboratory (United States), convened to reexamine the status of HCV genotype nomenclature, resolve conflicting genotype or subtype names among described variants of HCV, and draw up revised criteria for the assignment of new genotypes as they are discovered in the future. They also discussed how HCV sequence databases could introduce and facilitate a standardized numbering system for HCV nucleotides, proteins, and epitopes.

A comprehensive listing of all currently classified variants of HCV incorporates a number of agreed genotype and subtype name reassignments to create consistency in nomenclature. A consensus proposal was drawn up for the classification of new variants into genotypes and subtypes, which recognizes and incorporates new knowledge of HCV genetic diversity and epidemiology. The proposed numbering system was adapted from the Los Alamos HIV database, with elements from the hepatitis B virus numbering system. The system comprises both nucleotides and amino acid sequences and epitopes, and uses the full-length genome sequence of isolate H77 (accession number AF009606) as a reference. It includes a method for numbering insertions and deletions relative to this reference sequence.

Key words: HCV classification, HCV nomenclature, HCV numbering, HCV genomics.

Hengli Tang (ed.), *Hepatitis C: Methods and Protocols, Second Edition, vol. 510*
© 2009 Humana Press, a part of Springer Science+Business Media
DOI 10.1007/978-1-59745-394-3_4 Springerprotocols.com

1. Introduction

In the course of 2004, an initiative was started to provide standards for HCV nomenclature and numbering to facilitate HCV research. The initiative was supported by a broad group of experts in the field of HCV genetic variability and by those involved in development of HCV sequence databases, the Hepatitis Virus Database (Japan), euHCVdb (France), and the Los Alamos National Laboratory (United States). The purpose of the nomenclature initiative was to reexamine the status of HCV genotype nomenclature, to resolve conflicting genotype or subtype names of described variants of HCV, and to draw up revised criteria for the assignment of new genotypes as they are discovered in the future. The numbering initiative designed a standardized numbering system for HCV nucleotides, proteins, and epitopes. Both initiatives are actively supported by the HCV sequence databases.

2. The Nomenclature System

A standard classification system is important in studies of the epidemiology, evolution, and pathogenesis of HCV. Understanding genotype-specific differences in response to interferon-based and antiviral treatments has immediate clinical relevance. A new classification system must be robust, based on objective criteria, and able to accommodate new genetic variants and recombinant forms that are discovered in the future. The classification of HCV should therefore be based, as are other biological nomenclature systems, on its evolutionary history (as far as it is currently understood). First, we review current thoughts on the origins and epidemiology underlying the observed genetic diversity of HCV and address how these aspects could be incorporated into the proposed classification scheme.

2.1. HCV Sequence Variability

When the scale of genetic heterogeneity of HCV became clear in the early 1990s, a number of different, often incompatible, methods were used to classify variants, resulting in differences in the letters or numbers assigned to each recognized genetic group (1, 2). Progress toward resolving these uncertainties was made by publication of a consensus paper in 1994 (3), which proposed the classification of HCV by phylogenetic methods into six genotypes. An updated phylogenetic tree from this proposal is shown in **Fig. 4.1**. Each of the approximately equidistant genetic groups contains a variable number of more closely related,

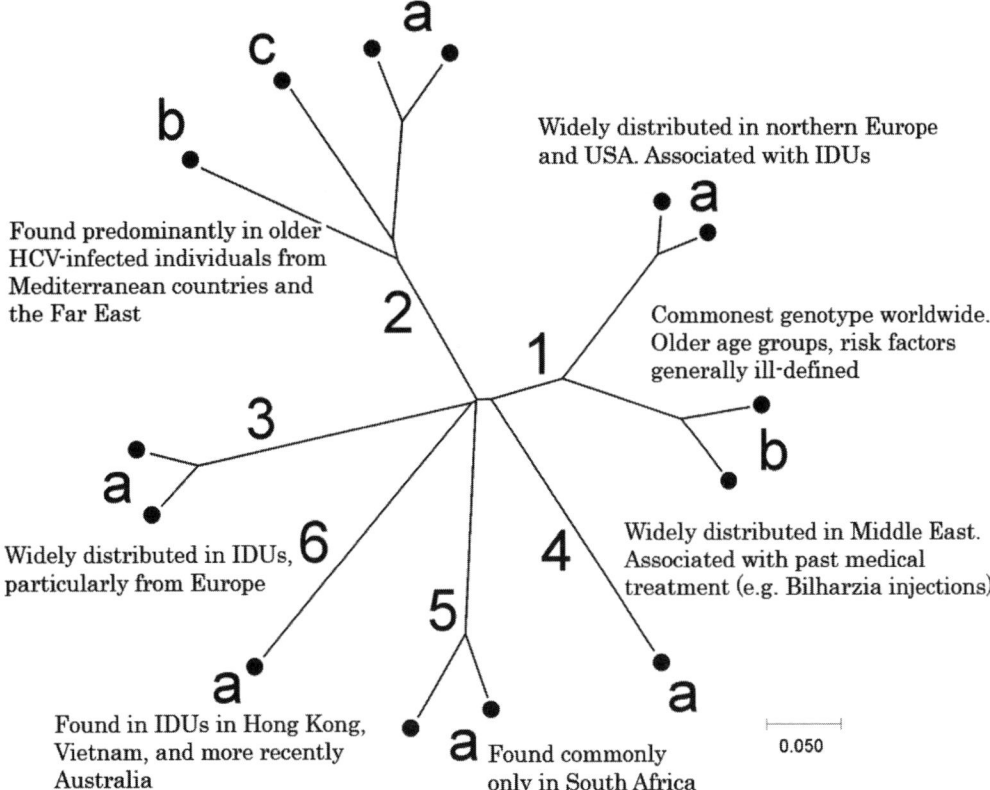

Fig. 4.1. Evolutionary tree of the principal genotypes of HCV found in industrialized countries. Phylogenetic analysis was carried out on complete genome sequences of genotypes of HCV found in the main identified risk groups for HCV infection: injection drug users (IDUs), recipients of unscreened blood or blood products, and those experiencing other parenteral exposures. These represent the main variants believed to have become prevalent over the course of the 20th century. The tree was constructed with the phylogeny program Neighbor-Joining implemented in the MEGA package, with Jukes-Cantor corrected distances.

genetically (and epidemiologically) distinct "subtypes." Genotypes differ from each other by 31–33% at the nucleotide level, subtypes by 20–25%.

Despite the sequence diversity of HCV, all genotypes share an identical complement of colinear genes of similar size in a large open reading frame, and the genetic inter-relationships of HCV variants are remarkably consistent throughout the genome (4). This consistency has allowed provisional classification of many of the currently recognized HCV variants, on the basis of partial sequences from subgenomic regions such as core/E1 and NS5B (5).

Subsequent molecular epidemiology studies have revealed great HCV diversity in certain regions of sub-Saharan Africa and in south and southeast Asia (**Fig. 4.2**). Most newly described variants originate from specific geographical regions. For example, infections in western Africa are predominantly genotype 2

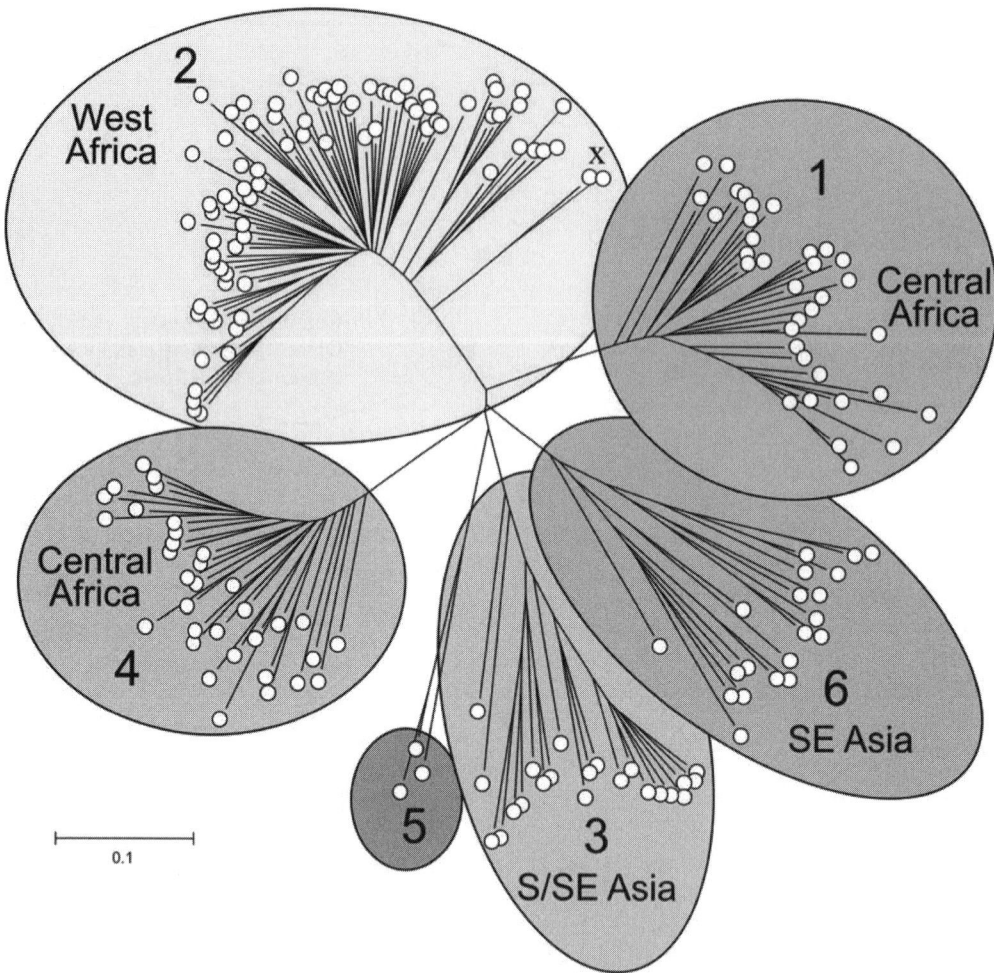

Fig. 4.2. Evolutionary tree of all known subtypes and genotypes HCV. This phylogenetic analysis of the NS5B region of all published sequences demonstrates that HCV variants still fall into six distinct genotypes, but each genotype includes numerous novel variants discovered in high-diversity areas in sub-Saharan Africa and southeast Asia. The tree was constructed with the phylogeny program Neighbor-Joining implemented in the MEGA package, with Jukes-Cantor corrected distances. More divergent members of genotype 2 are marked with an "x."

(6–8), whereas those in central Africa, such as the Democratic Republic of Congo and Gabon, are genotypes 1 and 4 (9–11). Genotypes 3 and 6 show similar genetic diversity in south and eastern Asia (12–14). This "endemic" pattern of sequence variability suggests that HCV has probably circulated in human populations in these parts of Africa and Asia for centuries, millennia, or longer.

In contrast, the most common variants found in Western countries (1a and 1b in genotype 1; 2a, 2b, and 2c in genotype 2) have become widely distributed over the past 50–70 years as a result of transmission through blood transfusion and other invasive medical procedures and of needle sharing by injection drug users (IDUs) *(15–17)*. They now represent the vast majority of infections encountered clinically in Western countries. Subtypes 1a, 1b, 2a, 2b, 3a, and 4a are likely to be the descendants of HCV variants from endemic areas that "seeded" these new, rapidly expanding transmission networks.

As discussed below, HCV classification should recognize the epidemiological associations of these "founder" viruses and incorporate their subtype names into the genotype nomenclature, while acknowledging that such labels are of little or no value in the description of HCV variants in high-diversity areas in sub-Saharan Africa and Southeast Asia.

2.1.1. Recombination

A recent discovery with implications for HCV classification is recombination between genotypes of HCV *(18–23)*. Homologous recombination in HCV could clearly be facilitated by the overlap in genotype distributions in many parts of the world. Because of the nature of HCV risk behavior, frequent exposures may surround the time of primary infection (e.g., repeated needle sharing by several IDUs), when protective immunity from reinfection has not yet been established. The true frequency of recombination may be underestimated because of the difficulty of detecting within-subtype recombination. In regions where HCV is highly diverse, such as west Africa, even intersubtype recombination will be difficult to detect. Finally, although a comparatively great number of complete genome sequences are available for common HCV genotypes, such as 1b, most studies of HCV variability in high-diversity areas are based on analysis of single subgenomic regions, such as NS5B or core/E1, making detection of potential recombinants unlikely.

2.1.2. HCV Genotype Identification

Because genotypes 1 and 4 are more resistant than genotypes 2 and 3 to the current standard of care, pegylated interferon and ribavirin combination therapy *(24)*, most treatment protocols require genotype information to tailor dose and duration of treatment. Genotyping assays are usually based on sequence analysis of an amplified segment of the genome, commonly the 5′ untranslated region (5′UTR). Although this region is highly conserved, a well-characterized set of polymorphisms predicts genotype and can be conveniently detected by probe hybridization *(25)*, changes in restriction sites *(26)*, or direct sequencing *(27)*. For prediction of treatment response and dose scheduling, currently used 5′UTR-based assays are acceptably accurate (more than 95% concordance with genotypes identified by nucleotide sequencing *(28-31)*), but their use for definitive genotype identification, for identification of subtypes, and more generally as an HCV classification tool is problematic for several reasons:

1. Even for genotype identification, the performance of genotyping assays is a property of the range of HCV variants tested. For example, some genotype 6 variants from southeast Asia have 5′UTR sequences identical to those of genotype 1a or 1b *(13,14,16,32)*. The currently used 5′UTR-based assays are unlikely to operate to the published level of accuracy in high-diversity areas.

2. Sequence differences between *subtypes* may be variable or nonexistent in the 5′UTR. For example, a sequence polymorphism at position 243 (numbered as in the H77 reference sequence, see below), frequently used to differentiate subtypes 1a and 1b, is unreliable. More generally, although some subtypes are identifiable in the 5′UTR (such as 2a and 2b), others, such as 2c, are not, even though all three subtypes are approximately equally divergent from each other elsewhere in the genome (**Fig. 4.1**).

3. All genotyping assays, whether based on the 5′UTR or elsewhere in the genome, involve the intrinsic assumption that the genotype inferred from one region reflects that of the genome as a whole. Although few recombinant forms have been described, the spread of HCV variants such as the 2k/1b recombinant and the generation of further hybrid viruses in multiply exposed individuals would further limit the accuracy of genotyping assays and attenuate their predictive value for treatment response.

Clearly, HCV identification is an activity distinct from HCV classification. Classification provides the framework within which the specificity and accuracy of genotyping assays can be assessed, and for this purpose, an agreed-upon and consistent set of classification criteria and a system of assigning genotype names are required.

2.2. Current Issues in HCV Classification

Several problems and uncertainties with current classification schemes for HCV cause inconsistencies with the nomenclature of HCV variants in published papers and create difficulties for the organization and retrieval of HCV sequences from the three databases.

2.2.1. Diversity Within Genetic Groups

The primary division of HCV variants into six genetic groups is evident from phylogenetic analysis (**Figs. 4.1 and 4.2**), but genetic diversity within groups 2, 3, and 6 is considerably greater than that among the originally classified subtypes 1a and 1b and 2a, 2b, and 2c *(34)*. In the past, more divergent variants within groups 3 and 6 had been treated as separate major genotypes of HCV, and "clades" subsumed "genotypes." At the first HCV classification meeting (Santa Fe, 1996), genetic group 6 was designated "clade 6," and its variants were labeled genotypes 6, 7, 8, 9, and 11. Similarly, "clade 3" contained variants classified as genotypes 3 and 10 *(2)*. This system compromised the simplicity of the original assignment of HCV genetic groups as genotypes. For example, both subtype 3b and the proposed new genotype 10a are in "clade" 3. Both are highly divergent from subtype 3a (10a 26%, 3b 13%), much more than other subtypes of genotype 3 (**Fig. 4.1**). This divergence is much lower than the 31–34% between variants in different genetic groups (such as between 1a and 2a). Divergence between the various proposed genotypes in group 6 is also consistently lower (mean, 27%; range, 21–29%) than between the originally classified genotypes.

Genetic group 2 may also contain more divergent sequences than other subtypes (marked "x" in **Fig. 4.2**). This situation might lead to the addition of further genotype designations. Apart from the difficulty in placing this further dividing line between genotype and clade, the resulting classification in a sub-type/genotype/clade hierarchy is geographically inconsistent. To many, the scheme has been confusing, because in some cases a clade contains only one genotype and the terms are interchangeable (e.g., genotype 1 and clade 1), whereas in others a clade may include five or more genotypes (e.g., clade 6, genotypes 6, 7, 8, 9, and 11). This confusion and lack of consensus has led to the continuing nomenclature differences between publications whenever variants from southeast Asia and elsewhere are described.

2.2.2. Conflicting Subtype Designations

Many examples exist of conflicting nomenclature within currently classified HCV variants. Most of these involve reference to two different subtypes by the same name, such as subtypes "4a" found in Egypt *(1)* and Zaire *(33)*. Conversely, the same variant is sometimes described with different subtype designation, such as VAT96, designated "2k" *(34)*, and RU169, designated "2j" *(35)*. These conflicts have been resolved in an agreed-upon catalogue of HCV variants *(3)*.

2.2.3. Recombination

Currently no method exists for classifying recombinant forms of HCV. For database retrieval and for cataloguing the occurrence of recombinant viruses, a nomenclature system that records the genotype composition and provides unique identifiers for the pattern of breakpoints would be of value. Such a system is in place for HIV-1 *(4)* and might be used as a model for HCV. In the HIV system, designation of intersubtype recombinant viruses as (circulating) recombinant forms (CRFs) requires detection and complete genome sequences of a recombinant virus from three or more independently infected individuals. To be a member of the same CRF, a recombinant must have breakpoints in the same positions in each sequence. Each CRF is then numbered sequentially in the order of discovery, and subtype identification letters are listed alphabetically to indicate their approximate composition. The HCV recombinant in St. Petersburg would therefore be designated CRF 01_1b2k.

2.3. Consensus Classification Proposals

2.3.1. Division of HCV into Clades/Genotypes

The primary division of HCV variants remains the six genetic groups, called genotypes, irrespective of the increased numbers of subtypes within these groups. Genotypes differ in level of within-group diversity and in within-group relationships, but varying degrees of diversity are becoming apparent in other genotypes too, for example, among the genotype 2 variants from west Africa. Specifying a degree of sequence divergence below which a subtype designation is made, and above which a new genotype is assigned, is difficult and arbitrary. The problem is epitomized by the difficulties with the classifications of 3b and 10a within genotype 3, discussed above.

The following points summarize the recommendations concerning the designation of HCV genotypes:

1. The primary division of HCV will be based on the six genetic groups apparent from **Figs. 4.1 and 4.2** and other published sequence analyses of HCV. This division is supported by the principal methods of phylogenetic analysis of the core/E1, NS5B, and complete genome sequences. For distance-based methods, greater than 70% of trees (actually invariably greater than 90%) support the primary division of HCV variants into the six genetic groups, whereas higher-level groupings are not consistently supported.

2. The genetic groups will be termed "genotypes." The previously proposed term "clade" might be regarded as an alternative, more descriptive term, but for consistency with previous classifications of HCV and current clinical usage, as well as to prevent confusion with the previously defined meaning of the term "clade," as distinct from "genotype," we recommend use of the term "genotype" in HCV sequence databases and publications.

3. Variants of HCV currently designated with genotype numbers above 6 will be renamed according to the genotype groups in which they fall and given the next available subtype designation. For example, genotype 10a will be reclassified as 3k, 7a as 6e, and so on. The proposed changes to the nomenclature are listed on the websites of the HCV databases.

4. The identification of new genotypes will henceforth require demonstration of a consistent independent phylogenetic grouping away from any of the currently classified genotypes of HCV (see below).

2.3.2. Classification and Nomenclature of Previously Described Subtypes of HCV

The group believed the existing nomenclature of HCV subtypes provides a valuable framework for ongoing studies of genetic variation. The following points summarize the group's decisions and recommendations for subtype designations:

Table 4.1
Numbering of H77 (AF009606) genomic regions under different systems

Region	Nucleic acid absolute numbering	Nucleic acid relative numbering	Amino acid absolute numbering	Amino acid relative numbering	Description
	1–341	1–341	na	na	5′ untranslated region
Core	342–914	1–573	1–191	1–191	Core protein
	915–1490	1–576	192–383	1–192	Envelope glycoprotein 1
E2	1491–2579	1–1089	384–746	1–363	Envelope glycoprotein 2
	2580–2768	1–189	747–809	1–63	Putative ion channel
NS2	2769–3419	1–651	810–1026	1–217	Autoprotease
NS3	3420–5312	1–1893	1027–1657	1–631	Serine protease and RNA-dependent RNA helicase
NS4A	5313–5474	1–162	1658–1711	1–54	NS3 cofactor
	5475–6257	1–783	1712–1972	1–261	NS4B protein
NS5A	6258–7601	1–1344	1973–2420	1–448	NS5A phosphoprotein
NS5B	7602–9377	1–1776	2421–3011	1–591	RNA-dependent RNA polymerase
3′UTR	9378–9646	1–269	na	na	3′ untranslated region

na, not applicable.

1. Existing designations, where they are consistent, will be retained, irrespective of the criteria agreed upon for the designation of new subtypes (**Table 4.1**).
2. Variants within genotypes 3 and 6 that have been redesignated subtypes (see previous section) will be incorporated into the updated list.
3. HCV variants with conflicting names in the literature have been redesignated on consultation with the originating authors.

2.3.3. Assignment of New Genotypes of HCV

Further variants of HCV will probably be discovered that merit designation as new genotypes, such as the candidate new genotype obtained from central Africa *(36,37)*. Their correct classification requires demonstration that they show no significant grouping within any of the existing genotypes, according to a rigorous phylogenetic analysis of a complete sequence of the coding region of the virus. This analysis will also confirm the absence of recombination with sequences from previously defined genotypes.

The following criteria were proposed for identification and designation of a variant of HCV as a new genotype:

1. *Provisional designation.* Provisional designation requires that one complete coding-region sequence be obtained, demonstration of a separate grouping from other genotypes by phylogenetic analysis, and an absence of recombination. The sequence of a candidate new genotype should be independently analyzed by submission to one of the HCV databases. The sequence will be analyzed by a variety of phylogenetic methods described previously. This analysis will allow the sequence to be assigned the next available genotype number and the subtype designation "a," for example, genotype 7a.

2. *Confirmed designation.* Confirmed designation requires coding sequences of two or more HCV variants from infections that are not directly linked epidemiologically. The sequences should further demonstrate a lack of grouping with current classified genotypes by the above methods. This further analysis, and any available sequences from subgenomic regions such as core/E1 and NS5B (see below), will provide valuable reference information on the genetic heterogeneity within the newly designated genotype, the existence of subtypes, the geographical origins of the variants, and their likely designation in 5′UTR-based genotyping assays.

The distinction between provisional and confirmed genotypes is also supported by the databases.

2.4. Assignment of New Subtypes of HCV

Different issues come up in the assignment of new subtypes. Some geographical regions contain so much diversity within genotypes that little is gained from their continued classification as subtypes. Elsewhere, however, subtype labels have epidemiological value and are used as genetic markers in studies of past and ongoing virus transmission in different risk groups. Therefore, a new subtype designation should only be provided if evidence supports its potential epidemiological value, that is, when the subtype is widely distributed, is spreading, or shows particular risk-group associations. The simplest method of fulfilling this criterion is to require evidence of infection with a new proposed subtype of HCV of several independently infected individuals. Therefore, the following criteria for assignment of new subtypes were adopted:

1. *Provisional subtype designation.* At least three infections with a new proposed subtype are required for subtype designation. Sequences are required from both the core/E1 region (sequence data from 90% or more of nucleotides corresponding to positions 869–1292 in the H77 reference sequences, accession number AF009606 *(33, 38, 39)* and the NS5B region (data from 90% or more of positions in the region 8276–8615 in H77 *(1, 40)*). A variety of distance-based, parsimony, and maximum-likelihood phylogenetic methods should provide consistent

evidence for clustering and for distinctness from other subtypes. Because currently classified subtypes of HCV differ in nucleotide sequence from each other by more than 15%, this level of divergence will be expected from other HCV variants within the genotype.

As described in the Introduction, the existence of subtypes is primarily an epidemiological phenomenon, associated with the recent spread of HCV. Because subtype designations are primarily epidemiological labels, a formal criterion for their assignment is not appropriate, and the varying degrees of sequence divergence of variants within different genotypes would make the development of such criteria extremely difficult.

Candidate subtypes will be provisionally assigned the next available subtype letter for the genotype on submission to one of the HCV databases. Single or pairs of variants of HCV that otherwise meet these criteria will not be assigned subtype letters in the database until a third sample of that subtype is found.

2. Confirmed subtype designation. One or more complete genome sequences will be required for confirmed designation. This requirement will allow the degree of sequence divergence from other subtypes over the whole genome to be assessed and the absence of recombination to be established.

2.4.1. Recombinant Forms of HCV

The ability of the classification scheme for HCV genotypes to incorporate recombinants is important, but because the number of confirmed recombinants currently described in the literature is small, their formal classification and the development of rules for their nomenclature were deemed to be of less immediate importance. Until review at a subsequent classification meeting, sequences with evidence of recombination will be annotated as such in the databases, and options will be provided to include them in or exclude them from downloads or analyses of sequences.

2.4.2. Interface with HCV Sequence Databases

The HCV sequence databases are in a unique position to support the effort to make HCV nomenclature more uniform. By assigning geno/subtypes to the sequences that people retrieve and download, they can influence the commonly used nomenclature of existing sequences, and they can have a coordinating role in assignment of new geno/subtypes and in keeping track of them, especially before journal publication. The databases are also committed to assisting in the naming of new geno/subtypes, by helping researchers name proposed new geno/subtypes, by checking existing names for consistency and correcting any inconsistencies, by helping the field to keep track of which geno/subtype names have already been assigned, and by providing tools for genotype or subtype identification and detecting recombinants. The HCV database websites will provide access to the criteria for assignment of new genotypes and subtypes of HCV presented here and will make HCV researchers, reviewers, and journal editors aware of these guidelines. They will provide the listing of current assigned subtypes and genotypes (based on **Table 4.1**). These will be automatically updated as sequence data are submitted, showing which designations exist in the databases and which have been assigned but not yet published. The distinction between "provisional" and "confirmed" designations will also be implemented in the databases.

3. The Numbering System

Inconsistent and inaccurate numbering of locations in DNA and protein sequences is a problem in the HCV scientific literature. Consistency in numbering is increasingly required for functional and clinical studies of HCV. For example, an unambiguous method for referring to amino acid substitutions at specific positions in NS3 and NS5B coding sequences associated with resistance to specific HCV inhibitors is essential in the investigation of antiviral treatment. Here, we provide a practical guide to help circumvent these problems in the future and to bring a common language into discussions in the field. The scope of the current system is limited to the HCV polyprotein and the UTRs; because of the controversial nature and extreme length variation of the alternate reading frame proteins, numbering for these proteins, if needed, will be decided at a later date.

We propose a numbering system adapted from the Los Alamos HIV database *(41)*, with elements used in the hepatitis B virus numbering system *(42)*. The system comprises both nucleotides and amino acid sequences and epitopes. It uses the full-length genome sequence of isolate H77 (accession number AF009606) as a reference sequence and includes a method for numbering insertions and deletions relative to this sequence. H77 was chosen because it is a commonly used reference strain for many different kinds of functional studies. Furthermore, RNA transcripts from this sequence are of demonstrated infectivity *(38, 43)*, providing evidence that the 5′ and 3′ ends of the sequence are complete. **Table 4.1** lists the boundaries of HCV genomic regions, and **Fig. 4.3** provides detailed nucleic acid and amino acid numbering over the complete AF009606 HCV genome sequence.

3.1. Protein and Polyprotein Numbering

Numbering of an amino acid sequence can be absolute, that is, based on the HCV polyprotein with numbering starting at the first amino acid of the core protein and continuing through the end of NS5B, or it can be relative, that is, starting

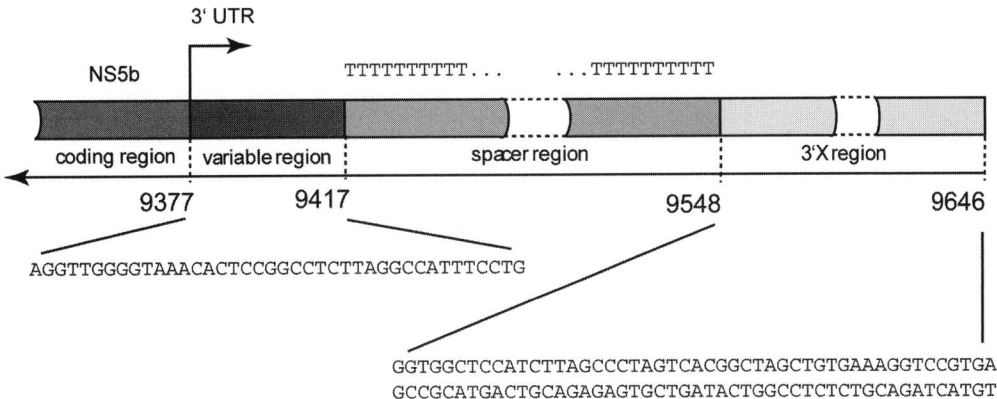

Fig. 4.3. Diagram of the 3′UTR, including detailed nucleic acid and amino acid numbering over the complete AF009606 HCV genome sequence.

over at every protein (core numbered from 1 to 191, then E1 from 1 to 192, etc.). Both of these numbering systems are used in HCV research. Although most epitopes are numbered absolutely, drug-resistance mutations tend to be numbered relatively, following the practice for HIV-1 and now HBV. All HCV sequence databases incorporate both numbering systems. In order to prevent confusion, the protein coordinates should be preceded by the protein name (e.g., NS3-A156T) when relative numbering is used. The relative and absolute numbering are easily interconverted by means of either the Sequence Locator tool, available at the American HCV database website (http://hcv.lanl.gov/content/sequence/LOCATE/locate.html), or the Number tool on the European HCV database website (http://euhcvdb.ibcp.fr/euHCVdb/).

3.2. Numbering of the 5′UTR

Two different methods are used to number the 5′UTR, each with advantages and disadvantages. One system starts numbering the polyprotein coding sequence at 1 and continues on through to the end of the 3′UTR; the 5′UTR is numbered in the reverse direction, starting at –1 and continuing on to –341. The other system simply starts at 1 for the first nucleotide of the 5′UTR and continues on to the last nucleotide of the 3′UTR. For reasons of simplicity, we have decided to adopt the numbering system that sets the start of the 5′UTR at 1. Nucleotide numbering continues uninterrupted into the coding sequence. In the case of AF009606, the AUG initiation codon would be numbered as 342–344. This system avoids the practical problem encountered in most sequence editors of numbering a sequence nonconsecutively (i.e., in the negative numbering system for the 5′UTR, the numbering at the 5′UTR/core junction should proceed ... – 3, –2, –1, +1, +2, +3 ..., whereas sequence editors tend to number the –1 base as zero).

3.3. Numbering of the 3′UTR

Numbering of the 3′ end of the HCV genome must accommodate a stretch of pyrimidine residues (the PPT) of highly variable length between the 40 alignable nucleotides at the beginning of the 3′UTR and the end of the 3′UTR, including the highly conserved 3′X-tail. Diagrammatically, the 3′UTR comprises the sequential stretches (AF009606 numbering) shown in **Fig. 4.3**.

The length of the PPT changes rapidly over time within an experimentally infected chimpanzee *(44)*; different clones of the original H77 isolate *(38, 43)* (accessions AF009069-77) are known to differ in PPT length. The actual numbering of the 3′ end of the 3′UTR sequence beyond the PPT is therefore of no significance. Furthermore, because the alignment of the PPT is arbitrary, so are any designation of insertions or deletions and any numbering attempt in this region.

An additional complication is caused by the choice of H77 as the reference sequence. The first three nucleotides of the 3′X region of H77 is AAT; this is not the case in most other isolates, in which the 3′X is only 98 nucleotides long. To prevent complications arising from this unusual feature of H77, the AAT of H77 should, for numbering purposes, be considered part of the PPT rather than the 3′X region.

With this exception, the 3′UTR follows the numbering of the reference sequence AF009606. If the PPT of the sequence under consideration is not longer than AF009606 (no sequence currently in the HCV database has a longer PPT), the numbering is straightforward. For a sequence with a longer PPT, we propose that the insertion be ignored for the purpose of numbering. The first nucleotide of the continuation of the alignment after the PPT, the region described in **ref.** *(38)* and numbered 9418 in their paper, will therefore be numbered 9549, regardless of whether that is its actual position. Positions after that will again be numbered consecutively.

3.4. Numbering of Mutations

Mutations are numbered according to their positions, for example, a NS5A:R217K would be used to indicate that the arginine in position 217 of the NS5A protein has mutated to a lysine. The codon involved changes from AGG to AAG (numbering 6906–6908), so the corresponding nucleotide mutation would be denoted G6907A.

3.5. Dealing with Insertions and Deletions Relative to the Reference Sequence

Insertions can be incorporated if the numbering is based on an alignment that includes all possible insertions or if a special numbering system is devised for insertions relative to a given numbering system. The second method is preferable for several reasons. First, sequences occasionally contain very long insertions, so defining an alignment that could accommodate all future insertions would be difficult, and changing the numbering system in a way that would invalidate all previous numbering would be very disruptive. This problem is circumvented by the method outlined below. Second, insertions in HCV are relatively rare, so this notation will not have to be used often.

Deletions are much simpler to deal with than insertions, and we outline a simple naming method for unequivocally designating them.

3.5.1. Insertion in Sequence Relative to the Reference Sequence

For insertions in sequence we propose a residue number/alphabet in which inserted bases or amino acids are indicated by lowercase letters following the nucleotide or amino acid position where they occur. For example, three inserted bases in an HCV variant inserted between positions 131 and 132 in the AF009606 reference sequence would be numbered 131a, 131b, and 131c. (For insertions longer than the length of the alphabet, numbering would proceed as 131x, 131y, 131z, 131aa, 131ab, 131ac, …, 131az, 131ba, 131bb, 131bc ….) A similar scheme has been used for numbering amino acids in the immunoglobulin complementarity-determining region (CDR) loops [e.g., **ref.** *(45)*].

For example, in the NS5A fragment shown in **Fig. 4.4**, the location in the variant of the inserted aspartate (D) between AF009606 residues 2412 and 2413 would be referred to as D2412a or NS5A-D440a.

Table 4.2 shows the numbering of each variant amino acid in this example.

3.5.2. Deletion in Sequence Relative to the Reference Sequence

Mentioning the deletions can be useful in cases where the length of the fragment is important or when the location of an amino acid located after a gap is referred to.

For example, in the following NS5A fragment (**Fig. 4.5**), the variant region would be numbered 2410-2415 (del 2413) or NS5A-438-443 (del 441) to make this explicit.

Table 4.3 shows the numbering of each variant amino acid in this example.

Mentioning the deletions would be useful in cases where the length of the fragment is important; in this example, if the deletion were not made explicit, the range 2410–2415 could be taken to refer to a peptide 6, rather than 5, amino acids long.

```
2411  2415 amino acids from start of H77 polyprotein
 |     |
GA-DTE  - AF009606
EADDTE  - variant sequence
```

Fig. 4.4. An example of an insertion variant.

Table 4.2
Example illustrating the numbering of insertions relative to the H77 sequence in an NS5A fragment (amino acids 211–215 from the beginning of H77 polyprotein, GA–DTE in AF009606), in which an aspartate (D) inserted between AF009606 residues 2412 and 2413 yields sequence EADDTE

H77 AF009606	G	A	—	D	T	E
Variant AA	E	A	D	D	T	E
AA absolute	2411	2412	2412a	2413	2414	2415
AA relative	439	440	440a	441	442	443
Variant NA	GAA	GCC	GAC	GAC	ACG	GAA
NA absolute	7572–7574	7575–7577	7577a–7577c	7578–7580	7581–7583	7584–7586
NA relative	1315–1317	1318–1320	1320a–1320c	1321–1323	1324–1326	1327–1329

AA = amino acid; NA = nucleic acid.

```
2410  2415 H77 AA position from start of polyprotein
 |     |
SGADTE - AF009606
EEA-TE - variant
```

Fig. 4.5. An example of a deletion variant.

Table 4.3
Example illustrating the numbering of deletions relative to the H77 sequence in an NS5A fragment (amino acids 2410–2415 from the beginning of H77 polyprotein, SGADTE), in a deletion of amino acid 2413 yields sequence EEA–TE

H77 AF009606	S	G	A	D	T	E
Variant AA	E	E	A	–	T	E
AA absolute	2410	2411	2412	–	2414	2415
AA relative	438	439	440	–	442	443
Variant NA	GAA	GAA	GCC	–	ACG	GAA
NA absolute	7569–7571	7572–7574	7575–7577	–	7581–7583	7584–7586
NA relative	1312–1314	1315–1317	1318–1320	–	1324–1326	1327–1329

AA = amino acid; NA = nucleic acid.

3.5.3. Synthetic
Sequences

A separate problem is that of artificially engineered sequences, like those that result when stretches of the HCV genome are replaced with extraneous genetic segments. The numbering system should be able to cope with this situation also, by treating the extraneous segment as a special type of HCV, which can be numbered according to the length of the insert. Finding the stretches that correspond to "real" HCV depends on pattern matching, which requires accurate location of the end and restart of HCV sequences on either side of the insert to retain correct numbering.

Using the same example (**Fig. 4.6**),

Table 4.4 shows the numbering of the insert in this example.

3.6. Numbering Positions in Other Genotypes

Although the numbering scheme is based on a genotype 1a reference sequence, both the Sequence Locator tool (U.S. database) and the Number tool (European database) are sufficiently flexible to accommodate other genotypes easily. The tools align sequences of all genotypes described to date unambiguously to the reference sequence, so the numbering will be uniform for other genotypes as well as for genotype 1.

```
2411              2415 amino acids from polyprotein start
  |                 |
  GADT-------E   - AF009606
  GADTYNTVATLE  - variant sequence
      =======
      insert
```

Fig. 4.6. An example of a synthetic sequence variant.

Table 4.4
Example illustrating the numbering of a sequence inserted after amino acid 2414 of the H77 sequence in an NS5A fragment (amino acids 211–215 from the beginning of H77 polyprotein, GA–DTE in AF009606), yielding sequence GADTYNTVATLE

H77 AF009606	G	A	D	T	— ... —	E
Variant AA	G	A	D	T	Y ... L	E
AA absolute	2411	2412	2413	2414	2414a ... j	2415
AA relative	439	440	441	442	442a ... j	443
Variant NA	GGA	GCC	GAC	ACG	TAT ... CTC	GAA
NA absolute	7572–7574	7575–7577	7578–7580	7581–7583	7583a–7583ad	7584–7586
NA relative	1315–1317	1318–1320	1321–1323	1324–1326	1326a–1326ad	1327–1329

AA = amino acid; NA = nucleic acid.

3.7. Summary and Conclusion

The numbering proposal presented here, using the AF009606 (isolate H77) sequence as a reference, should be able to number all possible mutations in HCV unequivocally, both natural and man-made. The HCV sequence databases (46) and the Los Alamos HCV immunology database (47) (as well as the Los Alamos HIV database) number positions and epitopes according to this system. Moreover, the database websites provide tools for finding stretches of sequence by their numbers, for assigning start and end coordinates to a sequence, and for converting among the various numbering systems.

HCV nucleotide sequences are numbered by analogy to H77. The first step is to align your sequence to H77. If length does not vary, the numbering is straightforward; nucleotide numbers run from 1 (start of 5′UTR) to 9646 (end of 3′UTR). Insertions relative to H77 are labeled with letters.

Protein numbering works like the nucleotide numbering but starts at the beginning of the polyprotein. The sequence databases will support both systems but use polyprotein numbering as a basis. Absolute numbering continues across the coding regions; relative numbering starts over at every coding region. Relative numbering is almost exclusively used for proteins, polyprotein numbering mainly in immunology, and protein numbering in drug-resistance research. The Los Alamos immunology database uses polyprotein numbering.

The 5′UTR numbering starts at 1 and ends at 341; the Core CDS starts at 342. The numbering of the 3′UTR starts at 9378 (after the stop codon), but complications arise because of the variable length of the PPT. The UTR consists of three elements: a variable 5′ region, the PPT, and a conserved 3′ region, often called X. The first region is numbered 9378–9410. The PPT consists almost entirely of T's and therefore cannot be meaningfully aligned; it is numbered according to its length in H77, 9411–9545. The X region starts at 9546 (regardless of its actual location, which depends on the length of the PPT) and ends at 9646.

Acknowledgments

For their contributions, we thank Jens Bukh, Christophe Combet, Gilbert Deléage, Nobuyuki Enomoto, Stephen Feinstone, Phillippe Halfon, Geneviève Inchauspé, Geert Maertens, Masashi Mizokami, Donald G. Murphy, Hiroaki Okamoto, Jean-Michel Pawlotsky, François Penin, Erwin Sablon, Tadasu Shin-I, Lieven J. Stuyver, Heinz-Jürgen Thiel, Sergei Viazov, Amy J. Weiner, and Anders Widell.

References

1. Simmonds, P., Holmes, E. C., Cha, T. A., Chan, S. W., McOmish, F., Irvine, B., et al. (1993) Classification of hepatitis C virus into six major genotypes and a series of subtypes by phylogenetic analysis of the NS-5 region. *J. Gen. Virol.* **74,** 2391–2399.

2. Cha, T. A., Beall, E., Irvine, B., Kolberg, J., Chien, D., Kuo, G., et al. (1992) At least five related, but distinct, hepatitis C viral genotypes exist. *Proc. Natl. Acad. Sci. USA* **89,** 7144–7148.

3. Simmonds, P., Bukh, J., Combet, C., Deléage, G., Enomoto, N., Feinstone, S., et al. (2005) Consensus proposals for a unified system of nomenclature of hepatitis C virus genotypes. *Hepatology* **42,** 962–973.

4. Robertson, B., Myers, G., Howard, C., Brettin, T., Bukh, J., Gaschen, B., et al. (1998) Classification, nomenclature, and database development for hepatitis C virus (HCV) and related viruses: proposals for standardization. International Committee on Virus Taxonomy. *Arch. Virol.* **143,** 2493–2503.

5. Simmonds, P., Holmes, E. C., Cha, T. A., Chan, S. W., McOmish, F., Irvine, B., et al. (1994) Identification of genotypes of hepatitis C virus by sequence comparisons in the core, E1 and NS-5 regions. *J. Gen. Virol.* **75,** 1053–1061.

6. Jeannel, D., Fretz, C., Traore, Y., Kohdjo, N., Bigot, A., Pê Gamy, E., et al. (1998) Evidence for high genetic diversity and long-term endemicity of hepatitis C virus genotypes 1 and 2 in West Africa. *J. Med. Virol.* **55,** 92–97.

7. Ruggieri, A., Argentini, C., Kouruma, F., Chionne, P., D¡ENT FONT=(normal text) VALUE=39¿'¡/ENT¿Ugo, E., Spada, E., et al. (1996) Heterogeneity of hepatitis C virus genotype 2 variants in West Central Africa (Guinea Conakry). *J. Gen. Virol.* **77,** 2073–2076.

8. Candotti, D., Temple, J., Sarkodie, F., and Allain, J.-P. (2003) Frequent recovery and broad genotype 2 diversity characterize hepatitis C virus infection in Ghana, West Africa. *J. Virol.* **77,** 7914–7923.

9. Mellor, J., Holmes, E. C., Jarvis, L. M., Yap, P. L., and Simmonds, P. (1995) Investigation of the pattern of hepatitis C virus sequence diversity in different geographical regions: implications for virus classification. The International HCV Collaborative Study Group. *J. Gen. Virol.* **76,** 2493–2507.

10. Ndjomou, J., Kupfer, B., Kochan, B., Zekeng, L., Kaptue, L., and Matz, B. (2002) Hepatitis C virus infection and genotypes among human immunodeficiency virus high-risk groups in Cameroon. *J. Med. Virol.* **66,** 179–186.

11. Fretz, C., Jeannel, D., Stuyver, L., Hervé, V., Lunel, F., Boudifa, A., et al. (1995) HCV infection in a rural population of the Central African Republic (CAR): evidence for three additional subtypes of genotype 4. *J. Med. Virol.* **47,** 435–437.

12. Tokita, H., Shrestha, S. M., Okamoto, H., Sakamoto, M., Horikita, M., Iizuka, H., et al. (1994) Hepatitis C virus variants from Nepal with novel genotypes and their classification into the third major group. *J. Gen. Virol.* **75,** 931–936.

13. Tokita, H., Okamoto, H., Luengrojanakul, P., Vareesangthip, K., Chainuvati, T., Iizuka, H., et al. (1995) Hepatitis C virus variants from Thailand classifiable into five novel genotypes in the sixth (6b), seventh (7c, 7d) and ninth (9b, 9c) major genetic groups. *J. Gen. Virol.* **76,** 2329–2335.

14. Tokita, H., Okamoto, H., Tsuda, F., Song, P., Nakata, S., Chosa, T., et al. (1994) Hepatitis C virus variants from Vietnam are classifiable into the seventh, eighth, and ninth major genetic groups. *Proc. Natl. Acad. Sci. USA* **91,** 11022–11026.

15. Marsh, M., Helenius, A., Pybus, O. G., Charleston, M. A., Gupta, S., Rambaut, A., et al. (2001) The epidemic behavior of the hepatitis C virus. *Science* **292,** 2323–2325.

16. Simmonds, P. The origin and evolution of hepatitis viruses in humans. *J. Gen. Virol.* **82,** 693–712.

17. Cochrane, A., Searle, B., Hardie, A., Robertson, R., Delahooke, T., Cameron, S., et al. (2002) A genetic analysis of hepatitis C virus transmission between injection drug users. *J. Infect. Dis.* **186,** 1212–1221.

18. Legrand-Abravanel, F., Claudinon, J., Nicot, F., Dubois, M., Chapuy-Regaud, S., Sandres-Saune, K., et al. (2007) New natural intergenotypic (2/5) recombinant of hepatitis C virus. *J Virol.* **81,** 4357–4362.

19. Tokita, H., Shrestha, S. M., Okamoto, H., Sakamoto, M., Horikita, M., Iizuka, H., et al. (2006) Serendipitous identification of natural intergenotypic recombinants of hepatitis C in Ireland. *Virol. J.* **3,** 95–95.

20. Kageyama, S., Agdamag, D. M., Alesna, E. T., Leaño, P. S., Heredia, A. M. L., Abellanosa-Tac-An, I. P., et al. (2006) A natural intergenotypic (2b/1b) recombinant of hepatitis C virus in the Philippines. *J. Med. Virol.* **78,** 1423–1428.

21. Colina, R., Casane, D., Vasquez, S., García-Aguirre, L., Chunga, A., Romero, H., et al.

(2004) Evidence of intratypic recombination in natural populations of hepatitis C virus. *J. Gen. Virol.* **85**, 31–37.

22. Kalinina, O., Norder, H., Mukomolov, S., and Magnius, L. O. (2002) A natural intergenotypic recombinant of hepatitis C virus identified in St. Petersburg. *J. Virol.* **76**, 4034–4043.

23. Noppornpanth, S., Lien, T. X., Poovorawan, Y., Smits, S. L., Osterhaus, A. D. M. E., and Haagmans, B. L. (2006) Identification of a naturally occurring recombinant genotype 2/6 hepatitis C virus. *J. Virol.* **80**, 7569–7577.

24. Hnatyszyn, H. J. (2005) Chronic hepatitis C and genotyping: the clinical significance of determining HCV genotypes. *Antiviral Ther.* **10**, 1–11.

25. Stuyver, L., Rossau, R., Wyseur, A., Duhamel, M., Vanderborght, B., Van Heuverswyn, H., et al. (1993) Typing of hepatitis C virus isolates and characterization of new subtypes using a line probe assay. *J. Gen. Virol.* **74**, 1093–1102.

26. Davidson, F., Simmonds, P., Ferguson, J. C., Jarvis, L. M., Dow, B. C., Follett, E. A., et al. (1995) Survey of major genotypes and subtypes of hepatitis C virus using RFLP of sequences amplified from the 5′ non-coding region. *J. Gen. Virol.* **76**, 1197–1204.

27. Ross, R. S., Viazov, S. O., Holtzer, C. D., Beyou, A., Monnet, A., Mazure, C., et al. (2000) Genotyping of hepatitis C virus isolates using CLIP sequencing. *J. Clin. Microbiol.* **38**, 3581–3584.

28. Simmonds, P., Rose, K. A., Graham, S., Chan, S. W., McOmish, F., Dow, B. C., et al. (1993) Mapping of serotype-specific, immunodominant epitopes in the NS-4 region of hepatitis C virus (HCV): use of type-specific peptides to serologically differentiate infections with HCV types 1, 2, and 3. *J. Clin. Microbiol.* **31**, 1493–1503.

29. Lau, J. Y., Mizokami, M., Kolberg, J. A., Davis, G. L., Prescott, L. E., Ohno, T., et al. (1995) Application of six hepatitis C virus genotyping systems to sera from chronic hepatitis C patients in the United States. *J. Infect. Dis.* **171**, 281–289.

30. Zheng, X., Pang, M., Chan, A., Roberto, A., Warner, D., and Yen-Lieberman, B. (2003) Direct comparison of hepatitis C virus genotypes tested by INNO-LiPA HCV II and TRUGENE HCV genotyping methods. *J. Clin. Virol.* **28**, 214–216.

31. Gault, E., Soussan, P., Morice, Y., Sanders, L., Berrada, A., Rogers, B., et al. (2003) Evaluation of a new serotyping assay for detection of anti-hepatitis C virus type-specific antibodies in serum samples. *J. Clin. Microbiol.* **41**, 2084–2087.

32. Mellor, J., Walsh, E. A., Prescott, L. E., Jarvis, L. M., Davidson, F., Yap, P. L., et al. Survey of type 6 group variants of hepatitis C virus in Southeast Asia by using a core-based genotyping assay. *J. Clin. Microbiol.* **34**, 417–423.

33. Bukh, J., Purcell, R. H., and Miller, R. H. (1993) At least 12 genotypes of hepatitis C virus predicted by sequence analysis of the putative E1 gene of isolates collected worldwide. *Proc. Natl. Acad. Sci. USA* **90**, 8234–8238.

34. Samokhvalov, E. I., Hijikata, M., Gylka, R. I., Lvov, D. K., and Mishiro, S. (2000) Full-genome nucleotide sequence of a hepatitis C virus variant (isolate name VAT96) representing a new subtype within the genotype 2 (arbitrarily 2k). *Virus Genes* **20**, 183–187.

35. Tokita, H., Okamoto, H,, Iizuka, H., Kishimoto, J., Tsuda, F., Lesmana, L. A., et al. (1996) Hepatitis C virus variants from Jakarta, Indonesia classifiable into novel genotypes in the second (2e and 2f), tenth (10a) and eleventh (11a) genetic groups. *J. Gen. Virol.* **77**, 293–301.

36. Depla, E., Maertens, G., De Nys, K., Blockx, H., Van Doorn, L. J., Quint, W., et al. (2003) A putative new clade 7 of hepatitis C virus containing at least one type and two subtypes, in *10th International Meeting on Hepatitis C Virus and Related Viruses*, Kyoto, Japan.

37. Murphy, D. G., Willems, B., Deschênes, M., Hilzenrat, N., Mousseau, R., and Sabbah, S. (2007) Use of sequence analysis of the NS5B region for routine genotyping of hepatitis C virus with reference to C/E1 and 5′ untranslated region sequences. *J. Clin. Microbiol.* **45**, 1102–1112.

38. Kolykhalov, A. A., Feinstone, S. M., and Rice, C. M. (1996) Identification of a highly conserved sequence element at the 3′ terminus of hepatitis C virus genome RNA. *J. Virol.* **70**, 3363–3371.

39. Bukh, J., Purcell, R. H., and Miller, R. H. (1994) Sequence analysis of the core gene of 14 hepatitis C virus genotypes. *Proc. Natl. Acad. Sci. USA* **91**, 8239–8243.

40. Enomoto, N., Takada, A., Nakao, T., and Date, T. (1990) There are two major types of hepatitis C virus in Japan. *Biochem. Biophys. Res. Comm.* **170**, 1021–1025.

41. Korber, B., Foley, B., Kuiken, C., Pillai, S., and Sodroski, J. (1998) Numbering positions in HIV relative to HXB2CG, in *Human Retroviruses and AIDS* (Korber. B. K., Foley, C. L., Hahn, B., McCutchan, F., Mellors, J. W., and

Sodroski, J., eds.), Los Alamos National Laboratory, Los Alamos, pp. 171–181.

42. Stuyver, L. J., Locarnini, S. A., Lok, A., Richman, D. D., Carman, W. F., Dienstag, J. L., et al. Nomenclature for antiviral-resistant human hepatitis B virus mutations in the polymerase region. *Hepatology* **33,** 751–757.

43. Yanagi, M., Purcell, R. H., Emerson, S. U., and Bukh, J. (1997) Transcripts from a single full-length cDNA clone of hepatitis C virus are infectious when directly transfected into the liver of a chimpanzee. *Proc. Natl. Acad. Sci. USA* **94,** 8738–8743.

44. Okamoto, H., Kojima, M., Okada, S., Yoshizawa, H., Iizuka, H., Tanaka, T., et al. (1992) Genetic drift of hepatitis C virus during an 8.2-year infection in a chimpanzee: variability and stability. *Virology* **190,** 894–899.

45. Lucas, A. H., Moulton, K. D., and Reason, D. C. (1998) Role of kappa II-A2 light chain CDR-3 junctional residues in human antibody binding to the *Haemophilus influenzae* type b polysaccharide. *J. Immunol.* **161,** 3776–3780.

46. Kuiken, C., Mizokami, M., Deleage, G., Yusim, K., Penin, F., Tadasu, S.-I., et al. (2006) Hepatitis C databases, principles and utility to researchers. *Hepatology* **43,** 1157–1165.

47. Yusim, K., Richardson, R., Tao, N., Dalwani, A., Agrawal, A., Szinger, J., et al. (2005) Los Alamos hepatitis C immunology database. *Appl. Bioinformat.* **4,** 217–225.

Chapter 5

A Real-Time Taqman® Method for Hepatitis C Virus Genotyping and Methods for Further Subtyping of Isolates

Kathryn J. Rolfe, Tim G. Wreghitt, Graeme J. M. Alexander, and Martin D. Curran

Abstract

The HCV genome is highly heterogeneous; more and more genotypes, each with several distinct subtypes, are being identified around the world. Knowledge of genotype is important for planning of treatment regimes, whereas subtype identification is useful in epidemiological studies and outbreak investigation. We describe HCV genotyping and subtyping assays, based on real-time PCR, that are sensitive, specific, and reliable. These assays provide fast, accurate, and convenient methods for HCV genotyping/subtyping to support clinical practice.

Key words: HCV, Taqman probes, genotype, Rotor-Gene, real-time PCR, subtype, NS5b, phylogenetic analysis.

1. Introduction

1.1. HCV Genotyping

Optimal therapy for HCV infection consists of pegylated interferon-α with ribavirin (combination therapy). Successful treatment results in a sustained viral response (SVR); defined as undetectable HCV RNA at the end of treatment and six months thereafter. The two most important predictors of achievement of SVR are pretreatment viral load and HCV genotype *(1–4)* Large, randomized, controlled trials have shown that SVR is achieved in a higher percentage of patients infected with genotype 2 or 3 (76–80%) than those with genotype 1 (46–52%) *(2, 5)*. As a result, therapy regimes are usually tailored according to genotype *(6–8)*. In the UK, for example, individuals with genotype 2 or 3

Hengli Tang (ed.), *Hepatitis C: Methods and Protocols, Second Edition, vol. 510*
© 2009 Humana Press, a part of Springer Science+Business Media
DOI 10.1007/978-1-59745-394-3_5 Springerprotocols.com

infection are treated for 24 weeks. Patients with genotype 1, 4, 5, or 6 are treated for 12 weeks initially and then for a further 36 weeks only if the 12-week treatment has reduced viral load to less than 1% of its initial level *(6, 7)*. Because combination therapy is expensive and can have severe side effects, correct identification of patients with HCV genotype 2 or 3 infection is crucial, allowing these patients to benefit from shorter courses of treatment, which result in significant savings and higher patient compliance.

HCV genotypes differ in geographical distribution. Genotypes 1, 2, and 3 appear to have worldwide distribution, but their prevalence differs in different regions *(9)*. Subtypes 1a, 1b, 2a, 2c, and 3a account for more than 90% of the HCV infections in North and South America, Europe, Russia, China, Japan, Australia, and New Zealand *(10)*. In Japan, subtype 1b is responsible for 73% of infections *(11)*. In northern and central Africa *(12)* and the Middle East *(13)*, genotype 4 predominates. In our region (eastern England), the predominant genotypes are 1 (48%), 3 (45%), and 2 (5%); genotypes 4 and 5 account for just 2% of infections (2005 data, unpublished).

Real-time PCR has become increasingly important in the diagnostic laboratory and has been used for HCV genotyping *(14–17)*. Recent advances in probe technology include the addition of $3'$ minor groove binding (MGB) groups and the incorporation of locked nucleic acids (LNA). These modifications stabilize the template probe hybrid and have allowed the design of shorter probes that are highly sequence specific over small regions and therefore ideal for genotyping assays. The ever-increasing array of fluorescent labels for these probes, along with the evolution of multichannel, real-time DNA amplification systems, permit the design of multiplex, rapid, and relatively inexpensive genotyping assays. We previously reported the development of a method for detecting the four most common HCV genotypes (1, 2, 3 and 4) based on Taqman probe technology using the Rotor-Gene^{TM} 3000 *(17)*. We have modified and improved this protocol and present here the updated version of our genotyping assay. This method is sensitive, specific, and reliable over a wide range of viral loads. Its target is the $5'$ untranslated region (UTR) of the HCV genome, which is highly conserved yet displays genotype-specific motifs. Nine type-specific Taqman probes, two for genotypes 1 and 2, four for genotype 3, and one for genotype 4, have been designed and incorporated into panels for detection and confirmation of each genotype. The first panel includes two probes for genotype 1 detection and a single probe each for genotypes 2 and 3. The second has a probe for confirmation of genotype 2, two probes for confirmation of genotype 3, and a probe for detection of genotype 4. Because previous experience has shown that subtype 3b isolates may be negative with the second-panel genotype 3 probes, a fourth genotype 3

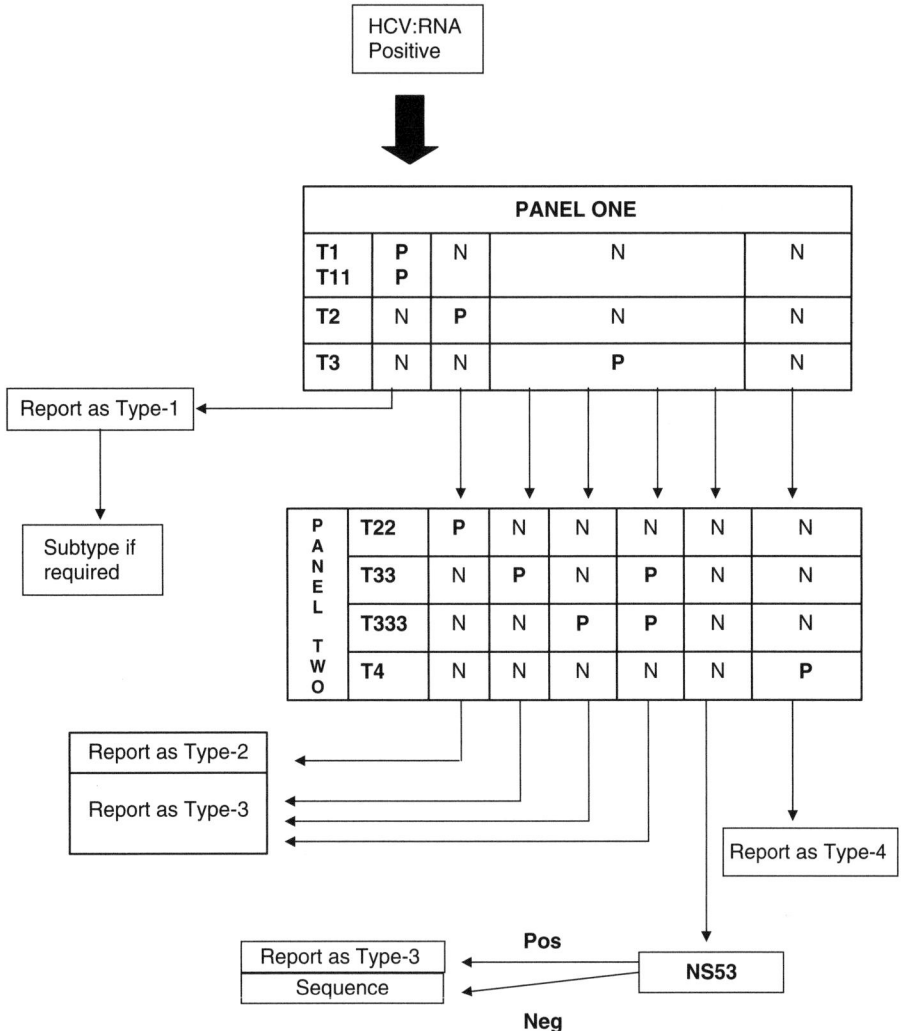

Fig. 5.1. An algorithm for Taqman HCV genotyping.

probe has been designed to confirm the identity of the small, yet significant, number of subtype 3b isolates. Genotypes are assigned and reported with an algorithm that covers all the permutations seen in our region (**Fig. 5.1**).

1.2. Generic HCV Subtyping

Because the RNA genome is so variable, each HCV genotype is further divided into a number of more closely related, genetically distinct subtypes. Knowledge of HCV subtype can be significant in epidemiological studies and in the investigation of infection transmission and outbreaks. Subtyping is problematic because the genomic region selected must be sufficiently conserved for maximum sensitivity but must be variable enough to allow discrimination between subtypes. The 5′ UTR has previously been used for subtyping but has proved unreliable, as discrimination between subtypes is often based on a single variation that may not occur consistently *(18)*. The nonstructural (NS) 5b-encoded RNA-dependent RNA polymerase region of the HCV genome shows more variability and provides better discrimination than the 5′ UTR, so it is superior for subtyping purposes *(19, 20)*.

An association between subtype 1b and disease progression has been suggested but remains controversial *(21–23)*. A link has also been made between infection with subtype 1b and poor response to interferon therapy *(24, 25)*. We have therefore designed a Taqman method with probes that further classify genotype 1 isolates into 1b and non-1b subtypes. To aid our epidemiological and outbreak studies, we have also developed a method using a generic primer pair to amplify a 383-bp fragment of the NS5b region. Sequencing this region can assign subtype to all HCV isolates by phylogenetic relatedness to known subtype reference strains.

2. Materials

2.1. Nucleic Acid Isolation

1. MagNA Pure LC automated nucleic acid extraction robot (Roche Diagnostics, UK).
2. Total nucleic acid isolation kit (Roche Diagnostics). Store at room temperature.

2.2. Reverse Transcription

1. Extracted RNA template.
2. pd(N)$_6$ 50 A$_{260}$ random hexamer (Amersham Bioscience). Store at −20°C.
3. 0.2 mL sterile, thin-wall tubes.
4. Thermal cycler.
5. Ice.
6. M-MLV reverse transcriptase (RT) 200 U/μL (Invitrogen, Paisley, UK). Store at −20°C.
7. dNTP mix (dATP, dCTP, dGTP, dTTP) at 10 mM (Invitrogen). Store at −20°C.
8. 10 × PCR buffer-II (200 mM Tris–HCl, pH 8.4, 500 mM KCl) (Invitrogen). Store at −20°C.
9. 50 mM MgCl$_2$ (Invitrogen, included with 10 × PCR buffer-II). Store at −20°C.
10. RNAse-free water.

2.3. Genotyping

1. Oligonucleotide primers: These are HCV3 (designed and identified as NCR4 by Garson et al. *(26)*) and HCV4 (designed and identified as OKA1 by Okamoto et al. *(27)*), whose targets are highly conserved sites in the 5′ UTR and which were selected for their robustness. These primers amplify 259 bp of the 5′ UTR. Slight modification of the

Table 5.1
Primers Used for HCV Genotyping and Subtyping

Genotyping primer	Sequence	Position[a]
HCV3	CACTCG[b]CA[b]AGCACCCTATCAGGCAGT	313–288
HCV4	TTCACGCAGAAAGCGTCTAG	63–82
Subtyping primer		
NS5bF2	TATGAYACCCGCTGYTTYGACTC	8256–8278
NS5bR	TACCTNGTCATRGCYTCCGTGAA	8638–8616

[a]Position of the 5′ base relative to HCV genomic sequence with reference to HCV type 1 sequence (GenBank accession number M62321). Y = T or C, R = G or A, N = A, T, C or G.
[b]Modifications to original (NCR4) primer, a T → G and a G → A change.

HCV3 primer from the original NCR4 improved its generic nature. See **Table 5.1** for primer sequence data. Primers are purchased from Metabion GmbH, Germany, in 100 μM concentrations and should be stored at −20°C; 20 μM working solutions are generated by mixing of 20 μL of stock solution with 80 μL of RNAse-free water. These should be stored at −20°C until use.

2. Oligonucleotide probes: Taqman probes are used for HCV genotyping. Each probe is labeled with a specific fluorescent dye at the 5′ end (FAM, Cy5, ROX, or JOE/VIC) and a quencher at the 3′ end (Tamra, BHQ-2, or NFQ). Probes T1, T11, T2, T3, and T333 are purchased from Metabion in 100 μM concentrations and should be stored, protected from light, at −20°C, as aliquots so that repetitive freeze-thaw cycles can be avoided; 10 μM working solutions are generated by mixing of 10 μL of stock solution with 90 μL of RNAse-free water. These should be stored, protected from light, at −20°C until use. Small amounts of working solution should be made regularly, as long-term storage of low concentrations is not recommended. The MGB probes (T33 and T4) are purchased from Applied Biosystems, Cheshire, UK, and should be prepared and stored as described above. The LNA probe (T22) is purchased from Sigma-Proligo France SAS at 100 μM concentration and should be stored, protected from light, at −20°C. A 10 μM working solution should be produced and stored as described above. See **Table 5.2** for probe sequence data.

3. LightCycler® FastStart DNA Master HybProbe kit (Roche Diagnostics): Store kit at −15°C to −25°C until expiration date. Prepare Master HybProbe (vial 1) by addition of vial 1b (reaction mix) to 1a (enzyme), and then store vial 1 at −15°C to −25°C for no more than 3 months. After thawing, store

Table 5.2
Taqman Probes Used for Genotyping and Subtyping

Genotyping probe	Reporter	Sequence	Position[a]	Quencher
Panel one				
T1	FAM	CGGAATTGCCAGGACGACCGGGTCCT	169–194	Tamra
T2	JOE	ATTRCCGGRAAGACTGGGTCCTTTC	173–197	Tamra
T3	ROX	CCCCGCRAGATCACTAGCCGAGT	237–259	BHQ-2
T11	Cy5	ACCCGCTCAATGCCTGGAGATTTGG	206–230	BHQ-2
Panel two				
T22 (LNA)	Cy5	ATA+AACCC+ACTC+TATGYCCG	202–221	BHQ-2
T33 (MGB)	VIC	CTCAATACCCAGAAATTTGG	211–230	NFQ
T333	ROX	CCCGGTCACCCCAGCGATTC	190–171	BHQ-2
T4 (MGB)	FAM	AGGCTGTACAACACTCATA	113–95	NFQ
Additional probe				
NS53	ROX	TGTRTTGCCRAAGCTGGTAGGCA	8477–8454	BHQ-2
Subtyping probes				
1a	FAM	GAYATCCGTACGGAGGAGGCAATCTA	8295–8320	BHQ-1
1b (MGB)	VIC	CCGAAGCCAGRCAG	8342–8355	NFQ

[a]Position of the 5′ base relative to HCV genomic sequence with reference to HCV type 1 sequence (GenBank accession number M62321). Y = T or C, R = G or A.
+ indicates locked nucleic acid (LNA) addition.

at +2°C to +8°C for no more than 1 week. Avoid repeated freezing and thawing.

4. 0.2 mL thin-wall flat-lid thermo tubes (ABgene, Surrey).

5. Rotor-Gene^{TM} 3000 real-time amplification system (Corbett Research Limited, Cambridge, UK).

2.4. Subtyping of Genotype 1 Isolates

1. Oligonucleotide primers: These are NS5bF2 and NS5bR, which amplify 383 bp of the NS5b region of all HCV genotypes. See **Table 5.1** for primer sequence data. These are purchased from Metabion GmbH and should be stored and working solutions produced as described for the genotyping primers.

2. Oligonucleotide probes: The FAM-labeled probe for subtype 1a detection is purchased from Metabion GmbH. The VIC-labeled MGB probe for subtype 1b detection is purchased from Applied Biosystems. Subtyping probes should be stored and working solutions produced as described for the genotyping probes. See **Table 5.2** for probe sequence data.

3. LightCycler® FastStart DNA Master HybProbe kit.
4. 0.2 mL thin-wall flat-lid thermo tubes.
5. Rotor-Gene™ 3000 real-time amplification system.

2.5. Generic HCV Subtyping

2.5.1. Amplification with SYBr Green

1. Oligonucleotide primers: These are NS5bF2 and NS5bR, as described above.
2. LightCycler® DNA Master SYBR Green I kit (Roche Applied Science): Store kit at −15°C to −25°C until expiration date. Keep vial 1b (reaction mix) away from light. Prepare master mix (vial 1) by addition of 1a (enzyme) to 1b. Keep vial 1 away from light and store at −15°C to −25°C for up to 3 months. After thawing, store vial 1 at +2°C to +8°C for no more than 1 week. Avoid repeated freezing and thawing.
3. 0.2 mL thin-wall flat-lid thermo tubes.
4. Rotor-Gene™ 3000 real-time amplification system.

2.5.2. Sequencing and Subtype Assignment

1. QIAQuick PCR purification kit (Qiagen Ltd., Crawley, UK)
2. Sequencing platform (e.g., Beckmann or ABI 3100)
3. DNASTAR Lasergene Software package (DNASTAR Inc., Madison, WI; *see* http://www.dnastar.com).
4. National Center for Biotechnology Information (NCBI) website (http://www.ncbi.nlm.nih.gov).

3. Methods

3.1. Nucleic Acid Extraction

1. These protocols require EDTA blood specimens, which should be stored at 4°C until use. These should be spun down in a centrifuge and the plasma separated and stored at −20°C until use.
2. Extract nucleic acid from 200 μL of serum with a Total Nucleic Acid Isolation kit using automated MagNA Pure extraction equipment and elute with 60 μL of elution buffer according to manufacturer's guidelines.
3. Store extracted nucleic acid at −20°C until use. Avoid repeated freezing and thawing.

3.2. Reverse Transcription

1. For each sample, add 1 μL (0.53 μg/reaction) pd(N)$_6$ 50 A$_{260}$ random hexamer to a labeled 0.2 mL tube.
2. Add 20 μL of extracted RNA to each tube, mix, and pulse spin in the centrifuge.
3. Place on a thermal cycler and heat to 70°C for 5 min to linearize the nucleic acid.
4. Meanwhile, prepare a master mix for reverse transcription. For each reaction you will need the following: 3.5 μL

of $10 \times$ PCR buffer-II, $3.5\,\mu L$ 50 mM $MgCl_2$, $1\,\mu L$ dNTP mix, $1\,\mu L$ M-MLV-RT, $5\,\mu L$ RNAse-free water.

5. After 5 min of thermal cycling, transfer the tubes containing linearized nucleic acid immediately to ice and incubate them for 2 min. Pulse spin in the centrifuge.

6. Add $14\,\mu L$ of prepared master mix to each tube, mix, and pulse spin. The total volume should be $35\,\mu L$.

7. Place tubes on the thermal cycler. Conditions for reverse transcription are as follows:

Incubation	25°C	2 min
Incubation	37°C	60 min
Denaturation	95°C	5 min

8. Pulse spin the prepared cDNA and store it at $-20°C$ until use.

3.3. Genotyping: Rotor-Gene PCR

3.3.1. Panel One

1. Prepare a master mix using LightCycler® FastStart DNA Master HybProbe. For each reaction you will need $2\,\mu L$ of prepared Master HybProbe (vial 1), $0.4\,\mu L$ of 25 mM $MgCl_2$ (to give a final concentration of 1.5 mM), 15 pmol of each primer (HCV3 and HCV4), 5 pmol of each of the type-specific probes T1, T2, T3, and T11, $11.1\,\mu L$ RNAse-free water.

2. Vortex the mix briefly and transfer $17\,\mu L$ aliquots into labeled, flat-lid 0.2 mL thermo tubes.

3. Thaw and pulse spin the prepared cDNA template. Add $3\,\mu L$ to each tube and mix well.

4. Include positive controls for each genotype (*see* **Note 1**) and a nontemplate control (water) with each run.

5. Pulse spin in a centrifuge and place tubes in the Rotor-Gene 3000 carousel.

6. Thermal cycling conditions on the Rotor-Gene are as follows:

Hold	10 min	95°C
45 cycles	8 s	95°C (not acquiring)
	45 s	65°C (acquiring on FAM, Cy5, JOE, and ROX)

7. Select "Calibration" to display the auto gain calibration set-up menu. Ensure that FAM, JOE, Cy5, and ROX channels are displayed. Set the temperature to 65°C and click "Start" to adjust gains for the four channels selected.

8. Close the calibration display and start the run. Edit the run worklist as appropriate. The run will take 104 min to complete.

9. Real-time measurements are taken and a threshold cycle (Ct) is calculated, determined by the point at which the fluorescence exceeds a threshold limit. After completion of the run, click on "Analysis" and then "Quantification." Click the linear scale to change it to log scale. Either manually set the threshold or use the "Auto-find threshold" tab. Each channel is analyzed separately. Genotype 1 isolates should appear in both FAM and Cy5 channels. Genotype 2 is detected on the JOE channel, and genotype 3 on the ROX channel. A nonexponential curve observed *only* in the Cy5 channel may indicate a genotype 4 or 5 isolate (*see* **Note 2**). A true exponential curve observed *only* in the Cy5 channel indicates a genotype 6 (*see* **Note 3**). Genotypes 2, 3, and 4 will require further confirmation with the second panel of Taqman probes. **Figure 5.2** shows a representative panel one run, demonstrating genotype 1, 2, and 3 positive isolates.

3.3.2. Panel Two

Isolates confirmed as genotype 1 by panel one need not be included on the panel two run.

1. Prepare a master mix using LightCycler® FastStart DNA Master HybProbe. For each reaction you will need $2 \mu L$ of prepared Master HybProbe (vial 1), $0.8 \mu L$ of 25 mM $MgCl_2$ (to give a final concentration of 2 mM), 15 pmol of each primer (HCV3 and HCV4), 5 pmol of each of the type-specific probes T22, T33, T333, and T4, and $10.7 \mu L$ RNAse-free water.
2. Vortex the mix briefly and transfer $17 \mu L$ aliquots into labeled, flat-lid 0.2 mL thermo tubes.

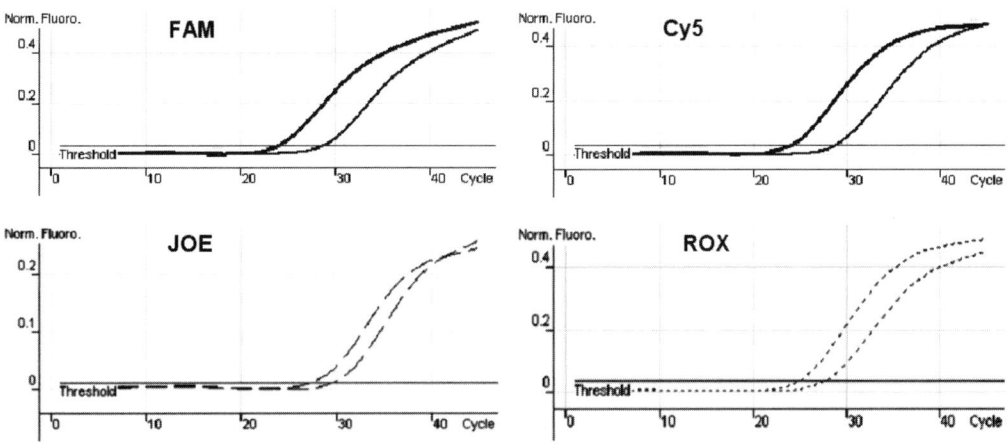

Fig. 5.2. A representative selection of clinical isolate results from panel one, demonstrating two genotype 1 isolates (positive on FAM and Cy5 channels), two genotype 2 isolates (positive on JOE channel), and two positive genotype 3 isolates (positive on ROX channel).

3. Thaw and pulse spin the prepared cDNA template. Add $3\,\mu L$ of cDNA to each tube and mix well.
4. Include positive controls for genotypes 2, 3, and 4 (*see* **Note 1**) and a nontemplate control (water) with each run.
5. Pulse spin in a centrifuge and place tubes in the Rotor-Gene 3000 carousel.
6. Thermal cycling conditions in the Rotor-Gene are as follows:

Hold	10 min	95°C
45 cycles	8 s	95°C (not acquiring)
	45 s	60°C (acquiring on FAM, Cy5, JOE, and ROX)

7. Set up the Rotor-Gene as described for panel one. Alter the settings so that calibration and cycling take place at 60°C. The run will take 109 min to complete.
8. Real-time measurements are taken as described above for panel one. Panel two confirms genotypes 2 and 3. It is a first and only line for genotype 4 isolates. Genotype 2 is detected on the Cy5 channel, genotype 4 on the FAM channel, and genotype 3 on the ROX and JOE channels.

3.3.3. Additional Probe

A very small number of isolates have been found to be positive with the T3 probe of panel one but negative with both genotype 3 probes of panel two. These have been found, by sequence analysis, to be subtype 3b isolates. An additional probe has been designed to confirm subtype 3b isolates demonstrating this probe pattern. Because no further genotype 3–specific motifs are available in the 5′ UTR, a probe has been designed around a genotype 3–specific motif in the NS5b region that is amplified with the subtyping primers (NS5bF2 and NS5bR). This probe detects all subtypes of genotype 3.

1. Prepare a master mix using LightCycler® FastStart DNA Master HybProbe. For each reaction you will need $2\,\mu L$ of prepared Master HybProbe (vial 1), $1.2\,\mu L$ of 25 mM $MgCl_2$ (to give a final concentration of 2.5 mM), 15 pmol of each primer (NS5bF2 and NS5bR), 3.5 pmol of NS53 probe, and $11.95\,\mu L$ of RNAse-free water.
2. Vortex the mix briefly and place $17\,\mu L$ aliquots in labeled, flat-lid 0.2 mL thermo tubes.
3. Thaw and pulse spin the prepared cDNA template. Add $3\,\mu L$ of cDNA to each tube and mix well.
4. Pulse spin the tubes in a centrifuge and place them in the Rotor-Gene 3000 carousel.

5. Set the thermal cycling conditions in the Rotor-Gene as follows:

Hold	10 min	95°C
45 cycles	8 s	95°C (not acquiring)
	45 s	60°C (acquiring on ROX)

6. Set up the Rotor-Gene as described for panel two, acquiring on ROX only. The run will take 91 min to complete. Any isolate positive with two of the four genotype 3 probes should be confirmed as genotype 3. *See* **Note 4**.

3.3.4. Reporting of Genotype Results

1. Genotyping results can now be reported as per the algorithm (**Fig. 5.1**).
2. Failure of samples to be amplified on either panel may be due to low HCV viral load. Check the samples' HCV viral-load result and report genotyping result as negative if viral load is $< \sim 6000$ IU/mL. *See* **Note 5**.

3.4. Subtyping of Genotype 1 Isolates: Rotor-Gene PCR

1. Prepare a master mix using LightCycler® FastStart DNA Master HybProbe. For each reaction you will need 2 μL of prepared Master HybProbe (vial 1), 0.8 μL of 25 mM MgCl₂ (to give a final concentration of 2 mM), 6 pmol of NS5bF2, 12 pmol of NS5bR, 2.5 pmol of each subtype-specific probe (1a and 1b), and 12.8 μL RNAse-free water.
2. Vortex the mix briefly and transfer 17 μl aliquots into labeled, flat-lid 0.2 mL thermo tubes.
3. Thaw and pulse spin the prepared cDNA template. Add 3 μL of cDNA to each tube and mix well.
4. Include positive controls for each subtype and a nontemplate control (water) with each run.
5. Centrifuge briefly and place in the Rotor-Gene 3000 carousel.
6. Thermal cycling conditions in the Rotor-Gene are as follows:

Hold	10 min	95°C
40 cycles	8 s	95°C (not acquiring)
	45 s	55°C (acquiring on FAM and JOE)

7. Select "Calibration" to display auto gain calibration set-up menu. Ensure that FAM and JOE channels are displayed. Set temperature to 55°C and click "start" to adjust gains for the two channels selected.
8. Close the calibration display and start the run. Edit the run worklist as appropriate. The run will take 92 min to complete.

9. Real-time measurements are taken as described above. Subtype 1a is detected on the FAM channel, and 1b on the JOE channel.

3.5. Generic HCV Subtyping

3.5.1. Rotor-Gene PCR

1. Prepare a reaction mix using LightCycler® DNA Master SYBR Green I. For each reaction you will need 2 μL of master mix (vial 1), 0.8 μL of 25 mM $MgCl_2$ (to give a final concentration of 2 mM), 6 pmol of each primer (NS5bF2 and NS5bR), and 13.6 μL RNAse-free water.
2. Vortex the mix briefly and transfer 17 μL aliquots into labeled, flat-lid 0.2 mL thermo tubes.
3. Thaw and pulse spin the cDNA template. Add 3 μL of cDNA template to each tube and mix well.
4. Include a nontemplate control (water) with each run.
5. Centrifuge briefly and place in the Rotor-Gene 3000 carousel.
6. Thermal cycling conditions in the Rotor-Gene are as follows:

Preincubation		10 min	95°C
Quantification (45 cycles)	(not acquiring)	10 s	95°C
	(not acquiring)	30 s	60°C
	(acquiring on FAM)	30 s	72°C
Melt curve	Hold	30 s	95°C
	Cool/hold	30 s	50°C
	Ramp		50–99°C rising in 1°C increments holding each increment for 5 s and continuously acquiring to MeltA on FAM

7. Select "Calibration" to display auto gain calibration set-up menu. Ensure only FAM channel is displayed. Set temperature to 60°C and click "Start" to adjust gains for the FAM channel.
8. Close the calibration display and start the run. Edit the run worklist as appropriate. The run will take 117 min to complete.
9. After completion of the run, click on "Analysis" and then "Quantification." Click the linear scale to change to log scale.
10. In the "Analysis" window, go to "Melt" to check the melt curve. The melting curves allow discrimination between primer-dimers and specific product. The melt should occur at 90–95°C. If melting temperature is lower (80–85°C), then amplification is nonspecific, and the resulting isolates should not be sequenced.

Table 5.3
National Center for Biotechnology Information Accession Numbers of Reference Isolates for HCV Subtyping (see www.ncbi.nlm.nih.gov). *See* **Note 6**

Subtype	Accession numbers
1a	D10749*, M62321*
1b	AY003957, AY003965
1c	AY051292*
1d	AF037233
1e	L38361
1g	AF271798
1h	AY257087
1i	L48495
1j	AY434113
2a	D00944*, AJ231479
2b	D10988*, AB030907*
2c	AJ231468, D50409*
2d	AB031663*
2e (2n)	L44602
2f (2p)	L44601
3a	D17763*, AY003970, D28917*
3b	AF279121
3k	D63821*
4a (4r)	AJ291282, Y11604
4d	AY743168
4f	AY743145
4k	AY743167
4o	AY743113, AF271815
4r	AY743159, AJ291282
5a	AJ291281, Y13184*
6a	AY973866
6b	D84262*
6d	D84263*

*Complete genome sequence.

3.5.2. Sequencing and
Subtype Assignment

1. Purify all Rotor-Gene PCR products with the correct melting curve using the QIAQuick PCR purification kit, following manufacturer's guidelines.
2. Sequence purified PCR products, using the NS5b primers described above at $0.2\,\mu M$ concentrations as sequencing primers. Use the sequencing kit and platform available within your laboratory.
3. Import the reference sequences using the accession numbers outlined in **Table 5.3** from the NCBI website to create a databank (*see* **Note 6**). Trim these to the correct size, to match the target sequence, using the software. Align the reference sequences using the MegAlign software in the DNASTAR Lasergene software package, using the Clustal W method. Import the sequences from the subtyping assay. Trim these, if necessary, using the software. Enter these sequences into the reference databank and align all using the Clustal W method. Go to "View" and then "Phylogenetic tree." Assign subtype according to phylogenetic relatedness

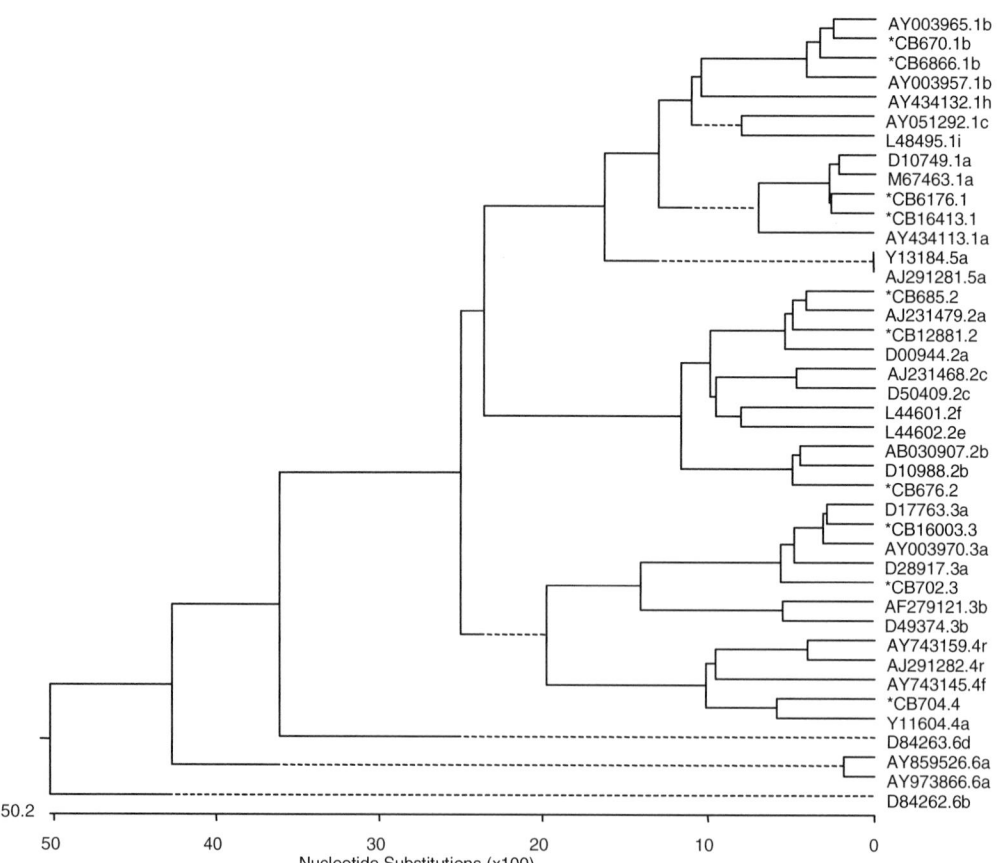

Fig. 5.3. An example phylogenetic tree for HCV subtyping. "Unknown" isolates are marked with asterisks.

to the reference isolates. A sample phylogenetic tree showing a subset of reference isolates and some "unknown" isolates is shown in **Fig. 5.3**.

4. Notes

1. Positive controls should be included for genotypes 1–4 in panel one and for genotypes 2–4 in panel two. If facilities exist, these should be created by production of plasmid clones with the TOPO TA Cloning® Kit for sequencing (Invitrogen) with the genotype-specific 5′ UTR amplicon for each clone. If plasmid clones cannot be produced, then the cDNA from a previous high-titer positive for each genotype can be used. In our laboratory, we produce plasmid clones and make serial dilutions of type-specific plasmid DNA controls in a polyA diluent, for long-term storage. A dilution of each plasmid DNA control that is reproducibly detected by the assay, but has too low a titer to pose a contamination threat, should be used. We have found 10^{-5} to 10^{-7} to be adequate, but the appropriate dilution will depend on the initial plasmid concentration.

2. Because of a weak hybridization with T11, genotype 4 and 5 isolates may be detected on the Cy5 channel of panel one with a shallow/nonexponential curve. We have seen this result with 60% of genotype 4 isolates. The genotype 4 identity of these isolates should still be confirmed on panel two, and they should be sequenced if negative on panel two (as probable genotype 5).

3. We have not observed any genotype 6 isolates while this assay has been in use. Genotype 6 isolates have 100% sequence identity with the T11 probe binding site, and we assume they would be positive *only* with this probe. Only sequencing can confirm the identity of these isolates.

4. Occasionally, the panel one genotype 3 probe is negative, but both panel two genotype 3 probes are positive. If this occurs, the isolate can still be confirmed as genotype 3. Subtype 3b isolates will not be detected with either of the panel two genotype 3 probes. Such an isolate, if positive with the panel one genotype 3 probe and the additional probe (NS53), can also be confirmed as genotype 3. The identities of isolates producing any other combination of genotype 3 probe patterns should be characterized by sequencing and genotype confirmed by phylogenetic analysis.

5. The genotyping assay may not detect isolates that have viral loads $<\sim 6000\,\text{IU/mL}$. Any loss of sensitivity at lower viral loads is likely to be a result of the optimization of cycling temperature and magnesium concentrations, which permitted maximum stringency and minimal nonspecific binding

of the probes but may decrease sensitivity. The genotypes of any samples that remain unidentified or show discrepant results with the Taqman genotyping assay and have viral load $>\sim 6000\,\text{IU/mL}$ should be ascertained by sequencing with both 5′ UTR and NS5b regions, with the sequencing platform in use within your laboratory.

6. The reference strains indicated in **Table 5.3** are only the most common seen. For a more complete reference database, refer to Simmonds et al. *(28)*, which gives reference sequences for HCV variants and provisionally assigned HCV subtypes. The trimmed, ready-to-use subtype reference sequences are available, on request, as a Fasta file from our laboratory (email request to kathryn.rolfe@addenbrookes.nhs.uk).

References

1. Manns, M. P., McHutchison, J. G., Gordon, S. C., Rustqi, V. K., Shiffman, M., Reindollar, R., et al. (2001) Peginterferon alfa-2b plus ribavirin compared with interferon alfa-2b plus ribavirin for the initial treatment of chronic hepatitis C: a randomised trial. *Lancet* **358**, 958–965.

2. Fried, M. W., Shiffman, N. L., Reddy, K. R., Smith, C., Marinos, G., Goncales, F. L., et al. (2002) Peginterferon alfa-2a plus ribavirin in patients with chronic hepatitis C virus infection. *N. Eng. J. Med.* **347**, 975–982.

3. Lee, S., Heathcote, E., Reddy, K., Zeuzem, S., Fried, M. W., Wright, T. L., et al. (2002) Prognostic factors and early predictability of sustained viral response with peginterferon alfa-2a. *J. Hepatol.* **37**, 500–506.

4. Lee, S. S. and Abdo, A. A. (2003) Predicting antiviral treatment response in chronic hepatitis C: how accurate and how soon? *J. Antimicrob. Chemoth.* **51**, 487–491.

5. Hadziyannis, S. J., Sette, H., Jr., Morgan, T. R., Balan, V., Diago, M., Marcellin, P., et al. (2004) Peginterferon-α2a and ribavirin combination therapy in chronic hepatitis C: a randomised study of treatment duration and ribavirin dose. *Ann. Intern. Med.* **140**, 346–355.

6. National Institute of Clinical Excellence (NICE) guidelines (2004) Interferon alfa (pegylated and non-pegylated) and ribavirin for the treatment of chronic hepatitis C. Technology Appraisal Guidance 75. Available at www.nice.org.uk.

7. NICE guidelines (2006) Peginterferon alfa and ribavirin for the treatment of mild chronic hepatitis C. Technology Appraisal Guidance 106. Available at www.nice.org.uk.

8. Strader, D. B., Wright, T., Thomas, D. L., and Seeff, L. B. (2004) Diagnosis, management, and treatment of hepatitis C. *Hepatology* **39**, 1147–1171.

9. Zein, N. N. (2000) Clinical significance of hepatitis C virus genotypes. *Clin. Microbiol. Rev.* **13**, 223–235.

10. Maertens, G., and Stuyver, L. (1997) Genotypes and genetic variation of hepatitis, in *The Molecular Medicine of Viral Hepatitis* (Harrison, T. J. and Zuckerman, A., eds.), John Wiley and Sons, Chichester, England, pp. 183–233.

11. Takada, N., Takase, S., Takada, A., and Date, T. (1993) Differences in the hepatitis C virus genotypes in different countries. *J. Hepatol.* **17**, 277–283.

12. Xu, L.-Z., Larzul, D., Delaporte, E., Bréchot, C., and Kremsdorf, D. (1994) Hepatitis C virus genotype 4 is highly prevalent in central Africa (Gabon). *J. Gen. Virol.* **75**, 2393–2398.

13. Chamberlain, R. W., Adams, N., Saeed, A. A., Simmonds, P., and Elliott, R. M. (1997) Complete nucleotide sequence of a type 4 hepatitis C virus variant, the predominant genotype in the Middle East. *J. Gen. Virol.* **78**, 1341–1347.

14. Bullock, G. C., Bruns, D.E., and Haverstick, D.M. (2002) Hepatitis C genotype determination by melting curve analysis with a single set of fluorescence resonance energy transfer probes. *Clin. Chem.* **48**, 2147–2154.

15. Schröter, M., Zöllner, B., Schäfer, P., Landt, O., Lass, U., Laufs, R., et al. (2002) Genotyping of hepatitis C virus types 1, 2, 3, and 4 by a one-step LightCycler method using three different pairs of hybridization probes. *J. Clin. Microbiol.* **40**, 246–250.

16. Lindh, M. and Hannoun, C. (2005) Genotyping of hepatitis C virus by Taqman real-time PCR. *J. Clin. Virol.* **34,** 108–114.

17. Rolfe, K. J., Alexander, G. J. M., Wreghitt, T. G., Parmar, S., Jalal, H., and Curran, M. D. (2005) A real-time Taqman method for hepatitis C virus genotyping. *J. Clin. Virol.* **34,** 115–121.

18. Chen, Z. and Weck, K. E. (2002) Hepatitis C genotyping: interrogation of the 5′ untranslated region cannot accurately distinguish genotypes 1a and 1b. *J. Clin. Microbiol.* **40,** 127–134.

19. Sandres-Sauné, K., Deny, P., Pasquier, C., Thibaut, C., Duverlie, G., and Izopet, J. (2003) Determining hepatitis C genotype by analyzing the sequence of the NS5b region. *J. Virol. Methods* **109,** 187–193.

20. Othman, S. B., Trabelsi, A., Monnet, A., Bouzgarrou, N., Grattard, F., Beyou, A., Bourlet, T., and Pozzetto, B. (2004) Evaluation of a prototype HCV NS5b assay for typing strains of hepatitis C virus isolated from Tunisian haemodialysis patients. *J. Virol. Methods* **119,** 177–181.

21. Amoroso, P., Rapicetta, M., Tosti, M. E., Mele, A., Spada, E., Buonocore, S., Lettieri, G., Pierri, P., Chionne, P., Ciccaglione, A. R., and Sagliocca, L. (1998) Correlation between virus genotype and chronicity rate in acute hepatitis C. *J. Hepatol.* **28,** 939–944.

22. Nakano, I., Fukuda, Y., Katano, Y., Toyoda, H., Hayashi, K., Kumada, T., and Nakano, S. (2001) Japan-specific subtype of hepatitis C virus genotype 1b, J subtype, has relatively low pathogenicity. *J. Med. Virol.* **65,** 45–51.

23. Wiese, M., Grungreiff, K., Guthoff, W., Lafrenz, M., Oesen, U. and Porst, H. (2005) Outcome in a hepatitis C (genotype 1b) single source outbreak in Germany—a 25-year multicenter study. *J. Hepatol.* **43,** 590–598.

24. Manesis, E. K., Papaioannou, C., Gioustozi, A., Kafiri, G., Koskinas, J., and Hadziyannis, S. J. (1997) Biochemical and virological outcome of patients with chronic hepatitis C treated with Interferon alfa-2b for 6 or 12 months: a 4-year follow-up of 211 patients. *Hepatology* **26,** 734–739.

25. Nakano, I., Fukuda, Y., Katano, Y., Toyoda, H., Hayashi, K., Kumada, T., and Nakano, S. (2001) Interferon responsiveness in patients infected with hepatitis C virus 1b differs depending on viral subtype. *Gut* **49,** 263–267.

26. Garson, J. A., Ring, C., Tuke, P., and Tedder, R. S. (1990) Enhanced detection by PCR of hepatitis C virus RNA. *Lancet* **336,** 878–879.

27. Okamoto, H., Okada, S., and Sugiyama, Y. F. (1990) Detection of hepatitis C virus RNA by a two stage polymerase chain reaction with two pairs of primers from the 3′ noncoding region. *Jpn. J. Exp. Med.* **60,** 215–222.

28. Simmonds, P., Bukh, J., Combet, C., Deleage, G., Enomoto, N., Feinstone, S., et al. (2005) Consensus proposals for a unified system of nomenclature of hepatitis C virus genotypes. *Hepatology* **42,** 962–973.

Chapter 6

Characterization of HCV Quasispecies in the Hypervariable Region 1 (HVR1) by In Vitro Translation and Mass Spectrometry

Carmen Yea, Melissa Ayers, and Raymond Tellier

Abstract

HCV infection provides a classic example of the phenomenon of quasispecies. Because several lines of investigation support the contribution of quasispecies to HCV's capacity to maintain a persistent infection, adequate characterization of the quasispecies is important. The hypervariable region 1 (HVR1) of the E2 glycoprotein has been particularly well studied in this regard. We present here a rapid method for characterizing the HVR1 quasispecies, based on in vitro coupled transcription/translation of the amplicons, followed by mass spectrometry of the resulting peptide mix.

Key words: HCV, quasispecies, HVR1, in vitro translation, MALDI-TOF.

1. Introduction

Like all RNA viruses, HCV encodes an RNA-directed RNA polymerase with a relatively low fidelity, leading to the generation of new mutants after replication. Because HCV typically causes a chronic infection, the accumulation of mutations over time leads to the genesis of a "quasispecies", a population of closely related but distinct mutants *(1, 2)*. The mutants' different degrees of fitness result in selective pressure on the HCV quasispecies; in particular, changes in the immune response are expected to be an important modulator of the quasispecies structure. Rapid changes in the mutant distribution in the quasispecies are thought to

Hengli Tang (ed.), *Hepatitis C: Methods and Protocols, Second Edition, vol. 510*
© 2009 Humana Press, a part of Springer Science+Business Media
DOI 10.1007/978-1-59745-394-3_6 Springerprotocols.com

contribute to the maintenance of a persistent HCV infection despite a strong immune response.

HCV quasispecies heterogeneity has been found within several regions of the genome *(3)*. By far the best studied of these has been the 27 amino acids at the amino terminus of the E2 glycoprotein, called "hypervariable region 1" (HVR1). Changes in amino acids in this region are associated with significant changes in the predicted secondary structure *(4)*. The high rate of mutation in this region probably results in part because this region is not essential for HCV replication and in vivo infection *(5)*, and the amino acid sequence therefore has great latitude for change. The HVR1 contains neutralizing epitopes *(6)*, and HVR1 mutants are often immune-escape mutants *(7–9)*. In a landmark study, Farci and colleagues *(10)* showed that a decrease in the diversity of HVR1 quasispecies at the amino acid level at the time of seroconversion was predictive of viral eradication. In patients with progressive hepatitis, the rate of nonsynonymous mutations in HVR1 was higher than that in patients with resolving hepatitis, but the synonymous mutation rates were comparable *(10)*. These observations strongly suggest that the nonsynonymous mutations are those that matter in HVR1, consistent with the idea that some HVR1 mutants are humoral immunity-escape mutants. In addition, other studies have shown that a decrease in diversity during interferon treatment is predictive of sustained response *(11)* and that, among vertically infected children, elevated ALT levels are associated with mono- or oligoclonal viral population, whereas in children with a normal or slightly elevated ALT level, the HVR1 quasispecies was shown to diversify gradually.

Laboratory characterization of the HVR1 quasispecies may therefore become an important part of the clinical follow-up of patients, necessitating a critical examination of the methods available for this characterization, from the point of view of the clinical laboratory. The most exhaustive method, cloning amplicons containing the HVR1 and sequencing many clones (≥ 20), is not practical for the clinical laboratory. Rapid methods, such as single-strand conformation polymorphism or heteroduplex assays, detect mutations in the nucleotide sequences in a way that does not allow differentiation between synonymous and nonsynonymous mutations. This situation was the starting point of the design of the method described here. Basically, amplification of the region containing the HVR1 on the HCV genome by RT-PCR is followed by in vitro transcription and translation of the amplicons, through a T7 promoter incorporated into one of the primers. The resulting peptide mixture is then submitted to mass spectrometry analysis, where the different HVR1 amino acid sequences are separated by mass. This method provides a rapid HVR1 quasispecies characterization that focuses exclusively on nonsynonymous mutations.

2. Materials

1. Trizol reagent (Invitrogen), chloroform, isopropanol, ethanol 70%, glycogen ($20 \mu g/\mu L$; Invitrogen).
2. Molecular grade water (free of RNase, DNase, and proteinase) (ddH_2O).
3. Dithiothreitol (DTT) 100 mM (Promega). Keep at $-20°C$.
4. RNasin 20–40 $U/\mu L$ (Promega). Keep at $-20°C$.
5. $10\times$ PCR buffer (Applied Biosystems). Keep at $-20°C$.
6. 25 mM $MgCl_2$ (Applied Biosystems). Keep at $-20°C$.
7. Avian myeloblastosis virus reverse transcriptase (AMV-RT) 5–10 $U/\mu L$ (Promega). Keep at $-20°C$.
8. AmpliTaq Gold (Applied Biosystems). Keep at $-20°C$.
9. dNTPs mix (10 mM each). Prepared in ddH_2O from 100 mM dNTP set (Amersham Biosciences). Keep at $-80°C$ in aliquots.
10. Mineral oil (Sigma).
11. Primers. Divide each primer into aliquots at a concentration of $10 \mu M$ in ddH_2O. Keep at $-20°C$.
12. Thin-wall PCR tubes (Stratagene).
13. Thermal cycler: Robocycler 40 (Stratagene).
14. In vitro coupled transcription–translation kit: TNT T7 Quick for PCR DNA (Promega).
15. Methionine solution (1 mM) (Promega).
16. Biotinylated anti-FLAG antibody M2 (Sigma).
17. Magnetic beads coated with streptavidin: Dynabeads M-280 (Dynal).
18. Phosphate-buffered saline (PBS)/bovine serum albumin (BSA).
19. Magnetic microfuge tube rack (Dynal).
20. Ammonium bicarbonate 25 mM (Sigma).
21. Trifluoroacetic acid 0.1% (Sigma).
22. Matrix solution: saturated solution of α-cyano-4-hydroxycinnamic acid (Sigma) in 50% acetonitrile and 0.1% trifluoroacetic acid. Prepare a fresh solution weekly; keep at $4°C$ in an opaque container.
23. Matrix-assisted laser desorption ionization–time of flight (MALDI-TOF) mass spectrometer Voyager-DE STR (Applied Biosystems).

3. Methods

3.1. Nested RT-PCR of HVR1-Containing Region (see Note 1)

1. Extract viral RNA with the TRIzol reagent according to the manufacturer's instructions, and resuspend the RNA pellet in 10 mM DTT and 5% (vol/vol) RNasin (20–40 $U/\mu L$) in ddH_2O (see **Note 2**).

2. Heat $10\,\mu L$ of RNA solution at $65°C$ for 2 min and put on ice; add $10.5\,\mu L$ of an RT master mix consisting of $0.5\,\mu L$ of RNasin, $2\,\mu L$ of $10 \times$ PCR buffer, $2\,\mu L$ of 25 mM $MgCl_2$, $2\,\mu L$ of dNTPs solution (10 mM each), $3\,\mu L$ of $10\,\mu M$ HCPEP-2 primer (5′-CTATTGATGTGCCAACTGCCG-3′), and $1\,\mu L$ of AMV-RT. Incubate the reaction at $42°C$ for 1 h.

3. Add the RT mixture to a PCR mix consisting of $8\,\mu L$ of 25 mM $MgCl_2$, $5\,\mu L$ of $10\,\mu M$ primer HCPEP-1 (5′-GGTTCTGATTGTGCTGCTACTATTTGC-3′), $5\,\mu L$ of $10\,\mu M$ primer HCPEP-2, $0.5\,\mu L$ of AmpliTaq Gold, and $53.5\,\mu l$ of ddH_2O. Perform the reactions in thin-wall PCR tubes overlaid with $100\,\mu L$ of mineral oil (Sigma) in a Robocycler 40 with the following parameters: one cycle of denaturation at $95°C$ for 10 min; annealing at $50°C$ for 1 min and elongation at $72°C$ for 1 min 30 sec; 35 cycles of denaturation at $95°C$ for 1 min, annealing at $50°C$ for 1 min, and elongation at $72°C$ for 1 min 30 sec. For the second round of the nested PCR, transfer $10\,\mu L$ of first-round PCR mixture into a thin-wall PCR tube containing $90\,\mu L$ of a master mix consisting of $9\,\mu L$ of $10\times$ PCR buffer, $9\,\mu L$ of 25 mM $MgCl_2$, $5\,\mu L$ each of $10\,\mu M$ HCFLAGALA and HCFLAG-4 primers (*see* **Note 3**), $2\,\mu L$ dNTPs solution, $0.5\,\mu L$ of AmpliTaq Gold, and $59.5\,\mu L$ of ddH_2O. Perform the reaction as above but change annealing temperature to $67°C$.

3.2. In Vitro Transcription and Translation

1. Carry out the reaction using a coupled transcription/translation rabbit reticulocyte lysate system (TNT T7 Quick for PCR DNA; Promega). Assemble the reaction mix as follows: $40\,\mu L$ of TNT T7 master mix (rabbit reticulocyte lysate), $7\,\mu L$ of second-round PCR mix, $2\,\mu L$ ddH_2O, and $1\,\mu L$ of 1 mM methionine.

2. Incubate the reaction mixture at $30°C$ for 1 h (*see* **Note 4**).

3.3. Magnetic-Bead Preparation

1. Wash 2.4 mg of streptavidin -coated magnetic beads with PBS containing 0.1% BSA, and resuspend them in $200\,\mu L$ of fresh PBS with 0.1% BSA.

2. Add 0.1 mg of biotinylated anti-Flag antibody M2 to the prewashed beads. Incubate the mixture at room temperature for 30 min; mix occasionally to prevent sedimentation of the beads.

3. After incubation, wash the beads five times with PBS/0.1% BSA, and then resuspend them in $250\,\mu L$ of fresh PBS/0.1% BSA.

4. Store the magnetic-bead preparation at $4°C$ until use. Prepare a fresh suspension every 2 months.

3.4. Peptide Purification

1. Add $10\,\mu L$ of anti-Flag magnetic-bead preparation to the in vitro transcription/translation mix in a microfuge tube. Incubate the tube at room temperature for 10 min, mixing occasionally.
2. Place the tube in a magnetic rack and remove the supernatant. Wash the beads three times with 25 mM ammonium bicarbonate and three times with ddH$_2$O.
3. To elute the bound peptides, resuspend the beads in $20\,\mu L$ of 0.1% trifluoroacetic acid and remove the beads using a magnet.
4. Lyophilize the eluted peptides in a vacuum concentrator (SpeedVac) and resuspend them in $5\,\mu L$ of matrix solution (*see* **Note 5**).

3.5. MALDI-TOF Mass Spectrometry

Analyze samples using a Voyager-DE STR mass spectrometer (Applied Biosystems) with a pulsed UV nitrogen laser (337 nm, 3-ns pulse) and a dual microchannel plate detector. Acquire mass spectra in linear DE mode; set the acceleration voltage to 20 kV, the grid voltage to 94% of the acceleration voltage, the guide-wire voltage at 0.05%, the delay time at 175 ns, and the low mass gate at 1000 Da. For sample analysis, apply approximately $1.5\,\mu L$ of sample solution to the sample plate. Record mass spectra after evaporation of the solvent and process them using the Data Explorer software (*see* **Note 6**).

4. Notes

1. We present the protocols in use in our laboratory for the convenience of the reader, but in principle any PCR method that yields amplicons with the specially designed primers for transcription/translation (*see* **Note 3**) should do as well.
2. In principle, other RNA extraction methods can be substituted for the method described here. Please note that, if the RNA is resuspended only in ddH$_2$O, adjustments in DTT must be made in the reverse transcription mix.
3. The inner primers used in the nested PCR are essential elements of the whole method. These primers are illustrated in **Fig. 6.1**, and several features are noteworthy. Primer HCFLA-GALA is the sense primer; the 20 nucleotides (nt) at the 3′ end are homologous to a region in the HCV genome just upstream of the HVR1; the sequence is a consensus sequence with strong homology to the corresponding sequence of HCV genotypes 1a, 1b, and 1c but not to other genotypes. To generate relatively short peptides (7000–7500 Da) that allow for single amino acid substitution detection in the HVR1, the primers must be close to the HVR1, and the sequences of the HCV genotypes are such that no consensus primer for all

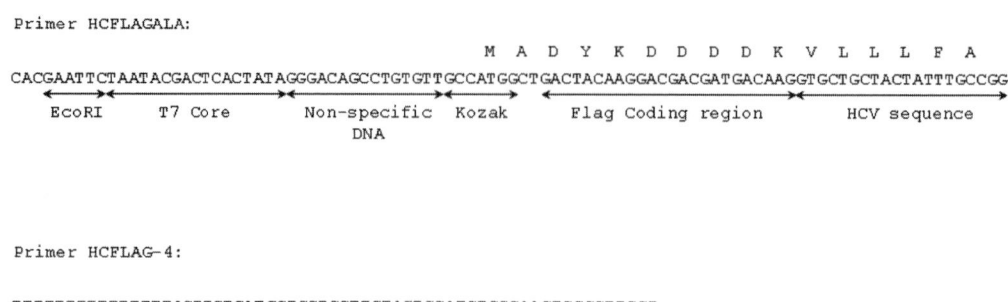

Fig. 6.1. Primers used in the synthesis of amplicons for in vitro coupled transcription–translation (*see* **Note 3**). Adapted from Ayers et al. *(14)*, with permission.

genotypes in this region is possible; consequently, the primer pair shown in **Fig. 6.1** is specific for genotype 1. Custom primers for other genotypes must be designed accordingly. Upstream of the HCV sequence on the HCFLAGALA primer is a sequence coding for the Flag epitope (DYKDDDDK), in frame with the HCV ORF and with an ATG codon embedded in a Kozak consensus sequence; a GCT codon immediately following the ATG codon encodes for an alanine residue and will direct the posttranslational, enzyme-mediated cleavage of the methionine *(12)*. Leaving the methionine residue leads to its oxidation in some, but not all, peptides *(13)* and therefore to a second peak 16 Da away in the MALDI-TOF spectrum *(14)*, which is undesirable. The HCFLAGALA primer also includes a T7 core promoter for RNA transcription, separated from the Kozak sequence by nonspecific "filler" DNA to allow for ribosomal scanning at the stage of translation. The T7 core promoter on an amplicon usually performs better with a 9–20 nt leader sequence upstream *(15)*.

Primer HCFLAG-4 is the antisense primer, with a sequence complementary to a part of the HCV genome just downstream of the HVR1; it is also specific for genotype 1. We found that peptides synthesized with Flag epitopes at both terminals were more stable *(14)*. The oligo-T stretch will result in a poly A tail for the transcribed RNA, which increases its stability.

Nested PCR may not be necessary if the viral load in the sample is high; in principle, many other outer pairs could be used for the first round of PCR, and the pair HCPEP-1 and HCPEP-2 is given here for the convenience of the reader. Please note that this outer pair is also specific for genotype 1 and that, because of the size of the bracketed regions and of the primers in **Fig. 6.1**, the amplicons synthesized from the second round of PCR are in fact larger than the amplicon from the first round (even though the HCV sequence in the second-round amplicon is a substring).

4. As alluded to in **Note 3**, peptides of the HVR1 flanked at both ends by Flag epitopes are quite stable, and translation in the presence of the protease inhibitor leupeptin (1 µg/mL) did not consistently improve the results, but experiments with other regions of the HCV genome did show marked improvements with the addition of leupeptin during the translation step.

5. Although our experience is still limited, in recent experiments we have eluted the peptides directly in 5 µL of matrix solution, avoiding the lyophilization step altogether, with good results.

6. MALDI-TOF mass spectrometry instruments are becoming widely available in universities and teaching hospitals, typically as part of core facilities.

Figure 6.2 illustrates the results obtained with the method from the H-77 strain, whose quasispecies in the HVR1 has been characterized in great detail *(8)*. The predicted masses of the peptides synthesized from the three most common mutants in the H-77 quasispecies are 7101.5, 7122.6, and 7055.5 Da, respectively. The frequencies of these three mutants in the H-77 quasispecies were estimated at 70/114, 6/114, and 5/114 clones, respectively *(8)*. The nominal precision of the Voyager instrument used in these studies is 0.1 %, approximately 7 Da in the mass range of the peptides we analyzed. Our experience has been that we usually get an even closer match with the predicted mass. In recent studies *(16)*, we took three measurements for each amplicon and took the average ± standard deviation as the measured value.

The calculation of the predicted mass must take into account the HVR1 amino acid sequence itself, the additional peptides encoded by the primers (including the two flag epitopes), and posttranslational modifications. For

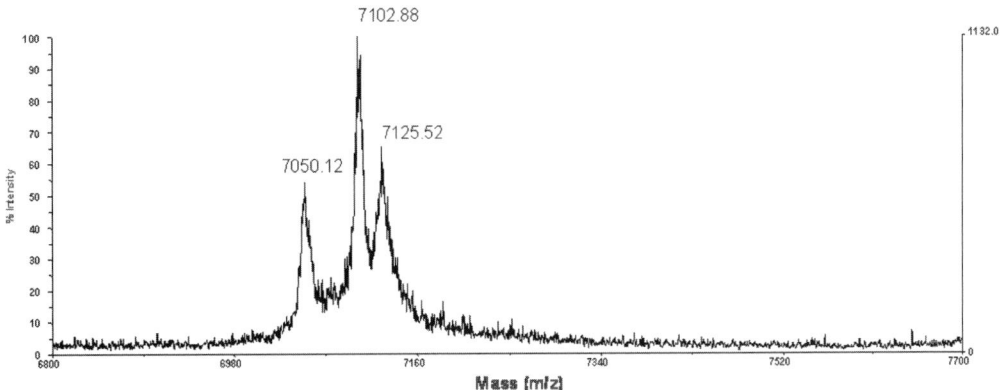

Fig. 6.2. Representative matrix-assisted laser desorption/ionization–time of flight (MALDI-TOF) spectrum obtained from the H77 strain (*see* **Note 6**).

example, the most frequent mutant in the H-77 quasispecies (which was used to construct the infectious clone pCV-H77C *(17)*) would yield, after the posttranslational cleavage of the methionine residue, the peptide "ADYKD-DDDKVLLLFAGVDAETHVTGGNAGRTTAGLVGLLTP-GAKQNIQLINTNGSWHIDYKDDDDK," whose predicted mass is 7059.5 Da. Enzyme-mediated posttranslational acetylation of the alanine residue at the amino terminal *(12)* adds 42 Da *(13)* for a resulting predicted mass of 7101.5 Da.

Acknowledgments

This work was supported by the Canadian Institutes of Health Research, The Canadian Liver Foundation, and the Department of Paediatric Laboratory Medicine, Hospital for Sick Children.

References

1. Holland, J. J., De La Torre, J. C., and Steinhauer, D. A. (1992) RNA virus populations as quasispecies. *Curr. Top. Microbiol. Immunol.* **176**, 1–20.
2. Eigen, M., and Biebricher, C. (1988) Sequence space and quasispecies distribution, in *RNA Genetics* (Domingo, E., ed.), CRC Press, Boca Raton, pp. 211–245.
3. Bukh, J., Miller, R. H., and Purcell, R. H. (1995) Genetic heterogeneity of hepatitis C virus: quasispecies and genotypes. *Semin. Liver Dis.* **15**, 41–63.
4. Taniguchi, S., Okamoto, H., Sakamoto, M., Kojima, M., Tsuda, F., Tanaka, T., et al. (1993) A structurally flexible and antigenically variable N-terminal domain of the hepatitis C virus E2/NS1 protein: implication for an escape from antibody. *Virology* **195**, 297–301.
5. Forns, X., Thimme, R., Govindarajan, S., Emerson, S. U., Purcell, R. H., Chisari, F. V., et al. (2000) Hepatitis C virus lacking the hypervariable region 1 of the second envelope protein is infectious and causes acute resolving or persistent infection in chimpanzees. *Proc. Natl. Acad. Sci. USA* **97**, 13318–13323.
6. Farci, P., Shimoda, A., Wong, D., Cabezon, T., De Gioannis, D., Strazzera, A., et al. (1996) Prevention of hepatitis C virus infection in chimpanzees by hyperimmune serum against the hypervariable region 1 of the envelope 2 protein. *Proc. Natl. Acad. Sci. USA* **93**, 15394–15399.
7. Farci, P., Alter, H. J., Wong, D. C., Miller, R. H., Govindarajan, S., Engle, R., et al. (1994) Prevention of hepatitis C virus infection in chimpanzees after antibody-mediated in vitro neutralization. *Proc. Natl. Acad. Sci. USA* **91**, 7792–7796.
8. Farci, P., and Purcell, R. H. (2000) Clinical significance of hepatitis C virus genotypes and quasispecies. *Semin. Liver Dis.* **20**, 103–126.
9. Shimizu, Y. K., Hijikata, M., Iwamoto, A., Alter, H. J., Purcell, R. H., and Yoshikura, H. (1994) Neutralizing antibodies against hepatitis C virus and the emergence of neutralization escape mutant viruses. *J. Virol.* **68**, 1494–1500.
10. Farci, P., Shimoda, A., Coiana, A., Diaz, G., Peddis, G., Melpolder, J. C., et al. (2000) The outcome of acute hepatitis C predicted by the evolution of the viral quasispecies. *Science* **288**, 339–344.
11. Farci, P., Strazzera, R., Alter, H. J., Farci, S., Degioannis, D., Coiana, A., et al. (2002) Early changes in hepatitis C viral quasispecies during interferon therapy predict the therapeutic outcome. *Proc. Natl. Acad. Sci. USA* **99**, 3081–3086.
12. Polevoda, B., and Sherman, F. (2000) N^{α}-terminal acetylation of eukaryotic proteins. *J. Biol. Chem.* **275**, 36479–36482.
13. Qin, J., and Chait, B. T. (1997) Identification and characterization of posttranslational modifications of proteins by MALDI ion trap mass spectrometry. *Anal. Chem.* **69**, 4002–4009.
14. Ayers, M., Siu, K., Roberts, E., Garvin, A. M., and Tellier, R.. (2002) Characterization of hepatitis C virus quasispecies by matrix-assisted laser desorption ionization-time of

flight (mass spectrometry) mutation detection. *J. Clin. Microbiol.* **40,** 3455–3462.

15. Logel, J., Dill, D., and Leonard, S. (1992) Synthesis of cRNA probes from PCR-generated DNA. *Biotechniques* **13,** 604–610.

16. Yea, C., Bukh, J., Ayers, M., Roberts, E., Krajden, M., and Tellier, R. (2007) Monitoring of hepatitis C virus quasispecies in chronic infection by matrix-assisted laser desorption ionization–time of flight mass spectrometry mutation detection. *J. Clin. Microbiol.* **45,** 1053–1057.

17. Yanagi, M,, Purcell, R. H., Emerson, S. U., and Bukh, J. (1997) Transcripts from a single full-length cDNA clone of hepatitis C virus are infectious when directly transfected into the liver of a chimpanzee. *Proc. Natl. Acad. Sci. USA* **94,** 8738–8743.

Part III

HCV Replication in vitro

Chapter 7

Purification and Crystallization of NS5A Domain I of Hepatitis C Virus

Joseph Marcotrigiano and Timothy Tellinghuisen

Abstract

The NS5A protein of HCV is an essential component of the viral RNA replication machinery and may also function in modulation of the host cell environment. The exact function of NS5A in these processes remains unknown. NS5A is a large hydrophilic phosphoprotein protein consisting of three domains. The amino-terminal domain, designated domain I, coordinates a single zinc atom that is required for virus replication. We have determined the X-ray crystallographic structure of the domain I region of NS5A, and the structure sheds some light on the previously reported RNA binding activity observed for NS5A and suggests that the protein functions as a dimer. Here we describe the bacterial expression, purification, crystallization, and structural determination of the amino-terminal domain I of NS5A. The methods described herein should be of use for the generation of domain I for biochemical studies as well as future crystallization studies as antiviral compounds directed against this region of NS5A become available.

Key words: Hepatitis C virus, NS5A, RNA replication, zinc metalloprotein, crystallography, RNA binding proteins.

1. Introduction

HCV NS5A protein is perhaps the most enigmatic component of the viral RNA replication machinery. Despite the absolute requirement for it in productive RNA replication *(1, 2)*, no specific function has been attributed to this phosphoprotein. NS5A has been a target of intense interest for many years, as it appears to interact with a large number of cellular proteins and may function to manipulate the host cell for productive virus replication *(3)*. Recent developments in NS5A characterization have provided some important clues to its possible function(s).

Hengli Tang (ed.), *Hepatitis C: Methods and Protocols, Second Edition, vol. 510*
© 2009 Humana Press, a part of Springer Science+Business Media
DOI 10.1007/978-1-59745-394-3_7 Springerprotocols.com

A three-domain organization model of NS5A has been proposed, on the basis of bioinformatic and biochemical analyses *(4)*. In this model, NS5A consists of a large, well-conserved amino-terminal domain (domain I) and two smaller, more variable domains (domains II and III) (**Fig 7.1A**). These domains are separated by low-complexity, repetitive sequences that are highly sensitive to proteolytic cleavage in vitro. The amino terminus of domain I contains an amphipathic α-helix that inserts into one leaflet of the membrane, tethering NS5A to the cytoplasmic side of the ER membrane *(5–7)*. After the membrane-anchor sequence, domain I contains a conserved four-cysteine zinc coordination site that binds a single metal ion *(4)*. This binding is essential for the role of NS5A in HCV replication. More recently, NS5A has been shown to bind to RNA, although the significance of this binding to RNA replication is not yet known *(8)*. The crystal structure of domain I provides a possible RNA

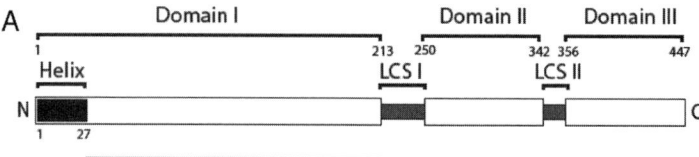

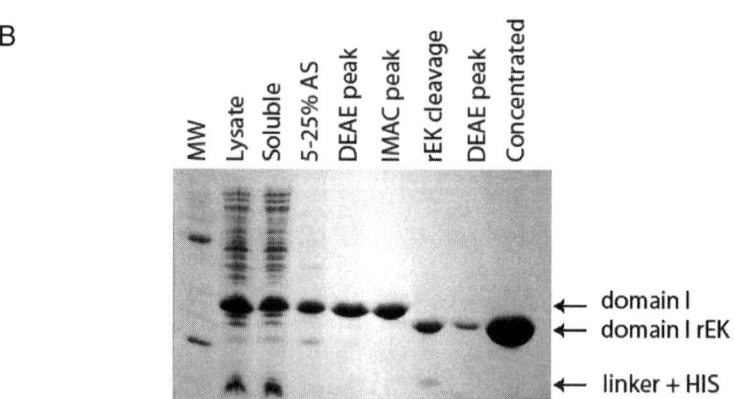

Fig. 7.1. Identification and purification of domain I Δ-h. **A.** Schematic of the full-length NS5A protein highlighting the location of the three domains and the interdomain connecting low-complexity sequences (LCS) defined previously. The location of the amino-terminal membrane anchor of NS5A is shown as a black bar labeled "helix." The black bar beneath the domain I region of the schematic highlights the region of NS5A corresponding to the domain I Δ-h construct. **B.** SDS-PAGE gel (12%, Coomassie Blue R-250 stained) showing the various purification steps of domain I Δ-h preparation discussed. The individual lanes of the gel correspond to the following steps in **Section 3.2**: Lysate (step 3), soluble (step 4), 5–25% ammonium sulfate (AS) (step 7), DEAE peak (step 9), IMAC peak (step 10), rEK cleavage (step 12), DEAE peak (step 13), and concentrated (step 17).

interaction site, a large basic groove generated by the interface of two domain I monomers in the dimeric structure *(9)*. Whether this groove is actually the site of RNA binding is not yet known. The remaining domains of NS5A are poorly understood. They appear to tolerate large heterologous insertions (such as green fluorescent protein) without loss of the RNA replication activities of NS5A *(1, 3, 4, 9)*. Much more research is clearly needed to elucidate fully the function(s) of NS5A in replication and cellular processes.

The development of an efficient system for the production of large amounts of heterologously expressed NS5A, or even portions of NS5A, has required considerable effort. The first real breakthrough came from the observation that fusion of the amino terminus of the NS5A coding sequence to ubiquitin allowed for expression of NS5A in bacteria *(10)*. Coexpression of this ubiquitin–NS5A fusion protein in the presence of an yeast ubiquitinase activity (encoded by the plasmid pCG1) allowed for the complete removal of the ubiquitin fusion from NS5A in a cotranslational manner, leading to an authentic, detergent-soluble NS5A protein *(10)*. Deletion of the amino-terminal membrane anchor of NS5A and fusion of the ubiquitin monomer to the region of NS5A following the anchor permitted production of soluble NS5A in the absence of detergents *(4)*. On the basis of these previous expression systems, and our observations made during the domain mapping of NS5A, we developed an expression construct with a ubiquitin monomer fused to a membrane-anchor-deleted domain I sequence that included a long flexible linker and a poly-histidine tag at the carboxyl terminus of domain I. The vector encoding this construct was called pET30b Ubi domain I Δ-h, and the expressed protein domain I Δ-h *(4)*. We evaluated a number of NS5A sequences from different HCV genotypes and found the Con1 isolate of genotype 1b to produce the highest yield of soluble domain I Δ-h in *E. coli*. The details of the construction of this plasmid have been published elsewhere *(4)*. A purification method was developed for extraction of domain I Δ-h and obtaining large quantities of very pure protein for crystallization *(9)*. The method for the expression and purification of domain I Δ-h are detailed in **Sections 3.1 and 3.2** of this chapter. Once purified domain I Δ-h was available, we screened a large number of crystallization trials with this protein and eventually developed the method of crystallization described in **Section 3.3**. Crystals produced in this manner were of sufficient quality to allow structural solution of domain I Δ-h *(9)*. The methods presented here provide the information necessary to express, purify, and crystallize domain I Δ-h. The methods presented represent further optimizations of our previously published work, which increase the reliability of the purification and crystallization of this protein.

2. Materials

2.1. Protein Expression

1. 2.8-L baffled glass Fernbach culture flasks.
2. Kanamycin (30 mg/mL in water) and choramphenicol (25 mg/mL in ethanol) antibiotic stocks.
3. Isopropyl β-D-1-thiogalactopyranoside (IPTG) stock, 1 M in water.
4. LB medium (10 g of tryptone, 10 g of NaCl, 5 g of yeast extract per liter), 4 L total, sterile.

2.2. Protein Purification

1. Buffer A: 25 mM Tris-HCl, pH 8.0, 25 mM NaCl, 20% glycerol, 0.22 μm filtered.
2. Buffer B: 25 mM Tris-HCl, pH 8.0, 1 M NaCl, 20% glycerol, 0.22 μm filtered
3. Buffer C: 25 mM Tris-HCl, pH 8.0, 250 mM imidazole, 25 mM NaCl, 20% glycerol, 0.22 μm filtered.
4. Protein storage buffer: 25 mM Tris-HCl, pH 8.0, 175 mM NaCl, 20% glycerol, 0.22 μm filtered.
5. Recombinant light-chain enterokinase (rEK) (New England Biolabs).
6. Benzonase nuclease, 90% purity (Novagen).
7. Ultra 4 10K molecular-weight cut-off centrifugal concentrators (Amicon).
8. Chromatography columns: 26/10 HiPrep desalting column, 16/10 HiPrep DEAE fast flow ion exchange column, 5 mL HiTrap IMAC metal chelate column (GE Healthcare).
9. Avestin Emulsiflex EC3 or EC5 air emulsifier cell-lysis system with water heat exchanger (Avestin).

2.3. Crystallization and Crystal Freezing

1. 24-well pregreased Linbro crystallization plates (Hampton Research).
2. 22-mm presiliconized circular glass coverslips (Hampton Research).
3. Crystallization-well solution: 100 mM HEPES, pH 7, 0.6 M trisodium citrate dihydrate, 13% (v/v) isopropanol.
4. 2 M nondetergent sulfobetaine 201 (Hampton Research).
5. 20-μm thread, 0.2–0.3 mm loop-size cryo loops mounted on appropriate crystal mounting bases (Hampton Research).
6. Cryo storage vials compatible with appropriate crystal loop mounting bases.
7. Crystal cryopreservation buffer: 100 mM HEPES, pH 7, 0.6 M trisodium citrate dihydrate, 13% (v/v) isopropanol, 30% glycerol.
8. CrystalCap vial holder/freezing apparatus (Hampton Research).

9. Laboratory-grade liquid propane (caution, flammable/explosive).

---

3. Methods

3.1. Protein Expression

1. Transform the plasmids pCG1 and pET30b Ubi domain I Δ-h into *E. coli* BL21(DE3) cells and plate transformations of LB agar plates containing $30\,\mu g/\mu L$ of kanamycin and $25\,\mu g/\mu L$ chloramphenicol (*see* **Note 1**).

2. After an overnight incubation at 37°C, pick a single colony from the bacterial plate with a sterile pipette tip and inoculate a 500-mL Erlenmeyer flask containing 100 mL of LB medium supplemented with $30\,\mu g/\mu L$ of kanamycin and $25\,\mu g/\mu L$ chloramphenicol. Grow this culture overnight at 37°C with 250 rpm shaking.

3. Inoculate 10 mL of the pET30b0 Ubi domain I Δ-h BL21(DE3) overnight culture into each of four 2.8-L baffled Fernbach flasks containing 1 L of LB medium. The LB medium should contain $30\,\mu g/\mu L$ of kanamycin and $25\,\mu g/\mu L$ chloramphenicol to ensure maintenance of the plasmids required for protein production.

4. Grow cells at 37°C with 250 rpm shaking until an optical density of 0.55 at 600 nm (OD_{600}) is reached. Typically, the time between inoculation and arrival at OD_{600} is approximately 3.5 h.

5. Once the cells reach the appropriate OD_{600}, chill the flasks at 4°C for 30 min by placing the flasks in a 4°C cold room or chilling them in an ice/water bath.

6. Add IPTG to a final concentration of 1 mM to induce expression of domain I Δ-h.

7. Incubate cells at 25°C with 180 rpm shaking for 5 h. The lower shaking speed is based on the empirical observation that slower shaking after induction produces a greater yield of soluble protein during purification (*see* **Note 2**).

8. Collect cells by centrifugation at 6000g for 10 min at 25°C. Pour off the used LB medium and save the cell pellet.

9. Resuspend cell-pellet material in a total of 80 mL of buffer A (20 mL/liter of culture volume) by vortexing the cell pellets in the presence of buffer A and then vigorously pipetting to disperse any cell-pellet clumps. Careful resuspension is important for proper loading of the cell suspension on the air emulsifier for lysis.

10. Either use resuspended cells directly in the method described in **Section 3.2** (protein purification) or flash freeze them in a liquid-nitrogen bath and store them at −80°C. For flash freezing, we place 20 ml aliquots of the resuspended cells

(equivalent to 1-L culture volumes) in 50-ml conical tubes and simply immerse the tubes in liquid nitrogen until the cell suspension is completely frozen. Frozen suspensions are then stored at −80°C until needed (*see* **Note 3**).

3.2. Protein Purification

1. If using a cell suspension frozen as in **Section 3.1**, thaw the suspension in a cold water bath. If using freshly resuspended cells, chill the material on ice for 15 min before proceeding with cell lysis.

2. Add 375 units of benzonase nuclease (Novagen, 90% purity grade enzyme) to the resuspended cell material (*see* **Note 4**).

3. Lyse the cells in the suspension by three passes through an Avestin air emulsifier at 10–15,000 psi (*see* **Note 5**).

4. Clarify the cell lysate by centrifugation at 25,000g for 30 min at 4°C.

5. Remove the soluble fraction and adjust the volume of this fraction to 100 mL with buffer A.

6. Add 2.65 g of ammonium sulfate to bring the soluble lysate to 5% saturation. Add ammonium sulfate slowly with stirring at 4°C. Stir lysate slowly at 4°C for 30 min after all of the ammonium sulfate has been added. Centrifuge the sample at 25,000g for 10 min at 4°C and save the soluble fraction.

7. Slowly add 10.8 g of ammonium sulfate to the supernatant fraction from step 6 to reach 25% saturation. Add ammonium sulfate slowly with stirring at 4°C. Stir the lysate slowly at 4°C for 30 min after all of the ammonium sulfate has been added. Centrifuge the sample at 25,000g for 10 min at 4°C and save the soluble fraction.

8. Discard the supernatant from the centrifugation and resuspend the pellet fraction in a total volume of 20 mL with buffer A. Desalt this material with 2 × 10 mL loadings onto a HiPrep 26/10 fast desalting column equilibrated with buffer A at a flow rate of 7 mL/min (*see* **Note 6**).

9. Load the pooled desalted material at a flow rate of 2.5 mL/min onto a HiPrep 16/10 DEAE fast-flow column equilibrated with buffer A. Wash the column with approximately 5 column volumes of buffer A or until the unbound material is completely through the column. Elute domain I Δ-h from the column with a linear salt gradient of 10 column volumes using buffer B. NS5A elutes as approximately 35% buffer B.

10. Load the pooled fractions from DEAE column elution directly onto a 5 mL HiTrap IMAC column charged with nickel and equilibrated with buffer A. Then wash the bound fraction on the column extensively (10 column volumes) with buffer A and then with buffer B for 4 column volumes. Then elute domain I Δ-h with a 25-mL linear gradient of buffer C.

11. Immediately desalt the IMAC elution fraction into buffer A using the same chromatographic method described in step 8 (*see* **Note 7**).

12. Remove the poly-histidine tag and linker region by overnight cleavage with rEK protease by adding 60 ng of rEK to the desalted domain I Δ-h and incubating overnight (approximately 14 h) at 25°C. After overnight digestion, we routinely check the completeness of proteolytic cleavage by SDS-PAGE.

13. Once complete cleavage has been obtained, pass the cleaved protein over a HiPrep 16/10 DEAE column as described in step 9. Domain I Δ-h binds to and elutes from this column as described in step 9, the cleaved linker and poly-histidine tag fragment is in the flow-through of the column.

14. Desalt the DEAE elution peak into protein storage buffer using the chromatographic method described in step 8.

15. Concentrate the purified protein with the Amicon ultra 4 centrifugal concentrators by centrifugation at 7000g at 20°C until target concentration is reached. Crystallization is conducted with the protein at a concentration of 26 mg/mL. A typical yield from this procedure is approximately 30 mg of protein at greater than 95% purity. We evaluate purity by silver-stained SDS-PAGE and determine protein concentration using the Bio-Rad protein assay with a bovine serum albumin standard concentration curve (*see* **Note 8**).

16. Purified domain I Δ-h can be stored at 4°C for several days or at −80°C indefinitely. Freezing the protein does not adversely affect later crystallization (*see* **Note 9**).

3.3. Crystallization and Crystal Freezing (Fig 7.2)

1. To each well of a pregreased 24-well Linbro crystallization plate add 1 mL of cold crystallization-well solution. Conduct this addition rapidly and on ice to minimize evaporation of isopropanol from the well solution.

2. To a 22-mm siliconized circular glass coverslip, add 1.5-μL of purified domain I Δ-h (26 mg/mL in protein storage buffer), 1.25 μL of crystallization-well solution, and 0.5 μL of 2 M nondetergent sulfobetaine 201. Invert the coverslip and place on the greased Linbro plate. Place gentle pressure on the coverslip to seal the well closed with the grease ring on the crystallization plate. Repeat these steps to load all the remaining wells of the Linbro plate (*see* **Note 10**).

3. Incubate the Linbro plate at 4°C in an area free of vibration and temperature variation (*see* **Note 11**).

4. After approximately 4–5 days, small crystals (0.05–0.07 mm) of a classic tetragonal habit should be apparent. These crystals will reach size usable for diffraction studies (0.2–0.3 mm) after 8–10 days of incubation. Longer incubation does not yield further crystal growth beyond 0.3 mm (*see* **Note 12**).

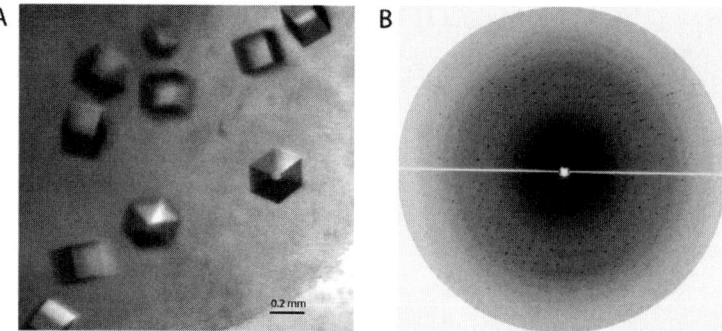

Fig. 7.2. Sample images of domain I Δ-h crystal quality. **A**. Light micrograph of domain I Δ-h protein crystals grown under conditions described here. Note the presence of amorphous precipitate in the drop. **B**. Sample diffraction image of frozen domain I Δ-h crystal (30-s exposure of 0.2-mm crystal at 1° oscillation with a 140-mm camera length). The edge of the detector plate corresponds to approximately 2.25 Å. Data collected at beam line X9A, Brookhaven National Labs, National Synchrotron Light Source.

5. Once maximal crystal growth has occurred, carefully remove the coverslip from the wells that have produced crystals of size and quality appropriate for diffraction studies. Using a 20-μm thread 0.3–0.4 mm nylon cryo loop, pick up individual crystals and transfer them to cryo buffer (essentially well solution supplemented with a total of 30% glycerol; *see* **Note 13**).

6. After a 10-min soak in cryo buffer, pick up individual crystals in nylon cryo loops mounted on cryo pin bases (Hampton Research) and transfer them to liquid propane in a 2-mL cryo vial appropriate to the cryo pin bases used. Using a Crystal-Cap holder, freeze the crystal and propane in a bath of liquid nitrogen (*see* **Note 14**).

7. Once the propane has frozen, transfer the vial to a liquid-nitrogen tank for long-term storage.

4. Notes

1. We have found the expression of domain I Δ-helix to be more reproducible in terms of protein yield when freshly transformed BL21(DE3) are used. Transformed cells on agar plates can be stored at 4°C for up to two weeks without loss of protein expression.

2. The lower temperature and slower shaking of cell cultures after IPTG induction are critical to the generation of soluble protein.

3. Cells can be frozen after induction for several months at −80°C without any noticeable effect on later protein

purification and crystallization. Flash freezing cells in liquid-nitrogen baths require proper cryogen safety precautions.

4. Benzonase treatment of the cell suspension greatly reduces the viscosity of the protein extract upon lysis of the cells. This treatment also increases the yield of soluble protein.

5. We have evaluated a number of protein lysis methods and determined that air emulsification produces the most soluble domain I Δ-h. Sonication or chemical lysis of cells is not recommended, as these method produce largely insoluble domain I Δ-h. Keeping the cell lysate cold during and immediately after lysis is crucial to solubility and protein stability. We use a recirculating water-flow heat exchanger and collect the lysate into prechilled tubes on ice.

6. The use of dialysis in place of column-based desalting is not recommended. Dialysis at this stage leads to protein degradation and/or aggregation.

7. The protein must be desalted into buffer A as soon as possible after elution from the IMAC column. Storage of protein for even a few hours in the presence of imidazole leads to aggregation of domain I Δ-h.

8. The limit of domain I Δ-h solubility in protein storage buffer is in excess of 60 mg/mL. Protein can be stored at high concentration and diluted in storage buffer to 26 mg/mL before crystallization. We have observed a light-brown precipitate at the bottom of the concentrators after the completion of concentration. This precipitate does not contain significant amounts of domain I Δ-h and can be removed by centrifugation of the concentrated protein for 1 min in a microcentrifuge at 10,000g.

9. We have successfully crystallized domain I Δ-h after more than five freeze-thaw cycles.

10. The mixture of protein, well solution, and additive will quickly turn white and show a significant amount of light-brown amorphous precipitate upon microscopic examination. This precipitation is normal, and protein crystals will grow from the precipitate. All of the crystallization buffers and solutions must be prepared with the highest quality of crystallization-grade materials. We use premade reagent stocks from Hampton Research. The use of nondetergent sulfobetaine 201 as an additive was determined by random additive screening. We have evaluated all of the commercially available sulfobetaine compounds as well as a number of detergents, and none produced the dramatic increase in crystal size and quality observed with nondetergent sulfobetaine 201.

11. Temperature variation and vibration can adversely affect crystallization. We routinely place our crystallization plates on foam-rubber supports to limit vibration. Trays are set in a

4°C cold room away from the door and air vents of the cold room. Low temperature is critical, and domain I Δ-h crystals will not grow at room temperature.

12. Crystals may not form in every drop. We regularly see crystals in about 18 of the 24 wells. If no crystals are observed, we suggest setting up a factorial grid-based screen in which sodium citrate and isopropanol concentrations are covaried.

13. When the coverslip is removed, the drop will quickly equilibrate with air and the contents of the drop will swirl rapidly. Although alarming in appearance, this phenomenon does not appear to affect crystal quality negatively. Simply wait for the swirling of the drop contents to stop and proceed with crystal processing.

14. Use appropriate precautions working with cryogenic liquids such as liquid propane and liquid nitrogen. In addition to being a cryogenic hazard, liquid propane is extremely flammable. Work with propane in a properly vented chemical hood approved for flammable gases and avoid open flames, sparks, and other sources of ignition. The choice of cyro pin and cryo vial depends largely on the goniometer that will be used for data collection and probably has little impact on the successful completion of these procedures.

References

1. Blight, K. J., Kolykhalov, A. A., and Rice, C. M. (2000) Efficient initiation of HCV RNA replication in cell culture. *Science* **290**, 1972–1974.

2. Lohmann, V., Korner, F., Dobierzewska, A., and Bartenschlager, R. (2001) Mutations in hepatitis C virus RNAs conferring cell culture adaptation. *J. Virol.* **75**, 1437–1449.

3. Macdonald, A. and Harris, M. (2004) Hepatitis C virus NS5A: tales of a promiscuous protein. *J. Gen. Virol.* **85**, 2485–2502.

4. Tellinghuisen, T. L., Marcotrigiano, J., Gorbalenya, A. E., and Rice, C. M. (2004) The NS5A protein of hepatitis C virus is a zinc metalloprotein. *J. Biol. Chem.* **279**, 48576–48587.

5. Brass, V., Bieck, E., Montserret, R., Wolk, B., Hellings, J. A., Blum, H. E., et al. (2002) An amino-terminal amphipathic α-helix mediates membrane association of the hepatitis C virus nonstructural protein 5A. *J. Biol. Chem.* **277**, 8130–8139.

6. Penin, F., Brass, V., Appel, N., Ramboarina, S., Montserret, R., Ficheux, D., et al. (2004) Structure and function of the membrane anchor domain of hepatitis C virus nonstructural protein 5A. *J. Biol. Chem.* **279**, 40835–40843.

7. Sapay, N., Montserret, R., Chipot, C., Brass, V., Moradpour, D., Deleage, G., et al. (2006) NMR structure and molecular dynamics of the in-plane membrane anchor of nonstructural protein 5A from bovine viral diarrhea virus. *Biochemistry* **45**, 2221–2233.

8. Huang, L., Hwang, J., Sharma, S. D., Hargittai, M. R., Chen, Y., Arnold, J. J., et al. (2005) Hepatitis C virus nonstructural protein 5A (NS5A) is an RNA-binding protein. *J. Biol. Chem.* **280**, 36417–36428.

9. Tellinghuisen, T. L., Marcotrigiano, J., and Rice, C. M. (2005) Structure of the zinc-binding domain of an essential component of the hepatitis C virus replicase. *Nature* **435**, 374–379.

10. Huang, L., Sineva, E. V., Hargittai, M. R., Sharma, S. D., Suthar, M., Raney, K. D., et al. (2004) Purification and characterization of hepatitis C virus non-structural protein 5A expressed in *Escherichia coli*. *Protein Expr. Purif.* **37**, 144–53.

Chapter 8

NS5A Phosphorylation and Hyperphosphorylation

Petra Neddermann

Abstract

NS5A phosphorylation can be studied in two ways: in living cells and in vitro. The former has several advantages: NS5A phosphorylation takes place in a cellular background and therefore might mimic more closely the real in vivo situation. Viral proteins and cellular kinases are in the correct cellular compartments, and dynamic processes like viral polyprotein processing and cellular signaling are in place. The disadvantage of this system is its great complexity, which makes limiting an observed effect to a single, well-defined agent, for example, a kinase, very difficult. NS5A phosphorylation in cells can easily be followed by metabolic labeling with either ^{35}S-methionine or ^{32}P-orthophosphate. The effect of a single, well-defined kinase on NS5A phosphorylation can be investigated in cells either by overexpression of this kinase in the presence of NS5A or by RNA interference of this kinase. If available, specific kinase inhibitors can be used to reveal the effect of this inhibition on NS5A phosphorylation. The problem with this approach is that only very few really specific kinase inhibitors are available. Biochemical in vitro experiments use purified components. This type of experiment allows direct investigation of the activity of a single kinase on NS5A as a substrate. In addition, the precise phosphorylation sites of a kinase can be mapped when NS5A-derived peptides are used instead of a full-length recombinant protein. Kinase inhibitors, which show a particular effect on NS5A phosphorylation in cells, can be retested in vitro on a particular kinase candidate. The problem with this approach is that purified components, like the purified NS5A substrate and the kinase of interest, are not always available.

Key words: Hepatitis C virus, NS5A, NS5A phosphorylation, NS5A hyperphosphorylation, CKI-α, cellular kinases, kinase inhibitors.

1. Introduction

The first indication that NS5A is a phosphoprotein came from the group of K. Shimotohno, who demonstrated that NS5A contains phospho-serine and to a small extent also phospho-threonine residues.

Hengli Tang (ed.), *Hepatitis C: Methods and Protocols, Second Edition, vol. 510*
© 2009 Humana Press, a part of Springer Science+Business Media
DOI 10.1007/978-1-59745-394-3_8 Springerprotocols.com

NS5A migrates as a protein with an apparent molecular weight of 56 kDa (NS5A-p56), but an additional slower migrating form of NS5A has been described that migrates with an apparent molecular weight of 58 kDa (NS5A-p58) *(1)*. Both forms are phosphorylated, and p58 is generally accepted to be a hyperphosphorylated form of NS5A-p56. In the case of the 1b-J strain, the generation of p58 requires the presence of NS4A *(2)* and, in other 1b strains, the presence of the nonstructural proteins NS3, NS4A, and NS4B together with NS5A *(3, 4)*. Mutagenesis studies identified three serine residues (S-2197, S-2201, and S-2204) in the central region of NS5A as phosphorylation sites important for the formation of p58 *(5)*, but none of these has been directly identified by phosphoamino acid analysis. To date only two serine residues have been unequivocally mapped as sites of phosphorylation in NS5A-p56, and they are thought to be involved in basal phosphorylation of NS5A *(6, 7)*.

Previous experiments depended on transient transfection experiments because no cell-culture system existed in which viral replication or viral infection could be reproduced. In many cases, the *Vaccinia* virus T7 infection/transfection system was used to increase protein expression. Protein expression is followed by metabolic labeling of the proteins with ^{35}S-methionine, whereas protein phosphorylation is followed by incubation of the cells with ^{32}P-orthophosphate. The protein of interest, in this case NS5A, is isolated from the cell extract by immunoprecipitation with NS5A-specific antibodies.

NS5A is phosphorylated at many different sites distributed throughout the entire protein, with the exception of the N-terminal part, and three different clusters of potential phosphorylation sites have been classified *(8)*. The existence of different basally and hyperphosphorylated forms has been confirmed by two-dimensional sodium dodecyl sulfate polyacrylamide gel electrophoresis (SDS-PAGE) *(9)*.

Coimmunoprecipitation and affinity-purification experiments have revealed a tight association of cellular kinase(s) with NS5A *(10–12)*. The effect of different protein kinase inhibitors on NS5A phosphorylation classified the kinase(s) as a member of the CMGC group of serine/threonine kinases *(12)*, and yeast two-hybrid experiments identified several kinases that interacted directly with NS5A-p56 in cells *(13)*.

In 1999, Lohmann et al. introduced the subgenomic replicon system *(14)*. the first cell-based system with which HCV RNA replication could be reproduced, and many studies of protein expression, modification, and function now use this system. Upon introduction of the subgenomic replicon, cell-culture adaptive mutations were observed to increase dramatically the replication efficiency of the Con1 HCV subgenomic RNA within the hepatic cell line Huh7. Interestingly, these mutations predominantly mapped to the NS5A-coding sequence, and notably, the most effective mutations were those that reduced the formation of hyperphosphorylated NS5A (NS5A-p58) *(15)*. Even though the functional relevance of the different phosphorylated forms for viral replication is still unknown, the importance of tight regulation of NS5A phosphorylation for the outcome of HCV RNA replication and/or infection has become more and more evident. Taking advantage of the tight association of cellular kinases with NS5A, we sought to identify kinase inhibitors that specifically inhibited the formation of hyperphosphorylated NS5A *(16)*. These inhibitors turned out to be a valuable tool for identifying at least one cellular kinase that is important for NS5A hyperphosphorylation: the α-isoform of protein kinase CKI (CKI-α). Overexpression and RNA interference of CKI-α in Huh7 cells confirmed the importance of this cellular kinase for NS5A hyperphosphorylation *(17)*. In addition, in vitro phosphorylation of NS5A by CKI resulted in the formation of two different forms of NS5A, which showed a remarkable similarity to NS5A-p56 and NS5A-p58 produced in cells *(18)*.

NS5A and especially its different phosphorylation variants therefore play a significant role in the outcome of HCV replication, but many questions remain open regarding the cellular kinases involved in basal and hyperphosphorylation, and too little is known about the precise phosphorylation sites in NS5A. Finally, the key question still remains to be answered: What is the role of NS5A-p56 and NS5A-p58 in HCV viral replication and/or infection?

2. Materials

2.1. Metabolic Labeling of NS5A in Huh7 Cells

1. Huh7 cells/Huh7 cells stably expressing the subgenomic replicon (Huh7-68). Store at −20°C.
2. *Vaccinia* virus stock vTF7-3. Store at −80°C.
3. Good quality DNA preparations. Here we used pcD-BLA-wt as NS5A expression plasmid *(17)*. Store at −20°C.
4. Fugene6 (Roche). Store at +4°C.
5. Optimem (Invitrogen). Store at +4°C.
6. Phosphate-buffered saline (PBS): 137 mM NaCl, 2.7 mM KCl, 10 mM Na_2HPO_4, and 1.8 mM KH_2PO_4, pH 7.4. Store at room temperature.
7. Kinase inhibitor H479 *(16)*.
8. Starvation medium: Dulbecco's modified Eagle's medium (DMEM) without methionine and cystine (MP Biomedicals). Store at +4°C.

9. Medium for ^{32}P-orthophosphate labeling: DMEM without phosphates (ICN). Store at +4°C.

10. [^{35}S]-labeled methionine (Promix, GE Healthcare). Store at −20°C.

11. [^{32}P]-orthophosphate (Perkin Elmer). Store at +4°C.

12. TEN: 40 mM Tris–HCl, pH 7.5, 1 mM EDTA, and 150 mM NaCl. Store at room temperature.

2.2. Immunoprecipitation of NS5A from Cell Extracts and Further Analysis by SDS-PAGE

1. Polyclonal or monoclonal antibodies suitable for NS5A-specific immunoprecipitation. Store at −20°C. Store small aliquots at +4°C.

2. Protein A-Sepharose (PAS, GE Healthcare). Store at +4°C.

3. IPB$_{150}$: 20 mM Tris–HCl, pH 8, 150 mM NaCl, 1% Triton-X100. Store at room temperature.

4. Lysis buffer: 25 mM Na$_2$HPO$_4$/NaH$_2$PO$_4$-buffer, pH 7.5, 20% glycerol, 1% Triton-X100, 150 mM NaCl, 1 mM EDTA, 2 mM dithiothreitol (DTT), 2 mM phenylmethylsulfonyl fluoride (PMSF). Add DTT and PMSF fresh. Store at −20°C.

5. 20% SDS. Store at room temperature.

6. 1 M DTT. Store at −20°C.

7. NDET/SDS: 1% Triton-X-100, 0.4% Na-deoxycholate, 66 mM EDTA, 10 mM Tris–HCl, pH 7.4, 0.3% SDS. Store at room temperature.

8. SDS sample buffer: 150 mM Tris–HCl, pH 6.8, 30% glycerol, 6% SDS, 0.3% bromphenol blue. Store at −20°C.

9. Standard materials for SDS-PAGE: Here mentioned specifically: 30% acrylamide stock solution, 29.2:0.8 ratio of acrylamide to *bis*-acrylamide (Invitrogen, Ultrapure) (*see* **Note 1**). Store at +4°C.

10. Protein-fixing solution: 10% glacial acetic acid, 20% methanol. Store at room temperature.

11. Amplify (GE Healthcare). Store at room temperature.

12. Films: Biomax MR (Kodak).

2.3. Phosphorylation of NS5A by Coimmunoprecipitated Cellular Kinases In Vitro

1. Huh7 cells or Huh7 cells stably expressing an HCV subgenomic replicon (Huh7-68, (*16*)). Store at −80°C.

2. NETN-buffer: 50 mM Tris–HCl, pH 7.5, 120 mM NaCl, 1 mM EDTA, 0.5% NP40. Store at +4°C.

3. Protein inhibitor cocktail Complete (Roche).

4. Protein A-Sepharose (PAS). Store at +4°C.

5. NS5A-specific antibody. Store at −20°C. Store small aliquots at +4°C.

6. Kinase buffer: 50 mM Tris–HCl, pH 7.5, 5 mM MnCl$_2$, 5 mM DTT, 50 mM NaF.

7. [$\gamma -^{33}$P]-ATP (GE Healthcare), Redivue. Store at +4°C.

8. SDS sample buffer. Store at −20°C.

2.4. Phosphorylation of NS5A by Purified Kinases In Vitro

1. Purified NS5A (Con1-1b strain). One example of NS5A purification protocol is described in Huang et al. *(19)*. Store the protein at −80°C.

2. Final NS5A dialysis buffer: 50 mM Hepes pH 7.5, 20% glycerol, 200 mM NaCl, 0.3% n-octyl-β-d-glucopyranoside (Inalco S.p.A.), 1 mM DTT. Store at +4°C.

3. Kinases: CKI-δ (BioLabs), protein kinase A (PKA) (Upstate). Store at −80°C.

4. PKA/CaMK inhibitor cocktail (Upstate). Store at −80°C.

5. ADBI: 20 mM 3-(*N*-morpholino)-propanesulfonic acid (MOPS), pH 7.2, 25 mM β-glycerolphosphate, 5 mM EGTA, 1 mM Na-orthovanadate, 1 mM DTT. This buffer can be considered twofold concentrated (2×). Store at −20°C.

6. ATP mix: 500 μM ATP, 75 mM MgCl$_2$ in ADBI. This buffer can be considered fourfold concentrated (4×). Store at −20°C.

7. SDS sample buffer. Store at −20°C.

8. Protein-fixing solution: 10% glacial acetic acid, 20% methanol. Store at room temperature.

9. SDS-PAGE.

10. Peptides: NS5A-derived peptides (Bio Synthesis Incorporated). Peptide 1: NH$_2$-RRKGSPPSLASSSASQLSAPSLK-NH$_2$; peptide 2: NH$_2$-RRKGSPPSLAS-pSer-SASQLSAP SLK-NH$_2$; pSer: indicates phosphorylated serine residue (*see* **Note 2**). Store at −20°C.

11. 3% phosphoric acid. Store at room temperature.

12. P30 Filtermat (Wallac).

13. Filter wash solutions: 75 mM phosphoric acid, 100% methanol. Store at room temperature.

14. Ready protein scintillation cocktail (Beckman Coulter). Store at room temperature.

2.5. Effect of Kinases on NS5A Phosphorylation in Cells

1. Cells that can be infected by *Vaccinia* virus and that express the desired kinase. In this case we use Huh7 cells. Store under liquid nitrogen.

2. High-quality DNA-expression plasmids. In this case we use pcD-BLA-wt and pCD-Flag-CKI-α *(17)*. Store at −20°C.

3. Material for DNA transfection, protein labeling, and immunoprecipitation as described in **Sections 2.1 and 2.2**.

4. PBS/DEPC: Incubate PBS with 0.1% diethylpyrocarbonate (DEPC) (Sigma) overnight at 37°C with continuous stirring. Autoclave the solution afterward. Store at room temperature.

5. High-quality siRNAs, double-stranded and annealed. *CKI −* α(*sense*): 5′-GAAACAUGGUGUCCGGUUUTT-3′. Store at −80°C.

6. Gene Pulser Cuvette 0.2 cm (BioRad).

7. Standard material for western blotting.
8. Antibodies: NS5A antibodies for immunoprecipitation; primary antibodies for western blot: anti-FLAG M2 monoclonal antibody, peroxidase conjugated (Sigma), anti-CKI-α (Santa Cruz Biotechnology, sc-6477). Secondary antibodies: horseradish peroxidase-linked secondary antibody (Pierce). Store at −20°C. Store small aliquots at +4°C.
9. ECL western blotting detection reagent (GE Healthcare). Store at +4°C.
10. Quantitative PCR:
   (a) RNeasy Mini Kit (Quiagen). Store at room temperature.
   (b) Quantitect Probe RT-PCR Kit (Quiagen). Store at −20°C.
   (c) Human β-actin endogenous control (Applied Biosystems). Store at −20°C.
   (d) Purified oligonucleotides for PCR amplification: CKI − α for: 5′-CATCTATTTGGCGATCAACATCA-3′; CKI − α rev: 5′-GCCTGGCCTTCTGAGATTCTA-3′. Store at −20°C.
   (e) DNA probes containing 5′-6FAM dye and 3′-TAMRA quencher: CKI − α probe: 5′-CAACGGCGA GGAAGTGGCAGTGA-3′. Store at −20°C in the dark.

3. Methods

3.1. Metabolic Labeling of NS5A in Huh7 Cells

In transient transfection experiments, the cellular backgrounds of different cells can be tested, and the effect of the differences on NS5A phosphorylation can be studied (*see* **Note 3**).

The subgenomic replicon is a more "physiological" system, because the Huh7 cells represent a cellular background in which the HCV RNA actively replicates, thus mimicking more closely the situation of in vivo replication.

3.1.1. Transient Transfection with the VacciniaT7 Infection/Transfection System

1. Plate 3.5×10^5 Huh7 cells in a six-well dish 24 h before the experiment in DMEM complemented with 10% FBS. The next day the cells should have grown to approximately 70–80% confluency.
2. Remove medium and incubate cells with 80 μl/well of the viral *Vaccinia* stock for 30 min at 37°C (*see* **Notes 4, 5**).
3. Add 2 ml of cell medium and incubate for an additional 30 min at 37°C.
4. In the meantime, prepare the DNA precipitate: Incubate 6 μl of Fugene6 with 94 μl of Optimem for 5 min at room temperature. Add the mixture dropwise to 2 μg of pcD-BLA-wt/well *(17)* and incubate for 20 min at room temperature.

After the incubation, add the DNA precipitate to the cells and incubate the cells for 4 h at 37°C.

5. Remove the medium and wash the cells once with 2 ml PBS.

   (a) For ^{35}S-methionine labeling: Incubate cells for 1 h at 37°C with starvation medium. After the starvation, incubate the cells with the starvation medium plus 100 μCi of ^{35}S-labeled methionine/ml of medium for 2–5 h at 37°C. Add kinase inhibitors during this starvation reaction and during the labeling reaction. In this case, we used the NS5A hyperphosphorylation-specific kinase inhibitor H479 at 5 μM final concentration (see **Note 6**).

   (b) For ^{32}P-labeling, a starvation reaction is not necessary. Wash the cells once with DMEM without phosphates and incubate the cells with 500 μCi of ^{32}P-orthophosphate/ml medium for 4 h at 37°C. Add kinase inhibitors directly during the labeling reaction. In this case, we used the NS5A hyperphosphorylation-specific kinase inhibitor H479 at 5 μM final concentration.

6. After the incubation, wash the cells once with PSB, collect the cells with TEN buffer, and freeze the cell pellet at −80°C until further processing.

7. A typical example of metabolic labeling using *Vaccinia* virus is shown in **Fig. 8.1**.

3.1.2. The Subgenomic Replicon

1. Plate 4.5×10^5 cells stably expressing the replicon in a six-well dish 24 h before the experiment.

2. Perform ^{35}S-methionine or ^{32}P-orthophosphate labeling as described above (see **Note 7**).

3.2. Immunoprecipitation of NS5A from Cell Extracts and Further Analysis by SDS-PAGE

1. Preparation of specific antibodies bound to Protein-A Sepharose (PAS): Take 50 μl of the PAS (25 μl packed volume)/point and equilibrate resin with IPB$_{150}$-buffer. Incubate 5 μl of polyclonal antibodies (rabbit serum) or $1 - 2$ μg of purified antibody with PAS in 300 μl IPB$_{150}$ for 1 h at 4°C. After the incubation, wash PAS with IPB$_{150}$ to remove unbound antibody.

2. Resuspend cell pellet in 110 μl of lysis buffer and leave on ice for 30 min.

3. Centrifuge lysate for 15 min at 4°C at maximum speed and retain the supernatant (discard the pellet).

4. Incubate 40 μl of cell extract with PAS in 300 μl of IPB$_{150}$ for 2 h at 4°C. For denaturing immunoprecipitation: Add 5 μl of 20% SDS and 5 μl of 1 M DTT to 40 μl of cell extract and boil the extract for 4 min at 95°C. Incubate the denatured extract with PAS in 500 μl of IPB$_{150}$ as above (see **Note 8**).

5. Centrifuge PAS for 30 s at maximum speed in Eppendorf centrifuge and remove the supernatant.

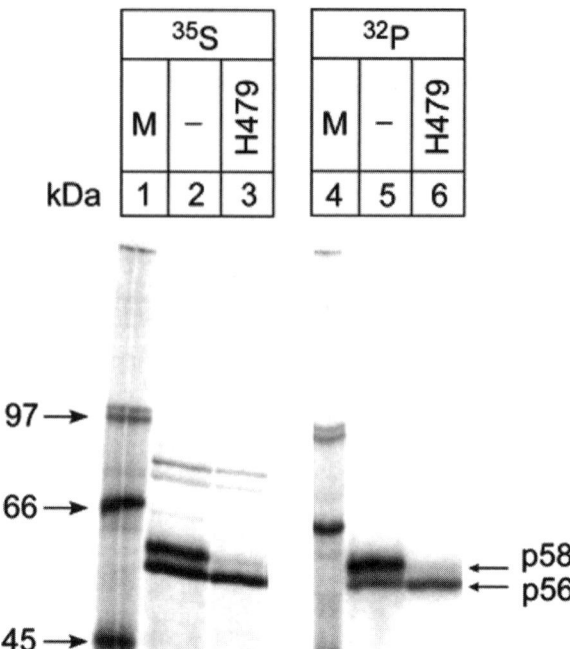

Fig. 8.1. Metabolic labeling of NS5A in cells. Viral proteins were expressed from the plasmid pcD-BLA-wt with the *Vaccinia* T7 infection/transfection system in the absence (lanes 2 and 5) and in the presence (lanes 3 and 6) of 5 µM of the kinase inhibitor H479. Proteins are labeled either with ^{35}S-methionine (lanes 2 and 3) or with ^{32}P-orthophosphate (lanes 5 and 6). Proteins were immunoprecipitated with a specific anti-NS5A antibody and separated on a 7.5% SDS-PAGE gel. Shown is an autoradiograph. M: protein standard (lanes 1 and 4). The positions of NS5A-p56 and NS5A-p58 are indicated at the *right*

6. Wash PAS once with 0.5 ml of IPB$_{150}$, once with 0.5 ml NDET/SDS, and once with 0.5 ml PBS. Finally, boil PAS with 50 µl of SDS sample buffer to denature the bound protein.

7. Load proteins onto 7.5% SDS-PAGE gels.

8. After the run, fix proteins with protein-fixing solution for 30 min.

9. In the case of ^{35}S-labeling, incubate the gel 30 min in Amplify.

10. Dry the gel and perform autoradiography.

3.3. Phosphorylation of NS5A by Coimmunoprecipitated Cellular Kinases In Vitro

In this type of experiment, NS5A phosphorylation occurs in vitro in the presence of cellular kinases associated with NS5A in cells.

1. Grow Huh7 or Huh7-68 cells to 50% confluency and prepare protein extract in NETN-buffer *(12)*. Briefly, lyse cells from a 15-cm-diameter dish in 500 µl of NETN-buffer supplemented with 5 mM DTT, 2 mM PMSF, 100 mM NaF, 20% glycerol, and protease-inhibitor cocktail Complete (Roche).

2. Incubate the lysate on ice for 30 min and centrifuge it for 15 min at maximum speed.

3. Perform immunoprecipitation as described above, but use only half of the above-mentioned quantities per point (2.5 µl of polyclonal antibody bound to 12 µl of PAS (PAS-antibody)). Incubate 30 µg of total protein with 12.5 µl of PAS antibody in 300 µl of NETN for 1 h at +4°C.

4. After binding, wash the resin once with 0.5 ml of NETN supplemented with 5 mM DTT and once with 0.5 ml of kinase buffer.

5. For the in vitro kinase reaction, add 36.5 µl kinase buffer to 12.5 µl of resin. Start the reaction by adding 1 µl of $[\gamma-^{33}P]$-ATP (2.5 µCi).

6. Incubate the reaction mixtures for 45 min at 37°C and terminate the reaction by adding 25 µl of SDS sample buffer.

7. Load the proteins onto a 7.5% SDS-PAGE gel and autoradiograph them.

8. A typical example of in vitro labeling of immunoprecipitated NS5A by associated cellular kinases is shown in **Fig. 8.2**.

3.4. In Vitro Phosphorylation of NS5A by Purified Kinases

In contrast to the above-mentioned experiment (**Section 3.3**), this kind of in vitro phosphorylation uses only purified components. The cellular kinases either come from a commercial source or must be purified in house. The kinase substrate NS5A can be offered either as purified protein or as peptides spanning certain regions of NS5A.

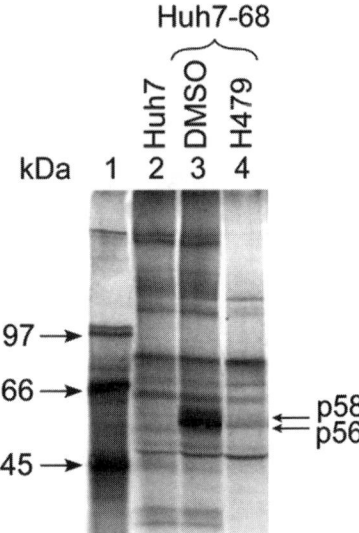

Fig. 8.2. In vitro phosphorylation of NS5A by associated cellular kinases. Proteins from Huh7 cells or Huh7-68 cells were immunoprecipitated with an anti-NS5A antibody. The isolated immunocomplexes were incubated with $[\gamma-^{33}P]$-ATP in the absence (lane 3) or in the presence (lane 4) of 5 µM of the kinase inhibitor H479. Proteins were separated on 7.5% SDS-PAGE gels and detected by autoradiography.

The assay conditions depend on the kinase used. In general, the best reaction conditions are provided by the manufacturer, if the kinases are available from a commercial source.

The specific activities of the kinases are usually different and are measured on either protein or peptide substrates, so a comparison of the activities of different kinases on NS5A as substrate can give only a qualitative result. Otherwise an accurate standardization of the specific activities of the single kinases is required. The kinase reaction should be linear with time, particularly if inhibition of the reaction is measured.

Below are listed some examples with different kinases either with NS5A protein purified from *E. coli* or with peptides, spanning a certain region of NS5A.

3.4.1. Purified Recombinant NS5A

1. NS5A (Con1-1b strain) was purified as described previously *(19)*. In our laboratory, the final buffer conditions were changed during dialysis. This change is important, because the NaCl concentration in the kinase reaction increases upon addition of the NS5A substrate, and some kinases are inhibited by the presence of salt.
2. The final volume of the reaction mix is 60 µl. The final concentration of NS5A substrate should be in the low micromolar range, in this case it is 2 µM. Below are given two examples using CKI-δ and PKA.
   (a) CKI-δ: Put 15 µl of ADBI buffer and 20 µl of purified NS5A (6 µM) in a reaction tube. Add 10 µl (12 ng/µl) of the kinase CKI-δ to this mix. Start the reaction by adding 15 µl of the ATP mix.
   (b) PKA: Put 15 µl of ADBI buffer and 20 µl of purified NS5A (6 µM) in a reaction tube. Add 5 µl (100 ng/µl) of PKA, 3 µl of the PKA/CaMK inhibitor cocktail (20-fold concentrated) as suggested by the manufacturer, and 2 µl of H_2O to adjust the volume to 60 µl. Start the reaction by adding 15 µl of the ATP mix.
3. Incubate for at least 4 h at room temperature.
4. Stop the reaction by adding 30 µl SDS sample dye.
5. Boil the samples for 5 min at 95°C and separate the proteins on a 10% SDS-PAGE gel.
6. Fix proteins by incubating the gel in protein-fixing solution.
7. Dry the gel and perform autoradiography.
8. A typical example of in vitro labeling of purified NS5A by purified kinases is shown in **Fig. 8.3A**.

3.4.2. NS5A-Derived Peptides

1. Assay conditions are similar as those mentioned for the recombinant protein as substrate.
2. The peptide-substrate concentration must be determined for each kinase separately. In general, the supplier provides a useful range to start with. The final reaction volume is 50 µl.

A

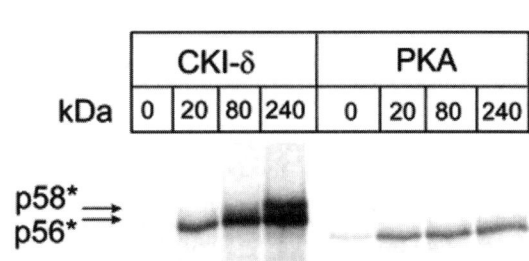

B

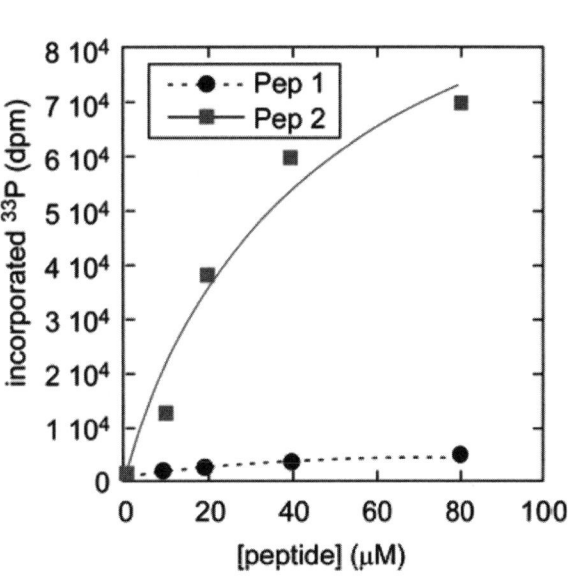

Fig. 8.3. In vitro phosphorylation of purified NS5A (**A**) or NS5A-derived peptides (**B**) by purified kinases. **A**. NS5A is incubated either with CKI-δ or with PKA for 0, 20, 80, or 240 min with [γ $-^{33}$ P]-ATP as described in Section 2. Proteins were separated on a 7.5% SDS-PAGE gel; the autoradiograph is shown. The positions of NS5A − p56* and NS5A − p58* are indicated on the *left*. **B**. NS5A-derived peptides 1 and 2 were incubated with CKI-δ for 30 min at room temperature and incorporated ^{33}P was measured as described in Section 2. ^{33}P incorporated into the peptides is shown as dpm on the *y*-axis at various concentrations of the substrate peptides (*x*-axis).

3. Below are given examples using CKI-δ and NS5A-derived peptides: Use the substrate peptide at the indicated concentration and 20 ng of enzyme/assay. Put 12.5 μl H$_2$O, 12.5 μl ADBI, and 10 μl of peptide (fivefold concentrated) in a reaction tube. Add 2.5 μl of CKI-δ (8 ng/μl). Start the reaction by adding 12.5 μl of ATP mix.

4. Incubate the reaction for 30 min at room temperature.

5. Stop the reaction by adding 8 μl of 3% phosphoric acid. Spot 20 μl of the reaction onto P30 Filtermat (Wallac). Let the

filter dry and wash it three times for 5 min with 75 mM phosphoric acid and once with methanol. Dry the filter and count the radioactivity in 5 ml of Ready Protein scintillation cocktail (Beckman Counter).

6. A typical example of in vitro labeling of NS5A-derived peptides by purified kinase is shown in **Fig. 8.3B**.

3.5. Effect of Kinases on NS5A Phosphorylation in Cells

Below are described two protocols with which the effect of any kinase on NS5A phosphorylation can be studied within the cell. In the first, the kinases are overexpressed together with NS5A and an expression plasmid is needed that contains the kinase gene of interest suitable for the expression in mammalian cells. The second protocol describes a method in which the expression of the kinases is knocked down and the effect of the absence of this kinase on NS5A phosphorylation is investigated. For this protocol, knowledge of the sequence of the gene of interest is needed for design of suitable siRNAs. In addition to the siRNA, other tools are needed to confirm successful overexpression or RNA interference of the kinase. In this case either suitable specific antibodies are required, or if these are not available, primers and probes must be designed that can be used for quantitative PCR of the kinase mRNAs *(17)*.

3.5.1. Overexpression of Kinases Together with NS5A with the Vaccinia Virus T7 Infection/Transfection System

Below is described an example of the effect of CKI-α overexpression on NS5A phosphorylation.

1. Plate 3.5×10^5 cells/35-mm-diameter dish the day before the experiment. Any cell line that can be infected by the *Vaccinia* virus is suitable. In this case we used Huh7 cells.

2. Follow the protocol for the *Vaccinia* T7 infection/transfection as already described in **Section 3.1**. The DNA precipitate should contain 2 μg of plasmid DNA expressing NS5A, either alone or in the context of the HCV polyprotein, together with 1 μg of the kinase expression plasmid. The ratio between DNA and Fugene6 should be 1:3 (9 μl Fugene6 in 91 μl Optimem). DNA plasmids used in this study are pcD-BLA-wt and pCD-Flag-CKI-α *(17)*.

3. Perform labeling and immunoprecipitation as described in **Sections 3.1 and 3.2**.

4. To control overexpression of the kinase, plate the cells in parallel on additional 35-mm-diameter dishes and perform *Vaccinia* T7 infection/transfection as above. In this case, protein labeling is not necessary.

5. The kinase is detected by western blot.
   (a) Prepare cell lysate as described in **Section 3.2**.
   (b) For CKI-α: Load 100 μg of total protein onto 10% SDS-PAGE gels and transfer proteins onto nitrocellulose.
   (c) The filter is blocked with Blocking buffer and washed three times with TBS 0.5% Tween 20. CKI-α is detected

A.   **B.**

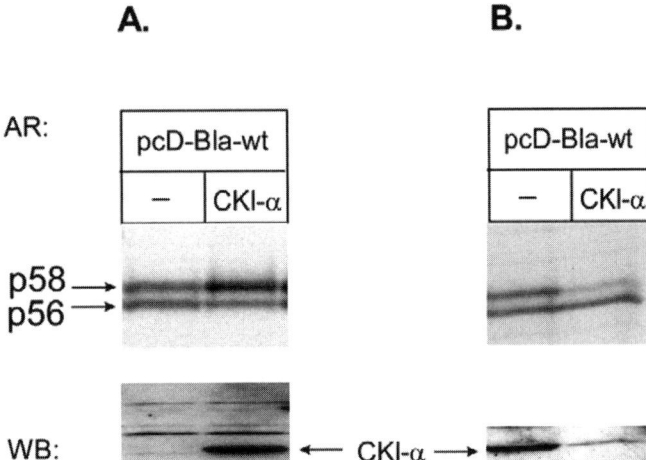

Fig. 8.4. Effect of overexpression (**A**) or RNA interference (**B**) of CKI-α on NS5A phosphorylation in cells. **A**. Plasmid pcD-Bla-wt was transfected together with plasmid pCD-Flag-CKI-α in Huh7 cells, and proteins were expressed with the *Vaccinia* T7 infection/transfection system. Proteins were labeled and NS5A was immunoprecipitated. Proteins were loaded onto a 7.5% SDS-PAGE gel; an autoradiogram (AR) is shown in the *upper panel. The lower panel* is a western blot (WB) of CKI-α. CKI-α was detected with the anti-FLAG antibody. NS5A and the kinases are indicated. **B**. CKI-α was silenced in Huh7 cells as described in Section 2. Forty-eight hours after siRNA transfection, pcD-Bla-wt was transfected, and proteins were expressed with the *Vaccinia* infection/transfection system. Proteins were labeled, and NS5A was immunoprecipitated as described above. The *upper panel* shows the autoradiogram. Silencing of CKI-α is shown in the western blot. CKI-α was detected with a CKI-α-specific antibody.

with the anti-FLAG antibody at a 1:1000 dilution in TBS, 0.05% Tween-20 over night at 4°C.

(d) Develop the filter using the ECL western blotting detection reagent.

6. A typical example of the effect of overexpressed CKI-α on NS5A phosphorylation is shown in **Fig. 8.4A**.

3.5.2. RNA Interference of Cellular Kinases

In contrast to the kinase overexpression, where NS5A is expressed together with the kinase, RNA interference and NS5A expression must be performed at different time points (*see* **Note 9**).

1. Split a confluent 150-mm-diameter dish of Huh7 cells 1:3 the day before siRNA electroporation.

2. The next day, typsinize the cells, detach the cells with 5 ml of DMEM medium, and fill a 50 ml Falcon tube with PBS/DEPC. From this time point on, keep the cells on ice.

3. Centrifuge the cells for 10 min at 300g at 4°C.

4. Remove the supernatant and resuspend the cells in 40 ml PBS/DEPC.

5. Count the cells and centrifuge them again for 10 min at 300g at 4°C.

6. Remove the supernatant and resuspend the cells at a concentration of 1.5–2×10^7 cells/ml.

7. Mix $10\,\mu M$ CKI-α siRNA with 1×10^6 cells in a final volume of $100\,\mu L$ of PBS/DEPC and add all to a Gene Pulser Cuvette 0.2 cm (BioRad).

8. Perform electroporation using a GenePulser XCell (BioRad) at $110\,V$ for 25 msec.

9. After electroporation, plate 4.5×10^5 cells on a 35-mm-diameter dish and incubate them for 2 days. After 2 days, perform protein expression and protein labeling of NS5A with the T7 *Vaccinia* system as described above (**Section 3.1**).

10. To control efficiency of RNAi, plate cells in parallel in an additional 35-mm-diameter dish and analyze them 2 days after electroporation.

    Western blot

    (a) Prepare cell lysate as described in **Section 3.2**.

    (b) Load $100\,\mu g$ of total protein onto a 10% SDS-PAGE gel and perform western blotting as described above. Detect endogenous CKI-α with an anti-CKI-α antibody.

    Quantitative PCR

    (a) Prepare RNA using the RNeasy Mini Kit (Quiagen) as described by the manufacturer. The final volume is $100\,\mu l$ of RNA/35-mm-diameter dish.

    (b) Use $100\,ng$ of RNA for quantitative RT-PCR using an ABI PRISM 7900 HT sequence detector (Applied Biosystems). Perform amplifications in duplicate using the Quantitect Probe RT-PCR Kit (Quiagen).

    (c) Use $0.2\,\mu M$ of oligonucleotide CKI-for, $0.8\,\mu M$ CKI-rev, and $0.15\,\mu M$ of CKI-α-probe for each reaction. The probe contains the 6FAM dye at the $5'$-end and the TAMRA quencher at its $3'$-end.

    (d) As endogenous standard we used the β-actin probe, containing the VIC-dye at its $5'$-end (Applied Biosystems).

    (e) Conduct reactions in three stages under the following conditions: stage 1, 30 min at $48°C$; stage 2, 10 min at $95°C$; and stage 3, 15 s at $95°C$ and 1 min at $60°C$, 40 cycles. The total volume of the reaction is $50\,\mu L$.

11. A typical example of the effect of RNA interference of CKI-α on NS5A phosphorylation is shown in **Fig. 8.4B**.

4. Notes

1. A good separation of p56 and p58 is important for the detection of effects on NS5A phosphorylation. The low percentage of *bis*-acrylamide helps to improve the resolution. In addition, perform electrophoresis using 7.5% PAA until NS5A nearly exits the gel. Loading a prestained molecular-weight marker helps to reveal when the electrophoresis should be stopped.

2. Peptide substrate, used here as an example, is dissolved in H_2O, but hydrophobic peptides might require solvents to be dissolved. In general, we use DMSO. Make sure that the final concentration of DMSO in the kinase reaction does not inhibit the enzyme.

3. Efficiency of NS5A hyperphosphorylation depends strongly on the context in which NS5A is expressed and on the HCV RNA sequence. Just remember that most of the adaptive mutations present in the subgenomic replicon significantly reduce NS5A hyperphosphorylation.

4. Be cautious when working with *Vaccinia* virus. Even though the viruses are unable to multiply in human and most mammalian cells, viral proteins are synthesized normally. Inactivate all liquids containing the virus with 1 N NaOH for several days before autoclaving them.

5. Infection with the *Vaccinia* virus is performed with small volumes (here 80 µl). Make sure that the virus is distributed well on the whole surface of the dish.

6. If the effect of kinase inhibitors is tested during metabolic labeling, the compound should be added during the starvation reaction and labeling reaction, if cellular uptake is fast enough. Addition of the inhibitor during DNA transfection might interfere with transfection efficiency.

7. Protein synthesis with the subgenomic replicon is much slower than with the *Vaccinia* virus system. Use as many cells as possible and longer labeling time. Do not use confluent cells, because HCV replication is shut down at cell confluency.

8. In the case of denaturing immunoprecipitation, make sure that the denatured cell extract is diluted at least 10 times with IPB_{150} before adding it to the PAS-bound antibodies. This dilution is important to dilute the SDS, which would otherwise denature the antibodies.

9. Efficiency of RNAi of cellular kinases depends on many factors. Many different methods of siRNA transfection can be applied, and the best method depends on the cell line used and on the kinase gene that will be silenced. For the CKI-α gene in Huh7 cells, the most efficient method turned out to be RNA electroporation. Also, the time course of kinase silencing is different for each protein. Make sure that NS5A is expressed when the kinase protein is at its minimum level.

References

1. Kaneko, T., Tanji, Y, Satoh, S., Hijikata, M., Asabe, S., Kimura, K., and Shimotohno, K. (1994) Production of two phosphoproteins from the NS5A region of the hepatitis C viral genome. *Biochem. Biophys. Res. Commun.* **205**, 320–326.

2. Asabe, S. I., Tanji, Y., Satoh, S., Kaneko, T., Kimura, K., and Shimotohno, K. (1997) The N-terminal region of hepatitis C virus-encoded NS5A is important for NS4A-dependent phosphorylation. *J. Virol.* **71**, 790–796.

3. Koch, J. O., and Bartenschlager, R. (1999) Modulation of hepatitis C virus NS5A hyperphosphorylation by nonstructural proteins NS3, NS4A, and NS4B. *J. Virol.* **73**, 7138–7146.

4. Neddermann, P., Clementi, A., and De Francesco, R. (1999) Hyperphosphorylation of the hepatitis C virus NS5A protein requires an active NS3 protease, NS4A, NS4B, and NS5A encoded on the same polyprotein. *J. Virol.* **73**, 9984–9991.

5. Tanji, Y., Kaneko, T., Satoh, S., and Shimotohno, K. (1995) Phosphorylation of hepatitis C virus-encoded nonstructural protein NS5A. *J. Virol.* **69**, 3980–3986.

6. Katze, M. G., Kwieciszewski, B., Goodlett, D. R., Blakely, C. M., Neddermann, P., Tan, S. L., and Aebersold, R. (2000) Ser(2194) is a highly conserved major phosphorylation site of the hepatitis C virus nonstructural protein NS5A. *Virology* **278**, 501–513.

7. Reed, K. E., and Rice, C. M. (1999) Identification of the major phosphorylation site of the hepatitis C virus H strain NS5A protein as serine 2321. *J. Biol. Chem.* **274**, 28011–28018.

8. Appel, N., Pietschmann, T., and Bartenschlager, R. (2005) Mutational analysis of hepatitis C virus nonstructural protein 5A: potential role of differential phosphorylation in RNA replication and identification of a genetically flexible domain. *J. Virol.* **79**, 3187–3194.

9. Hirota, M., Satoh, S., Asabe, S., Kohara, M., Tsukiyama-Kohara, K., Kato, N., Hijikata, M., and Shimotohno, K. (1999) Phosphorylation of nonstructural 5A protein of hepatitis C virus: HCV group-specific hyperphosphorylation. *Virology* **257**, 130–137.

10. Ide, Y., Tanimoto, A., Sasaguri, Y., and Padmanabhan, R. (1997) Hepatitis C virus NS5A protein is phosphorylated in vitro by a stably bound protein kinase from HeLa cells and by cAMP-dependent protein kinase A-alpha catalytic subunit. *Gene* **201**, 151–158.

11. Kim, J., Lee, D., and Choe, J. (1999) Hepatitis C virus NS5A protein is phosphorylated by casein kinase II. *Biochem. Biophys. Res. Commun.* **257**, 777–781.

12. Reed, K. E., Xu, J., and Rice, C. M. (1997) Phosphorylation of the hepatitis C virus NS5A protein in vitro and in vivo: properties of the NS5A-associated kinase. *J. Virol.* **71**, 7187–7197.

13. Coito, C., Diamond, D. L., Neddermann, P., Korth, M. J., and Katze, M. G. (2004) High-throughput screening of the yeast kinome: identification of human serine/threonine protein kinases that phosphorylate the hepatitis C virus NS5A protein. *J. Virol.* **78**, 3502–3513.

14. Lohmann, V., Korner, F./, Koch, J. O., Herian, U., Theilmann, L., and Bartenschlager, R. (1999) Replication of subgenomic hepatitis C virus RNAs in a hepatoma cell line. *Science* **285**, 110–113.

15. Blight, K. J., Kolykhalov, A. A., and Rice, C. M. (2000) Efficient initiation of HCV RNA replication in cell culture. *Science* **290**, 1972–1974.

16. Neddermann, P., Quintavalle, M., Di Pietro, C., Clementi, A., Cerretani, M., Altamura, S., Bartholomew, L., and De Francesco, R.(2004) Reduction of hepatitis C virus NS5A hyperphosphorylation by selective inhibition of cellular kinases activates viral RNA replication in cell culture. *J. Virol.* **78**, 13306–13314.

17. Quintavalle, M., Sambucini, S., Di Pietro, C., De Francesco, R., and Neddermann, P. (2006) The α isoform of protein kinase CKI is responsible for hepatitis C virus NS5A hyperphosphorylation. *J. Virol.* **80**, 11305–11312.

18. Quintavalle, M., Sambucini, S., Summa, V., Orsatti, L., Talamo, F., De Francesco, R., and Neddermann, P. (2007) Hepatitis C virus NS5A is a direct substrate of casein kinase I-α, a cellular kinase identified by inhibitor affinity chromatography using specific NS5A hyperphosphorylation inhibitors. *J. Biol. Chem.* **282**, 5536–5544.

19. Huang, L. Y., Sineva, E. V., Hargitta, M. R. S., Sharma, S. D., Suthar, M., Raney, K. D., and Cameron, C. E. (2004) Purification and characterization of hepatitis C virus non-structural protein 5A expressed in *Escherichia coli*. *Protein Expr. Purif.* **37**, 144–153.

Chapter 9

Preparation and Handling of Hepatitis C Viral Proteins NS3 and NS5B for Structural Studies

Stefania Di Marco and Andrea Carfí

Abstract

HCV is a small positive-strand RNA virus responsible for a considerable proportion of acute and chronic hepatitis in humans. Although all HCV enzymes are, in theory, equally appropriate for therapeutic intervention, the NS3-NS4A serine protease and the NS5B RNA-dependent RNA polymerase are the most popular targets from a drug-discovery perspective. A number of active-site inhibitors of the NS3 protease as well as allosteric inhibitors of the NS5B polymerase are being developed. We determined the crystal structures of complexes of NS3/NS4A/active-site inhibitor as well as NS5B/allosteric inhibitor to permit structure-based drug design and the efficient optimization of leads. The methods for obtaining such structures by crystal soaking procedures are described.

Key words: Hepatitis C virus, NS3 protease, NS5B polymerase, crystallization, soaking, drug design.

1. Introduction

HCV affects approximately 2% of the world population. It causes chronic liver disease in 50% of infected patients, and in the United States is the number one indication for liver transplant surgery. The currently available therapeutic regimens, based on the use of different forms and combinations of interferon-α (INF-α), are generally poorly tolerated, contraindicated in a large number of patients, and effective in controlling the disease only in a fraction of the individuals who are eligible for therapy. The need is therefore urgent for more effective and better-tolerated treatments.

Hengli Tang (ed.), *Hepatitis C: Methods and Protocols, Second Edition, vol. 510*
© 2009 Humana Press, a part of Springer Science+Business Media
DOI 10.1007/978-1-59745-394-3_9 Springerprotocols.com

HCV is a small positive-strand-RNA virus and is classified as the genus *Hepacivirus* in the *Flaviviridae* family. HCV encodes for a polyprotein of approximately 3000 amino acids *(1)* that is subsequently proteolytically cleaved into four structural proteins (core, envelope proteins E1 and E2, and the ion channel p7) and six nonstructural proteins (NS2, NS3, NS4A, NS4B, NS5A, and NS5B). All cleavages of the nonstructural proteins downstream of NS3 are mediated by a chymotrypsin-like viral serine protease, encoded within the N-terminal portion of the NS3 protein *(2)*. The NS3 protease domain is necessary but not sufficient for the efficient processing of the HCV polyprotein. Another protein, NS4A, is required as a cofactor, the active protease being a heterodimer consisting of both NS3 and NS4A. Deletion mutagenesis experiments have shown that residues 20–34 (out of a total of 54), corresponding to the extended, hydrophobic region of NS4A, are sufficient for the full activation of NS3 protease *(3)*. The system composed of a 20-kDa NS3 protease domain and a synthetic 14-mer peptide as a NS4A cofactor mimic, therefore, represents a good in vitro system for the development of new antiviral agents. The crystal structure of the HCV NS3 protease in complex with a truncated NS4A cofactor (residues 21–34) revealed a flat and featureless substrate binding cleft, which lacks cavities, holes, and flaps that could facilitate the design of small-molecule inhibitors *(4, 5)*. Nevertheless, potent and selective inhibitors of the NS3/NS4A catalytic activity have recently been described.

Another important HCV target is NS5B, the RNA-dependent RNA polymerase responsible for the replication of the viral genome *(1)*. The crystal structure of NS5B polymerase, like those of other polymerases, can be envisioned as shaped like a right hand and divided into three domains: the palm, containing the active site of the enzyme; the fingers and the thumb domains, which modulate the interaction with the RNA chain and show the presence of an extension in the fingers; and the so-called fingertip subdomain, which anchors the fingers to the thumb. As a result, the polymerase has a relatively closed and spherical appearance, and the active-site cavity, to which the RNA-template and the NTP substrates have access via two positively charged tunnels, is completely encircled *(6, 7)*.

A number of competitive inhibitors of the NS3 protease, as well as nucleoside analogues and nonnucleoside inhibitors of the NS5B polymerase, are being developed, and their efficacy has been demonstrated in recent proof-of-concept clinical trials *(1)*. Structure-based design has greatly contributed to the development of these potential drugs. The methods for obtaining inhibited structures of NS3/NS4A and NS5B by crystal soaking procedures are described.

2. Materials

2.1. NS3 Protein

1. M9 modified minimal medium: 5 g/L glucose, 1 g/L ammonium sulfate, 5 μM biotin, 7 μM thiamine, 0.5% casamino acids, 0.5 mM MgSO$_4$, 0.5 mM CaCl$_2$, 100 mg/L ampicillin, in 100 mM potassium phosphate buffer at pH 7.0 (*see* **Note 1**).
2. 50 μM ZnCl$_2$.
3. 400 μM isopropyl-β-thiogalactopyranoside (IPTG).
4. Microfluidizer (Model 110-S, Microfluidics).
5. Lysis buffer A: 50 mM sodium acetate, pH 6.3, 20% glycerol, 50 mM NaCl, 0.1% *n*-octyl-β-d-glucopyranoside, 10 mM dithiothreitol (DTT), 0.1 mM phenylmethylsulfonyl fluoride (PMSF), Complete protease-inhibitor cocktail tablets (Boehringer), and 0.02% NaN$_3$.
6. Sorvall SS34 centrifuge rotor.
7. Source 15-S column (Pharmacia).
8. Superdex-75 High Load 26/60 column (Pharmacia).
9. Gel filtration buffer: 50 mM sodium acetate buffer, pH 6.3, 10% glycerol, 500 mM NaCl, 0.1% *n*-octyl-β-d-glucopyranoside, 10 mM DTT, and 0.02% NaN$_3$.
10. FPLC chromatographic system (Pharmacia).
11. CENTRIPREP3 ultrafiltration system (Amicon).
12. Spectrophotometer DU640 (Beckman).
13. Liquid nitrogen.

2.2. NS4A Peptide

1. Pep4A: a peptide mimicking the central portion of NS4A, with Lys residues added to the termini to assist aqueous solubility, purchased from Peptides International, Inc., H-KGSVVIVGRIILSGRK-OH-TFA salt.

2.3. NS5B Protein

1. 0.4 mM isopropyl-1-thio-β-d-galactopyranoside.
2. LB medium: 10 g/L tryptone, 5 g/L yeast extract, 5 g/L NaCl. Adjust the pH to 7.2 with 1 M NaOH. Autoclave at 121°C for 20 min.
3. Lysis buffer B: Buffer A-Hep plus Complete EDTA-free protease-inhibitor cocktail tablets (Boehringer).
4. Buffer A-Hep: 20 mM Tris-HCl, pH 7.5, 300 mM NaCl, 20% glycerol, 0.2% β-octylglucoside, 1 mM EDTA, 10 mM DTT.
5. Cell disruptor (Costant System).
6. DNase I.
7. 10 mM MgCl$_2$.
8. Heparin column (Pharmacia).
9. Ion-Exchange Resource S column (Pharmacia).

10. A-S buffer: 20 mM Tris-HCl, pH 7.5, 150 mM NaCl, 20% glycerol, 0.2% β-octylglucoside, 1 mM EDTA, and 10 mM DTT.

11. G75 26/60 gel filtration column (Pharmacia).

12. Buffer C: 20 mM Tris-HCl, pH 7.5, 500 mM NaCl, 20% glycerol, 0.2% β-octylglucoside, 1 mM EDTA, and 10 mM DTT.

13. CENTRIPREP10 ultrafiltration system (Amicon).

2.4. Inhibitors

1. All inhibitors were dissolved in 100% dimethylsulfoxide (DMSO) at a concentration of 0.1 M and stored in 1 μl aliquots at –20°C.

2.5. NS3 and NS5B Crystallization

1. 24-well crystallization plates, VDX Plate with sealant (HR3-172), for hanging-drop or sitting-drop vapor diffusion crystallization (Hampton Research). These plates are stackable with optically clear plastic wells with raised covers (to allow room for cover slides). Each well is individually sealed with a 22 mm × 0.22 mm siliconized circle cover slide (HR3-231), **Fig. 9.1A–C.**

2. Solid glass sitting drop rods, each with a concave depression in the top and that can hold up to 50 μL and fit into the well of a 24-well VDX Plate. These rods are 14 mm tall and 10 mm in diameter (HR3-146) and were used for soaking experiments.

3. Micro-Bridges (HR3-310), small bridge-shaped devices designed to hold sitting drops for vapor diffusion crystallization and/or inhibitor soaks inside a VDX Plate, **Fig. 9.2A–B.**

4. Kit crystallization buffers (Crystal Screen kit HR2-110, Crystal Screen 2 kit HR2-112, Crystal Screen Lite kit HR2-128, PEG/Ion Screen kit HR2-126, SaltRx kit HR2-108, Index Kit HR2-144) and detergents (Detergent Screen 1-3, HR2-410, HR2-411, HR2-412), all from Hampton. All buffers were stored at 4°C; detergents were stored at –20°C.

5. Ultrafree-MC centrifugal filter devices, for volume less than 0.5 mL (Millipore).

6. Minimal NS5B dilution buffer: 20 mM HEPES, pH 7.5, 1 mM EDTA, and 10 mM DTT.

7. NS3/NS4A crystallization buffer: 3.4 M NaCl, 4.8 mM cyclohexyl-pentyl-β-d-maltoside, 5 mM DTT, and 0.02% NaN_3 in 0.1 M sodium citrate buffer, pH 5.1.

8. NS5B crystallization buffer: 100 mM sodium citrate, pH 6.0, 5–8% polyethylene glycol 8.000, 5–8% 2-propanol, and 10 mM DTT.

2.6. Soaking of Crystals with Inhibitors

1. Synchrotron facility: European Radiation Synchrotron Facility (ESRF, Grenoble) beamlines ID14.

A

B

DROP =
1 vol. **PROTEIN** +
1 vol. PRECIPITANT

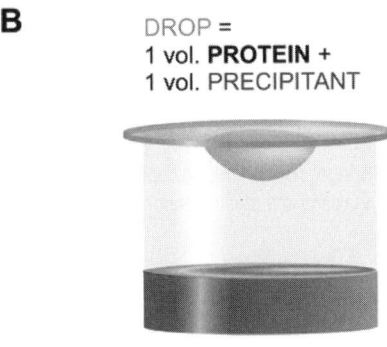

PRECIPITANT

C

GREASE POINT
TO SEAL

SILICONISED
COVERSLIP

HANGING DROP

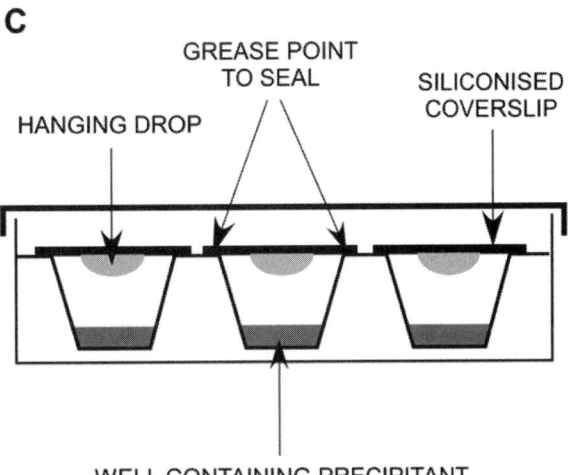

WELL CONTAINING PRECIPITANT

DROP = **PROTEIN** + PRECIPITANT

Fig. 9.1. Vapor diffusion crystallization in hanging drops. (**A**) Photograph of a 24-well VDX crystallization plate. (**B**) Schematic representation of hanging-drop vapor diffusion process in a single well. (**C**) Schematic representation of hanging-drop vapor diffusion process along the *y*-axis of the crystallization plate.

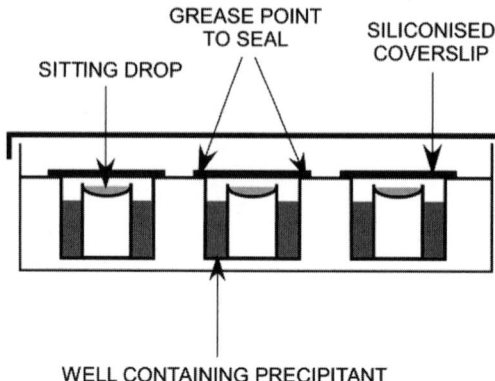

DROP = **PROTEIN** + PRECIPITANT

Fig. 9.2. Vapor diffusion crystallization in sitting drops. (**A**) Photograph of the Micro-Bridge. (**B**) Schematic representation of sitting-drop vapor diffusion process along the *y*-axis of the crystallization plate shown in Fig.9.1A.

2. Stabilizing solution (NS3/NS4A crystals): 4.5 M NaCl, 10 mM DTT, 0.1 M citrate buffer, pH 5.1.
3. Soaking solution A (NS3/NS4A crystals): 4.5 M NaCl, 10 mM DTT, 0.1 M citrate buffer, pH 5.1, 5 mM inhibitor, 5% DMSO.
4. Cryogenic solution (NS3/NS4A crystals): 30% glycerol, 4.0 M NaCl, 10 mM DTT, 0.1 M citrate buffer, pH 5.1.
5. Nylon Cryo-loops (Hampton, HR4-617).
6. Soaking solution B (NS5B crystals): 100 mM 2-morpholinoethanesulfonic acid (MES), pH 6.0, 14% polyethylene glycol 8.000, 14% 2-propanol, 10 mM DTT, 10 mM $MnCl_2$, 2.5% DMSO, and 2.5 mM inhibitor.
7. Cryoprotectant solution (NS5B crystals): 100 mM MES, pH 6.0, 14% polyethylene glycol 8.000, 14% 2-propanol, 18% 2-methyl-2,5-pentanediol, 10 mM DTT, 5 mM $MnCl_2$, and 1.25 mM inhibitor.

3. Methods

3.1. NS3 Protein

1. Obtain a cDNA fragment encoding the serine protease domain of NS3 J, subtype 1b (amino acids 1–187, UniProt

reference P26662 1027-1213), by PCR and clone it down-stream of the T7 promoter of the pT7-7 vector, in frame with the first ATG of the protein of gene 10 of the T7 phage.

2. Express the NS3 protein in *Escherichia coli* BL21 (DE3) according to the following protocol. Grow a 1-L culture at 37°C to $A600\,nm = 0.8$ (as determined here and in all chromatographic steps on the FPLC system) in M9 modified minimal medium. Cool the culture to 18°C, add $50\,\mu M\,ZnCl_2$, and induce it with $400\,\mu M$ IPTG for 22 h at 18°C. Perform all subsequent operations at 4°C unless otherwise instructed.

3. Harvest cells and disrupt them with the Microfluidizer in lysis buffer A. Pellet the insoluble material at 27,000g for 30 min in the Sorvall SS34 rotor.

4. Load the clarified NS3 J supernatant, which should contain about 70% of the recombinant protein, onto the Source 15-S column, preequilibrated in lysis buffer A and elute it with a 0.05–0.5 M NaCl linear gradient. Detect the protein peak, in fractions containing approximately 0.30 M NaCl, and pool the fractions.

5. Size-fractionate the protein on the Superdex-75 column, equilibrated with gel filtration buffer. The pure protein (> 95% as judged by SDS-PAGE) should elute as a monomer after 170 mL of buffer. The final yield of protein should be about 20 mg from 1 L of cell culture.

6. Pool protein fractions for concentration by ultrafiltration at about $1\,mg/mL$ ($48\,\mu M$, MW 20,697) in the gel filtration buffer, as quantified by UV spectroscopy at a theoretical calculated molar absorption coefficient of $\varepsilon_{280} = 18,200\,M^{-1}\,cm^{-1}$. Freeze the concentrated protein in liquid nitrogen and store in $100\,\mu L$ aliquots at $-70°C$ until use (*see* **Note 2**).

3.2. NS4A Peptide

1. Dissolve peptide Pep4A in gel filtration buffer to a final concentration of 2 mM and store it in $10\,\mu L$ aliquots at $-20°C$.

3.3. NS5B Protein

1. Obtain a cDNA fragment encoding the NS5B protein, genotype 1b and strain BK, lacking the C-terminal 55 amino acids (amino acids 1–536, UniProt reference P26663, 2420–2955), by PCR and clone it downstream of the T7 promoter of the pT7-7 vector, in frame with the first ATG of the protein of gene 10 of the T7 phage.

2. Overexpress the protein by induction of mid-log phase BL21 DE3 *E. coli* cells with 0.4 mM isopropyl-1-thio-β-d-galactopyranoside for 22 h at 18°C in LB medium.

3. Resuspend harvested cells in ice-cold lysis buffer B and lyse them with the cell disruptor. Incubate the lysate with DNase I (5 U/mL) in the presence of $10\,mM\,MgCl_2$ and clarify it by centrifugation at 35,000g for 1 h in the Sorvall SS34 rotor.

4. Load the supernatant into a Heparin column preequilibrated with buffer A-Hep at 4°C. Subject the column to a linear gradient of buffer A-Hep containing 1 M NaCl; the protein should elute at 700 mM NaCl.

5. Pool the protein fractions, and dilute them with buffer A-Hep without NaCl to reach 150 mM NaCl, and load them onto an ion-exchange Resource S column equilibrated in A-S buffer. Wash the column in buffer with 300 mM NaCl and subject the protein to a linear gradient from 300 mM to 1 M NaCl. The protein should elute at approximately 500 mM NaCl.

6. Load the protein fractions onto a G75 26/60 gel filtration column equilibrated in buffer C. The pure protein (> 95% as judged by SDS-PAGE) should elute as a monomer after 145 mL of gel filtration buffer. The final yield of protein should be about 3 mg from 1 L of cell culture.

7. Pool fractions for concentration by ultrafiltration in CENTRIPREP10 at about 10 mg/mL (167 μM, MW 59,804) in gel filtration buffer, as quantified by UV spectroscopy with a theoretical calculated molar absorption coefficient of $\varepsilon_{280} = 72,970\,M^{-1}\,cm^{-1}$. Freeze the concentrated protein in liquid nitrogen and store in 100 μL aliquots at −70°C until use.

3.4. Inhibitors

1. Dissolve all inhibitors in 100% (DMSO) at a concentration of 0.1 M and store them in 1 μL aliquots at −20°C.

3.5. NS3 and NS5B Crystallization

1. Thaw ten 100 μL aliquots of purified NS3 protease, 48 μM, slowly on ice and mix 952 μL slowly with 48 μL of NS4A peptide (2 mM, dissolved in gel filtration buffer) to a NS3:NS4A-peptide molar ratio of approximately 1:2 (*see* **Notes 3, 4**).

2. Incubate the sample overnight in an ice basket at 4°C and then dilute it 20-fold with gel filtration buffer and concentrate it by ultrafiltration with CENTRIPREP3 to 215 μM.

3. By ultrafiltration, remove the excess of NS4A peptide. Confirm 1:1 stoichiometry of the NS3/NS4A peptide complex by electrospray ionization (ESI) mass spectrometry *(8)*.

4. Filter the NS3/NS4A peptide complex using 0.1 μM centrifugal filter devices and store it in 50 μL aliquots at −70°C until use (*see* **Note 5**).

5. After thawing the NS3/NS4A peptide complex on ice, centrifuge it at 10,000g for 15 min and filter it with 0.1 μM centrifugal filter devices. Use aliquots of the protein solution immediately for crystallization experiments.

6. Thaw two 100 μL aliquots of purified NS5B 167 μM slowly on ice in gel filtration buffer and dilute them 20-fold in minimal NS5B dilution buffer.

7. Bring the sample volume down to 200 μL by ultrafiltration using CENTRIPREP10 and then dilute it again 20-fold in the minimal NS5B dilution buffer to bring it down to a final NS5B concentration of 67 μM.

8. Filter the sample with the 0.1 μM centrifugal filter devices. Use the concentrated protein immediately for crystallization experiments.

9. Crystallization experiments on the NS3/NS4A peptide complex and NS5B should be performed with the "hanging drop" and "sitting drop" vapor diffusion methods at room temperature and at 4°C, respectively, for NS3/NS4A and NS5B.

10. To use the hanging-drop technique (**Fig. 9.1B–C**), place a small droplet (1–3 μL) of the sample, mixed with the same volume of crystallization buffer, on a siliconized glass cover slide, and then invert the cover slide and put it on the top of the well in vapor equilibration with 600 μL of the reservoir containing the crystallization buffer. Seal each individual well of the 24-well plate with high-vacuum grease to prevent evaporation. The initial reagent concentration in the droplet should therefore be half than that in the reservoir. Over time, the reservoir will pull water from the droplet in a vapor phase until equilibrium is reached between the drop and the reservoir. During this equilibration process the sample is also concentrated, so the relative supersaturation of the sample in the drop will increase (*see* **Note 6**).

11. To use the sitting-drop technique (**Fig. 9.2A–B**), place a drop on a Micro-Bridge inserted into each well of a 24-well VDX Plate, and then seal each well with a greased glass coverslip. Sitting drops can be up to 10 μL. The principle of vapor diffusion is the same as for the hanging-drop technique, but the sitting-drop technique is suitable for larger drops (*see* **Note 7**).

12. Perform the crystallization experiments and optimization with Hampton Research Buffer kits (*see* **Note 8**).

13. NS3/NS4A peptide crystals, with a maximum size of $0.6 \times 0.3 \times 0.2 \, mm^3$, should have formed by both hanging- and sitting-drop methods after two weeks at room temperature in NS3/NS4A crystallization buffer (**Fig. 9.3A**).

14. NS5B crystals, with a maximum size of $1 \times 0.2 \times 0.2 \, mm^3$, should have formed by the hanging-drop technique after 1 week at 4°C in NS5B crystallization buffer (**Fig. 9.3B**).

3.6. Soaking of Crystals with Inhibitors

3.6.1. NS3/NS4A

NS3/NS4A peptide crystals belong to spacegroup $P6_1$ with unit cell dimensions of $a = b = 93.49 \, \text{Å}$ and $c = 80.24 \, \text{Å}$, with two molecules per asymmetric unit, and diffract to maximum 2.1 Å resolution on a Synchrotron system at cryogenic temperature (100 K). The crystal packing makes these crystals

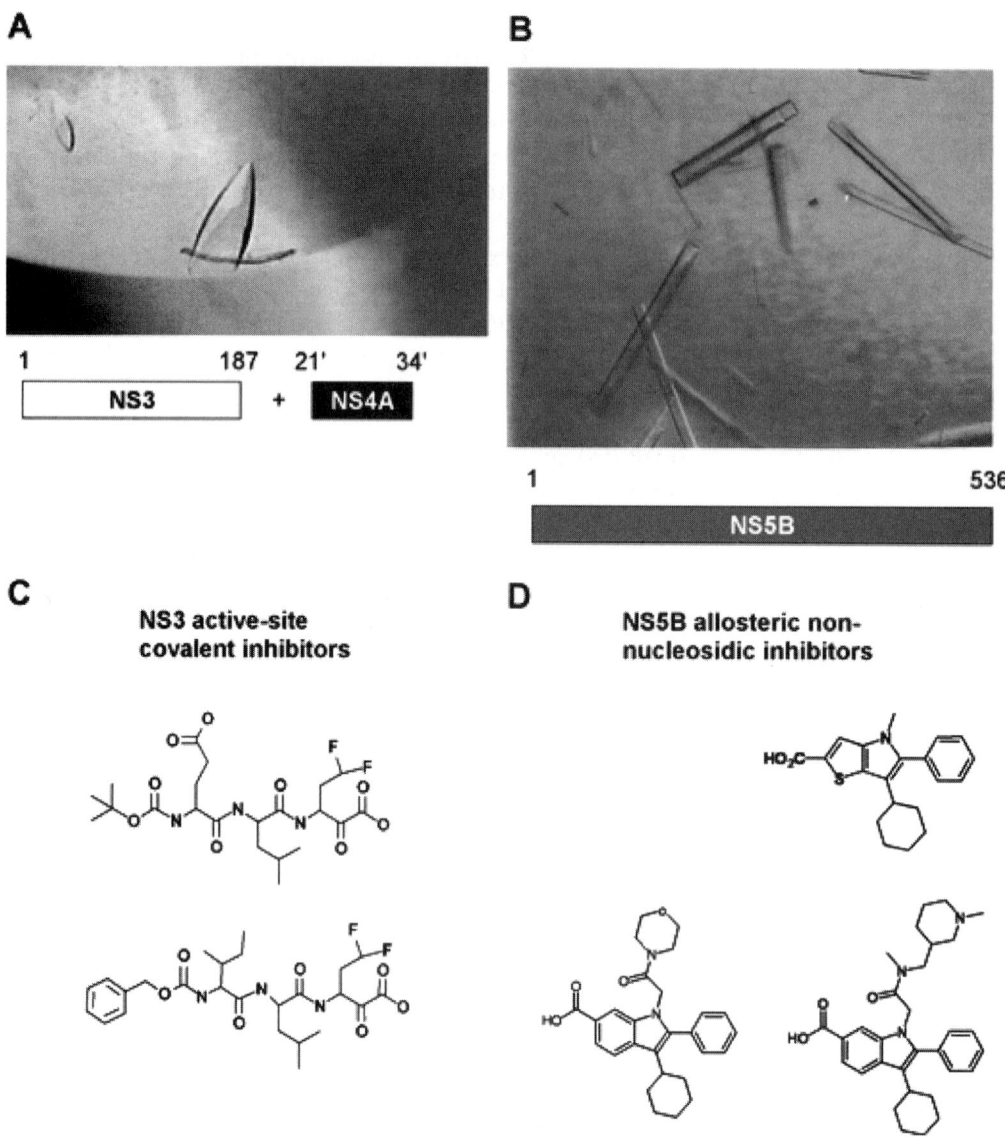

Fig. 9.3. Crystals of NS3/NS4A and NS5B proteins and structures of the inhibitors. (**A**) Hexagonal NS3/NS4A crystals. (**B**) Orthorhombic NS5B crystals. (**C**) Small-molecule inhibitors of NS3/NS4A protease used for soaking experiments. (**D**) Small-molecule inhibitors of NS5B polymerase used for soaking experiments.

suitable for soaking experiments with active-site covalent inhibitors (**Fig. 9.3C**). These crystals grow in the presence of high NaCl concentration, which promotes the interaction between the protease and the peptide cofactor, but the situation is reversed for multiply charged inhibitors, which in the presence of high concentrations of counter-ions should not easily bind inside the NS3/NS4A crystals, because of the solvent-exposed and flat active-site cavity *(4, 5)*.

1. Slowly remove a coverslip bearing 5–10 NS3/NS4A crystals, of good morphology and free from defects, from the well of the VDX plate and place it, with the drop facing upward, under a dissecting microscope at a magnification 10–100× (*see* **Note 9, 10**).

2. Immediately add 10 µL of stabilizing solution to the drop.

3. Transfer crystals, using a capillary or a Gilson-200 tip (**Fig. 9.4A–B**), from the drop to a Micro-Bridge containing 30–40 µL of soaking solution A, seal the well with a new coverslip, and leave to soak at room temperature for several days (*see* **Notes 11, 12**).

4. Follow inhibitor binding inside the crystals by multiple reaction monitoring mass spectrometry, a very sensitive (fmol) technique that can check the presence of covalent inhibitors inside soaked crystals. Briefly, transfer NS3/NS4A crystals soaked with the inhibitor at days 5, 10, and 24 to new Micro-Bridges containing only the stabilizing solution for crystal washing to eliminate unspecific inhibitor binding. Then transfer crystals to a Micro-Bridge containing 10–20 µL HPLC-grade water and analyze them by mass spectrometry. Inhibitor is typically found inside the crystals soaked in high salt/low pH only after 24 days of soaking. Crystals are not damaged by this long soaking time.

5. After 24 days of soaking at room temperature, transfer NS3/NS4A crystals to a new Micro-Bridge containing cryogenic solution (30% glycerol, 4.0 M NaCl, 10 mM DTT, 0.1 M citrate buffer, pH 5.1), fishing them individually with the nylon Cryo-loop, and freeze them immediately in liquid nitrogen for X-ray data collection at the Synchrotron beam line using a stream of gaseous nitrogen at 100 K.

3.6.2. NS5B

NS5B crystals belong to spacegroup $P2_12_12_1$ with unit cell dimensions of $a = 67.06\,\text{Å}$, $b = 96.89\,\text{Å}$, and $c = 194.43\,\text{Å}$, with two molecules per asymmetric unit, and diffract to maximum 2.0 Å resolution on a Synchrotron system at cryogenic temperature (100 K). We were interested in having the structure in complex with allosteric inhibitors (**Fig. 9.3D**), of which we did not know the real binding site.

1. Perform soakings of NS5B crystals (**Fig. 9.3B**), after transferring them from the crystallization drop directly to a Micro-Bridge containing 40 µL of soaking solution B. In this case, use soaking times of less than 1 h, because crystals do not survive longer soaking times.

2. Transfer crystals to cryoprotectant solution. After a 1 min soak in this solution, plunge crystals directly into liquid nitrogen for X-ray data collection at 100 K with synchrotron

A

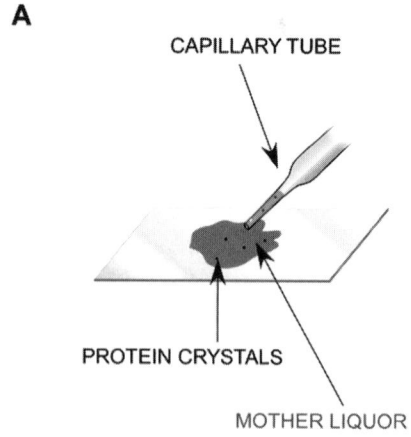

CAPILLARY TUBE

PROTEIN CRYSTALS

MOTHER LIQUOR

B

PROTEIN CRYSTAL in the stabilizing
solution plus inhibitor

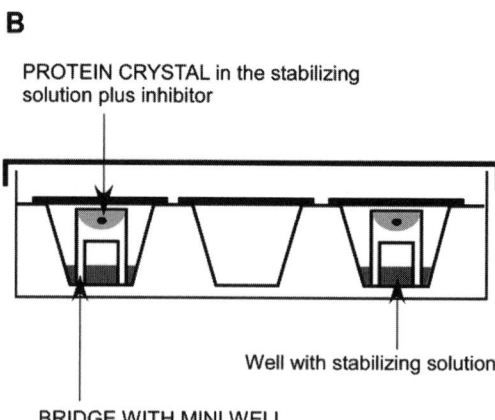

Well with stabilizing solution

BRIDGE WITH MINI WELL

Fig. 9.4. Soaking procedure. (**A**) Crystals are picked up from a drop to be transferred to sitting drops for soaking experiments. (**B**) Schematic representation of the soaking technique in sitting drops.

radiation. Perform both soaking and cryofreezing at 4°C.

3. Inhibited NS3/NS4A and NS5B crystal structures are described in *(9–11)*.

4. Notes

1. The water used for all buffers should be double distilled (MilliQ) and filtered with $0.22\,\mu$M filters.

2. Repeated freezing and thawing of the protein sample as well as of the inhibitor sample should be avoided. Divide the sample into aliquots in multiple small Eppendorf tubes. Make the aliquots small enough that the entire aliquot can be consumed in one experiment after thawing.

3. When thawing a sample be gentle and avoid foam that can be a sign of protein denaturation.

4. Avoid combining different purification batches for crystallization experiments. Especially during the crystal optimization process, the same batch should be used.

5. For NS3/NS4A complex, control experiments had shown that this freezing procedure does not affect the NS3 protease specific activity and/or complex stability.

6. The advantages of the hanging-drop technique include the ability to view the drop under a microscope, through the glass, without the optical interference from plastic, and easy access to the drop. In addition, multiple drops can be pipetted onto the same cover slide. This technique can be used to screen different drop sizes and ratios against the same reservoir and also for additive/detergent screening, in which each drop contains a different additive/detergent. The disadvantage is that a little extra time is required for set-up.

7. Each Micro-Bridge contains a smooth, concave depression at the center of its top region of the bridge that can hold up to 40 μL drop. The Micro-Bridge can be removed from the plate for crystal manipulation and observation. Micro-Bridges are designed as disposable devices. They cannot be siliconized or autoclaved.

8. Optimal conditions for crystal nucleation and growth are difficult to predict, and screening is a very efficient and effective tool for determining the initial crystallization conditions of biological macromolecules. The variables influencing crystal growth are too large to allow an exhaustive search. A highly effective approach is the use of a sparse matrix method of trial conditions that is biased and selected from known crystallization conditions for macromolecules. We tested a wide range of buffers and pH and three categories of precipitants: salts, polymers, and organic solvents. In addition, we use three different detergent kits from Hampton. The same experiment was performed at room temperature as well as at 4°C.

9. To access the drop and/or reservoir from the VDX Plate, grasp the edge of the cover slide with forceps, twist and pull gently.

10. Crystals useful for X-ray diffraction analysis are typically single, larger than 20–50 μm, free of cracks and defects, birefringent, and with feature edges.

11. Take great care when transporting and viewing the crystallization plates to facilitate crystal growth.

12. If a crystal adheres to the crystallization surface, withdrawing liquid from the drop and gently ejecting it onto a chosen crystal will often dislodge it.

References

1. De Francesco, R., and Migliaccio, G. (2005). Challenges and successes in developing new therapies for hepatitis C. *Nature* **436**, 953–960.
2. Urbani, A., Bazzo, R., Nardi, M. C., Cicero, D. O., De Francesco, R., Steinkuhler, C., et al. (1998). The metal binding site of the hepatitis C virus NS3 protease. A spectroscopic investigation. *J. Biol. Chem.* **273**, 18760–18769.
3. Bianchi, E., Urbani, A., Biasiol, G., Brunetti, M., Pessi, A., De Francesco, R., et al. (1997). Complex formation between the hepatitis C virus serine protease and a synthetic NS4A cofactor peptide. *Biochemistry* **36**, 7890–7897.
4. Kim, J. L., Morgenstern, K. A., Lin, C., Fox, T., Dwyer, M. D., Landro, J. A., et al. (1996). Crystal structure of the hepatitis C virus NS3 protease domain complexed with a synthetic NS4A cofactor peptide. *Cell* **87**, 343–355.
5. Yan, Y., Li, Y., Munshi, S., Sardana, V., Cole, J. L., Sardana, M., et al. (1998). Complex of NS3 protease and NS4A peptide of BK strain hepatitis C virus: a 2.2 Å resolution structure in a hexagonal crystal form. *Protein Sci.* **7**, 837–847.
6. Bressanelli, S., Tomei, L., Roussel, A., Incitti, I., Vitale, R. L., Mathieu, M., et al. (1999). Crystal structure of the RNA-dependent RNA polymerase of hepatitis C virus. *Proc. Natl. Acad. Sci. USA* **96**, 13034–13039.
7. Lesburg, C. A., Cable, M. B., Ferrari, E., Hong, Z., Mannarino, A. F., and Weber, P.C. (1999). Crystal structure of the RNA-dependent RNA polymerase from hepatitis C virus reveals a fully encircled active site. *Nat. Struct. Biol.* **6**, 937–943.
8. Orsatti, L., Di Marco, S., Volpari, C., Vannini, A., Neddermann, P., and Bonelli, F. (2002). Determination of the stoichiometry of non-covalent complexes using reverse-phase high-performance liquid chromatography coupled with electrospray ion trap mass spectrometry. *Anal. Biochem.* **309**, 11–18.
9. Di Marco, S., Rizzi, M., Volpari, C., Walsh, M. A., Narjes, F., Colarusso, S., et al. (2000). Inhibition of the hepatitis C virus NS3/4A protease. The crystal structures of two protease-inhibitor complexes. *J. Biol. Chem.* **275**, 7152–7157.
10. Ontoria, J. M., Di Marco, S., Conte, I., Di Francesco, M. E., Gardelli, C., Koch, U., et al. (2004). The design and enzyme-bound crystal structure of indoline based peptidomimetic inhibitors of hepatitis C virus NS3 protease. *J. Med. Chem.* **47**, 6443–6446.
11. Di Marco, S., Volpari, C., Tomei, L., Altamura, S., Harper, S., Narjes, F., et al. (2005). Interdomain communication in hepatitis C virus polymerase abolished by small molecule inhibitors bound to a novel allosteric site. *J. Biol. Chem.* **280**, 29765–29770.

Chapter 10

Structural and Functional Analysis of the HCV p7 Protein

Nathalie Saint, Roland Montserret, Christophe Chipot, and François Penin

Abstract

The p7 membrane polypeptide from HCV is essential for virus infection. It exhibits ion-channel activity reported to be specifically blocked by various compounds. These properties make p7 an attractive candidate target for antiviral intervention to combat viral hepatitis C infection. In this context, in vitro functional analyses of isolated p7 coupled to structural characterization are critical for further understanding of the molecular mechanisms of p7 ion-channel activity and for the development of new antiviral drugs. We present here in vitro assays designed to purify synthetic p7 by RP-HPLC, to investigate its ion-channel properties by means of planar lipid-bilayer assays and patch-clamp recordings after reconstitution into liposomes, and to analyze its structural features by circular dichroism (CD), nuclear magnetic resonance (NMR), and molecular dynamics (MD).

Key words: Ions-channel activity, viroporin, RP-HPLC, circular dichroism (CD), nuclear magnetic resonance (NMR), molecular dynamics (MD).

1. Introduction

The p7 membrane polypeptide from HCV is essential for virus infection in vivo *(1)* and for production of infectious HCV particles *in cellulo* but is not necessary for RNA replication *(2)*. P7 localizes primarily to the endoplasmic reticulum as an integral membrane protein and displays a topology in which both N- and C-termini point toward the endoplasmic reticulum lumen *(3)*. Structure predictions suggest that p7 comprises two transmembrane (TM) helices connected by a short cytoplasmic loop *(3)*. Recent in vitro studies indicate that p7 oligomerizes *(4, 5)* and can conduct ions across artificial membrane systems in a

Hengli Tang (ed.), *Hepatitis C: Methods and Protocols, Second Edition, vol. 510*
© 2009 Humana Press, a part of Springer Science+Business Media
DOI 10.1007/978-1-59745-394-3_10 Springerprotocols.com

cation-selective manner *(4, 6, 7)*. Because of these properties, p7 has been tentatively included in the family of viroporins *(8)*. Ion-channel activity has been reported to be specifically blocked by various compounds *(4, 6, 7)*. These properties make p7 an attractive candidate target for antiviral intervention. In addition, it was recently shown that p7 variants from different isolates differ substantially in their capacities to promote virus production, suggesting that p7 is an important virulence factor that may modulate fitness and in turn virus persistence and pathogenesis *(9)*.

In this context, in vitro functional analyses of isolated p7 coupled to structural characterization are critical for further understanding of the molecular mechanisms of p7 ion-channel activity and for the development of new antiviral drugs to combat viral hepatitis C. Production of p7 by chemical synthesis as well as its purification by RP-HPLC was successful despite its size (63 amino acids) and hydrophobicity. P7 peptide was found to form cation-selective ion channels in planar lipid bilayer and at the single-channel level by the patch-clamp technique. Structure analyses in the presence of membrane-mimetic (trifluoroethanol–water mixtures, detergents, and lyso-lipids) systems by circular dichroism (CD), nuclear magnetic resonance (NMR), and molecular dynamics (MD) simulation in a membrane bilayer allowed us to propose a three-dimensional (3D) structure model of the monomer form of p7 in the membrane environment.

2. Materials

2.1. RP-HPLC

1. HPLC apparatus (Waters) consisting of two M510 pumps, a U6K injector including a 2 mL injection loop, and a 991 Photodiode Array Detector including data-processing software M990.
2. Columns: C8 semipreparative column (reverse-phase Vydac C8 208TP1010, 10×250 mm, 300 Å, particle size 10 μm) equipped with a guard column (Brownlee Aquapore C8 RP-300, 4.6×30 mm, particle size 7 μm) (*see* **Note 1**).
3. Solvents: MilliQ water, acetonitrile (HPLC grade), trifluoroacetic acid (TFA). Solution A: degassed water containing 0.1% TFA (v:v); solution B: degassed acetonitrile–water mixture (80:20, v:v) containing 0.08% TFA (*see* **Note 2**).

2.2. Planar Lipid-Bilayer Assays

1. Lipids: L-α-phosphatidylcholine type IV S from soybean (also called L-α-lecithin or azolectin) available from Sigma (*see* **Note 3**).
2. Organic solvents: hexane (98% HPLC grade), chloroform, and methanol (ACS reagent, $\geq 99.8\%$) from Sigma-Aldrich.

3. A bilayer chamber composed of two glass cells clamped together (volume of each cell: 4 mL) designed by NHverre (Puéchabon, France).

4. High-frequency spark tester PPM Mk3 (Girovac LTD, U.K.).

5. Teflon septum (Goodfellow, U.K.), thickness 10 μm.

6. Coating agent for Teflon septum: hexadecane (Fluka).

7. Home-made Ag/AgCl electrodes: Ag wire (A-M Systems, Carlsborg, WA) and AgCl (Sigma-Aldrich).

8. PC-S3 electrode holders (Bio-Logic, France).

9. Electrolyte: 500 mM KCl, 10 mM HEPES, pH 7.4.

10. Home-made Faraday cage made of aluminum plates 5 mm thick (dimensions: 50 cm × 50 cm × 50 cm).

11. Amplifier BLM 120 (Bio-Logic, France), amplifier BBA-01 (Eastern Scientific, Rockville, MD).

12. Oscilloscope TDS 3012 (Tektronix, Beaverton, OR).

13. CD recorder PDR-W739 Pioneer (Bio-Logic, France).

14. Biotools, WinEDR, and Scope softwares (Bio-Logic, France) for data analysis.

15. Solution for cleaning glass cells: 200 g of KOH in 70% MeOH and 30% H_2O.

2.3. Reconstitution into Liposomes and Patch-Clamp Recordings

1. Lipids: same as described in **Section 2.2**.

2. Polystyrene beads: Bio-Beads SM2 adsorbent (Bio-Rad).

3. Electrolyte: 500 mM KCl, 10 mM HEPES, pH 7.4.

4. Patch chamber RC-25F (Warner Instrument, Hamden, CT).

5. Patch pipettes from borosilicate glass capillaries (Harvard Apparatus, U.K.).

6. Patch pipette puller PA-10 (Bio-Logic, France).

7. Three-axis hydraulic micromanipulator (Narishige Scientific Instrument, Japan).

8. Microscope Axiovert 135 (Zeiss, Germany).

9. Antivibration table "Newport" (Bio-Logic, France) (*see* **Note 4**).

10. Ag/AgCl electrodes as described in **Section 2.2**.

11. PC-S3 electrode holder with suction (Bio-Logic, France).

12. Visual patch 500 amplifier (Bio-Logic, France).

13. Biotools software (Bio-Logic, France).

2.4. Circular Dichroism

1. Detergents and lyso-lipid: 0.7 M stock solution of dode-cylphosphocholine (DPC, Avanti Polar Lipids, AL); 0.8 M stock solution of sodium dodecyl-sulfate (SDS, from Sigma), 1 M stock solution of *n*-dodecyl-β-d-maltoside (DM, from Boehringer). DPC, DM, and SDS stock solutions can be stored frozen for several months. 10% freshly prepared stock solution of α-lyso-phosphatidylcholine (LPC, from Sigma).

2. Trifluoroethanol (TFE, analytical grade, Sigma).

3. Sodium phosphate buffer 100 mM, pH 7.4.

4. Jobin Yvon CD6 spectrometer calibrated with ammonium d10-camphorsulfonate and equipped with CD6 program for acquisition and treatment of CD data.

5. Flat quartz cuvette of 0.1 cm path length equipped with a PTFE stopper (Stopper Top cell, Hellma).

2.5. Nuclear Magnetic Resonance

1. NMR solvents, reagents, and tube: D_2O; deutered trifluoroethanol (TFE-d_2, 2,2,2-trifluoroethyl-1,1-$d2$ alcohol); DSS (2,2-dimethyl-2-silapentane-5-sulfonate); precision NMR tube 5 mm OD (Wilmad-Labglass).

2. NMR spectrometers: Varian Unity-*plus* 500 MHz and Varian *Inova* 800 MHz equipped with ultraNMR shims and using a 5 mm triple resonance ^1H-^{13}C-^{15}N probe with a self-shielded z-gradient coil.

3. NMR softwares: Vnmr (Varian) and SPARKY (T. D. Goddard and D. G. Kneller, SPARKY 3, University of California, San Francisco; http://www.cgl.ucsf.edu/home/sparky/).

4. Software for structure calculation, analysis, and visualization: X-PLOR *(10)*, PROCHECK-NMR *(11)*, and VMD program *(12)*, respectively.

3. Methods

Expression of recombinant p7 in bacteria is limited because of its cell toxicity. In contrast, despite its size (63 amino acids) and hydrophobicity, p7 has been successfully produced by chemical synthesis by Clonestar Peptide Service Ltd. Studies of the structural and functional features of membrane protein domains such as p7 require highly purified peptides, especially when they are obtained by chemical synthesis. Hydrophobic peptides indeed exhibit detergent-like properties when reconstituted with lipids, and the presence of irrelevant, side-product peptides from chemical synthesis can generate unexpected ion channel–like activities, which disturb the recording and analysis of electrophysiology experiments. The purification of p7 was achieved by RP-HPLC, which is generally used for the preparation of highly purified peptides, but which could be inefficient in the case of membrane peptides. Indeed, the handling of such peptides requires careful attention to prevent aggregation. The preparation of samples is therefore an essential step in obtaining relevant information about p7's structure and function. Its ion-channel properties were investigated with planar lipid-bilayer assays and patch-clamp recordings after reconstitution into liposomes. The characterization of p7 by CD in cosolvents (TFE–water mixtures), detergents (SDS, DPC), or lyso-phospholipid (LPC) rapidly provides information about the conformational preferences and the secondary structure of p7 in these membrane-mimetic environments. The CD analyses are also essential to evaluation and preparation of samples for NMR studies using relevant physicochemical conditions. NMR is one of the three main approaches (together with X-ray crystallography and cryoelectron microscopy) for determining the 3D structure of individual proteins *(13)*. NMR is used to study the atomic structure of relatively small molecules (10–12 kDa), but labeling of proteins with ^{15}N and ^{13}C isotopes allows study of larger proteins. NMR is particularly useful for structural analysis of membrane protein domains for which the crystallization is very challenging *(14)*. Liquid-state NMR in membrane-mimetic media, like detergents or organic solvents, is routinely used for membrane-peptide structure determination. Studies in organic solvents such as TFE–water mixtures, however, are only useful for probing membrane-peptide conformational preferences. On the other hand, studies in detergents are limited by the broadening of NMR signals due to the large molecular size of the peptide-detergent complexes that are not amenable to proton liquid-state NMR. Solid-state NMR, which is still under development and requires ^{15}N- and ^{13}C-labeled proteins, is a particularly promising technique for the study of membrane proteins in phospholipid environments rather than in membrane-mimetic ones. Nevertheless, experimentally determined 3D structures of membrane peptides by NMR in membrane mimetics do not allow a detailed understanding of protein–phospholipid interactions. In this context, numerical simulations of membrane peptides in lipid membranes by molecular dynamics contribute to understanding of membrane-bound protein structure and function. Combined synergistically with experimental structure data from NMR and molecular modeling, simulations by molecular dynamics allowed us to propose a 3D model of the monomer form of p7 associated to the phospholipid bilayer.

3.1. Purification of Synthetic p7 by RP-HPLC

The 63-amino-acid-long p7 peptide was synthesized by standard solid-phase synthesis by Clonestar Ltd (*see* **Note 5**). Analysis of crude lyophilized material by electrospray mass spectrometry is required for rough estimation of content in the expected peptide (*see* **Note 6**). As a rule, the expected peptide should represent at least 20% of crude product peptides, for efficient purification by RP-HPLC.

1. To prepare a 1 mg/mL peptide sample, dissolve 10 mg of crude synthesis product in 50 μL of pure TFA (the solution generally becomes yellow-brownish), and then dilute to 6 mL with pure acetonitrile and to 10 mL with water (the final solution is colorless).
2. Equilibrate the RP-HPLC column with 30% solvent B and 70% solvent A at 1.5 mL/min for 15 min (i.e., until the baseline is stable) (*see* **Note 7**).

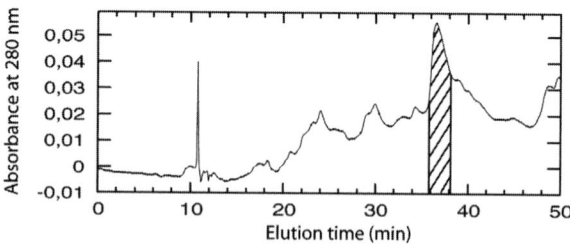

Fig. 10.1. Purification of p7 from crude synthesis peptide product by reversed-phase chromatography. The shading indicates the collected fraction containing purified p7. The unretained products eluted at about 11 min correspond to the void volume of the column. This time corresponds approximately to the delay between the programmed gradient and the effective gradient.

3. Run an analytical sample of 50 μL (50 μg) to test the whole chromatographic system. The typical elution gradient of solution B in solution A used linear gradient steps performed as follows: 30% solution B at initial time, 70% solution B at 5 min, 90% solution B at 35 min, 100% solution B at 40 min, and 100% solution B at 42 min. A decreasing linear gradient down to 30% solution B is then performed in 5 min, followed by a minimum of 15 min column equilibration at this concentration before the next run. P7 is typically eluted at 36–38 min retention time (i.e., about 80% solution B, i.e., 64% acetonitrile) (see **Note 8**). A typical elution profile is shown in **Fig. 10.1** (see **Note 9**).

4. Run a series of semipreparative samples of 2 mg (2 mL) and collected the p7-containing fractions (see **Note 10**). Pool the p7 eluted fractions and dilute to 10% acetonitrile for freezing at −80°C and lyophilization.

5. Store lyophilized p7 at 4°C, either dry or dissolved in TFE 50%. Estimate protein concentration by UV absorbance at 280 nm. The overall yield of purification from the crude synthesis product is about 20%, and the purity is over 95%, as attested by electrospray mass spectroscopy and NMR analyses.

3.2. Preparation of p7 Samples

Knowledge of the features of hydrophobic membrane peptide is essential for their correct handling and for acquisition of relevant structural and functional data. Membrane peptides are indeed difficult to solubilize and tend to aggregate. Direct solubilization of lyophilized membrane peptide with an aqueous solution of detergent is often not efficient and/or does not yield monodispersed solution of peptide-detergent micelles. In contrast, membrane peptides are generally quite soluble in TFE, provided that some water (5–10%) is present. Indeed, TFE is very hydrophobic and prevents aggregation of peptides by hydrophobic, nonspecific interactions in an aqueous environment, but the presence of some water is required to prevent peptide aggregation by polar and ionic interactions of polar and/or charged peptide groups. A highly soluble, monodispersed concentrated solution of p7 (e.g., 20 mg/mL) can then be diluted to 50% TFE with water, and this stock solution can be stored for months at $-20°C$.

1. To prepare a solution of p7 in any detergent (e.g., DM or DPC), add the required volume of p7 stock solution in 50% TFE to a water solution of concentrated detergent and mix carefully. Dilute the sample with water down to 10% TFE, freeze at $-80°C$, and lyophilize to remove any trace of TFE.

2. To dissolve the previous lyophilized sample, add some water and wait a few minutes, without shacking. Then mix and add concentrated buffer solution to the required final concentrations (e.g., 50 μM p7 in 100 mM DPC, 10 mM sodium phosphate, pH 7.4; see **Section 3.5.** for examples). Note that the direct addition of the buffer prevents dissolution of the lyophilizate, probably because a salt effect disturbs the dissolution of peptide-detergent micelles in water.

3. To reconstitute p7 in lipids, mix the required quantity of p7 stock solution into 50% TFE with phospholipids (e.g., egg yolk phosphatidylcholine) dissolved in 1:1 chloroform:methanol. The molar ratio of p7 to phospholipid must be at least 1:100 to prevent a peptide-dependent detergent effect. If the resulting solution is opalescent, add some TFE or chloroform to obtain a limpid solution attesting a homogenous mixture. Then remove the solvents under a nitrogen flux and lyophilize to remove any trace of solvent and water. The dry film can then be used to prepare liposomes.

3.3. Reconstitution into Planar Lipid Bilayer

Virtually solvent-free planar lipid bilayers were formed by the technique outlined by Montal and Mueller *(15)* in which two monolayers are apposed within a hole in a thin Teflon septum.

1. Make a hole of 100–200 μm across a 10-μm-thick Teflon film using a high-frequency spark tester. The tester generates an electrical arc that penetrates the film and therefore makes a hole.

2. Check the size and the shape of the hole with a dissecting microscope.

3. Sandwich the Teflon film between two glass cells and coat the hole in the septum with 2 μL of a solution of hexadecane in hexane (1:40, v:v).

4. After hexane evaporation (about 1 h later), fill the two compartments of the bilayer chamber with the electrolyte solution and spread 15 μL of lipids on the air–water interface of the aqueous buffer. Prepare lipids from azolectin IV S from soybean dissolved in hexane at a concentration of 0.5% (w:v).

At the air–water interface, lipids will form structured mono-layers.

5. Thirty minutes after lipid addition (the time necessary to ensure hexane evaporation), lower the Ag/AgCl electrodes into both cells of the chamber. Connect one chamber of the bilayer, called the *trans* side, to the ground. Connect the other electrode, on the *cis* side, to the amplifier.

6. Lower the level of the electrolyte solution in the compartment of the chamber below the hole, and then raise the level of the buffer over the hole to form the lipid bilayer by apposition of two monolayers within the hole in the Teflon septum.

7. Monitor the formation of the bilayer by the capacitance response, as measured on the oscilloscope.

8. Reconstitute the p7 protein by adding 5×10^{-9} to 5×10^{-8} M of purified protein dissolved in 1% DM in the aqueous subphase in front of the azolectin bilayer.

9. Monitor p7 insertion into the bilayer by measuring the membrane current after subjecting the membrane to slow ramps of potential (10 mV/s). Feed the transmembrane currents into the amplifier and store the current–voltage curves on computer and analyze them with the Scope program (**Fig. 10.2**).

10. Conduct all experiments at room temperature and stir the solution in the chamber until currents are observed.

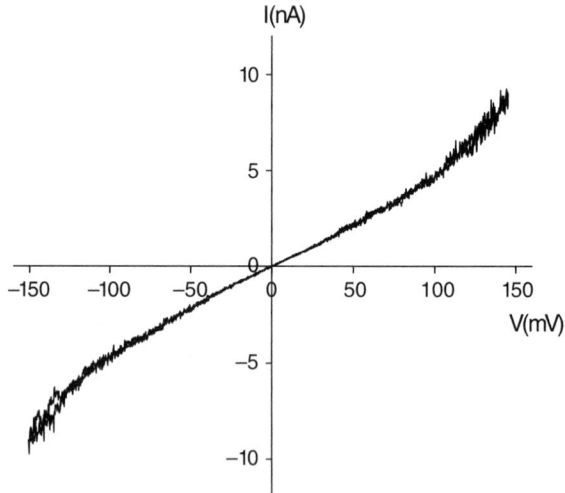

Fig. 10.2. Macroscopic current-voltage (I-V) curve from p7 protein inserted into an azolectin bilayer. This I-V curve is useful for testing the ability of p7 protein to induce channel activity in planar lipid bilayers. In these experiments, several hundreds or thousands of p7 channels can be incorporated into bilayers submitted to slow voltage ramps (10 mV/s). The resulting I-V curve is linear between –100 mV and +100 mV and shows a small rectification above ±100 mV. Nonlinearity of the current–voltage curve above ±100 mV indicates weak voltage-dependent channel formation.

3.4. Reconstitution into Liposomes and Patch-Clamp Recordings

Patch-clamp experiments *(16)* were performed on blisters induced from giant liposomes containing the purified protein.

1. Dissolve soybean azolectin in 500 mM KCl, 10 mM HEPES, pH 7.4, by sonication to a final concentration of 10 mg/mL.

2. Add an aliquot of purified p7 dissolved in 1% DM to azolectin vesicles to get a defined protein-to-lipid ratio of 1:500–1:5000 (w:w). Incubate the mixture for 30 min at room temperature and then add 80 mg Bio-Beads per mL of protein solution in 1% DM (*see* **Note 11**).

3. After 4 h, remove the Bio-Beads by sedimentation and ultracentrifuge the liposomes for 20 min at 200 000 × g at 4°C (70 Ti Rotor, Beckman).

4. Resuspend the resulting pellet in 30 μL of 10 mM HEPES (pH 7.4). Subject aliquots to a dehydration/rehydration procedure to obtain giant proteoliposomes: Place three spots of 10 μL each on a glass slide and dehydrate them for 4 h at 4°C; then rehydrate them with 10 μL of 500 mM KCl, 10 mM HEPES, pH 7.4, overnight at 4°C.

5. Place 1–2 μL of proteoliposomes in the patch chamber containing recording solution (500 mM KCl, 10 mM HEPES, pH 7.4).

6. Examine membrane patches, obtained from unilamellar blisters of collapsing proteoliposomes, using the standard patch-clamp technique. Clamp the voltage across the membrane at different values using the VP500 amplifier.

7. Record the resulting currents on a PC computer and analyze them with the Biotools program (**Fig. 10.3**).

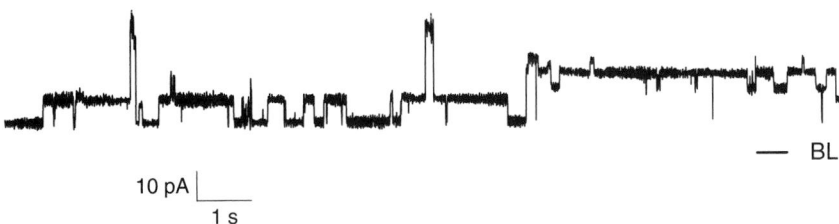

10 pA

1 s

— BL

Fig. 10.3. Selected current recordings from p7 ions channels. Giant liposomes of azolectin doped with p7 protein were studied by the patch-clamp technique. The electrolyte solution used was 10 mM HEPES (pH 7.4), 0.5 M KCl. The current recordings obtained under these conditions with a protein-to-lipid ratio of 1:500 resulted in observation of multiple channels with conductance levels of 35, 61, 123, and 244 pS. BL represents the baseline at zero pA level.

3.5. Circular Dichroism Recordings

CD measurements *(17)* are performed at room temperature with solutions of p7 ranging from 10 to 80 μM. Spectra are recorded in the 190–250 nm wavelength range, with 0.5 nm increments, 1 s integration time. For noisy samples, accumulation of two or four scans increases the signal-to-noise ratio.

1. For CD spectrum of p7 in TFE 50%, first add 196 μL of TFE 50% to the cuvette (0.1 cm path-length quartz cuvette equipped with a PTFE stopper) and then record the baseline spectrum in the Delta-Absorption (ΔA) mode.

2. Add 4 μL of 2 mM p7 stock solution in 50% TFE, mix carefully, and record the assay spectrum (*see* **Note 12**).

3. Baseline-correct the assay spectrum by subtracting the baseline spectrum, and smooth it by using a third-order least-squares polynomial fit. Then convert the ΔA corrected spectrum to molar ellipticity per residue by taking into account p7 protein concentration and number of p7 residues (i.e., 63).

4. For CD spectra of p7 in each of the detergents (SDS, DM, DPC) or in lyso-lipid (LPC), add successively to the cuvette 156 μL of water, 20 μL of 1 M detergent or 10% LPC stock solutions, and 4 μL of 2 mM p7 stock solution in 50% TFE; mix well and then add 20 μL of 100 mM sodium phosphate buffer, pH 7.4. The final concentrations are 100 mM detergent (or 1% LPC), 10 mM sodium phosphate, pH 7.4, and 2% TFE (*see* **Note 13**). Samples for baseline correction are treated in the same way, without p7 but including 2% TFE final concentration.

5. CD spectra of p7 in TFE 50%, detergents, or lyso-lipid exhibit the typical shape of proteins mainly folded into an α-helix, that is, two minima around 208 and 222 nm and a maximum at 192 nm (**Fig. 10.4**). Assuming that the residue molar

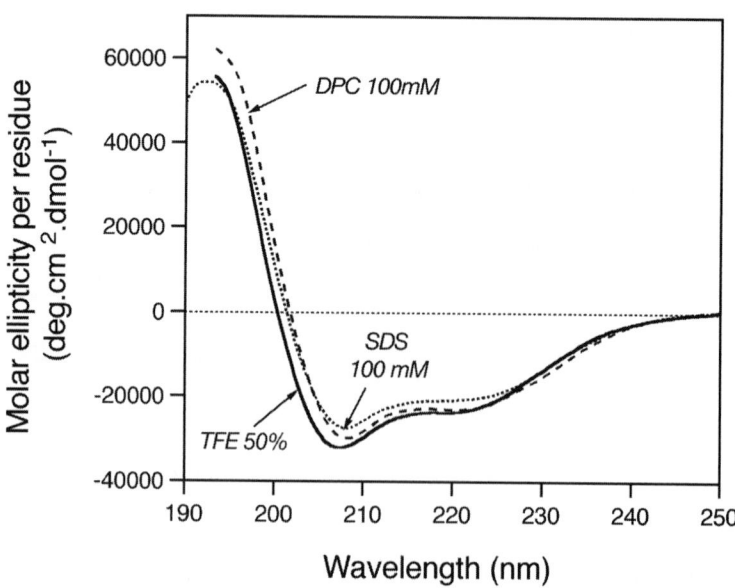

Fig. 10.4. Far UV CD analyses of synthetic p7 in various membrane-mimetic environments. CD spectra were recorded either in 50% TFE (*solid line*) or in 100 mM SDS (*dotted line*) or 100 mM dodecylphosphocholine (DPC, *dashed line*).

ellipticity at 222 nm is exclusively due to α-helix, the α-helical content is estimated according to Chen et al. *(18)* according to the following equation: % helix = $100[\theta]_{obs}/[\theta]_H$, where $[\theta]_{obs}$ is the observed mean residue ellipticity at 222 nm, and $[\theta]_H$ is the maximum mean residue ellipticity for a helix of finite length. $[\theta]_H$ was calculated using the empirical equation for helix length dependence proposed by Chen et al.: $[\theta]_H = [\theta]_{H\infty}(f_H - ik/N)$, Where f_H is the fraction of helix (i.e., $f_H = 1$ when assuming helical folding only), i is the number of helical segments (3 in this case, see below), k is a wavelength-dependent constant (2.57 at 222 nm), N is the number of residues (63 for p7), and $[\theta]_{H\infty}$ is the maximum mean residue ellipticity for a helix of infinite length (–39500 at 222 nm). On this basis, the $[\theta]_H$ for p7 in a 100% helical conformation was calculated to be –34665 deg cm^2 dmol^{-1}. A similar content of α-helix (ranging from 59 to 68%) is observed in the various membrane-mimetic conditions (**Fig. 10.4**).

3.6. Overview of NMR Analysis of p7

The protein solution is placed in a strong permanent magnetic field, and the magnetic properties of ^{1}H, ^{13}C, and ^{15}N isotopes are measured. The signals (resonances) emitted by individual atoms and interactions between pairs of atoms are captured in multidimensional NMR spectra and are ascribed to each atom in each amino acid composing the protein. Distances between hydrogen atoms (≤ 5 Å) can be deduced from magnetic interactions between pairs of atoms that produce cross-peaks at the resonances of the two atoms (nuclear Overhauser effect). Large collections of known interatom distances are used to impose structural constraints when 3D protein structure is modeled *(13)*.

Preliminary NMR spectra of p7 dissolved in deutered detergents (DPC, SDS) showed broad NMR signals because of the large size of the peptide-detergent micelle and the expected oligomerization of p7. Such complex could not be analyzed by proton NMR and would require the preparation of ^{15}N- and ^{13}C-labeled p7, which has not yet been accomplished. In contrast, well-resolved proton NMR spectra were obtain from p7 dissolved in 50% TFE (**Fig. 10.5**). As CD experiments showed that a similar α-helix content was observed for p7 in TFE 50% and in various detergent, TFE 50% appears to be an appropriate medium for analysis of the structure of p7 in membrane-like environments. Because of its hydrophobic solvent properties, however, TFE does not allow p7 to oligomerize, and only the secondary structure of p7 could be describe in this medium (*see* **Note 14**).

1. About 5.5 mg of purified p7 was dissolved in 650 μL of 50% deutered trifluoroethanol in MilliQ H$_2$O (p7 final concentration, 1.26 mM). DSS (2,2-dimethyl-2-silapentane-5-sulfonate) was added to the NMR sample as an internal ^{1}H chemical shift reference (0.01% final concentration).

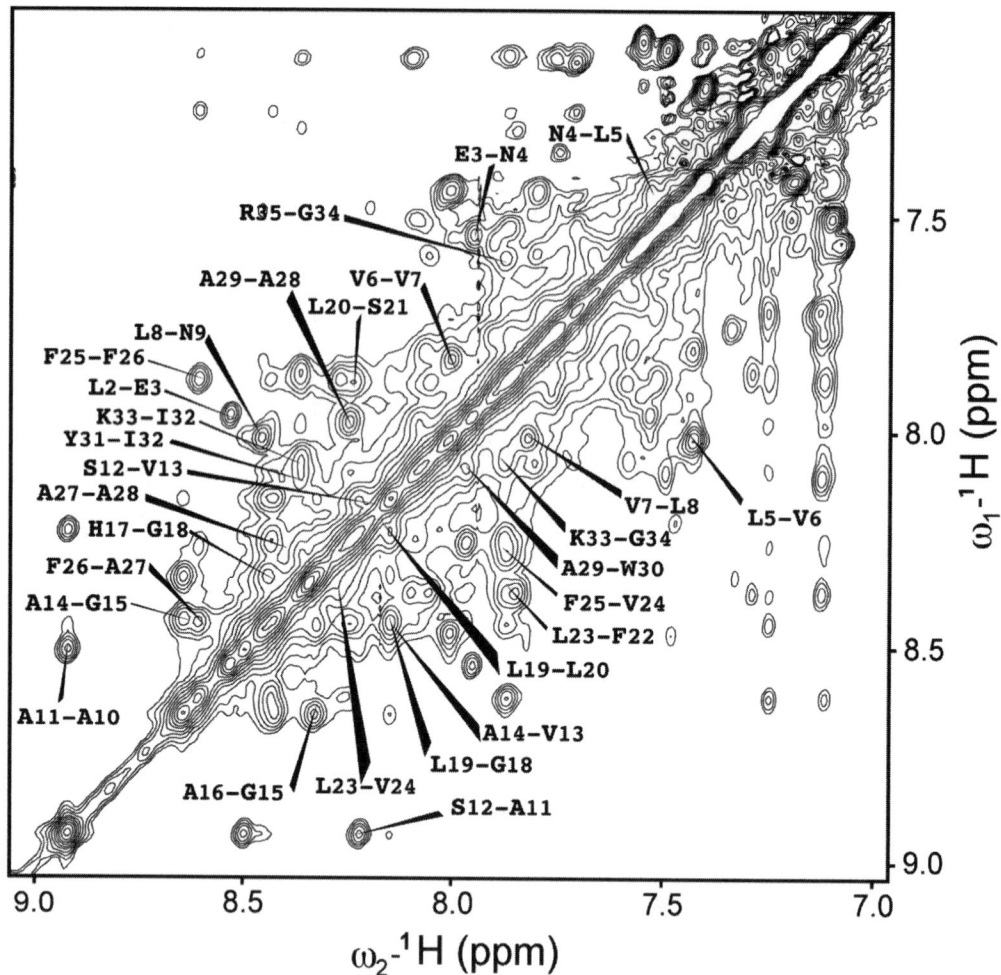

Fig. 10.5. NMR analysis of synthetic p7. Extract of NOESY spectra of 1.3 mM p7 in 50% TFE. NH-NH region, where sequential resonance assignments are traced out. NOESY spectra were recorded at 20°C with a mixing time of 150 ms.

2. A set of conventional multidimensional NMR spectra were acquired, including 2D homonuclear ^{1}H experiments (phase sensitive DQF-COSY, clean-TOCSY with isotropic mixing time of 80 ms, and NOESY with mixing times between 100 and 250 ms) and 2D heteronuclear ^{1}H -^{13}C HSQC and HSQC-TOCSY in ^{13}C natural abundance (19, 20).

3. Spectra were processed and analyzed with the Vnmr and SPARKY programs. The resonances of protons were ascribed by the conventional assignment method (13). In short, intraresidue backbone resonances and aliphatic side chains were identified from 2D DQF-COSY and TOCSY spectra, with the help of ^{1}H -^{13}C HSQC and HSQC-TOCSY spectra at various temperatures (20 and 25°C), to resolve the overlap of resonances. Sequential assignments were determined by correlation of intraresidue assignments with interresidue cross-peaks observed in 2D ^{1}H-NOESY.

4. NOE intensities obtained from the NOESY spectra were partitioned into three categories of intensities that corresponded to distances ranging from a common lower limit of 1.8 Å to upper limits of 2.6, 3.8, and 5 Å, for strong, medium, and weak intensities, respectively. Three-dimensional structures were generated from NOE distances with X-PLOR *(10)* using the standard force fields and default parameter sets. Sets of 50 structures were calculated and analyzed for NOE and dihedral angle violations. The secondary-structure elements and Ramachandran plots of the final set of selected structures were analyzed according to the Kabsch–Sander definition rules, as incorporated into PROCHECK-NMR *(11)*.

5. Combined analysis of homo- and heteronuclear NMR spectra allowed the complete sequence-specific backbone resonance assignment of p7 (*see* **Note 15**) and identification of its secondary-structure elements. P7 includes three α-helices, namely the N-terminal helix 2–16, the TM1 helix 19–33, and the TM2 helix 40–56. The N-terminal helix is connected to the TM1 helix by a turn, and TM1 is connected to TM2 by a small loop. The C-terminus region of p7 (57–63) appears to be unstructured. As TFE break the expected hydrophobic interactions between TM1 and TM2 helices and imperfectly mimic the phospholipid environment, the reconstruction of the 3D structure of p7 monomer was done by molecular modeling and numerical simulation in a palmitoyloleylphosphatidylcholine (POPC) membrane model by molecular dynamics.

3.7. Overview of Molecular Dynamics Simulation on the p7 Monomer in Membrane

On the basis of the structural information acquired from both CD experiments in membrane mimetics and NOESY signals from NMR in TFE-water solutions, a model of the complete monomeric protein was built. Because the latter experiments only supply fragments of the complete 3D structure, a number of assumptions ought to be made when the segments of known structure are combined with those of undefined structure. The p7 topology with both N- and C-termini pointing toward the ER lumen *(3)* indicates that the TM1 and TM2 α-helices probably form a hairpin motif that spans the entire hydrophobic core of the membrane.

1. Construction of this motif implies pairing the TM segments, which is usually achieved by means of protein–protein molecular docking. Here, use was made of the GRAMM program *(21)* to explore α-helix association through rigid-body translation and rotation of the individual helical units. An optimized scoring function was employed, the target of which was membrane proteins. Exploration was restrained to the formation of antiparallel α-helix bundles. Among the plethora of solutions proposed by GRAMM, only those corresponding to strictly antiparallel helical segments facing each other were

retained. Selection was further narrowed by examination of the interhelical contacts, which eliminated unphysical spatial arrangements (**Fig. 10.6b** and Color Plate 1).

2. The two TM α-helices were subsequently bridged into a hairpin by means of a short, four-residue loop domain (**Fig. 10.6c** and Color Plate 1) using the VMD program *(12)*. The N-terminal α-helical domain was then pasted to the TM1 segment (**Fig. 10.6d** and Color Plate 1), before attachment of the unstructured C-terminal fragment at the tail end of the TM2 α-helix, completing the model of the protein (**Fig. 10.6e** and Color Plate 1). Backbone dihedral angles were checked against rotamer libraries and corrected whenever necessary. A rapid energy minimization was then performed, with the NAMD molecular dynamics program *(22)* with the CHARMM27 macromolecular force field *(23)* to remove spurious intramolecular contacts.

3. Monomeric p7 protein was finally immersed in a membrane environment consisting of a fully hydrated POPC lipid bilayer. Molecular dynamics simulations in this environment

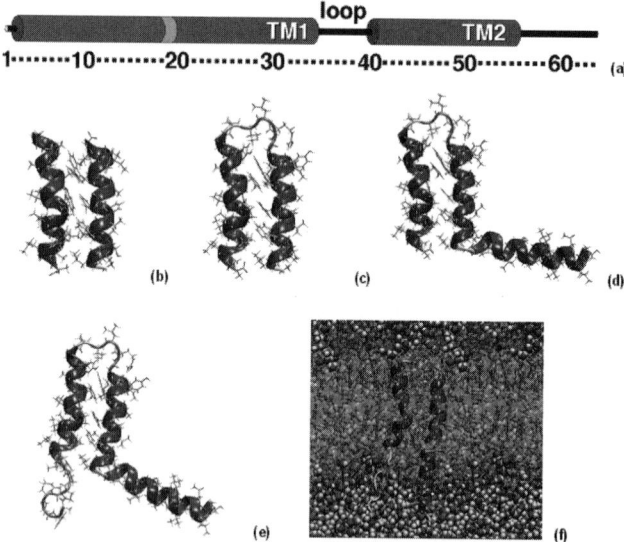

Fig. 10.6. Sequential construction of the monomeric p7 protein. (**a**) Cartoon representation of the protein sequence and secondary-structure elements. (**b**) Protein–protein molecular docking of the TM1 and TM2 α-helices. (**c**) Inclusion of the loop domain, yielding the hairpin motif. (**d**) The N-terminal α-helix is added to the hairpin. (**e**) The model is completed by insertion of the remaining unstructured C-terminal segment. (**f**) Incorporation of the protein in a fully hydrated POPC phospholipid bilayer shown in semitransparent van der Waals spheres. The N-terminal α-helix and the unstructured C-terminal part protrude into the aqueous medium of the ER lumen, whereas the loop region interacts with the head groups of the membrane. (*see* Color Plate 1)

Color Plate

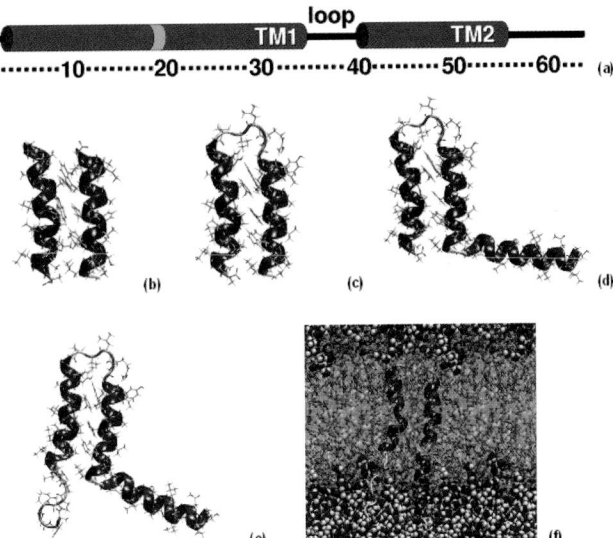

Color Plate 1. Sequential construction of the monomeric p7 protein. (**a**) Cartoon representation of the protein sequence and secondary-structure elements. (**b**) Protein–protein molecular docking of the TM1 and TM2 α-helices. (**c**) Inclusion of the loop domain, yielding the hairpin motif. (**d**) The N-terminal α-helix is added to the hairpin. (**e**) The model is completed by insertion of the remaining unstructured C-terminal segment. (**f**) Incorporation of the protein in a fully hydrated POPC phospholipid bilayer shown in semitransparent van der Waals spheres. The N-terminal α-helix and the unstructured C-terminal part protrude into the aqueous medium, whereas the loop region interacts with the head groups of the membrane. (*See* discussion on p. 138)

not only appraise the relevance of the protein model but also dissect the participating protein–protein interactions and shed new light on the subtle protein–lipid interplay.

4. Notes

1. A higher yield of purification for hydrophobic, membrane peptide is obtained when C8 (or C4) RP-HPLC, rather than C18, columns are used. The use of media with pore size smaller than 300 Å often decreases the yield of purification for larger peptides such as p7.

2. For efficient degassing of water under vacuum (e.g., water-tap pump), put the water container in a sonifier bath filled with about 1.5 cm of water. Degassing is completed when no more small air bubbles are observed (10–15 min). Degassing of acetonitrile is very rapid under such conditions (a few seconds). To prepare both solution A and solution B, use the following protocol: (1) degas water and acetonitrile separately, (2) add 1 mL of TFA 100% in 1 L of water under gentle magnetic agitation to prepare solution A (as TFA is highly volatile, immerse the pipette tip in the water to inject TFA), (3) add 100 mL of solution A to 400 mL of degassed acetonitrile under gentle magnetic agitation, and (4) then add 0.32 mL TFA 100% as above to obtain solution B. To avoid a drift of the baseline of the chromatogram at 220 nm during running of the acetonitrile gradient, as both acetonitrile and TFA absorb at 220 nm, the amount of TFA in solution B must be adjusted to yield the same final absorbance as in solution A. As a rule, this result is obtained by addition of 0.08% of TFA in solution B rather than 0.1%.

3. The major phospholipids contained in the azolectin type IV S preparation are an average of 55% phosphatidylcholine and 20% phosphatidylethanolamine as assayed by thin liquid chromatography. Other trace components such as carbohydrates and other lipids (phosphatidylinositol and phosphatidylserine) are also present in this preparation. We used azolectin to form our lipid bilayers because mixtures of phospholipids with various polar groups, acyl chain lengths, and degrees of saturation generally form stable planar bilayers rather than single types of phospholipids (24).

4. Usually the price of such tables is quite high. A cheaper solution is to build a table from a heavy stone plate placed on four tennis balls. This apparatus alone provides quite good vibration isolation. Alternatively, a partially inflated inner tube from a vehicle tire can be used to support the stone plate.

5. p7 sequence from genotype 1b, used in this study (strain J, accession number D90208), contains one cysteine at position

27. This cysteine was changed to alanine to facilitate peptide synthesis and to prevent unexpected disulfide-bond formation, which is difficult to maintain reduced in a hydrophobic environment. Note that Val, Ala, or Thr residues have been reported at position 27 for HCV strains of genotype 2, including JFH1.

6. To prepare a sample of any crude synthesis peptide product for electrospray mass spectroscopy analysis, weight out 0.3–0.5 mg of the product, add 3–5 μL of pure TFA, evaporate excess TFA under nitrogen flux, dissolve the product in 20 μL of pure formic acid, add 90 μL of methanol and then 100 μL of water. This solution is subsequently diluted in formic acid 1% in 50% methanol to the concentration required for direct injection on the electrospray loop. For highly insoluble material, electrospray sample containing up to 50% formic acid in 90% methanol can be successfully used.

7. Equilibration of RP-HPLC columns is a critical step for efficient peptide separation. As the equilibration of acetonitrile–water mixtures between the pores of RP-column media and the mobile phase is a quite slow diffusion process, too-rapid change of solvent concentrations must be avoided. We therefore recommend use of only increasing and decreasing gradients of linear steps. In addition, before any RP-HPLC experiment with a stored column is started, the column must be cycled carefully. So long as the column is stored in a 50:50 (v:v) acetonitrile/water mixture, the following cycling program of linear gradient steps can be used: 50% solution B at initial time, 100% B at 10 min, 100% B at 12 min, 10% B at 18 min, 10% B at 25 min, 100% B at 50 min, and 30% B at 55 min. Then equilibrate the column at 30% B for 15 min, until a stable baseline is reached. Then run a blank or a control sample with the required gradient program.

8. Fractions eluted after 38 min mainly contain p7 with few various contaminants. These fractions can be pooled and lyophilized for further purification according to the same protocol. The presence of p7 in various eluted fractions is in keeping with its properties of oligomerization.

9. The easiest way to identify the eluted peptides is to analyze the various fractions by mass spectroscopy after lyophilization and dissolution in a small volume of 1% formic acid in 50% methanol (or TFE). The recording of the chromatogram at several wavelengths (e.g., 220 and 280 nm) and the UV spectra of the eluted fractions are helpful for detecting the relevant fractions containing peptides to be analyzed.

10. As the capacity of RP-HPLC column is several tens of milligrams, a common practice for purification is to inject a large sample amount of crude product (e.g., 50–100 mg) and

then to analyze the various eluted fractions by analytical RP-HPLC to identify the fractions containing the pure peptide. This approach works well for common small peptides when the content of the required peptide in the crude synthesis product is high. In contrast, a semipreparative procedure like that reported here for p7 allows the optimization of both purity and yield of purification, especially for membrane peptides, which are often present as relatively low amounts in crude synthesis product. In addition, better peptide separation and yield are obtained when the acetonitrile concentration in the equilibration solution is close to that of peptide elution. This condition greatly decreases the column capacity but prevents the peptide aggregation at the top of the column that often occurs when the acetonitrile concentration in the equilibration solution is too low.

11. Bio-Beads are polystyrene beads used to remove detergents by adsorption (25). Bio-Beads can be stored at 4°C in distilled water. Before use, the beads are blotted with tissue paper, which removes excess moisture. The desired number of moist beads are weighed and added directly to the solution containing proteins, detergent, and lipids.

12. For titration experiments with TFE, add either water to dilute TFE (e.g., up to 200 μL in the present cuvette, i.e., 25% TFE final concentration) or TFE 100% (e.g., up to 200 mL, i.e., 75% TFE final concentration). To reach upper and lower TFE concentrations, start with 90% and 20% TFE sample solution, respectively. Note that p7 aggregates if diluted in water solution containing less than 20% TFE.

13. As the pH of the peptide solution is generally acidic (about 4.0), probably because of the presence of trace amounts of TFA, we recommend verification of the pH of the final solution (e.g., with pH paper). The presence of 2% TFE has no detectable effect on p7 folding in the presence of detergent. This possibility could easily be checked by lyophilization of the CD samples to remove all traces of TFE and addition of water up to the initial volume (200 μL) to reconstitute the sample for CD recording.

14. As a rule, any peptides dissolved in 50% TFE yield well-resolved NMR spectra that are therefore quite easy to analyze. Moreover, the assignment of resonances is often helpful for the analysis of peptide dissolved in detergents, which generally exhibit less well-resolved NMR spectra than those in TFE. Recording NMR spectra of membrane peptides dissolved in 50% TFE is therefore always interesting and, in addition, constitutes a required step for the preparation of peptide samples in detergent (see **Section 3.2**).

15. The leucine-rich region 50–57 of p7 was resolved, thanks to a synthetic peptide including residues 34 to 63 of p7.

Acknowledgments

This work was supported by the French Centre National de la Recherche Scientifique (CNRS) and Université de Lyon, by grants QLK2-CT1999-00356 and QLK2-CT2002-01329 from the European Commission, and by the French National Agency for Research on AIDS and Viral Hepatitis (ANRS). CD experiments were performed on the platform "Production et Analyse de Protéines" from the IFR 128 BioSciences Gerland Lyon-Sud.

References

1. Sakai, A., St. Claire, M., Faulk, K., Govindarajan, S., Emerson, S. U., Purcell, R. H., et al. (2003) The p7 polypeptide of hepatitis C virus is critical for infectivity and contains functionally important genotype-specific sequences. *Proc. Natl. Acad. Sci. USA* **100**, 11646–11651.

2. Lohmann, V., Körner, F., Koch, J. O., Herian, U., Theilmann, L., and Bartenschlager, R. (1999) Replication of subgenomic hepatitis C virus RNAs in a hepatoma cell line. *Science* **285**, 110–113.

3. Carrère-Kremer, S., Montpellier-Pala, C., Cocquerel, L., Wychowski, C., Penin, F., and Dubuisson, J. (2002) Subcellular localization and topology of the p7 polypeptide of hepatitis C virus. *J. Virol.* **76**, 3720–3730.

4. Griffin, S. D. C., Beales, L. P., Clarke, D. S., Worsfold, O., Evans, S. D., Jaeger, J., et al. (2003) The p7 protein of hepatitis C virus forms an ion channel that is blocked by the antiviral drug, Amantadine. *FEBS Lett.* **535**, 34–38.

5. Clarke, D., Griffin, S., Beales, L., Gelais, C. S., Burgess, S., Harris, M., et al. (2006) Evidence for the formation of a heptameric ion channel complex by the hepatitis C virus p7 protein in vitro. *J. Biol. Chem.* **281**, 37057–37068.

6. Pavlovic, D., Neville, D. C. A., Argaud, O., Blumberg, B., Dwek, R. A., Fischer, W. B., et al. (2003) The hepatitis C virus p7 protein forms an ion channel that is inhibited by long-alkyl-chain iminosugar derivatives. *Proc. Natl. Acad. Sci. USA* **100**, 6104–6108.

7. Premkumar, A., Wilson, L., Ewart, G. D., and Gage, P. W. (2004) Cation-selective ion channels formed by p7 of hepatitis C virus are blocked by hexamethylene amiloride. *FEBS Lett.* **557**, 99–103.

8. Gonzalez, M. E. and Carrasco, L. (2003) Viroporins. *FEBS Lett.* **552**, 28–34.

9. Steinmann, E., Penin, F., Kallis, S., Patel, A. H., Bartenschlager, R., and Pietschmann, T. (2007) Hepatitis C Virus p7 Protein Is Crucial for Assembly and Release of Infectious Virions. *PLoS Pathog.* **3**, e103.

10. Brünger, A. T. (1992) *Xplor, a System for Crystallography and NMR*, Yale University Press, New Haven, CT.

11. Laskowski, R. A., Rullmann, J. A. C., MacArthur, M. W., Kaptein, R., and Thornton, J. M. (1996) AQUA and PROCHECK-NMR: programs for checking the quality of protein structures solved by NMR. *J. Biomol. NMR* **8**, 477–486.

12. Humphrey, W., Dalke, A., and Schulten, K. (1996) VMD: visual molecular dynamics. *J. Mol. Graph.* **14**, 27–28.

13. Wuthrich, K. (ed.) (1986) *NMR of Proteins and Nucleic Acids.* John Wiley & Sons, New York.

14. Opella, S. J. and Marassi, F. M. (2004) Structure determination of membrane proteins by NMR spectroscopy. *Chem. Rev.* **104**, 3587–3606

15. Montal, M. and Mueller, P. (1972) Formation of bimolecular membranes from lipids monolayers and a study of their electrical properties. *Proc. Natl. Acad. Sci. USA* **69**, 3561–3566.

16. Hamill, O. P., Marty, E., Neher, E., Sakmann, B., and Sigworth, F. J (1981) Improved patch-clamp techniques for high resolution current recording from cells and cell-free membrane patches. *Pflugers Arch.* **391**, 85–100.

17. Kelly, S. M., Jess, T. J., and Price, N. C. (2005) How to study proteins by circular dichroism. *Biochim Biophys Acta.* **1751**, 119–39.

18. Chen, Y.-H., Yang, J. T., and Chau, K. H. (1974) Determination of the helix and β form of proteins in aqueous solution by circular dichroism. *Biochemistry* **13**, 3350–3359.

19. Penin, F., Geourjon, C., Montserret, R., Böckmann, A., Lesage, A., Yang, Y. S., et al. (1997) Three-dimensional structure of the

DNA binding domain of the fructose repressor from *Escherichia coli* by ^{1}H and ^{15}N NMR. *J. Mol. Biol.* **270,** 496–510.

20. Favier, A., Brutscher, B., Blackledge, M., Galinier, A., Deutscher, J., Penin, F., et al. (2002) Solution structure and dynamics of Crh, the *Bacillus subtilis* catabolite repression HPr. *J. Mol. Biol.* **317,** 131–144.

21. Vakser, I. A., Matar, O. G., and Lam, C. F. (1999) A systematic study of low-resolution recognition in protein-protein complexes. *Proc. Natl. Acad. Sci. USA* **96,** 8477–8482.

22. Phillips, J. C., Braun, R., Wang, W., Gumbart, J., Tajkhorshid, E., Villa, E., et al. (2005) Scalable molecular dynamics with NAMD. *J. Comput. Chem.* **26,** 1781–1802.

23. MacKerell, A. D., Jr., Bashford, D., Bellott, M., Dunbrack Jr., R. L., Evanseck, J. D., Field, M. J. et al. (1998) All-atom empirical potential for molecular modeling and dynamics Studies of proteins. *J. Phys. Chem. B* **102,** 3586–3616.

24. Schindler, H. (1989) Planar lipid-protein membranes: strategies of formation and of detecting dependencies of ion transport functions on membrane conditions. *Meth. Enzymol.* **171,** 225–253

25. Rigaud, J. L., Mosser, G., Lacapere, J. J, Olofsson, A., Levy, D., and Ranck, J. L. (1997) Bio-Beads: an efficient strategy for two-dimensional crystallization of membrane proteins. *J. Struct. Biol.* **118,** 226–235.

Chapter 11

HCV Replicons: Overview and Basic Protocols

Volker Lohmann

Abstract

Subgenomic replicons have been the first efficient cell-culture system for HCV and still are a valuable tool for studying different aspects of RNA replication. A variety of replicons based on different viral isolates and vector designs have been established. Here, I give a brief overview of viral isolates, applicable host-cell lines, replicon structures, and general considerations regarding replicon experiments, supplemented by basic protocols for in vitro transcription, electroporation, selection of replicon cells, transient replication assays, and northern hybridization.

Key words: HCV, replicon, Con1, JFH1, electroporation, transient replication, in vitro transcription, northern hybridization.

1. Introduction

HCV isolates taken from patients usually replicate poorly, if at all, in cell culture. Selectable subgenomic replicons first overcame this problem when hepatoma cells were generated that harbor these autonomously replicating RNA molecules persistently (1). Analysis of the viral sequences in these so called replicon cell lines revealed the presence of a variety of adaptive mutations (2–5), which are necessary for efficient RNA replication in cell culture for most HCV isolates. Although the whole viral life cycle can now be studied in cell culture on the basis of the exceptional isolate JFH1, HCV replicons are still a valuable tool for studying viral RNA replication and drug screening. Here, I provide a brief overview of different aspects of the replicon technology to aid in the choice of appropriate tools for particular experimental designs, but making no claim to completeness. The basic

Hengli Tang (ed.), *Hepatitis C: Methods and Protocols, Second Edition, vol. 510*
© 2009 Humana Press, a part of Springer Science+Business Media
DOI 10.1007/978-1-59745-394-3_11 Springerprotocols.com

protocols provided can be accommodated for all types of replicons and host cells.

1.1. Structure of HCV Replicons

The structure of a "classical" HCV bicistronic subgenomic replicon is shown in **Fig. 11.1A**, consisting of the 5′ nontranslated region (NTR), the first 16 codons of core (which are required for optimal activity of the HCV internal ribosomal entry site (IRES)), a reporter gene (rg) or a gene encoding a selection marker (sm), the IRES of the encephalomyocarditis virus directing translation of the HCV nonstructural proteins 3 to 5B, and the HCV 3′NTR *(1)*. Nonstructural proteins 3 to 5B are necessary and sufficient for autonomous RNA replication, but the portion of the HCV coding sequences can be extended by NS2 *(1)* to contain the NS2/3 protease (**Fig. 11.1B**), or even the entire HCV open reading frame can be included to yield full-length replicons (**Fig. 11.1C**) *(6, 7)*. A variety of selection markers like bsr^r, hygro^r, and zeo^r have been used in these configurations to select for replicon cell clones, among which neo^r is the most widespread and efficient. Possible options for reporter genes include firefly and renilla luciferases, β-lactamase *(8)* and green fluorescent protein (GFP). GFP can also be inserted at certain positions within the NS5A coding sequence, allowing the direct visualization of NS5A in living cells *(9)*. A combination of a reporter gene and neo^r fused by ubiquitin (**Fig. 11.1D**) allows

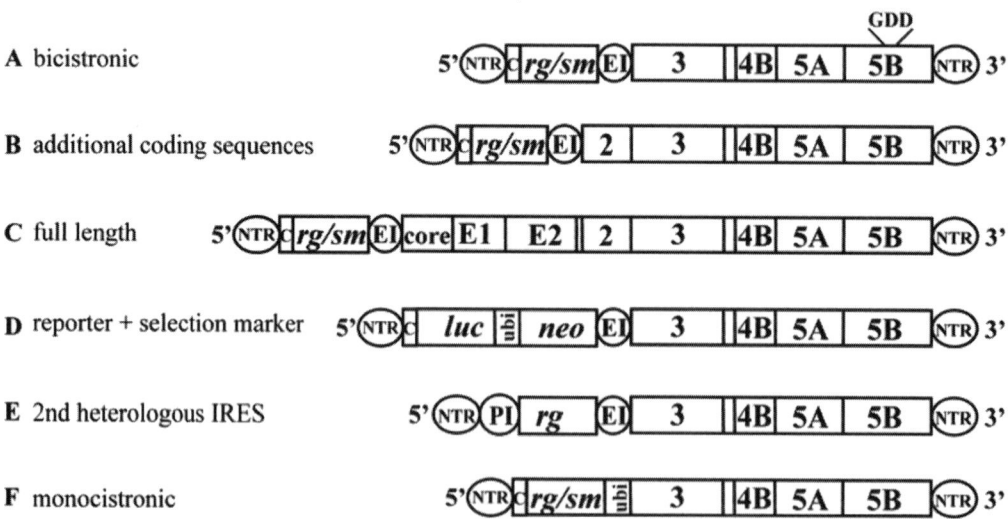

Fig. 11.1. Schematic structure of different subgenomic replicons. NTR, nontranslated region; rg, reporter gene; sm, selection marker; luc, firefly luciferase; neo, gene encoding neomycin phosphotransferase; PI, poliovirus IRES; EI, encephalomyocarditis virus IRES; ubi, ubiquitin; c, N-terminal portion of the core protein (nt 342–389) required for optimal IRES activity; GDD, amino acid motif in the active center of the polymerase; core, E1, E2, coding sequence for structural proteins, 2, 3, 4B, 5A, 5B, sequences encoding nonstructural proteins. P7 and NS4A are not labeled. Coding sequences are marked by rectangles; noncoding regions are indicated by oval forms.

the generation of replicon cell clones persistently expressing firefly luciferase upon HCV replication, providing a convenient tool particularly for study of the role of antivirals *(10)*. A variation of the bicistronic replicon contains a poliovirus IRES to control translation of the firefly luciferase (**Fig. 11.1E**). Because the HCV 5'NTR is not involved in translation, this type of replicon is particularly useful for study of the role of the 5'NTR in RNA replication *(11)*. Furthermore, the translation of the reporter gene is more efficient and RNA replication is enhanced for some unknown reasons in these replicons *(5)*, so they are a preferred option for sensitive transient replication assays (**Fig. 11.2**, see **Section 1.6**). Replicons resembling most closely the structure of viral genomes are monocistronic (**Fig. 11.1F**). In this case, reporter gene or selection marker and the HCV proteins are expressed as one polyprotein under translational control of the HCV IRES. Cleavage between marker and HCV proteins can be achieved by insertion of ubiquitin, which is cleaved C-terminally by a cellular enzyme, or by picornavirus 2A. Because ubiquitin (or 2A) remains at the C-terminus of the marker protein, neomycin phosphotransferase is not active in this configuration. Monocistronic reporter replicons harboring a firefly-luciferase gene are slightly less efficient than their bicistronic counterparts (unpublished data).

1.2. Available Viral Isolates

HCV Con1 was the first viral isolate successfully used to generate subgenomic replicons efficiently replicating in cell culture *(1)*. The genome is a specific consensus sequence cloned from the total liver RNA of an infected patient and is classified genotype 1b. The wild-type isolate barely replicates in Huh-7 cells, as compared to a negative control (**Fig. 11.2A**, Con1wt versus Con1 GND). To produce efficient replication in cell culture with Con1-based replicons, cell-culture-adaptive mutations must be included, which were originally identified in selected replicon cell clones. Commonly used adaptive mutations conferring efficient replication in cell culture are S2204I *(2)* and the combinations E1202G, T1280I, S2197P, designated NK5.1 *(3)*, and E1202G, T1280I, K1846T, designated "ET" *(5)*. Efficient adaptive mutations increase replication levels several orders of magnitude (**Fig. 11.2A**). The mechanisms underlying cell-culture adaptation are still widely enigmatic, but a reduction of hyperphosphorylation of NS5A is correlated to some extent with the adaptive phenotype of NS4B and NS5A mutations *(12, 13)*. The mode of cell-culture adaptation seems to be conserved within HCV genotype 1, however, because the mutations identified in Con1 also confer the adaptive phenotype on other HCV isolates, and similar mutations were found in replicon cell clones of other genotype 1 isolates. At least four other genotype 1b isolates were successfully used to generate subgenomic replicons of genotype

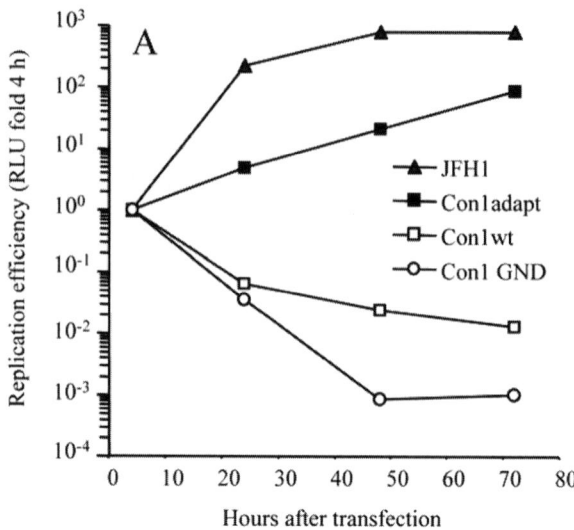

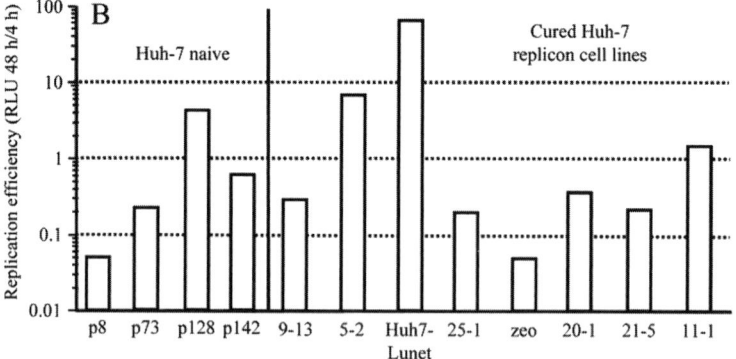

Fig. 11.2. Impact of viral isolate and host cells on HCV replication efficiency. **A**. Transient replication efficiency of Con1wt, adapted Con1, and JFH1 compared to a replication-deficient control. Reporter replicons as shown in Fig. 1E, harboring firefly-luciferase genes, were electroporated into Huh-7-Lunet cells. Con1adapt contained adaptive mutations E1202G, T1280I, and K1846T. Luciferase activity was determined in cell lysates prepared at 4, 24, 48, and 72 h after transfection. **B**. Relative replication efficiency of replicon Con1-ET in different passages of naïve Huh-7 cells and in various cured replicon cell clones. Luciferase activity was measured at 4 and 48 h after electroporation. Data were normalized for transfection efficiency among identical replicons as determined by measurement of the luciferase activity 4 h after transfection. Note the logarithmic scale of the ordinate.

1b: HCV-N *(7)*, HCV-BK *(14)*, 1B-1 *(15)*, and 1B-2R1 *(16)*. The initial replication capacity of replicons derived from HCV-N was highest among the genotype 1b isolates because of a unique four-amino-acid insertion in NS5A (SSYN at position 2219), which was shown to be required for efficient replication in cell culture *(7)*, mimicking cell-culture adaptation.

Only one isolate of genotype 1a, designated H77, has been established to replicate in cell culture *(17)*. Its wild-type isolate is not replication competent at all but requires two mutations for detectable replication levels (P1496L and S2204I) and three

additional alterations for highly efficient replication, allowing
transient reporter assays *(18)*.

The most outstanding HCV isolate analyzed so far is
termed JFH1 and grouped with genotype 2a *(19)*. JFH1
replicates to extremely high levels without requirement for
adaptive mutations (**Figs. 11.2A,   11.3**). JFH1 reaches far
higher peak replication levels than Con1-ET at only 24–
48 h after transfection; at later time points, replication seems
to be limited by the host cell, and slight cytopathic effects
are even observed. JFH1 is currently the only isolate giv-
ing rise to efficient production of virus particles (*(20)*. see
Chapters 23, 24 for related protocols).

Isolates from genotypes other than 1 and 2a that replicate efficiently in cell culture have not yet been identified. The choice of
isolate will very much depend on the experimental requirements.
Selected replicon clones will provide very similar persistent replication levels, irrespective of the isolate. For transient replication
assays, JFH1 provides by far the most robust and reliable replication capability, but because of its peculiar properties, it may
or may not properly reflect HCV replication in general. Therefore validation might be required for basic studies performed with
JFH1 with a genotype 1 isolate, especially whenever host-cell factors are implied. Because of their medical importance, genotype 1
isolates should be the first choice for the development of antiviral drugs, because compounds can show very different potency
on different genotypes. Judging which genotype 1 isolate is
most efficient is difficult since no careful side-by-side comparison
with replicons established in different laboratories has yet been

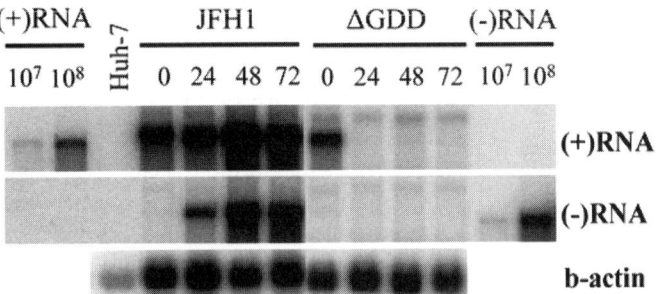

Fig. 11.3. Analysis of transient replication of a JFH1-based replicon compared to a
replication-deficient control by northern hybridization. Huh-7 cells were transfected by
electroporation and analyzed for positive- and negative-strand RNA synthesis 0–72 h
after transfection as described in **Sections 3.3 and 3.5**. β-actin RNA levels were used
to normalize for loading in each lane. Ten μg of total RNA was analyzed by northern hybridization with a [^{32}P]-labeled negative- or positive-sense riboprobe to reveal
HCV positive-strand RNA ((+)RNA) or negative-strand RNA ((–)RNA), respectively, both
encompassing the NS5A to 3′NTR region (nt 6273–9678), or with a negative-sense
riboprobe specific for β-actin as given on the right. 10^7 or 10^8 in vitro transcripts of
positive-sense ((+)RNA)) or negative-sense ((–)RNA) orientation were loaded for quantification of HCV RNA, as given above the two outer left and right lanes, respectively.

performed. The major drawback of genotype 1 replicons lies in the enigmatic role of adaptive mutations, especially important for reverse genetic studies on HCV proteins harboring adaptive mutations (e.g., NS5A). The most widely used isolates for reverse genetic studies have been Con1 and HCV-N. Because of its high replication levels and the possibility of studying the whole viral life cycle, the importance of JFH1 is increasing.

1.3. Importance of the Host Cell

An important determinant of efficient HCV replication is the host cell. Huh-7 cells support HCV replication most efficiently and are the cell line of choice for all kinds of assays, but transient HCV replication assays revealed more than 100-fold differences in permissiveness between passages of naive Huh-7 cells generated from the same stock. Highly permissive naive cells are found preferentially among higher passages, but the phenotype is not stable (**Fig. 11.2B**, Huh-7 naive (5)). More permissive host-cell lines were generated by treatment of replicon cell clones with interferons or selective drugs (8, 21), among which so called Huh-7.5 are the most widely used (21). Only some "cured" Huh-7 clones are more permissive for HCV replication than naive cells (**Fig. 11.2B**). In our hands, Huh-7-Lunet is by far the most efficient cured Huh-7 clone for HCV RNA replication, but because of low surface expression levels, cd81 must be ectopically expressed in Huh-7-Lunet to yield efficient HCV infection (22).

Comparing the permissivenesses of naive Huh-7 cells and cured replicon clones used in different laboratories is very difficult, because the data rely on various assays, but the host cells, as well as the usage of appropriate adaptive mutations, are both critical determinants of the establishment of replicon cell clones and transient replication assays. In contrast, once a replicon cell clone is established, the levels of persistent replication in different cell clones, replicons carrying various isolates (23), and adaptive mutations are rather similar and seem to be regulated by a complex interplay between the host cell and the replicon.

Besides Huh-7 cells, Con1 replicon cell clones have been generated with very low efficiency in HeLa cells, mouse hepatoma cells (24), HEK 293 cells (25), and HuH-6 cells (26). The impact of Huh-7 permissiveness on JFH1 replication is far less pronounced than that on genotype 1 replicons. JFH1 also reaches easily detectable replication levels in poorly permissive Huh-7 cells, albeit with delayed kinetics (M. Binder, V. Lohmann, unpublished data). In addition to Huh-7, JFH1 has been shown to replicate in HuH-6 cells (26), HepG2 cells, IMY-N9 cells (27), HEK293 cells, HeLa cells (28), mouse embryonic fibroblasts (29), and mouse hepatocyte and fibroblast cell lines (30).

The host factors limiting HCV RNA replication to particular cell lines are not known, nor are the determinants for Huh-7 permissiveness. Recent results indicate that the RIG-I/IRF-3 pathway is not significantly involved in permissiveness *(31)* as has been previously suggested for Huh 7.5 cells.

1.4. Dependence of HCV Replication on Host-Cell Growth

At least in Huh-7 cells, levels of HCV replication vary drastically with the state of the cell, being highest in actively proliferating cells *(32)*. In resting cells, that is, cell monolayers that are confluent and left in a confluent state for several days, HCV RNA-amounts are significantly lower. This important observation must be considered in several experimental settings: (i) Substances and experimental manipulations that interfere with cell growth will reduce the amount of replicon RNA. This reduction does not necessarily indicate a specific inhibition of HCV RNA replication but can simply be due to interference with host-cell metabolism. Therefore, cytostatic and cytotoxic effects must be monitored in every experimental setup. (ii) Whenever a cell population becomes overconfluent in the course of any experiment, HCV replication will apparently decrease.

The link between cell growth and HCV replication is probably due to intracellular nucleotide concentrations *(33)* and is not observed in HuH6-cells *(26)*.

1.5. Replicon Cell Lines

Selected replicon cell clones show high, persistent HCV replication in almost all cells of a given population, which is maintained by constant selection with G418. Replication can be monitored by many robust assays, like northern blot (**Section 3.5**) or quantitative RT-PCR for RNA detection and western blot or immunofluorescence for antigen detection. Particular cell lines based on replicons containing combinations of a luciferase reporter gene and npt have been established and allow monitoring of persistent HCV replication by firefly-luciferase activity (**Fig. 11.1D**).

Replicon cell clones are the optimal tool whenever a constantly high level of replication is required without the need for manipulation of the HCV sequence. Therefore the major topics addressed with selected replicon cell clones are screening and evaluation of antiviral drugs, the effect of interferon on HCV replication, biochemical analyses of the viral replicase complex, ultrastructural analyses, and virus host-cell interactions. Selected replicon cell clones are available for all cell types and isolates described above. The choice of appropriate negative controls will depend strongly on the experiment and can range from naïve cells to cells harboring a selection marker identical to that of the replicon cells. In some cases, the most appropriate control will be a subset of the replicon cells cured by interferon or HCV-specific drugs. Particular protocols addressing some of the above matters can be found in **Chapters 11, 12, 13, 14, 15, 16, 17, and 18**. Protocols for the establishment of replicon cell clones are provided in **Sections 3.2 and 3.3** of this chapter.

1.6. Transient Replication Assays

Transient replication assays have been established for quantitative analysis of the impact of changes in the viral sequence on RNA replication efficiency. The critical determinants for all transient replication assays are the viral isolate, the permissiveness of the Huh-7 cells, and the method used to quantify replication. As a control for all types of transient replication assays, a replication-deficient RNA of the same composition as that of the replicon to be analyzed must be transfected in parallel to define the background of the assay. Most widely used are point mutants or small deletions in the GDD motif of NS5B, which is essential for RNA synthesis (e.g., GND instead of GDD, **Fig. 11.1A**). True HCV replication is defined to reach significantly higher signals in a replication assay than the appropriate negative control.

The simplest "transient" replication assay uses selectable replicons, and the efficiency of colony formation after transfection and selection is quantified as the number of selected colonies per microgram of transfected RNA (**Section 3.2 and 3.3**). This kind of assay has the advantage of being relatively sensitive and robust. The main disadvantages are the time required for selection of colonies, typically 3–4 weeks, and the quantification by colony counting, which is time consuming and error prone.

Another type of transient replication assay is based on the detection of HCV RNA soon after transfection by northern blot (**Section 3.5**), which eliminates the need for any foreign sequences like selectable markers or reporter genes and allows even analyses of in vitro transcribed full-length genomes. The drawbacks of northern blot assays are limited sensitivity and the hands-on time required for sample preparation and analysis.

The easiest and fastest approach to measuring HCV replication soon after transfection is based on replicons with reporter genes instead of selectable markers. A variety of replicons with different reporters have been established; a detailed description of them goes beyond the scope of this report. Examples for transient assays using reporter replicons harboring a firefly-luciferase gene are shown in **Fig. 11.2**; the experimental setup is described in protocols 3.1, 3.2, and 3.4.

Transient HCV replication assays have been involved in several studies, including the analysis of adaptive mutations, the characterization of different HCV isolates, the role of the HCV NTRs, the structure-function analysis of HCV NS5B and NS5A, and the permissiveness of Huh-7 cells. In contrast to replicon cell clones, which are extremely robust and show high levels of persistent replication, transient assays are in general technically more difficult and require careful optimization, because the total readout depends on many determinants, among which transfection efficiency, HCV isolate and adaptive mutations used, permissiveness of the Huh-7 cells, and type of transient assay are the most important.

2. Materials

2.1. Cell Culture

1. DMEM complete: Dulbecco's modified Eagle medium (with 4.5 g/l glucose, without pyruvate) supplemented with 2 mM L-glutamine, penicillin (100 IU/mL)/streptomycin (100 µg/mL), 1× nonessential amino acids (all from Invitrogen, Karlsruhe, Germany), and 10% fetal calf serum (PAA, Pasching, Germany).
2. Solution of trypsin 0.05%-EDTA (Invitrogen).
3. Phosphate-buffered saline (PBS): 8 mM Na_2HPO_4, 2 mM NaH_2PO_4, 140 mM NaCl, sterilize by autoclaving.
4. G418 ("Geneticin," Invitrogen): concentrations are given as weight per volume of the original substance. Specific activity of a typical batch is ca. 700 µg/mg as stated by the manufacturer. This value does not necessarily reflect the biological activity in our system. Therefore each new batch of G418 should be tested individually, for example, in an electroporation experiment using different selection conditions (0.25–1 mg/mL).

2.2. In Vitro Transcription, Electroporation, and Northern Hybridization

1. High-yield transcription buffer (5×): 400 mM HEPES, pH 7.5, 60 mM $MgCl_2$, 10 mM spermidine, 200 mM DTT; all chemicals should be of the highest available quality.
2. NTP solution: 100 mM solutions of ATP, CTP, GTP, and UTP (all Roche, Mannheim, Germany) mixed in equal volumes to yield a mix containing 25 mM for each nucleotide.
3. RNasin (40 u/µL) and T7 RNA Polymerase (20 u/µL) (Promega, Mannheim, Germany).
4. Cytomix (34): 120 mM KCl, 0.15 mM $CaCl_2$, 10 mM K_2HPO_4/KH_2PO_4, pH 7.6, 25 mM HEPES (1 M stock solution, cell-culture grade, Invitrogen), 2 mM EGTA, 5 mM $MgCl_2$. Adjust pH to 7.6 with KOH, sterilize by filtration, and store at room temperature. Add 2 mM ATP (100 mM stock) and 5 mM reduced glutathione (250 mM stock) just before use. pH of ATP and glutathione stock solutions must be adjusted to 7.6 with KOH; stocks should be sterilized by filtration and stored in aliquots at −20°C.
5. 100 mM sodium phosphate, pH 7.0: 5.77 ml 1 M Na_2HPO_4, 4.23 ml 1 M NaH_2PO_4, water to 100 ml.
6. Solution of glyoxal, 40% (Sigma, Taufkirchen, Germany), deionized by mixing with an ion-exchange resin (BioRad AG501-X8, BioRad, Munich, Germany) until the pH is > 5. Store in aliquots at −20°C.
7. Glyoxal-loading dye: 0.25 mg/mL bromophenol blue, 0.25 mg/mL xylene cyanol, 10 mM sodium phosphate, pH 7, 50% glycerol.

8. 20× SSC: 3 M NaCl, 0.3 M Na$_3$ citrate; adjust pH to 7.0 with HCl.
9. 100× Denhardt solution: 10 g Ficoll 400, 10 g polyvinylpyrrolidone, 10 g bovine serum albumin, fraction V; add water to 500 mL, sterilize by filtration, and store at −20°C.
10. Formamide hybridization solution: 5× SSC, 5× Denhardt solution, 50% (w:v) formamide, 1% SDS.

3. Methods

3.1. In Vitro Transcription

1. Preparation of DNA for in vitro transcription (*see* **Note 1**): To remove RNases, extract DNA twice with equal volumes of phenol/TE and once with chloroform, ethanol-precipitate, wash with 70% ethanol and dissolve in 20–50 μL of RNase-free water before in vitro transcription.
2. For 100 μL in vitro transcription reaction, add 20 μL 5× high-yield transcription buffer, 12.5 μL NTP solution, 100 u RNasin, 5 μg purified replicon plasmid DNA, Rnase-free water to 100 μL total volume, and 80 u T7-RNA-Polymerase.
3. Incubate the reaction for 2 h at 37°C, add an additional 40 u T7-RNA-Polymerase, and incubate it for an additional 2 h at 37°C.
4. Add 10 u DNase (RNase free, Promega) and incubate for 30 min at 37°C.
5. Add 60 μL 2 M sodium acetate (pH 4.5), 440 μL water, and 400 μL water-saturated phenol, pH < 5; vortex vigorously, incubate for 10 min on ice, and centrifuge 10 min at 4°C at 18,000g.
6. Transfer 400–500 μL of the upper phase into a fresh tube, extract it with one volume of chloroform, and spin it for 3 min at room temperature.
7. Transfer the upper phase into a fresh tube, precipitate RNA by adding 0.7 volume of isopropanol equilibrated to room temperature, and centrifuge 15 min at room temperature at maximum speed.
8. Wash the pellet once with 70% ethanol, quantitatively remove any liquid, air dry very briefly, and dissolve the pellet in 50–100μL water at least 5 min at 37°C with shaking. Transcript yield is typically 0.3–1 μg RNA/μL transcription reaction (*see* **Note 2**).

3.2. Electroporation of Huh-7 Cells

1. Seed 15 cm dishes with the appropriate numbers of cells 3–4 days before the electroporation experiment. Cells should be confluent at the time of harvest. A 15 cm dish of Huh-7 cells typically contains 1–2 × 10^7 Huh-7 cells and is sufficient for

2–5 electroporations. Carry out all steps at room temperature unless otherwise instructed.

2. Prepare 0.001–10 μg of in vitro transcribed replicon RNA, depending on the type of experiment and replicon, in a sterile 1.5 ml reaction tube for each electroporation and store the tubes on ice (*see* **Sections 3.3 and 3.4** and **Note 3**).

3. Remove the cell-culture medium from the 15 cm dishes, wash the cells once with 10 mL PBS, add 5 mL of trypsin to each plate, incubate the plates briefly at room temperature, remove the trypsin, and incubate the plates a further for 5 min at 37°C, until all the cells detach from the plate. Harvest cells by rinsing each plate with 10 mL DMEM complete medium. Pass the cells several times through a serological pipette with a sterile 1 mL blue tip attached to obtain a suspension of single cells.

4. Remove a small aliquot for counting and centrifuge the remainder 5 min at 700 rpm. Wash once with ∼10 mL PBS per 15 cm dish; centrifuge 5 min at 700 rpm.

5. Resuspend the cells to 1×10^7/ml in Cytomix by pipetting up and down several times with a Gilson P1000 pipette. If serological pipettes are used, attach a sterile 1 mL blue tip to ensure that the cells are efficiently separated.

6. Transfer 400 μL of the cell suspension into the reaction tube containing the nucleic acid and mix by gently pipetting up and down 5–6 times. Transfer the mixture immediately into the electroporation cuvette with 0.4 cm gap width (BioRad, Munich, Germany) and pulse once with a Gene pulser II (BioRad; settings: 975 μF; 270 V; expected time constant ∼20 ms; *see* **Note 4**).

7. Transfer cells immediately into 7–24 mL DMEM complete medium with a pasteur pipette and rinse the cuvette with medium to remove all cells (*see* **Note 5**). Seed the cells according to the experimental requirements (**Sections 3.3, 3.4, and 3.5**).

3.3. Selection for G418-Resistant Colonies

This protocol can be used to generate cell lines harboring persistent replicons for further analysis (**Section 1.5**) or to count the number of colonies raised from different replicons when this number is taken as a semiquantitative measure of HCV replication capability (**Section 1.6**).

1. Transfect Huh-7 cells by electroporation (*see* 3.2), using 1–500 ng of in vitro transcripts (*see* **Note 6**); add carrier RNA to 10 μg (Huh-7 total RNA or tRNA) to prevent degradation of low amounts of transcripts and to keep experimental conditions constant. Suspend the cells to 24 mL of DMEM complete medium, and seed three 10 cm dishes with 8 mL each. Include a replication-deficient RNA harboring an npt

gene and electroporate the same amounts of in vitro transcripts in parallel.

2. Change the medium of the 10 cm dishes to 8 mL DMEM complete containing 0.5 mg/mL G418 24 h after electroporation (*see* **Note 7**).

3. Change the medium once a week and wait for colonies ~2–3 weeks, until the cells harboring the negative control RNA are detached from the plate. During the first week the medium might have to be changed twice, if it turns yellow (this point depends on the amount of in vitro transcripts used).

4. To generate cell lines, harvest colonies about 4–5 weeks after transfection by removing the medium, adding 20 μL of trypsin solution to the colonies of choice, incubating briefly at room temperature, detaching the cells by gentle pipetting, and transferring them into 24-well dishes with 1 mL of selection medium per well. Further expand cells whenever the dish is confluent. If a large number of colonies appear and a clonal cell line is not necessary or desired, all colonies from a plate can be passaged at once to produce a more variable cell pool. Check cell lines for the presence of HCV positive- and negative-strand RNA by northern blot (*see* **Section 3.5** and **Note 8**).

5. For a quantitative assay, remove the medium, wash the colonies once with PBS, and fix and stain them for 5 min with coomassie solution (250 mg/L coomassie R-250, 50% methanol, 10% acetic acid). Rinse the colonies intensively with water and count them. Data from several independent electroporations, in which different amounts of in vitro transcripts were used, must be combined to produce conclusive results. Data can be expressed as number of colony forming units per microgram in vitro transcribed RNA (cfu/μg).

6. In case colonies appear after transfection of replication-deficient RNAs, G418 resistance by integration of contaminating plasmid DNA is the source, and more care must be taken to remove the template DNA after transcription (*see* **Note 2**).

3.4. Transient Replication Assays Using Luciferase Reporter Replicons

Besides the intrinsic replication capacity of the replicons involved, the most important parameters for a successful transient assay with reporter replicons are the permissiveness of the host cells and the choice of the reporter construct (*see* **Sections 1.1, 1.2, 1.3 and 1.6**). This protocol is based on reporter replicons harboring a firefly-luciferase gene (rg in **Fig. 11.1A, B, E and F**) and Con1 sequences with adaptive mutations or JFH1 (**Fig. 11.2**).

1. Transfect Huh-7 cells by electroporation (*see* above), using 1–5 μg of in vitro transcripts. Transfer the cells into 12 mL of DMEM complete medium and seed individual wells of six-well plates with 1 mL aliquots of the cell suspension and 1 mL

of DMEM complete (*see* **Note 9**). Duplicate electroporations for each transcript are recommended to produce a conclusive set of data. Electroporate in parallel identical amounts of replication-deficient HCV RNA harboring a firefly-luciferase gene (e.g., a GND mutant, *see* **Section 1.6**).

2. Harvest cells at 4, 24, 48, and 72 h after transfection. Wash them once with 4 mL of PBS and lyse them by adding 500 μL of luciferase-lysis buffer (*see* Chapter 26: Full-length infectious HCV chimeras, 2.4). Either scrape the cells, transfer the lysate to a 1.5 mL reaction tube, and store it at −70°C or freeze the whole six-well plate at −70°C.

3. After cells have been harvested at all time points, thaw the cell lysates on ice and measure luciferase activity in 100 μL of lysate (detailed protocol in Chapter 26: Full-length infectious HCV chimeras, 3.4). Each lysate should be measured twice.

4. Reporter-gene activity 4 h after transfection reflects mainly translation of the input RNA and should be used to normalize for transfection efficiency. This normalization can most easily be accomplished if relative light units (RLU) obtained 4 h after transfection are set to 1 or 100% and all other time points are expressed relative to these data (*see* **Fig. 11.2** and **Note 10**).

3.5. Northern Hybridization

Northern hybridization is the most accurate method for detecting HCV full-length RNA in persistent replicon cells or after electroporation. It furthermore allows the specific detection of positive- and negative-strand RNA by the choice of strand-specific probes *(35)*.

1. For preparation of total RNA from persistent replicon cells, harvest one confluent 6 cm or 10 cm culture dish per sample (corresponding to $10^6 - 5 \times 10^6$ cells). For detection of HCV RNA after electroporation, use 5–10μg of in vitro transcribed replicon RNA per electroporation and combine several electroporations for each replicon, depending on the number of time points that must be analyzed. Seed cells from one electroporation on a 10 cm dish per time point of harvest within the first 24 h after transfection and half the number of cells on the same area for later time points up to 72 h after electroporation. Harvest one aliquot of cells, corresponding to one electroporation, immediately after electroporation by centrifugation at 700 rpm to determine input RNA levels. Total RNA of sufficient quality can be obtained with commercially available systems or by means of standard single-step RNA-isolation methods with guanidine.

2. Prepare sufficient amounts of 1% agarose in 10 mM sodium phosphate, pH 7, and pour a gel in an appropriate device for horizontal electrophoresis. The gel should be 0.5–1 cm thick and each well large enough to hold 25 μL of sample.

Submerge the gel in 10 mM sodium phosphate, pH 7, place it on magnetic stirrers, and put a magnetic stir bar in each buffer tank.

3. Adjust 5–20 μg of total RNA to 10 μL; add 4.1 μL of 100 mM sodium phosphate, pH 7, 6 μL of 40% glyoxal, and 20.5 μL of dimethyl sulfoxide; denature RNA 1 h at 50°C; snap cool on ice and spin briefly. Set up samples with defined amounts of in vitro transcribed HCV RNA (10^7 to 10^9 copies per load) mixed with 2–5 μg Huh-7 total RNA and denature these samples in parallel to obtain positive controls for quantification.

4. Add 10.9 μL glyoxal-loading dye to each sample and load 25 μL onto the agarose gel. Run the gel at 4 V/cm until the bromophenol blue dye has migrated two-thirds the length of the gel. Start magnetic stirrers after the samples have entered the gel (ca. 15 min) to allow efficient recirculation of the buffer during electrophoresis.

5. Cut a positively charged nylon membrane (e.g., Hybond N+, GE Healthcare, Braunschweig, Germany) slightly larger than the gel, wet the membrane in 10 mM sodium phosphate, pH 7, place it on the rubber mask of a vacuum blotting device (von Keutz, Reiskirchen, Germany), cover the membrane with 10 mM sodium phosphate, pH 7, and carefully put the gel on the membrane, avoiding formation of air bubbles.

6. Cover the gel with 50 mM sodium hydroxide and transfer the RNA to the membrane for 1 h at 0.3 bar negative pressure. Be sure that the surface of the gel does not run dry during transfer.

7. Disassemble the blot, rinse the membrane briefly in 20 mM Tris-HCl, pH 8, remove liquid, and crosslink RNA with UV light (twice auto-crosslink in a "Strata-linker," Stratagene, Amsterdam, The Netherlands).

8. Stain the membrane for 1 min in 0.03% (w:v) methylene blue in 0.3 M sodium acetate, pH 5.2; destain by rinsing with water until bands of the 28S and 18S RNA become clearly visible. Cut the membrane ca. 1 cm below the 28S RNA. Use the upper part for the detection of HCV RNA and the lower part for probing against β-actin mRNA.

9. Prehybridize both pieces of the membrane in separate bottles, each in 10 mL formamide hybridization solution with 100 μg/mL denatured salmon sperm DNA for 15 min to several hours at 68°C in a hybridization oven (GFL, Burgwedel, Germany).

10. Probe synthesis: Cut the plasmid containing the desired probe sequence under control of a T7 or T3 promoter with a restriction enzyme appropriate to ensure runoff-transcription of probes with desired length and orientation. We have successfully used probes from 0.7 to 6 kb in length. Purify DNA

by phenol/chloroform extraction and ethanol precipitation; dissolve it in water to $0.5\,\mu g/\mu L$. Mix $0.5\,\mu g$ DNA, $4\,\mu L$ transcription buffer (Promega), $10\,mM$ DTT, $10\,\mu M$ CTP, $0.5\,mM$ each of ATP, GTP, UTP, $20\,u$ RNasin (Promega), $50\,\mu Ci$ of $[\alpha^{32}P]$ CTP (GE-Healthcare, $3000\,Ci/mmol$), and $20\,u$ T7 or T3 RNA polymerase (Promega) in a total volume of $20\mu L$ and incubate for 1 h at $37°C$. Add $1\,u$ DNase (Promega) and incubate for 15 min at $37°C$. Equilibrate a NAP-5 column (GE-healthcare) 3 times with $3\,mL$ water, add the probe, add $480\mu L$ water, and discard flow through. Elute probe with $1\,mL$ water.

11. Discard prehybridization solution, add $10\,mL$ of new for-mamide hybridization solution to each piece of the blot, an HCV-specific probe to the upper part of the blot, and a β-actin specific probe to the lower part of the blot. Hybridize overnight at $68°C$.

12. Discard the hybridization solution and wash the blots twice with $10\,mL$ of $2\times$ SSC/0.1% SDS and twice with $10\,mL$ $0.2\times$ SSC/0.1% SDS; carry out all washes at $68°C$ for 15 min. Remove the liquid, wrap the blots in Saran Wrap, and subject them to autoradiography (e.g., with Biomax MR film, Kodak, Stuttgart, Germany). Generally, $10^7 - 10^{10}$ HCV RNA molecules can be detected with this method; an example is shown in **Fig. 11.3**.

4. Notes

1. Depending on the way the $3'$ end of the HCV genome is generated, the plasmid must be digested with appropriate restriction enzymes. *Sca*I is used for most Con1 repli-cons, making use of a reported isolate variation in the X-region *(36)*. Because *Sca*I cuts supercoiled plasmid DNA rather poorly, prelinearization with an appropriate restric-tion enzyme is recommended (e.g., *Ase*I, cutting the amp^r gene in the plasmid backbone). *Xba*I is used for many JFH1 replicons as well as for replicons and genomes from other isolates. In the case of *Xba*I digests, the DNA must be treated with an exonuclease (e.g., mung bean nuclease) to generate runoff transcripts with an authentic HCV $3'$ end; otherwise three extra bases will be added to the transcript. These extra sequences do not to interfere with RNA replication, how-ever. If the $3'$ end of the replicon is fused to a ribozyme of the hepatitis delta virus, followed by a T7-terminator sequence, linearization of the plasmid is not required *(14,26)*. Clean-up of DNA before in vitro transcription is also possible with commercially available systems (e.g., Nucleospin extract, Macherey-Nagel, Düren, Germany).

2. Isopropanol precipitation at room temperature removes unincorporated nucleotides. RNA can be quantified by OD measurement. The transcripts must be checked by agarose-gel electrophoresis and should contain a single band with only minimal smear. DNase digestion and extraction with water-saturated phenol will help to remove most of the plasmid DNA. Residual plasmids will not interfere with most experimental procedures, but traces of DNA will be detectable by PCR and might lead to the selection of drug-resistant colonies because of recombination with the cellular genome and expression of the selection marker protein. Therefore, if PCR-based detection techniques are intended or drug-resistant colonies appear after transfection of negative control RNA (**Section 3.3**), transcripts must be further purified, for example, by repetitive cycles of the above DNase-digestion/purification protocol. An alternative to phenol extraction for efficient clean-up of the in vitro transcripts is use of commercially available spin-column-based systems for the purification of total RNA, used according to the manufacturer's protocol (e.g., RNeasy, Qiagen).

3. If less than $1\,\mu g$ of in vitro transcripts are used, $10\,\mu g$ of carrier RNA (Huh-7 total RNA or tRNA) can be added to protect the transcripts from degradation.

4. These settings and conditions are also a good starting point for other cell lines.

5. To reuse the electroporation cuvettes, collect them in 70% ethanol, clean them intensively with water to remove all debris, rinse them with 70% ethanol, and let them dry under a sterile laminar-flow bench. Cuvettes can be used 20–30 times without negative effects on the results.

6. The amount of in vitro transcript used in an electroporation experiment is critical to the number of colonies that will be obtained. Because the neomycin-phosphotransferase seems to be very stable, cells gain resistance to G418 for several days to weeks after transfection from the transfected RNA coding for this enzyme. This protective effect of the transfected RNA is visible if more than $100\,ng$ of in vitro transcribed replicon RNA is used in an electroporation experiment. Above $500\,ng$ RNA, cells get very confluent under selection with the given protocol even in the absence of RNA replication, and this might be a problem for colony formation, because HCV only replicates well in actively dividing cells (see above). In general, therefore, the amount of RNA used should be as low as possible. The Con1 wild-type replicon yields about 20–40 colonies per microgram, so $100–500\,ng$ of RNA should be used for an electroporation, whereas for the adapted replicons 1–10 ng are sufficient.

7. A critical parameter for colony formation is the G418 concentration. The lower the G418 concentration the higher the number of colonies obtained, because a cell is more likely at low G418 concentrations to reach the level of HCV replication necessary for resistance before it is killed by the drug, but low G418 concentrations increase the risk of overconfluence, which leads to formation of fewer colonies. 0.1–1 mg/mL G418 can be used for selection. 0.5 mg/mL G418 is a good starting point. Once a cell line is established from a colony, the G418 concentration is less critical but may influence the replication level of the replicon.

8. Further convincing evidence for true HCV replication is provided by the decrease in HCV RNA levels upon treatment with interferons or specific drugs abrogating HCV replication. The presence of short HCV RNA fragments detected by RT-PCR is not necessarily due to HCV replication but might also be caused by intracellular transcription from integrated plasmid DNA contaminating the in vitro transcripts (*see also* **Note 2**).

9. The appropriate number of cells will depend on the general electroporation conditions and on the growth rate of the Huh-7 cells and must therefore be optimized. Cells should not become overconfluent within the time scale of the experiment (typically 4–72 h after transfection), but low cell numbers will result in insufficient growth.

10. Different luciferase assays and luminometers will vary widely in absolute numbers of RLU. Typical yields in our system are in the range of 10^5 and 10^6 RLU per 100 µL of cell lysate at 4 h after electroporation; the assay background is less than 10^3 RLU.

Acknowledgments

This work was funded within the National Genome Research Network, NGFN, by the German Ministry for Research and Education.

References

1. Lohmann, V., Körner, F., Koch, J. O., Herian, U., Theilmann, L., and Bartenschlager, R. (1999) Replication of subgenomic hepatitis C virus RNAs in a hepatoma cell line. *Science* **285,** 110–113.
2. Blight, K. J., Kolykhalov, A. A., and Rice, C. M. (2000) Efficient initiation of HCV RNA replication in cell culture. *Science* **290,** 1972–1974.
3. Krieger, N., Lohmann, V., and Bartenschlager, R. (2001) Enhancement of hepatitis C virus RNA replication by cell culture—adaptive mutations. *J. Virol.* **75,** 4614–4624.
4. Lohmann, V., Körner, F., Dobierzewska, A., and Bartenschlager, R. (2001) Mutations in hepatitis C virus RNAs conferring cell culture adaptation. *J. Virol.* **75,** 1437–1449.

5. Lohmann, V., Hoffmann, S., Herian, U., Penin, F., and Bartenschlager, R. (2003) Viral and cellular determinants of hepatitis C virus RNA replication in cell culture. *J. Virol.* **77,** 3007–3019.

6. Pietschmann, T., Lohmann, V., Kaul, A., Krieger, N., Rinck, G., Rutter, G., et al. (2002) Persistent and transient replication of full-length hepatitis C virus genomes in cell culture. *J. Virol.* **76,** 4008–4021.

7. Ikeda, M., Yi, M., Li, K., and Lemon, S. M. (2002) Selectable subgenomic and genome-length dicistronic RNAs derived from an infectious molecular clone of the HCV-N strain of hepatitis C virus replicate efficiently in cultured Huh7 cells. *J. Virol.* **76,** 2997–3006.

8. Murray, E. M., Grobler, J. A., Markel, E. J., Pagnoni, M. F., Paonessa, G., Simon, A. J., et al. (2003) Persistent replication of hepatitis C virus replicons expressing the β-lactamase reporter in subpopulations of highly permissive Huh7 cells. *J. Virol.* **77,** 2928–2935.

9. Moradpour, D., Evans, M. J., Gosert, R., Yuan, Z., Blum, H. E., Goff, S. P., et al. (2004) Insertion of green fluorescent protein into nonstructural protein 5A allows direct visualization of functional hepatitis C virus replication complexes. *J. Virol.* **78,** 7400–7409.

10. Frese, M., Schwärzle, V., Barth, K., Krieger, N., Lohmann, V., Mihm, S., et al. (2002) Interferon-γ inhibits replication of subgenomic and genomic hepatitis C virus RNAs. *Hepatology* **35,** 694–703.

11. Friebe, P., Lohmann, V., Krieger, N., and Bartenschlager, R. (2001) Sequences in the 5′ nontranslated region of hepatitis C virus required for RNA replication. *J. Virol.* **75,** 12047–12057.

12. Appel, N., Pietschmann, T., and Bartenschlager, R. (2005) Mutational analysis of hepatitis C virus nonstructural protein 5A: potential role of differential phosphorylation in RNA replication and identification of a genetically flexible domain. *J. Virol.* **79,** 3187–3194.

13. Evans, M. J., Rice, C. M., and Goff, S. P. (2004) Phosphorylation of hepatitis C virus nonstructural protein 5A modulates its protein interactions and viral RNA replication. *Proc. Natl. Acad. Sci. USA* **101,** 13038–13043.

14. Grobler, J. A., Markel, E. J., Fay, J. F., Graham, D. J., Simcoe, A. L., Ludmerer, S. W., et al. (2003) Identification of a key determinant of hepatitis C virus cell culture adaptation in domain II of NS3 helicase. *J. Biol. Chem.* **278,** 16741–16746.

15. Kishine, H., Sugiyama, K., Hijikata, M., Kato, N., Takahashi, H., Noshi, T., et al. (2002) Subgenomic replicon derived from a cell line infected with the hepatitis C virus. *Biochem. Biophys. Res. Commun.* **293,** 993–999.

16. Kato, N., Sugiyama, K., Namba, K., Dansako, H., Nakamura, T., Takami, M., et al. (2003) Establishment of a hepatitis C virus subgenomic replicon derived from human hepatocytes infected *in vitro*. *Biochem. Biophys. Res. Commun.* **306,** 756–766.

17. Blight, K. J., McKeating, J. A., Marcotrigiano, J., and Rice, C. M. (2003) Efficient replication of hepatitis C virus genotype 1a RNAs in cell culture. *J. Virol.* **77,** 3181–3190.

18. Yi, M. and Lemon, S. M. (2004) Adaptive mutations producing efficient replication of genotype 1a hepatitis C virus RNA in normal Huh7 cells. *J. Virol.* **78,** 7904–7915.

19. Kato, T., Date, T., Miyamoto, M., Furusaka, A., Tokushige, K., Mizokami, M., et al. (2003) Efficient replication of the genotype 2a hepatitis C virus subgenomic replicon. *Gastroenterology* **125,** 1808–1817.

20. Wakita, T., Pietschmann, T., Kato, T., Date, T., Miyamoto, M., Zhao, Z., et al. (2005) Production of infectious hepatitis C virus in tissue culture from a cloned viral genome. *Nat. Med.* **11,** 791–796.

21. Blight, K. J., McKeating, J. A., and Rice, C. M. (2002) Highly permissive cell lines for subgenomic and genomic hepatitis C virus RNA replication. *J. Virol.* **76,** 13001–13014.

22. Koutsoudakis, G., Herrmann, E., Kallis, S., Bartenschlager, R., and Pietschmann, T. (2007) The level of CD81 cell surface expression is a key determinant for productive entry of hepatitis C virus into host cells. *J. Virol.* **81,** 588–598.

23. Miyamoto, M., Kato, T., Date, T., Mizokami, M., and Wakita, T. (2006) Comparison between subgenomic replicons of hepatitis C virus genotypes 2a (JFH-1) and 1b (Con1 NK5.1). *Intervirology* **49,** 37–43.

24. Guo, J. T., Bichko, V. V., and Seeger, C. (2001) Effect of alpha interferon on the hepatitis C virus replicon. *J. Virol.* **75,** 8516–8523.

25. Ali, S., Pellerin, C., Lamarre, D., and Kukolj, G. (2004) Hepatitis C virus subgenomic replicons in the human embryonic kidney 293 cell line. *J. Virol.* **78,** 491–501.

26. Windisch, M. P., Frese, M., Kaul, A., Trippler, M., Lohmann, V., and Bartenschlager, R. (2005) Dissecting the interferon-induced

inhibition of hepatitis C virus replication by using a novel host cell line. *J. Virol.* **79,** 13778–13793.

27. Date, T., Kato, T., Miyamoto, M., Zhao, Z., Yasui, K., Mizokami, M., et al. (2004) Genotype 2a hepatitis C virus subgenomic replicon can replicate in HepG2 and IMY-N9 cells. *J. Biol. Chem.* **279,** 22371–22376.

28. Kato, T., Date, T., Miyamoto, M., Zhao, Z., Mizokami, M., and Wakita, T. (2005) Nonhepatic cell lines HeLa and 293 support efficient replication of the hepatitis C virus genotype 2a subgenomic replicon. *J. Virol.* **79,** 592–596.

29. Chang, K. S., Cai, Z., Zhang, C., Sen, G. C., Williams, B. R., and Luo, G. (2006) Replication of hepatitis C virus (HCV) RNA in mouse embryonic fibroblasts: protein kinase R (PKR)-dependent and PKR-independent mechanisms for controlling HCV RNA replication and mediating interferon activities. *J. Virol.* **80,** 7364–7374.

30. Uprichard, S. L., Chung, J., Chisari, F. V., and Wakita, T. (2006) Replication of a hepatitis C virus replicon clone in mouse cells. *Virol. J.* **3,** 89.

31. Binder, M., G. Kochs, R. Bartenschlager, and V. Lohmann (2007) Hepatitis C virus escape from the interferon regulatory factor 3 pathway by a passive and active evasion strategy. *Hepatology* **46,** 1365–1374.

32. Pietschmann, T., Lohmann, V., Rutter, G., Kurpanek, K., and Bartenschlager, R. (2001) Characterization of cell lines carrying self-replicating hepatitis C virus RNAs. *J. Virol.* **75,** 1252–1264.

33. Nelson, H. B. and Tang, H. (2006) Effect of cell growth on hepatitis C virus (HCV) replication and a mechanism of cell confluence-based inhibition of HCV RNA and protein expression. *J. Virol.* **80,** 1181–1190.

34. van den Hoff, M. J., Moorman, A. F., and Lamers, W. H. (1992) Electroporation in "intracellular" buffer increases cell survival. *Nucleic Acids Res.* **20,** 2902.

35. Binder, M., D. Quinkert, O. Bochkarova, R. Klein, N. Kezmic, R. Bartenschlager, and V. Lohmann (2007) Identification of determinants involved in initiation of hepatitis C virus RNA synthesis by using intergenotypic replicase chimeras. *J. Virol.* **81**:5270–5283.

36. Kolykhalov, A. A., Feinstone, S. M., and Rice, C. M. (1996) Identification of a highly conserved sequence element at the 3′ terminus of hepatitis C virus genome RNA. *J. Virol.* **70,** 3363–3371.

Chapter 12

Reverse Transcription PCR-Based Sequence Analysis of Hepatitis C Virus Replicon RNA

Timothy L. Tellinghuisen and Brett D. Lindenbach

Abstract

Since the advent of efficient cell-culture methods for HCV replication and, more recently, infection, there has been a need to efficiently sequence the viral RNA in these systems. This need is especially urgent in light of the error-prone nature of HCV RNA replication, which leads to a variety of interesting mutations. The adaptation of hepatitis C replicons to cell culture, which greatly increased their replication capacity, and the subsequent identification of viral point mutations responsible for this adaptation are prime examples of the type of phenotype–genotype connection that viral RNA sequencing methods can provide. More recently, researchers have used similar sequencing methods to identify changes in replicons that represent viral adaptation to engineered mutations, adaptation to a variety of host cells, and viral evasion of antiviral compound susceptibility. Here, we describe the cloning and isolation of HCV replicon-bearing cells, the extraction of total RNA, the generation of cDNA, and the amplification of specific HCV replicon sequences for sequence analysis. The methods we describe permit rapid and robust determination of HCV RNA sequences from cell culture.

Key words: HCV, RNA, reverse transcription, sequence analysis, cell cloning, PCR, primers.

1. Introduction

The last decade has seen an explosion in our ability to manipulate HCV genetically in a cell-culture environment. Since the development of the first infectious clone for HCV 10 years ago, we have been able to introduce directed mutations into the HCV genome and to analyze the fitness of the resultant virus in chimpanzees (1). The development of subgenomic replicons of the genotype 1b Con1 isolate in 1999 permitted the study of RNA replication in a more tractable cell-culture environment (2). Modifications of

Hengli Tang (ed.), *Hepatitis C: Methods and Protocols, Second Edition, vol. 510*
© 2009 Humana Press, a part of Springer Science+Business Media
DOI 10.1007/978-1-59745-394-3_12 Springerprotocols.com

the original replicon system, and hepatoma cell line used in replicon studies, have greatly increased the efficiency and utility of this system *(3–5)*, enabling many labs to perform reverse genetics on HCV. This system and others based on the same general theme have allowed an explosion in our understanding of HCV RNA replication. Within half a decade, the complete HCV life cycle, including production of infectious virus, has become susceptible to study in cell culture *(6–8)*.

Despite all these novel research tools, at least one thing has remained constant: the need to isolate and sequence HCV RNA from the cell-culture environment. A general method such as RNA sequence analysis is at once an experimentally simple technique and an incredibly powerful method for tracking genetic variations in a replicating HCV RNA population in response to a researcher's manipulations. It is general enough to preclude a detailed list of all its potential applications, so we will highlight only a few examples of the method's impact on HCV research. Perhaps the most important observation garnered from RT-PCR sequencing in HCV research was the identification of adaptive mutations in the HCV protein coding sequence that lead to vastly increased efficiency of RNA replication in cell culture when engineered into the parental RNA sequence *(3,5)*. A similar adaptation of virus to host has been observed in the infectious HCV cell-culture system, although the magnitude of increased HCV fitness is lower *(9)*. Nonetheless, through RT-PCR sequencing, scientists have observed the adaptation of HCV to the host cell. Similar research using replicons in cell lines other than the archetypal Huh-7 lineage has identified changes in HCV that lead to adaptation to other human and mouse cell lines *(10, 11)*. RT-PCR sequencing has also been useful in identifying replicons and viruses exhibiting reversions or second site mutations that suppress the effect of an initial input mutation *(12)*. Thus, forward genetic selection can be integrated with reverse genetic manipulation. Another key contribution of this method to HCV research has been the identification of mutations that confer resistance to various antiviral compounds in cell culture. These mutations have proven useful in validating the putative site of action of some compounds and aided in the understanding of resistance to antiviral drugs relatively early in compound development (see **ref.** *(13)* for review). These results collectively demonstrate that analysis of the sequences that come out of an HCV genetic experiment can be as important as the mutation engineered into the original input RNA.

Here, we describe a combination of commercial reagents and protocols that, with minor modifications, allow for the efficient isolation of total cellular RNA, conversion to cDNA, and PCR amplification of HCV-specific sequences for sequence analysis.

Although the methods presented are focused on the genotype 1b subgenomic replicon *(2)*, the same methods have been successfully applied to the genotype 2a J6/JFH-1 infectious cell-culture system RNA *(6)* with only modification of the sequence of the HCV-specific primers used in amplification.

2. Materials

2.1. Colony Cloning and Expansion

1. Phosphate-buffered saline (PBS) (Invitrogen Dulbecco's PBS without calcium or magnesium, catalog # 14190-250).
2. Trypsin/EDTA (Invitrogen catalog # 25300-062).
3. Cloning cylinders (Pyrex 10 × 10 mm glass cloning cylinders, catalog # 3166-10).
4. Vacuum grease (Dow Corning high-vacuum grease, catalog # 1597418).

2.2. RNA Extraction

1. TRIzol (Invitrogen catalog # 15596).
2. 1.5-mL Eppendorf microfuge tubes (Brinkmann catalog # 022364111).
3. Isopropanol.
4. Chloroform.
5. 75% Ethanol.
6. 1 mM sodium citrate, pH 6.4.
7. RNase-free water.

2.3. Reverse Transcription

1. Superscript II reverse transcriptase (Invitrogen catalog # 18064-022).
2. 5× first-strand buffer (provided with Superscript II enzyme).
3. 0.1 M DTT (provided with Superscript II enzyme).
4. RNase-free water.
5. Random hexamers (Invitrogen catalog # 48190-011).
6. 10 mM dNTP mix (Invitrogen catalog # 18427-013).
7. RNase Out RNase inhibitor (Invitrogen catalog # 10777-019).
8. RNase H (Invitrogen catalog # 18021-014).
9. Qiagen QIAquick PCR cleanup kit (Qiagen catalog # 28104).

2.4. PCR Amplification of HCV-Specific Sequences

1. Phusion DNA polymerase (New England Biolabs catalog # F540S).
2. Phusion buffer (provided with Phusion polymerase).
3. 10 mM dNTP stock (Invitrogen catalog # 18427-013).
4. HCV-specific PCR primers (*see* **Table 12.1**).

Table 12.1
Oligonucleotide primers for the amplification of specific HCV sequences

Primer	Position* (nt)	Orientation	Sequence (5′–3′)
1	1624–1644	F	CTCTCCTCAAGCGTATTCAAC
2	2487–2507	R	GTGGCGGCGACGGACGGGTTC
3	2392–2412	F	TTCCAGGTGGCCCATCTACAC
4	3040–3060	R	GGTAAAGCCCGTCATTAGAGC
5	2983–3003	F	GGCCTTGATGTATCCGTCATA
6	3852–3872	R	GGGAGGTGTGAGGCGCACTCT
7	3750–3770	F	GACAACAGGCAGCGTGGTCAT
8	4533–4553	R	CGTGCTGCAGCGTCGCTCTCA
9	4469–4489	F	GGATGAACCGGCTGATAGCGT
10	5265–5285	R	TTAGCCGTCTCCGCCGTAATG
11	5170–5190	F	GTCGGGCTCAATCAATACCTG
12	6779–6799	R	TCTGCCCTTTAGAATTAGTCA
13	5959–5979	F	GCTAGTGAGGACGTCGTCTGC
14	6732–6752	F	ACAGGCCATAAGGTCGCTCAC
15	7615–7595	R	ACTGGGACGCAGCCGGGATTG
16	7378–7398	F	CTCCATGGCCTTAGCGCATTT
17	7781–7801	R	AAAAACAGGATGGCCTATTGG

*Positions are relative to the I377/NS3 − 3′ UTR replicon *(2)*, accession number AJ242652.

3. Methods

3.1. Colony Cloning and Expansion

1. After replicon transfection and drug selection *(2,3)*, identify a well-isolated colony at least 1–2 mm in diameter for cloning and circle the clone with a black marking pen on the outside of the plate bottom.
2. Remove the growth medium from the plate by aspiration and wash the plate with 10 mL of PBS (acceptable volume for 100- to 150-mm plates).
3. Remove the PBS and aseptically place a presterilized 0.5-cm glass cloning cylinder greased on one end with silicon vacuum grease on the plate such that the colony of interest is completely surrounded by the cylinder wall (*see* **Note 1**). Using

sterile forceps, gently press down on the cloning cylinder to create a watertight seal of vacuum grease on the culture surface of the plate.

4. Add 100 μL of trypsin to the cylinder center and incubate the plate at 37°C for 5 min to release the cells in the colony from the plate.

5. Add 100 μL of complete growth medium to the well and mix well by pipetting to dislodge and separate the cells from the colony.

6. Transfer the cell suspension to a fresh 24-well plate containing 500 μL of complete growth medium.

7. Once the 24-well plate is 70–80% confluent, split the cells and expand to a 12-well plate and continue to expand the cells until cell density is sufficient to seed two wells of two separate six-well plates.

8. Proceed to RNA isolation as detailed in **Section 3.2** with one six-well plate. The other can be passaged as a backup plate or used to freeze cells for long-term storage.

3.2. RNA Extraction

1. Remove growth medium from the well of a six-well plate of cells at 80% confluence by aspiration. Wash the cell monolayer by adding 2 mL of PBS per well and then remove the PBS by aspiration. Repeat the PBS wash with an additional 2 mL of PBS and remove it.

2. Add 0.75 mL of TRIzol reagent per well to lyse the cells. Rock the plate briefly to wet the monolayer completely with the reagent. Lysis with TRIzol occurs rapidly and is complete within 1–2 min (*see* **Note 2**).

3. Homogenize the TRIzol lysates by pipetting the lysates 10 times. Use of filter tips prevents cross-contamination if multiple samples are to be processed.

4. Transfer lysates to a clean, unautoclaved 1.5 mL Eppendorf microcentrifuge tube (*see* **Note 3**).

5. Lysates can be stored at −80°C or processed immediately. If lysates are stored at −80°C, be sure to thaw the samples before proceeding to step 6. Allow the thawed or fresh samples to sit at room temperature for 5 min before proceeding to step 6 (*see* **Note 4**).

6. Add 150 μL of chloroform to the TRIzol lysates and shake the tubes vigorously by hand for 15 s (*see* **Note 5**).

7. Allow the tube to sit at room temperature for 3 min.

8. Centrifuge the TRIzol/chloroform mixture at $12,000 \times g$ (maximum speed in a microfuge) at 4°C for 20 min to separate aqueous and organic phases.

9. After centrifugation, carefully remove the clear, colorless aqueous (upper) layer and transfer it to a new microfuge tube. The red organic (bottom) layer can be discarded in an appropriate hazardous-waste container.

10. Add 0.5 mL of isopropanol to the aqueous layer and incubate at room temperature for 10 min to precipitate total cellular RNA.

11. Centrifuge the RNA precipitation reaction at $12,000 \times g$ (maximum speed in a microfuge) at 4°C for 10 min.

12. Carefully remove the supernatant by slow aspiration (*see* **Note 6**).

13. Wash the remaining RNA pellet by adding 1 mL of 75% ethanol and then vortex briefly (~ 20 s).

14. Centrifuge the sample at $7500 \times g$ at 4°C for 5 min.

15. Carefully remove the supernatant by slow aspiration. Allow the RNA pellet to air dry at room temperature for 5–10 min. Overdrying the sample (as in a speed-vac or a similar device) will make RNA resuspension very difficult.

16. Dissolve the RNA pellet in 25 μL of 1 mM sodium citrate, pH 6.4, or RNase-free water (*see* **Note 7**).

17. Dilute a 1-μL aliquot of RNA (1:20 to 1:100) and determine the concentration of RNA (*see* **Note 8**).

18. Bring the final concentration of the purified RNA to $0.25\,\mu g/\mu L$ by dilution in RNase-free water. RNA can be stored at −80°C at this step (*see* **Note 9**).

3.3. Reverse Transcription

1. Thaw RNA samples (if frozen) and reagents for reverse transcription.

2. Set up the reverse transcription reaction with the following volumes of components (*see* **Note 10**): 1 μL of random hexamer, 1 μL of 10 mM dNTP mix, 4 μL of total RNA, $0.25\,\mu g/\mu L$ (from **Section 3.2**, step 18), and 6 μL of RNase-free water.

3. Heat the reverse transcription reaction at 65°C for 5 min, chill on ice for 1 min, and centrifuge briefly to collect any condensation from the tube cap from heating.

4. Add the following components to the reaction mix: 4 μL of 5× first-strand buffer, 2 μL of 0.1 M DTT, and 1 μL of RNase inhibitor.

5. Incubate the reaction at 42°C for 2 min, cool to room temperature, and incubate for an additional 2 min at room temperature.

6. Add 1 μL of Superscript II reverse transcriptase to the reaction and mix well.

7. Incubate the reaction at 50°C for 1 h to allow cDNA synthesis to occur.

8. Heat the reaction to 75°C for 15 min to inactivate the Superscript II enzyme.

9. After allowing the reaction to cool to room temperature, add 1 μL of RNase H, mix well, and incubate the reaction at 37°C for 20 min.

10. Purify the cDNA product from the random hexamers and other components of the reverse transcription reaction *before* proceeding to the PCR amplification procedure described in **Section 3.4**. Although any method of cDNA purification can be used, we routinely use the Qiagen QIAquick spin purification kit. We briefly outline below the modified version of the QIAquick purification method we use.

11. Add 79 μL of water to the cDNA sample and mix.

12. Add 500 μL of buffer PB to the cDNA sample and mix.

13. Apply the PB/cDNA mixture to a QIAquick spin column placed in a 2-mL microfuge tube and centrifuge for 1 min at $12,000 \times g$.

14. Discard the column flow-through fraction and place the spin column back in the 2-mL microfuge tube.

15. Add 0.75 mL of buffer PE to the column and centrifuge for 1 min at $12,000 \times g$.

16. Discard the column flow-through and place the spin column back in the 2-mL microfuge tube.

17. Centrifuge for 1 min at $12,000 \times g$.

18. Discard the column flow-through and the 2-mL microfuge tube.

19. Transfer the spin column to a new 1.5-mL microfuge tube and centrifuge for an additional 2 min (*see* **Note 11**).

20. Discard the flow-through and the 1.5-mL microfuge tube.

21. Place the spin column in a new 1.5-mL microfuge tube and add 50 μL of buffer EB to the center of the resin bed of the spin column.

22. Centrifuge the spin column for 1 min at $12,000 \times g$ to collect the eluted cDNA.

23. Discard the spin column and save the eluted cDNA in the 1.5-mL microfuge tube. Proceed to the HCV-specific PCR amplification described in **Section 3.4** (*see* **Note 12**).

3.4. PCR Amplification of HCV-Specific Sequences

1. Thaw cDNA samples (if frozen) and the reagents for PCR amplification.

2. Set up the PCR amplification reactions. The choice of primers is dictated by the region of the HCV polyprotein from which sequence information is desired (*see* **Tables 12.1 and 12.2** and **Notes 13 and 14**). The PCR amplification mixture contents are as follows: 10 μL 5 μL Phusion PCR buffer, 23.75 μL water, 1 μL 10 mM dNTP mix, 2.5 μL 10 μM forward primer, 2.5 μL 10 μM reverse primer, 10 μL, cDNA sample (from **Section 3.3**, step 23), and 0.25 μL Phusion DNA polymerase.

3. Perform PCR amplification using the following conditions (*see* **Note 15**): 98°C for 30 s, 98°C for 10 s, 54°C for 20 s,

Table 12.2
HCV-specific amplicons for sequence analysis

PCR	Primers	Position*	Size (bp)	Replicon region
A	1 + 2	1624–2507	883	EMCV IRES–NS3
B	3 + 4	2392–2412	668	NS3
C	5 + 6	2983–3872	889	NS3–NS4A
D	7 + 8	3750–4553	803	NS4A–NS4B
E	9 + 10	4469–5285	816	NS4B–NS5A
F	11 + 12	5170–6799	1629	NS5A–NS5B
G	12 + 13	6779–5979	800	NS5A–NS5B
H	14 +15	6732–7595	883	NS5B
I	16 + 17	7378–7801	423	NS5B–3′ UTR

*Positions are relative to the I377/NS3-3′ UTR replicon *(2)*, accession number AJ242652.

72°C for 1 min 45 s, repeat steps 2 through 4, 29 more times, 72°C for 2 min, end, and incubate at 4°C.

4. Gel purify the appropriate PCR products (*see* **Table 12.2**) to remove primers and other reagents from the amplification reaction. We regularly use the Qiagen QIAquick gel extraction kit and protocol (catalog # 28704, used without modification of the manufacturer's instructions).

5. At this point, the end user must decide what information is desired from the sequence analysis. For simple population sequencing, a determination of the dominant RNA sequence(s) in the cell, the PCR products can be directly sequenced at this stage. For an estimation of the diversity of sequence variants in the original pool of HCV RNA, PCR products can be cloned into a PCR cloning vector, and individual clones can be isolated for sequence analysis (*see* **Note 16**). The number of clones to be sequenced is again to be decided by the end user, but we typically sequence 10–20 clones to provide a picture of the original RNA population (*see* **Note 17**).

6. The HCV-specific PCR products (or clones) are now ready for DNA sequence analysis according to the method of choice for DNA sequence analysis in the end user's laboratory. The primers used to generate the PCR products (*see* **Tables 12.1 and 12.2**, **Fig. 12.1**) are suitable primers (forward and reverse orientation) for sequence analysis (*see* **Note 18**).

PCR Reaction

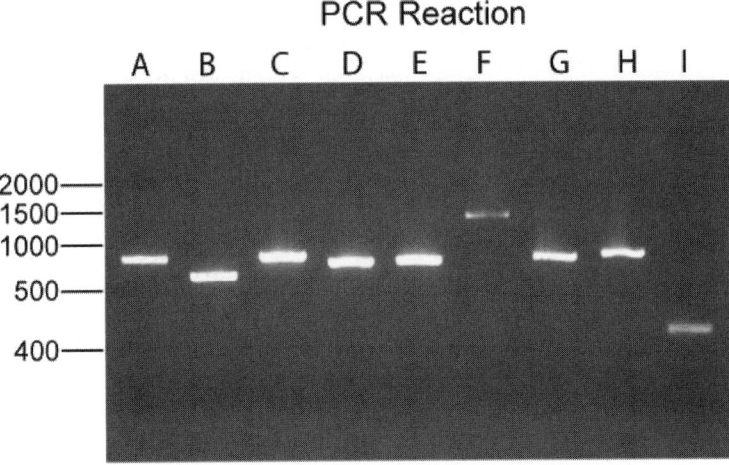

Fig. 12.1. HCV-specific PCR amplicons after gel purification. A 1% agarose gel showing the relative sizes of the nine individual PCR amplicons described in **Table 12.2**. Approximate migration of a dsDNA size-standard ladder is shown in units of base pairs.

4. Notes

1. We prepare cloning cylinders by placing sufficient vacuum grease in a Pyrex glass-covered 150-mm petri dish to form a layer of grease about 5 mm thick. We then place the cloning cylinders, standing upright, in the grease, cover, and autoclave the plate.

2. TRIzol is a hazardous chemical and should be handled according to the manufacturer's instructions.

3. We have found that Eppendorf 1.5-mL Snap-Cap microfuge tubes (part number 022364111) are sufficiently RNase free for use in this procedure. Autoclaving tubes can add moisture/condensation residue to the tubes and should be avoided.

4. We have amplified and sequenced HCV from TRIzol lysates stored at −80°C for more than 1 year.

5. Chloroform is a hazardous chemical. Take appropriate precautions when working with, and disposing of, this material.

6. Use caution to avoid accidental aspiration of the RNA pellet. The pelleted material may be located in the bottom and on the side of the tube and may be clear, colorless, and therefore difficult to see. Be sure to take this problem into account when resuspending the pellet fraction.

7. The choice of 1 mM sodium citrate, pH 6.5, or RNase-free water is the user's. We prefer to use citrate buffer, as the low pH and metal-chelating properties of citrate promote long-term RNA stability. The method works equally well with either resuspension medium.

8. Typical total RNA yield from this procedure is approximately 10–20 μg per well of a six-well plate. The quality of RNA can be monitored by standard agarose gel electrophoresis and analysis of the integrity of ribosomal RNAs if desired. An ethidium bromide–stained 1% agarose gel should show discrete (not smeared) bands of mRNA greater than a 7 kb DNA marker: two prominent bands, 28S and 18S ribosomal RNA, which co-migrate with 5 kb and 2 kb DNA markers, and low molecular weight RNAs of 100–500 bases (tRNAs and 5S RNA).

9. We have amplified and sequenced HCV from purified RNA stored at −80°C for more than 1 year.

10. If complete coverage of the replicon open reading frame is desired, two reverse transcription reactions are required to generate sufficient cDNA.

11. Removal of residual ethanol (present in the PE wash buffer) is important. We have added an additional 2-min spin in a clean 1.5-mL microfuge tube to dry the QIAquick spin column fully. Failure to remove the ethanol may significantly reduce the yield of PCR products in later steps.

12. We have stored cDNA at −80°C for several months with no adverse effects on subsequent PCR amplification.

13. The region of the HCV replicon to be sequenced is chosen by the end user. **Tables 12.1 and 12.2** provide oligo pairs for PCR and information about the regions amplified.

14. The procedure is written for Phusion DNA polymerase amplification. We have also successfully used Pfu turbo and Klentaq polymerases with appropriate modifications to the amplification conditions. The choice of polymerase is the end user's, although a proofreading polymerase should be used to maintain accurate amplification.

15. Remember to perform a negative control (no cDNA) for the PCR reaction to ensure that the reagents are not contaminated. Contaminants are a common source of failure for this procedure, so take care to ensure the PCR represents an amplification of the input cDNA and not contaminating DNA.

16. Select a destination vector for the PCR products appropriate for the type of polymerase used for amplification (polymerases generating blunt ends versus those leaving overhangs).

17. The user should be aware of bias introduced in this, or any, PCR amplification/cloning experiment and how it might affect the end results.

18. Using the PCR primers (**Table 12.1**) as sequencing primers should provide sufficient sequence coverage to span each PCR fragment.

References

1. Kolykhalov, A. A., Agapov, E. V., Blight, K. J., Mihalik, Feinstone, S. M., and Rice, C. M. (1997) Transmission of hepatitis C by intrahepatic inoculation with transcribed RNA. *Science* **277**, 570–574.

2. Lohmann, V., Korner, F., Koch, J. O., Herian, U., Theilmann, L., and Bartenschlager, R. (1999) Replication of subgenomic hepatitis C virus RNAs in a hepatoma cell line. *Science* **285**, 110–113.

3. Blight, K. J., Kolykhalov, A. A., and Rice, C. M. (2000) Efficient initiation of HCV RNA replication in cell culture. *Science* **290**, 1972–1974.

4. Blight, K. J., McKeating, J. A., and Rice, C. M. (2002) Highly permissive cell lines for hepatitis C virus genomic and subgenomic RNA replication. *J. Virol.* **76**, 13001–13014.

5. Lohmann, V., Korner, F., Dobierzewska, A., and Bartenschlager, R. (2001) Mutations in hepatitis C virus RNAs conferring cell culture adaptation. *J. Virol.* **75**, 1437–1449.

6. Lindenbach, B. D., Evans, M. J., Syder, A. J., Wolk, B., Tellinghuisen, T. L., Liu, C. C., et al. (2005) Complete replication of hepatitis C virus in cell culture. *Science* **309**, 623–626.

7. Wakita, T., Pietschmann, T., Kato, T., Date, T., Miyamoto, M., Zhao, Z., et al. (2005) Production of infectious hepatitis C virus in tissue culture from a cloned viral genome. *Nat. Med.* **11**, 791–796.

8. Zhong, J., Gastaminza, P., Cheng, G., Kapadia, S., Kato, T., Burton, D. R., et al. (2005) Robust hepatitis C virus infection in vitro. *Proc. Natl. Acad. Sci. USA* **102**, 9294–9299.

9. Zhong, J., Gastaminza, P., Chung, J., Stamataki, Z., Isogawa, M., Cheng, G., et al. (2006) Persistent hepatitis C virus infection in vitro: coevolution of virus and host. *J. Virol.* **80**, 11082–11093.

10. Kato, T., Date, T., Miyamoto, M., Zhao, Z., Mizokami, M., and Wakita, T. (2005) Nonhepatic cell lines HeLa and 293 support efficient replication of the hepatitis C virus genotype 2a subgenomic replicon. *J. Virol.* **79**, 592–596.

11. Zhu, Q., Guo, J. T., and Seeger, C. (2003) Replication of hepatitis C virus subgenomes in nonhepatic epithelial and mouse hepatoma cells. *J. Virol.* **77**, 9204–9210.

12. Lindenbach, B. D., Pragai, B. M., Montserret, R., Beran, R. K., Pyle, A. M., Penin, F., Rice, C. M. (2007) The C terminus of hepatitis C virus NS4A encodes an electrostatic switch that regulates NS5A hyperphosphorylation and viral replication. *J. Virol.* **81**, 8905–8918.

13. Tomei, L., Altamura, S., Paonessa, G., De Francesco, R., and Migliaccio, G. (2005) HCV antiviral resistance: the impact of in vitro studies on the development of antiviral agents targeting the viral NS5B polymerase. *Antivir. Chem. Chemother.* **16**, 225–245.

Chapter 13

Studying HCV RNA Synthesis In Vitro with Replication Complexes

Wengang Yang and Mingjun Huang

Abstract

HCV replication complexes are well-organized protein and lipid structures responsible for HCV RNA replication. The nonstructural protein NS5B, an RNA-dependent RNA polymerase, is the catalytic subunit of these replication complexes. After being isolated from HCV replicon-containing cells as a crude membrane fraction, these replication complexes have been shown to remain active in synthesis of viral RNA under proper assay conditions. Under the improved assay conditions presented here, we recently showed that isolated replication complexes are able to synthesize two species of nascent viral RNA, one double stranded and the other single stranded. NS5B nucleoside inhibitors block synthesis of both species, whereas nonnucleoside inhibitors inhibit mostly single-stranded RNA synthesis. Our results support the discoveries with recombinant NS5B biochemical assay that nucleoside inhibitors and nonnucleoside inhibitors block viral RNA synthesis by different mechanisms.

Key words: HCV replication complex, HCV replicase, NS5B, NS5B inhibitors.

1. Introduction

The nonstructural protein 5B (NS5B) of HCV is an RNA-dependent RNA polymerase (1). It is one of the most extensively explored drug targets because it fits traditional criteria for development of antivirals, that is, is essential for viral replication, specific for the virus, historically validated as an antiviral target, and well enough characterized biochemically and biophysically to make rational drug design feasible.

Biochemical assays using recombinant NS5B have provided insight into the enzymology of NS5B as well as tools for discovering inhibitors of NS5B, but these assays may not reflect

Hengli Tang (ed.), *Hepatitis C: Methods and Protocols, Second Edition, vol. 510*
© 2009 Humana Press, a part of Springer Science+Business Media
DOI 10.1007/978-1-59745-394-3_13 Springerprotocols.com

authentic replication. Recombinant NS5B copies a wide range of template RNA and DNA without preference. Initiation of nascent viral RNA synthesis by recombinant NS5B could occur by primer-dependent or de novo RNA synthesis. As a consequence, the inhibitory potency of nonnucleoside compounds on NS5B, when evaluated in biochemical assays, varies widely depending on the form of recombinant NS5B used and/or the concentration of primer and/or template used (2, 3). The availability of HCV replicon–containing cells (4) and most recently HCV-infected cells (5) offers the opportunity to determine the correlation of the inhibitory effects of nonnucleoside inhibitors in vitro and in vivo and thereby to elucidate further the nature of recombinant NS5B biochemical assays. Unfortunately, compound penetration of cells, toxicity to cells, and stability during the culture period often complicate such analyses.

Here, we describe an assay for the study of viral RNA synthesis as well as the effect of compounds on viral RNA synthesis with crude replication complexes (CRCs) prepared from replicon-containing cells. As well documented, the viral proteins NS3, NS4A, NS4B, NS5A, and NS5B are sufficient to support HCV RNA replication (5). These nonstructural proteins assemble on specialized intracellular membranes into a multimolecular complex, the replication complex or replicase. This well-organized machinery supplies the structural bases responsible for various features of viral RNA replication, such as template specificity, de novo initiation, and ratio of plus- to minus-stranded RNA. The replication complexes, after they are isolated from HCV replicon–containing cells as a crude membrane fraction, remain

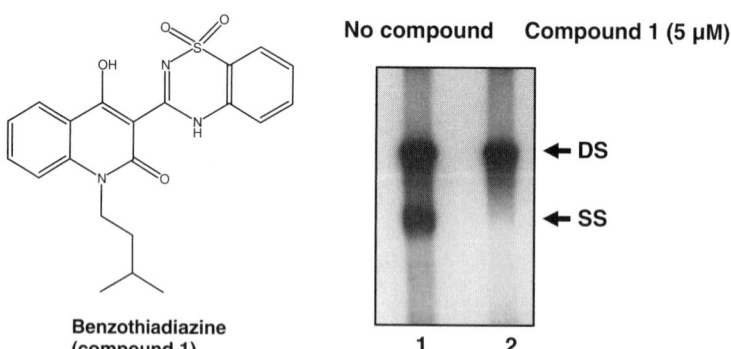

Fig. 13.1. Inhibition of nascent single-stranded RNA synthesis by a benzothiadiazine-based compound (compound 1) with CRCs prepared from replicon-containing cells (Huh-9-13), under the conditions described here. Total RNAs are extracted with TRIzol reagent, dissolved in water and resolved on a 1% agarose gel in × TBE buffer. Two nascent viral RNA species are visualized in the untreated control (lane 1), one double stranded (DS) and the other single stranded (SS). NS5B nucleoside inhibitors inhibit both synthesis (data not shown), whereas NS5B nonnucleoside inhibitors inhibit mainly SS synthesis as shown in lane 2 in the presence of compound 1, a nonnucleoside inhibitor (15).

active, under proper conditions, in incorporation of radiolabeled nucleotides into the nascent viral RNA molecules with HCV RNA templates within the complexes it(*6–13*). NS5B nucleoside inhibitors, in their triphosphate form, have been shown to block the nascent viral RNA synthesis in the system *(7, 8, 10, 11)*. Although the matter was controversial for a while *(7, 11, 13)*, we have also demonstrated unequivocally that NS5B nonnucleoside inhibitors can exert their inhibitory effect in the system *(14)* (**Fig. 13.1**). As a result, the assay described here can now be used to study viral RNA synthesis under in vitro conditions much more closely than is possible with recombinant NS5B, and at the same time, the complexities associated with the cell-based assay can be avoided.

2. Materials

1. Replicon-containing cells: Of the various types of replicon-containing cell lines, we have used cell line Huh-9-13 to prepare CRCs. This line was established by transfection of a subgenomic replicon, I377/NS3-3′, as described by Lohmann et al. *(4)*. Direct sequencing of the replicon RNA molecules isolated from the cell line repeatedly revealed an adaptive mutation at NS5A (S2204I) and sometimes an additional adaptive mutation at NS3 (E1202G).

2. Culture media: Dulbecco's modified Eagle's medium (DMEM) (HyClone, Ogden, UT) supplemented with 10% fetal bovine serum (FBS; Gemini Bio-Products, West Sacramento, CA), 1× nonessential amino acids (Mediatech, Herndon, VA), penicillin (100 IU/mL), streptomycin (100 μg/m), and 0.5 mg/mL G418 (Mediatech, Herndon, VA).

3. Hyptonic buffer: 10 mM Tris-HCl, pH 7.5 (Mediatech, Herndon, VA), and 10 mM NaCl.

4. Storage buffer: Hypotonic buffer supplemented with 15% glycerol.

5. Reaction buffer: 1 M HEPES, pH 7.3 (Gibco-Invitrogen, Carlsbad, CA), 1 M KCl, 1 M MgCl$_2$, 40 units/μL RNAse inhibitor (Promega, Madison, WI), 10 mg/mL actinomycin D (Sigma-Aldrich, St. Louis, MO), 3.3 mM ATP, 3.3 mM GTP, 3.3 mM UTP (Promega, Madison, WI), [α^{32}P] CTP (10 μCi/μL, 800 Ci/mmol) (PerkinElmer, Waltham, MA), and 30 mM MnCl$_2$. The reaction buffer should be freshly prepared.

6. RNA loading buffer: 50% glycerol, 1 mM EDTA, pH 8.0, 0.25% bromophenol blue (Shelton Scientific, Shelton, CT),

and 0.25% xylene cyanol FF (Sigma-Aldrich, St. Louis, MO). This solution can be stored at 4°C.

7. 10 × TBE (Invitrogen, Carlsbad, CA).
8. Agarose (Invitrogen, Carlsbad, CA).
9. Yeast tRNA (Invitrogen, Carlsbad, CA).
10. TRIzol LS reagent (Invitrogen, Carlsbad, CA).
11. 10 × PBS (Mediatech, Herndon, VA).
12. RNAse-free H_2O (Mediatech, Herndon, VA).
13. Other general chemicals (Sigma-Aldrich, St. Louis, MO).

3. Methods

3.1. Preparation of Crude Replication Complexes (CRCs)

1. Maintain cells in the media at 37°C in an atmosphere of 5% CO_2 (see **Note 1**).
2. Plate the cells in the media into 150-mm culture dishes at a density of 2×10^6 cells per dish.
3. Incubate the dishes at 37°C in an atmosphere of 5% CO_2 until the cells reach confluence, usually 3–5 days.
4. Wash the cell layer once with 10 mL cold 1 × PBS.
5. Add 1 mL of cold hypotonic buffer to each dish. Scrape off the cells and transfer the cell suspension to an Eppendorf tube or a 10-mL conical tube, depending on the number of dishes processed.
6. Place the tube on ice for 15–20 min.
7. Use a 10-mL Wheaton Dounce homogenizer with a tight A pestle to break the cells with 40 strokes of the pestle.
8. Spin the lysate at $900 \times g$ for 5 min at 4°C in a microcentrifuge or a Sorvall high-speed centrifuge equipped with a SA-600 rotor.
9. Transfer the supernatant to a 3-mL ultracentrifuge tube and spin in a Beckman ultracentrifuge equipped with SW 55Ti rotor at $15,000 \times g$ for 20 min at 4°C.
10. Decant the supernatant and resuspend the pellet (containing HCV RC) with the storage buffer at a density of 1×10^8 cells/mL.
11. Divide the suspension into aliquots (CRCs) for storage at −80°C (see **Note 2**).

3.2. In Vitro Reaction Using CRCs

1. Set up a cell-free complex reaction in an Eppendorf tube. We always use an Eppendorf tube with O-ring to prevent any leaks of radioactive materials. Each reaction volume is 60 µL, which usually yields enough signal for detection by autoradiography. The volume of reaction can be scaled up accordingly (see **Notes 3, 4, and 5**).

Stock concentration	Added volume/reaction (μL)	Final concentration
1 M HEPES, pH 7.3	3	50 mM
1 M KCl	0.6	10 mM
1 M MgCl$_2$	0.6	10 mM
40 units/μL RNAse inhibitor	1.5	1 unit/μL
10 mg/mL actinomycin D	0.06	10 μg/mL
3.3 mM ATP, GTP, UTP	9	0.5 mM
[α^{32}P] CTP (10μCi/μL, 800 Ci/mmol)	1	0.2 μM
30 mM MnCl$_2$	0.6	0.3 mM
CRCs	6	
H$_2$O	37.6	

2. Incubate the reaction tube at 30°C for 1–2 h.
3. Add 190 μL H$_2$O containing 5 μg of carrier yeast tRNA to each tube.
4. Add 750 μL TRIzol LS reagent (Gibco-Invitrogen) to each tube and mix well with a pipette. Keep the tube at room temperature for 5 min for a complete dissociation of protein–nucleic acid complexes.
5. Add 0.2 mL chloroform to each tube. Close the tube tightly. Shake vigorously for 15 s. Keep the tube at room temperature for another 10 min.
6. Spin the tube in a microcentrifuge at 12,000 × g for 15 min at 4°C.
7. Transfer 0.5 mL of the upper layer to a new Eppendorf tube with O-ring containing 0.5 mL of isopropanol and mix.
8. Spin the tube in a microcentrifuge at 20,000 × g for 15 min at 4°C.
9. Decant the supernatant. At this stage, a small white pellet at the bottom of the tube is usually observed.
10. Add 1 ml 75% ethanol to wash the pellet.
11. Decant the supernatant. Carefully remove the residual ethanol with a pipette. Air dry the pellet for 5 min at room temperature and finally dissolve the pellet with 20 μL RNAse-free H$_2$O. The RNA solution can be used directly for the next step or stored at −80°C for later use.

3.3. Nondenaturing Gel Electrophoresis

1. Prepare a 1% agarose gel in 1 × TBE buffer. The BIO-RAD gel system is used in our laboratory. The size of the gel is 10 × 15 cm. The thickness of the gel is about 0.8 cm (*see* **Notes 6 and 7**).

2. Mix the RNA solution with 1/10 volume loading buffer.
3. Load 20- to 25-μL RNA samples onto the gel.
4. Run the gel in 1 × TBE. For better resolution of RNA bands, a constant voltage of 2 V/cm is suggested. Running time is about 4–6 h.
5. Fix the gel with 10% acetic acid at room temperature for 30 min and soak the gel in absolute ethanol for another 30 min.
6. Dry the gel. We use a BIO-RAD gel dryer (Model 583). It takes 2 h at 58°C.
7. Expose the gel to an X-ray film. To increase sensitivity, the BioMax MS film (Kodak) can be used. Nascent RNA bands can usually be visualized in less than 2 days.

4. Notes

1. The health of the cells is an important factor in obtaining CRCs with good activity. Never grow cells beyond confluence. We usually split cells 1:6 twice a week.
2. CRC stocks can be stored at −80°C for up to 6 months without any apparent loss of activity. Discard any leftover CRCs after thawing, because activity is usually impaired after one more round of freezing and thawing.
3. To maintain CRCs at their optimal state of the activity, keep the tubes on ice before mixing the contents.
4. For study of the effect of an inhibitor on the activity of CRCs, a 30-min preincubation without addition of any NTP substrates is employed to assure that the inhibitor is in extensive contact with the HCV replication complexes in the preparation. Because the stock solutions of inhibitors are usually made in dimethylsulfoxide (DMSO), the volume of a stock inhibitor added to a reaction should result in a final concentration of DMSO ≤ 2%. A solvent control, for example, DMSO, should also be included. After this preincubation, the substrates including [α^{32}P] CTP are added into the reaction mixture to initiate the incorporation.
5. Mn^{2+} affects the yield of nascent RNA species synthesized in vitro by CRCs. Initially, we examined several preparations of the crude membrane fractions under the reaction mixture as described but without Mn^{2+} (**Fig. 13.2**). Although two radiolabeled RNA bands, one double stranded (DS) and the other single stranded (SS), are readily detected in the reactions containing the crude membrane fractions prepared from replicon-containing cells, the intensity and mobility of the SS-RNA band varies from one preparation to another (**Fig. 13.2A**, lanes 1–7). After addition of 0.3 mM Mn^{2+} to the reaction

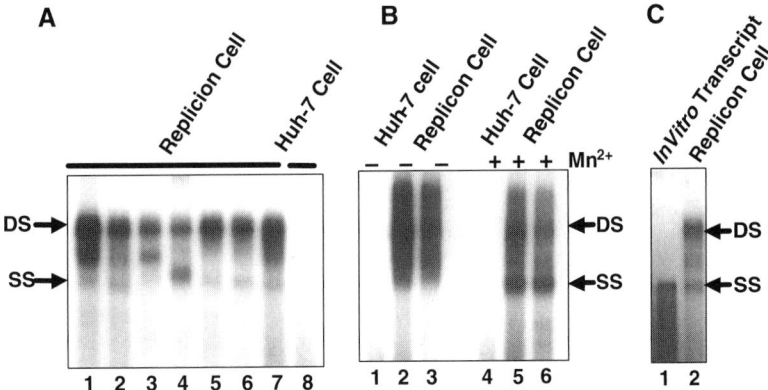

Fig. 13.2. The effect of Mn^{2+} on nascent HCV RNA synthesis catalyzed by CRCs in vitro. (**A**) Nascent RNA synthesis with CRCs prepared from replicon-containing cells (Huh-9-13) at different times (lane 1–7), in the absence of Mn^{2+}. Total RNAs are extracted with TRIzol reagent, dissolved in water, and resolved on 1% agarose gel in ×TBE buffer. Two readily detected radiolabeled RNA bands, one double stranded (DS) and the other single stranded (SS), are present only in the reactions containing the crude membrane fractions prepared from replicon-containing cells and not in those from the parental Huh-7 cells, confirming the identities of these radiolabeled RNA products as HCV RNAs. The DS band is consistently observed in all membrane preparations, but the intensity and mobility of the SS band varies from one preparation to another (lanes 1–7). (**B**) Nascent HCV RNA synthesis in the presence and absence of Mn^{2+}. Two different preparations of CRCs (preparation 1: lanes 2 and 5; preparation 2: lanes 3 and 6) are used in reactions in which 0.3 mM Mn^{2+} is either included (+) or not included (−). Preparation 1 is the same preparation used in lane 1 of Fig. 13.1, and preparation 2 is the same preparation used in lane 3 of Fig. 13.2A. RNA is isolated and analyzed as described for Fig. 13.1. Clearly, the addition of Mn^{2+} results in a consistent increase in the yield as well as the size uniformity of SS species (compare lanes 2 and 3 to lanes 5 and 6). (**C**) Size of the nascent HCV RNAs synthesized by CRCs in vitro. Radiolabeled subgenomic replicon RNA synthesized from the plasmid pI377/N3-3′ by T7 RNA polymerase is used as the size marker (lane 1). Nascent HCV RNA is synthesized by CRCs in the presence of Mn^{2+} (lane 2). RNA is resolved on 1% agarose gel in ×TBE buffer. Two major RNA products appear, one DS and the other SS. The SS RNA migrates to the same position as the unit-length replicon RNA.

mixture, however, a consistently higher yield, as well as size uniformity of the SS-RNA species, is observed, as illustrated with two CRC preparations in **Fig. 13.2B** (compare lanes 2 and 3 to lanes 5 and 6). We therefore run all the reactions in the presence of 0.3 mM Mn^{2+}.

6. Although no special steps are taken to treat the reagents and apparatus, a gel apparatus is reserved for electrophoresis so that no other samples, especially samples containing RNAse, will be run on the same apparatus.

7. The samples can also be resolved on denaturing gels, where all nascent RNA species migrate as one band with a size of a subgenomic unit length. In studies of nucleotide inhibitors that block synthesis of all nascent RNA species, resolving the samples in either gel will reveal a clear inhibitory effect, whether all nascent RNA species migrate as one band on a denaturing gel or whether they are separated into two major bands on a nondenaturing gel. Because nonnucleoside

inhibitors block only SS-RNA synthesis, however, no apparent inhibitory effect will be revealed unless the samples are resolved on a nondenaturing gel.

References

1. Lindenbach, B. D., and Rice, C. M. (2001) Flaviviridae: the viruses and their replication, in *Fields' Virology* (Fields, B. N., Knipe, D. M., Howley, P. M., and Griffin, D. E., eds.), Lippincott/Williams and Wilkins, Philadelphia, PA, pp. 991–1041.
2. McKercher, G., Beaulieu, P. L., Lamarre, D., LaPlante, S., Lefebvre, S., Pellerin, C., et al. (2004) Specific inhibitors of HCV polymerase identified using an NS5B with lower affinity for template/primer substrate. *Nucleic Acids Res.* **32**, 422–431.
3. Tomei, L., Altamura, S., Bartholomew, L., Bisbocci, M., Bailey, C., Bosserman, M., et al. (2004) Characterization of the inhibition of hepatitis C virus RNA replication by nonnucleosides. *J. Virol.* **78**, 938–946.
4. Lohmann, V., Korner, F., Koch, J., Herian, U., Theilmann, L., and Bartenschlager, R. (1999) Replication of subgenomic hepatitis C virus RNAs in a hepatoma cell line. *Science* **285**, 110–113.
5. Lindenbach, B. D., Evans, M. J. A., Syder, J., Wolk, B., Tellinghuisen, T. L., Liu, C. C., et al. (2005) Complete replication of hepatitis C virus in cell culture. *Science* **309**, 623–626.
6. Ali, N., Tardif, K. D., and Siddiqui, A. (2002) Cell-free replication of the hepatitis C virus subgenomic replicon. *J. Virol.* **76**, 12001–12007.
7. Hardy, R. W., Marcotrigiano, J., Blight, K. J., Majors, J. E., and Rice, C. M. (2003) Hepatitis C virus RNA synthesis in a cell-free system isolated from replicon-containing hepatoma cells. *J. Virol.* **77**, 2029–2037.
8. Klumpp, K., Leveque, V., Le Pogam, S., Ma, H., Jiang, W. R., Kang, H., et al. (2006) The novel nucleoside analog R1479, 4′-azidocytidine, is a potent inhibitor of NS5B dependent RNA synthesis and HCV replication in cell culture. *J. Biol Chem.* **81**, 3793–3799.
9. Lai, V. C., Dempsey, S., Lau, J. Y., Hong, Z., and Zhong, W. (2003) *In vitro* RNA replication directed by replicase complexes isolated from the subgenomic replicon cells of hepatitis C virus. *J. Virol.* **77**, 2295–2300.
10. Lu, G., Aizaki, H., He, J., and Lai, M. M. (2004) Interactions between viral nonstructural proteins and host protein hVAP-33 mediate the formation of hepatitis C virus RNA replication complex on lipid raft. *J. Virol.* **78**, 3480–348.
11. Ma, H., Leveque, V., De Witte, A., Li, W., Hendricks, T., Clausen, S. M., et al. (2005) Inhibition of native hepatitis C virus replicase by nucleotide and non-nucleoside inhibitors. *Virology* **332**, 8–15.
12. Migliaccio, G., Tomassini, J. E., Carroll, S. S., Tomei, L., Altamura, S., Bhat, B., et al. (2003) Characterization of resistance to non-obligate chain terminating ribonucleoside analogs which inhibit HCV replication *in vitro*. *J. Biol. Chem.* **278**, 49164–49170.
13. Quinkert, D., Bartenschlager, R. and Lohmann, V. (2005) Quantitative analysis of the hepatitis C virus replication complex. *J. Virol.* **79**, 13594–13605.
14. Yang, W., Sun, Y., Phadke, A., Deshpande, M., and Huang, M. (2007) Hepatitis C virus (HCV) NS5B nonnucleoside inhibitors specifically block single-stranded viral RNA synthesis catalyzed by HCV replication complexes *in vitro*. *Antimicrob. Agents Chemother.* **51**, 338–342.
15. Dhanak, D., Duffy, K. J., Johnston, V. K., Lin-Goerke, J., Darcy, M., Shaw, A. N., et al. (2002) Identification and biological characterization of heterocyclic inhibitors of the hepatitis C virus RNA-dependent RNA polymerase. *J. Biol. Chem.* **277**, 38322–38327.

Chapter 14

Proteomics Study of the Hepatitis C Virus Replication Complex

Kyungsoo Chang, Tianyi Wang, and Guangxiang Luo

Abstract

RNA replication of HCV occurs in the multiprotein complexes associated with the endoplasmic reticular (ER) membranes. The HCV NS3 to NS5B proteins are necessary and sufficient for HCV RNA replication in the cell, but cellular proteins in the HCV replication complex (RC) have not been determined. Several methods have been used to isolate the HCV RC, including crude cell extract preparation, subcellular fractionation, and affinity purification. The components of the HCV RC can be separated by two-dimensional electrophoresis and then determined by proteolytical digestion and mass spectrometry analysis in conjunction with peptide/protein database search and immunobiochemistry and functional genomic studies.

Key words: HCV, replication complex, RNA replication, cellular proteins, 2-D electrophoresis, proteomics, mass spectrometry.

1. Introduction

Viral RNA replication is an orchestrated biological process occurring in a multiprotein replication complex consisting of viral and cellular proteins acting through protein–protein and protein–RNA interactions. Like that of many other positive-strand RNA viruses, replication of HCV RNA occurs in a membrane-bound protein complex containing HCV RNAs and viral and cellular proteins *(1)*. Replication of a subgenomic HCV RNA encoding the NS3 to NS5B proteins demonstrated that HCV NS3–NS5B proteins are sufficient for HCV RNA replication in the cell *(2)*. A number of studies also demonstrated that most HCV proteins interact with each other in the cell. In addition, numerous cellular

Hengli Tang (ed.), *Hepatitis C: Methods and Protocols, Second Edition, vol. 510*
© 2009 Humana Press, a part of Springer Science+Business Media
DOI 10.1007/978-1-59745-394-3_14 Springerprotocols.com

proteins have been identified that specifically bind to HCV RNAs or proteins. Cellular proteins of the HCV replication complex are believed to play important roles in the control of HCV RNA replication *(3)*, but the biochemical compositions and properties of the HCV replication complex have not been determined. As a consequence, the roles of viral and cellular proteins of the HCV replication complex in HCV RNA replication and virion assembly have not been defined.

Recent advances in functional genomics and proteomics provide powerful technologies for separation and identification of components of protein complexes that can be concentrated by affinity chromatography or fractionation methods. Proteins and peptides can then be separated either by two-dimensional polyacrylamide gel electrophoresis (2D PAGE) or by liquid chromatography *(4, 5)*. Proteins are separated on the basis of their differences in isoelectric point (pI) and molecular mass by 2D PAGE. Protein spots are visualized by staining with various protein dyes. The identity of the protein of interest excised from the gel can be determined by proteolytical digestion and mass spectrometry analysis using either a solid phase-based matrix-assisted laser desorption ionization (MALDI) or a liquid phase-based electrospray ionization (ESI) approach in conjunction with HPLC/tandem mass spectrometry (MS/MS). The proteomic technologies are routinely available at the mass spectrometry facilities of most of institutions or commercially.

2. Materials

2.1. Equipment

1. Immobiline DryStrip kit.
2. Reswelling cassette.
3. Multiphor II electrophoresis unit (Amersham Biosciences, Piscataway, NJ).
4. Multitemp III thermostatic circulator.
5. HydroTech gel drying system.
6. Protean II stretch kit.
7. Protean II xi multicell.
8. Protean II xi multicell 2D conversion kit
9. Plate washer for Protean IIxi glass plates.
10. Model 395 gradient former.
11. Protean IIxi multicasting chamber.
12. PowerPac 200, PowerPac 3000.
13. Trans Blot SD system.
14. Peristaltic pump.
15. Consort 6000-V power supply.
16. Storm 860 imager (Molecular Dynamics, Sunnyvale, CA) or Typhoon Imager (Bio-Rad).

17. Automated spot picker (Amersham Biosciences) or gel cutter (Bio-Rad).
18. MassPREP Station automated protein digestion system (Micromass, UK).
19. CapLC capillary LC system (Micromass, UK).

2.2. Buffers and Supplies

2.2.1. Isolation of the HCV Replication Complex

1. Hypotonic buffer: 20 mM HEPES, pH 7.4, 10 mM KCl, 1.5 mM MgCl$_2$, 1 mM DTT, 1 mM phenylmethylsulfonyl fluoride (PMSF), and 2 μg/mL of leupeptin.

2.2.2. Better Separation and Purification of the HCV Replication Complex

1. Immobilized linear pH gradient (IPG) strips: (1) Immobiline DryStrip, 7 cm, pI 4–7 and pI 3–10; (2) Immobiline DryStrip, 11 cm, pI 4–7 and pI 3–10; (3) Immobiline DryStrip, 18 cm, pI 4–7 and pI 3–10 (Bio-Rad Laboratories, Hercules, CA; Applied Biosciences; or Amersham Biosciences).
2. Pharmalyte 3-10 (Bio-Rad).
3. Isoelectric focusing (IEF) lysis buffer: 7 M urea, 2 M thiourea, 4% CHAPS, 5 mM tributylphosphine.
4. IEF rehydration buffer: 7 M urea, 2 M thiourea, 2% CHAPS, 65 mM DTT, and 2% IPG buffer (Amersham Biosciences), pH 3–10.
5. Equilibration buffer 1: 50 mM Tris-HCl, pH 8.5, 6 M urea, 20–30% glycerol, 1–2% SDS, and 1–2% (w:v) DTT.
6. Equilibration buffer 2: 50 mM Tris-HCl, 6 M urea, 20–30% glycerol, 1–2% SDS, and 2.5–5% (w:v) iodoacetamide.
7. Staining solutions: SYPRO Ruby dye (Bio-Rad or Molecular Probes), 0.025% coomassie brilliant blue R250 in 40% methanol and 10% acetic acid, 0.1% bromophenol blue in 20% ethanol, or 0.1% Amido Black 10B in 7% acetic acid.

3. Methods

3.1. Isolation of the HCV Replication Complex

Several independent groups have demonstrated that the crude HCV replication complex (RC) isolated from Huh-7 cells that contain a persistently replicating subgenomic or genomic HCV RNA is able to synthesize HCV RNA in vitro (6, 7) The crude HCV RC is prepared from HCV RNA–replicating cells by cell lysis and removal of unlysed cells and nuclei.

1. Wash HCV replicon–containing Huh-7 cells twice with PBS, detach them by scraping, and collect them by centrifugation at 500–1000g for 10 min at 4°C.
2. Resuspend the cell pellet in hypotonic buffer and hold the suspension on ice for 20 min.
3. Lyse the cells further using a Dounce homogenizer.

4. Remove unlysed cells and nuclei by centrifugation at 1000g for 5–10 min at 4°C.

5. Concentrate the membrane-bound HCV RC in the supernatant by ultracentrifugation at 15,000g or 68,500g for 20 min to 1 h at 4°C

3.2. Better Separation and Purification of the HCV Replication Complex

Better separation and purification are desirable for identification of unique cellular proteins assembled in the HCV replication complex. The association of the HCV RC with intracellular ER–originated membrane structures allows it to be separated from other subcellular organelles on the basis of their differences in density *(1)*.

1. As schematically depicted in **Fig. 14.1**, apply the crude cytoplasmic extracts to the bottom of a density gradient (e.g., sucrose, glycerol, or iodixanol).

2. Spin at 100,000–150,000g for 16 h at 4°C. The membrane-bound HCV RC, having lower density, will float to the top of the gradient, whereas the soluble proteins will remain at the bottom.

3. Concentrate the membrane-bound HCV RC by ultracentrifugation as described above.

4. Use the resulting HCV RC to determine cellular proteins of the HCV RC by 2D PAGE and mass spectrometry (*see* **Note 1**).

The HCV RC can also be separated and purified by membrane-flotation centrifugation. Alternatively, the HCV RC can be purified by affinity chromatograph using tagged HCV RNA complementary to HCV genome or the N-terminus of NS3 or NS5A

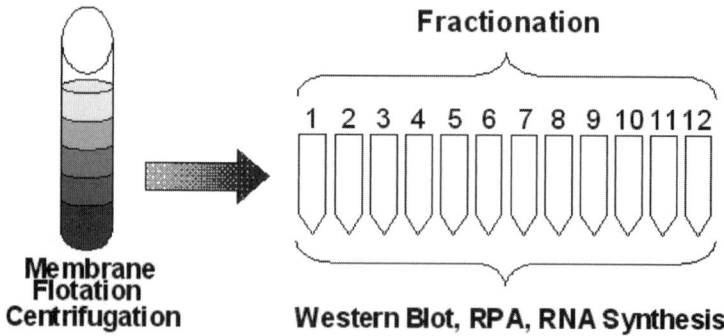

Fig. 14.1. Subcellular fractionation analysis. The HCV replicon-harboring Huh-7 cells or parental Huh-7 (control) cells are homogenized, and cell extracts are applied to the bottom of the gradient. Subcellular fractionation is carried out by density-gradient centrifugation. Fractions are collected either from top to bottom of the gradient or vice versa. HCV RNA and proteins in each fraction are determined by RNase protection assay (RPA) (or northern blot) and western blotting, respectively. The HCV RC in each fraction is determined by in vitro assay for its ability to synthesize HCV RNA in vitro. The HCV RC in the membrane fraction is concentrated by ultracentrifugation.

protein *(8, 9)*. Cross-linking reagents can also be used to stabilize the HCV RC before its isolation and separation. These methods warrant further investigation.

3.3. Separation and Identification of Unique Cellular Proteins in the HCV RC by 2D PAGE (Small Scale)

1. Dissolve proteins of the isolated and purified HCV replication complex in IEF lysis buffer containing a complete protease inhibitor cocktail.
2. Apply 200 μg of proteins to an immobilized pH gradient (IPG) strip.
3. Carry out IEF using pH 3–10 carrier ampholytes on a Multiphor II 2D apparatus for a total of 28,000 V h (starting with 200 V and gradually raising the voltage to 3500 V).
4. Equilibrate the first-dimension gel in equilibration buffer 1 and 2, and then lay it onto the second-dimension 10–20% gradient Criterion gel (Bio-Rad).
5. Run the second-dimension gel at 200 V until the dye reaches the bottom of the gel (*see* **Note 2**).
6. Visualize the protein spots by staining with a fluorescent dye, SYPRO Ruby.
7. Determine the fluorescent intensity of the protein spots with a Typhoon Imager and analyze them with PDQuest gel analysis software (Bio-Rad). Protein spots can be excised from the gel with a Bio-Rad spot cutter. Cellular proteins of the HCV RC purified by the above-described approaches can be determined by proteolytical digestion and mass spectrometry.

3.4. Separation and Identification of Unique Cellular Proteins in the HCV RC by 2D PAGE (Fig. 14.2) (Various Scale)

1. Dilute 100, 200, or 500 μg of purified HCV RC to 200 μL with IEF rehydration buffer [2D lysis buffer adjusted to 1% ampholytes (IPG buffer) or 0.5% Pharmalytes 3–10, 0.4% DTT, and a trace amount of bromophenol blue]. The volume of rehydration buffer depends on the size of the strip, as follows: 7 cm, 125 μL; 11 cm, 185 μL; 17 cm, 300 μL; 18 cm, 315 μL; and 24 cm, 410 μL.

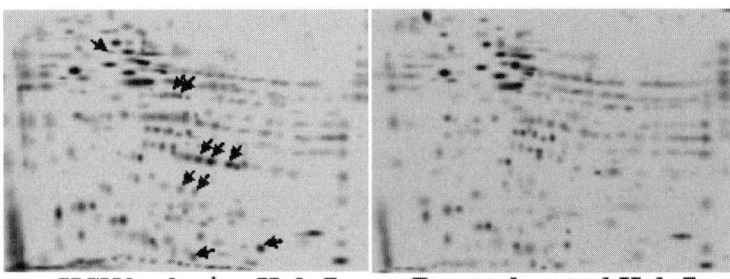

HCV-harboring Huh-7   Parental control Huh-7

Fig. 14.2. Two-dimensional PAGE analysis of protein complexes. The HCV RC was isolated from a subgenomic HCV RNA-harboring Huh-7 cell line by subcellular fractionation methods (**Fig. 14.1**). The protein complexes from naïve Huh-7 cells were used as a control. Arrows indicate unique or significantly up-regulated proteins associated with the HCV RC.

2. Rehydrate a 7, 11, or 18-cm immobilized pH gradient (IPG) strip containing a pH 3–10 gradient overnight (11–16 h) at room temperature with HCV RC proteins in 2–3 mL of mineral oil.

3. Run IEF on a Multiphor II flatbed apparatus cooled to 20°C, following the manufacturer's instructions, for a total of 28,000 Vh, starting at 200 V and gradually raising the voltage to 3500 V.

4. Gently blot the strips while draining off excess oil after IEF.

5. Equilibrate the strip for 10–15 min in equilibration buffer 1.

6. Equilibrate the strip for 10–15 min in equilibration buffer 2 before loading it onto SDS-polyacrylamide slab gels for second-dimension electrophoresis of proteins based on their molecular mass.

7. Load the strip onto the top of a 10–20% polyacrylamide gradient Criterion gel or onto SDS-polyacrylamide slab gels (12.5%).

8. Place 1% low-melting agarose (trace of bromophenol blue) overlay in the IPG well of the gel and wait for 5 min to allow agarose gel solidification before electrophoresis.

9. Conduct electrophoresis at 200 V until the dye reaches the bottom of the gel. The necessary length of time depends on the length of the strip: 7 cm, 40 min; 11 cm, 65 min; 17 cm, 5.5 h; and 18 or 24 cm, 6 h.

10. Stain the gel with SYPRO Ruby fluorescent dye according to the manufacturer's instructions to reach optimal detection of proteins with lower abundance.

11. Acquire images of proteins on the gel with the Storm 860 imager or Typhoon Imager.

3.5. Image and Statistical Analysis

Gel images are analyzed with PDQuest software. The positions of molecular weight standards used for electrophoresis are used as references to calibrate the 2D gel y-axis for determining unknown protein spot mass. The pH gradient is determined in the horizontal dimension with PDQuest, on the assumption that the physical ends of the IPG gel bed define the boundaries of the pH gradient and that the gradient is linear within the strip. For determining protein spot intensities and x and y coordinates, a digitized Gaussian master gel image containing all spots is generated with PDQuest.

1. Collect SYPRO Ruby-dye images using a Storm 860 gel scanner or a Typhoon Imager.

2. Normalize gel images by adjusting the exposure times according to the average pixel values observed.

3. Quantify the intensity of protein spots and calibrate gel images using PDQuest gel analysis software, which was developed specifically for 2D-DIGE, using the experimental design described above.

4. Merge two gel images derived from the HCV RC and parental cell control to determine proteins of interest by visualizing the intensities of protein spots.

5. Calculate the spot volume (sum of pixel intensities) in each gel and then normalize according to the corresponding standard gel spot volume (*see* **Note 3**)

6. Compare the intensity of protein spots across all gel images for each matched spot.

7. Identify unique or significantly up-regulated protein spots upon analysis of gel images.

8. Locate protein spots on the gel stained with SYPRO Ruby dye.

9. Excise protein spots of interests from preparative gels by means of an automated spot picker or gel cutter.

3.6. Identification of Viral/Cellular Proteins in the HCV RC by Tryptic Digestion and Mass Spectrometry

Mass spectrometry (MS) can determine the mass of a protein and the masses of its digested peptides. The identity of cellular proteins assembled in the HCV RC can be identified if the mass spectrum of the unique protein spots in the HCV RC, as visualized by 2-D PAGE analysis, is first determined. MALDI and MS/MS have been developed for proteomic studies of proteins and protein complexes. Mass spectrometry is the method of choice for rapid identification of proteins because of its high sensitivity. It requires much less protein than the classical amino acid sequence determination by Edman degradation. It can also accurately determine the amino acid sequence with modification *(5)*. The mass spectrum of digested protein can be used to search databases for identification of protein candidates. The unique cellular proteins identified by MS should be further confirmed by means of antibodies specific to identified proteins in conjunction with functional studies. Database searches using the known peptide sequences will reveal the identity of the cellular protein. In addition, MS/MS can be used to identify the amino acids of cellular proteins in case database search does not yield positive or clearcut results. The MS and MS/MS technologies are sufficient to identify cellular proteins of the HCV RC. Proteins can be identified by MASCOT MS/MS Ions Search at the website http://www.matrixscience.com/cgi/search_form.pl? FORMVER=2&SEARCH=MIS.

3.6.1. In-Gel Trypsin Digestion

In-gel trypsin digestion is performed with an automated protein digestion system, MassPREP Station.

1. Wash the gel slices twice with 50 μL of 25 mM NH_4HCO_3 and 50 μL of acetonitrile.

2. Reduce the cysteine residues by adding 50 μL of 10 mM dithiothreitol at 57°C and alkylate them by adding 50 μL of 55 mM iodoacetamide.

3. After dehydration with acetonitrile, digest proteins in the gel with 8 μL of 12.5 ng/μL of modified porcine trypsin (Promega, USA) in 25 mM NH_4HCO_3 at room temperature overnight (*see* **Note 4**).

4. Extract trypsin-digested peptides with 60% acetonitrile in 5% formic acid.

3.6.2. Mass Spectrometry Analysis

The resulting peptide mix is initially used for matrix-assisted laser desorption ionization time-of-flight mass spectrometry (MALDI-TOF MS) analysis. The resulting mass spectrum is then used for database research through Mascot or other search engines against SWISS-PROT and TrEMBL databases.

Further confirmation of the identity of putative proteins suggested by the MALDI-TOF MS analysis and database searches can be obtained by nanoscale capillary liquid chromatography–tandem mass spectrometry (LC–MS/MS) analysis carried out for trypsin-digested proteins with a CapLC capillary LC system (Micromass, UK) in conjunction with a hybrid quadrupole orthogonal acceleration time-of-flight tandem mass spectrometer (Q-TOF II, Micromass) (*see* **Note 5**).

3.6.3. Amino Acid Sequence Alignment

Amino acid sequences retrieved from Swiss-Prot and/or TrEMBL databases are aligned with either Dialign 2.2.1 or ClustalW Software, both of which are available at http://www.expasy.org.

4. Notes

1. The purity of the isolated membrane-bound HCV RC is critical and can be checked by detection of contaminant proteins from other intracellular organelles by use of known resident proteins of specific organelles as markers. The other important factor for 2D separation of the HCV RC components is the solubility of the proteins in the lyses buffer, which could be problematic. Undissolved precipitates should be removed.

2. IPG strip hydration and IEF running are critical to the success of a 2D proteomics analysis. For details, *see* pages 31–37 of the document at http://www.bio-rad.com/cmc_upload/0/000/013/012/4006164B.pdf. In addition, a detailed description of in-gel rehydration can be found in the instruction on immobilized linear pH gradient pH 3–10 IPG strip and a horizontal electrophoresis apparatus, Multiphor II, APB. IEF is performed on IPG strips with nonlinear pH gradients for alkaline protein samples and linear pH gradients for acidic protein samples.

3. Comparison of normalized protein expression intensities within each gel will give a standardized expression ratio.

4. For in-gel tryptic digestion, just enough buffer is typically added to cover the gel slice so that the local concentration of trypsin is high. Depending on the amount of the protein in each gel band, different amounts of trypsin can be tested so that a complete digestion can be achieved with the minimal amount of trypsin. This procedure will reduce the peptides generated by autolysis of trypsin. In addition, detergents such as SDS or NP-40 are incompatible with most of the mass spectrometric analysis and should therefore be avoided. If detergent must be used in dissolve protein samples, or in other steps, the gel should be thoroughly destained before in-gel digestion. For a typical coomassie-stained gel band, the gel slice should turn completely transparent after destaining, indicating the removal of SDS from the gel.

5. The mass spectrometric analysis will probably be handled in a core facility. Please follow the standard guidelines set by your facility. MALDI-based and ESI-based mass spectrometers have been shown to analyze some peptides preferentially, depending on their hydrophobicity. Complementary use of the two ionization sources will yield higher coverage of peptide identification.

References

1. El-Hage, N. and Luo, G. 2003. Replication of hepatitis C virus RNA occurs in a membrane-bound replication complex containing nonstructural viral proteins and RNA. *J. Gen. Virol.* **84,** 2761–2769.

2. Lohmann, V., Korner, G., Koch, J., Herian, U., Theilmann, L., and Bartenschlager, R. 1999. Replication of subgenomic hepatitis C virus RNAs in a hepatoma cell line. *Science* **285,** 110–113.

3. Luo, G. 2004. Molecular virology of hepatitis C virus. Birkhauser, Basel.

4. Andersen, J. S. and Mann, M. 2006. Organellar proteomics: turning inventories into insights. *EMBO Rep.* 7,874–879.

5. Yates, J. R., III. 2000. Mass spectrometry. From genomics to proteomics. *Trends Genet.* **16:**5–8.

6. Hardy, R. W., Marcotrigiano, J., Blight, K. J., Majors, J. E., and Rice, C. M. 2003. Hepatitis C virus RNA synthesis in a cell-free system isolated from replicon-containing hepatoma cells. *J. Virol.* 77, 2029–2037.

7. Quinkert, D., Bartenschlager, R., and Lohmann, V. 2005. Quantitative analysis of the hepatitis C virus replication complex. *J. Virol.* 79, 13594–13605.

8. Huang, H., Sun, F., Owen, D. M., Li, W., Chen, Y., Gale, M., Jr., et al. 2007. Hepatitis C virus production by human hepatocytes dependent on assembly and secretion of very low-density lipoproteins. *Proc. Natl. Acad. Sci. USA* **104,** 5848–5853.

9. Waris, G., Sarker, S., and Siddiqui, A. 2004. Two-step affinity purification of the hepatitis C virus ribonucleoprotein complex. *RNA* **10,** 321–329.

Chapter 15

Investigation of the Hepatitis C Virus Replication Complex

Volker Brass, Rainer Gosert, and Darius Moradpour

Abstract

Formation of a membrane-associated replication complex, composed of viral proteins, replicating RNA, altered cellular membranes, and other host factors, is a hallmark of all positive-strand RNA viruses. In the case of HCV, RNA replication takes place in a likely endoplasmic reticulum–derived membrane alteration referred to as the "membranous web." In vitro transcription-translation, membrane extraction and flotation analyses, immunofluorescence microscopy, fluorescent in situ hybridization, and RNA metabolic labeling followed by confocal laser scanning microscopy have yielded insights into the structure and function of the HCV replication complex. We describe these techniques and highlight selected results.

Key words: Replication complex, membrane flotation, in situ hybridization, membranous web, RNA-dependent RNA polymerase.

1. Introduction

Genome replication of all positive-strand RNA viruses investigated so far takes place in membrane-bound replication complexes containing viral proteins, replicating RNA, and altered cellular membranes [reviewed in **refs.** *(1, 2)*]. Depending on the virus, replication may occur on altered membranes derived from the endoplasmic reticulum (ER), Golgi apparatus, mitochondria, or even lysosomes. The role of membranes in viral RNA synthesis is not well understood. It may include (i) the physical support and organization of the RNA replication complex *(3)*, (ii) the compartmentalization and local concentration of viral products *(4)*, (iii) tethering of the viral RNA during unwinding, (iv) provision of lipid constituents important for replication, and (v) protection

Hengli Tang (ed.), *Hepatitis C: Methods and Protocols, Second Edition, vol. 510*
© 2009 Humana Press, a part of Springer Science+Business Media
DOI 10.1007/978-1-59745-394-3_15 Springerprotocols.com

of the viral RNA from 5′ triphosphate or double-strand RNA-mediated host defenses or RNA interference.

A specific membrane alteration, designated the membranous web, has been identified as the site of RNA replication in Huh-7 cells containing subgenomic HCV replicons *(5)*. Formation of the membranous web was induced by nonstructural protein 4B (NS4B), a polytopic membrane protein, and this web was very similar to the "sponge-like inclusions" previously found by electron microscopy in the liver of HCV-infected chimpanzees *(6)*. Incorporation of the other nonstructural proteins depends on specific membrane segments located at the N-terminus of NS4A *(7)* and NS5A *(8, 9)* as well as the C-terminus of NS5B *(10,11)* [reviewed in **ref.** *(12)*]. The membranous web is currently believed to be derived from ER membranes. Ongoing studies are aimed at characterizing the host factors and cellular processes involved in formation of the HCV replication complex.

Interestingly, the mechanism of membrane association of NS4A, NS5A, and NS5B appears to be different from that of the vast majority of cellular membrane proteins, as membrane association occurs by means of a posttranslational and presumably signal-recognition particle (SRP)-independent pathway *(8, 11)*. Detailed structural and functional analyses of the membrane segments of these viral nonstructural proteins suggest complex interactions within the phospholipid bilayer that may be required for the assembly of a functional replication complex *(9, 10)*. Recent studies have allowed the identification of a number of host factors required for efficient HCV RNA replication. These include, among others, human vesicle-associated membrane protein–associated protein A (hVAP-A) *(13)*, F-box and leucine-rich repeat protein 2 (FBL-2) *(14)*, and cyclophilin B *(15)*.

Once the multiprotein replication complex is functionally established, RNA replication is carried out by the NS5B RNA-dependent RNA polymerase. During this process, positive-strand RNA serves as template for a negative-strand replicative intermediate that is subsequently used as template for positive-strand genome synthesis. The mechanisms regulating translation, replication, and packaging of the viral genome are not well understood. Current evidence indicates that NS5A might function as a "molecular switch" between replication and assembly [reviewed in **ref.** *(16)*]. In this context, the three-dimensional structure of the N-terminal domain I of NS5A revealed that this protein dimerizes and forms a basic cleft at the surface of the membrane *(17)*. This cleft could perfectly accommodate single- or double-stranded RNA. According to one hypothesis, multiple NS5A dimers form a two-dimensional array on intracellular membranes, thereby creating a "basic railway" that would tether the viral RNA onto intracellular membranes and coordinate its different fates during HCV replication.

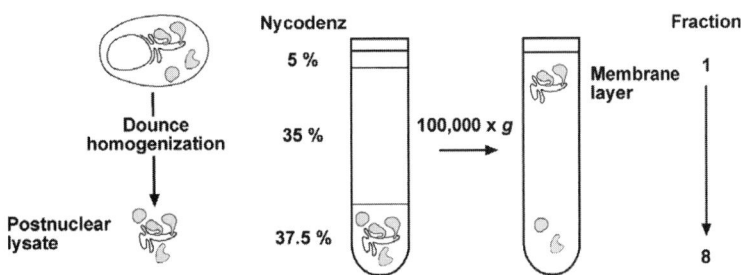

Fig. 15.1. Schematic diagram of the membrane flotation procedure. After hypotonic lysis, postnuclear lysates are loaded at the bottom of a 37.5 to 5% Nycodenz gradient. During equilibrium centrifugation at 100,000g for 20 h, membrane components float to the upper, low-density fractions. A cloudy membrane layer can be visible. Fractions of equal volume are collected from the top to the bottom and analyzed by immunoblotting. Membrane-bound proteins are found in the low-density fractions. Unbound cytosolic and aggregated proteins remain in the high-density fractions.

Here, we describe selected techniques used to investigate the structure and function of the HCV replication complex (**Fig. 15.1**). The first step in this process is the membrane association of newly synthesized nonstructural proteins that can be studied by in vitro transcription-translation (IVTT) as well as membrane extraction and flotation analyses. We also discuss techniques for identifying and visualizing the membrane-bound replication complex, that is, immunofluorescence, fluorescent in situ hybridization, and RNA metabolic labeling. For description of electron microscopy techniques used to analyze the ultrastructure of the replication complex, the reader is referred to a comprehensive review on this technique *(18)*.

2. Materials

2.1. In Vitro Transcription-Translation

1. RNAse-free water.
2. TNT T7-Coupled Rabbit Reticulocyte Lysate System (Promega, Madison, WI): TNT buffer (50×); TNT amino acid mix without methionine; TNT T7 polymerase; TNT reticulocyte lysate (TNT RRL, *see* **Note 1**). See also the manufacturer's recommendations.
3. RNAsin (Promega, Madison, WI).
4. 0.8 mCi/mL [^{35}S]methionine (Amersham Biosciences, Little Chalfont, Buckinghamshire, UK).
5. Canine microsomal membranes (e.g., Promega, Madison, WI). Store membrane preparations at −80°C (*see* **Note 2**).
6. Puromycin (Sigma, St. Louis, MO). Store a 12.5 mM stock solution in water at −20°C.

7. Gel fixation buffer 1: 50% methanol, 10% glacial acetic acid.
8. Gel fixation buffer 2: 7% methanol, 7% glacial acetic acid, 1% glycerol.

2.2. Hypotonic Cell Lysis

1. Hypotonic lysis buffer: $2\,mM\,MgCl_2$, $10\,mM$ Tris-HCl, pH 7.5, protease inhibitor (e.g., Complete Protease Inhibitor Cocktail, Roche Diagnostics, Mannheim, Germany).
2. Dounce homogenizer (e.g., Tissue Grinder Size 20, Kontes Glass Co., Vineland, NJ).

2.3. Membrane Extraction

1. NTE buffer: $100\,mM$ NaCl, $10\,mM$ Tris-HCl pH 7.0, $0.1\,mM$ EDTA.
2. High-salt extraction buffer ($2\times$): $2\,M$ NaCl in NTE.
3. Alkaline extraction buffer ($2\times$): $200\,mM\,Na_2CO_3$, pH 11.5.
4. Urea extraction buffer ($2\times$): $8\,M$ urea.
5. Membrane dissolving buffer ($2\times$): 2% Triton-X 100 in NTE.

2.4. Membrane Sedimentation

No additional reagents are required for membrane sedimentation analyses.

2.5. Membrane Flotation

1. 75% (w:v) Nycodenz (AXIS-SHIELD, Oslo, Norway) in $2\times$ PBS (*see* **Note 3**).
2. 35% (w:v) Nycodenz in PBS.
3. 5% (w:v) Nycodenz in PBS.

2.6. Indirect Immunofluorescence

1. 2% paraformaldehyde (PFA, Sigma, St. Louis, MO) in PBS (*see* **Note 4**).
2. 0.05% saponin (Sigma, St. Louis, MO) in PBS (*see* **Note 5**).
3. 0.05% saponin, 3% bovine serum albumin (BSA, Sigma, St. Louis, MO) in PBS (*see* **Note 6**).
4. Primary antibody.
5. Secondary antibody.
6. Antifading reagent (e.g., SlowFade, Molecular Probes, Eugene, OR).

2.7. Fluorescent In Situ Hybridization of Viral RNA

1. 4% PFA (Sigma, St. Louis, MO) in PBS.
2. 4% PFA (Sigma, St. Louis, MO)/0.2% Triton-X100 in PBS.
3. $0.5\,M$ NH_4Cl in PBS.
4. $10\times$ transcription buffer, $100\,mM$ NaCl.
5. $10\,mM$ ATP, CTP, GTP, and $6.5\,mM$ UTP (Roche, Diagnostics, Mannheim Germany).
6. $10\,mM$ FITC (or Texas Red)-UTP (Roche Diagnostics, Mannheim, Germany).
7. RNAse inhibitor $40\,IU/\mu L$ and DNAse $5\,IU/\mu L$ (Roche Diagnostics, Mannheim, Germany).
8. SP6 or T7 RNA polymerase (Roche Diagnostics, Mannheim, Germany).

9. Mini Biospin columns 6 and 30 (BioRad, Hercules, CA).

10. 400 mM $NaHCO_3$, 600 mM Na_2CO_3.

11. Hybridization buffer: 50% formamide, 10 mM Tris-HCl, pH 7.4, 600 mM NaCl, 10% dextran sulfate, 10 mM DTT, 0.05% BSA (DNAse free, RNAse free, Sigma, St. Louis, MO), 0.1% SDS, 200 μg/μL salmon sperm DNA (Sigma, St. Louis, MO), 100 μg/μL yeast tRNA (Sigma, St. Louis, MO).

12. 1.0 M sodium acetate (NaOAc), pH 6.0.

13. Acetic acid (1:10 diluted).

14. Ethanol, absolute.

15. $0.1 \times$ SSC.

2.8. Bromode-oxyuridine (BrU) Labeling of Nascent Viral RNA (See Note 7)

1. 5-bromouridine 5′-triphosphate (BrUTP, Sigma, St. Louis, MO). Prepare a stock solution at 400 mM in RNAse-free water. It can be stored for several weeks at −20°C.

2. Lipofectin (Invitrogen, Carlsbad, CA).

3. Actinomycin D (Merck Sharp and Dohme, Whitehouse Station, NJ). To produce a stock solution, dissolve actinomycin D at a concentration of 500 μg/mL in RNAse-free water.

4. MAbs against BrU, available from several companies (e.g., Calbiochem, La Jolla, CA; Santa Cruz Biotechnologies, Santa Cruz, CA; Invitrogen, Carlsbad, CA).

5. RNAse inhibitor 40 IU/μL (Roche Diagnostics, Mannheim, Germany).

6. Fixation buffer: 4% PFA in PBS.

7. Permeabilization buffer: 4% PFA with 0.2% Triton-X100 in PBS.

3. Methods

IVTT is often used to study the membrane association of proteins. Coupled transcription-translation systems are particularly useful in this regard. Canine microsomal membranes mimic the ER and can be added either during or after IVTT, allowing distinction of co- from posttranslational membrane association. Most cellular membrane proteins are targeted to membranes by means of an SRP-dependent pathway. The SRP interacts with the signal sequence of nascent polypeptide chains and directs the translation complex to the ER membrane. Therefore, SRP-mediated direction occurs only cotranslationally.

Membrane association of a given protein can be assessed and quantified by membrane sedimentation or flotation (19). The latter method is more complex and time consuming but more specific, as it allows distinction of aggregated from truly membrane-associated material. The nature and strength of membrane association can be analyzed by extraction studies. High-salt conditions (1 M NaCl) shield charges and interfere with ionic interactions that bind peripheral proteins to membranes either directly or indirectly through other membrane proteins. Extractions with 100 mM sodium carbonate, pH 11.5, are expected to release peripheral and intraluminal proteins by transforming microsomes into membrane sheets. The application of strongly denaturing conditions (4 M urea) is expected to break protein interactions of peripheral membrane proteins, so only integral membrane proteins are able to resist the latter extraction procedures.

Immunofluorescence microscopy is the method of choice for study of the subcellular localization of proteins. A typical fine reticular staining pattern with sparing of the nucleus and labeling of the nuclear membrane is an obvious indication of membrane association of the investigated protein. Double-label immunostaining and colocalization experiments with known cellular marker proteins allow further definition of the cellular compartment. When expressed in the context of the HCV polyprotein, nonstructural proteins accumulate in "dot-like structures" representing membranous webs. Fluorescent in situ hybridization (FISH) is used to study the subcellular localization of positive-strand RNA. HCV-specific positive-strand RNA was found as bright-fluorescing dots distributed over the cytoplasm with accumulation in the perinuclear region.

BrU labeling of nascent viral RNA allows determination of the site of HCV RNA replication. To this end, cells harboring replicating HCV RNA (i.e., cells transfected with an HCV replicon or infected with cell culture–derived HCV) are treated with actinomycin D to block cellular DNA-dependent RNA transcription and transfected with BrUTP, which is incorporated into nascent RNA produced by the viral RNA-dependent RNA polymerase. Incorporated BrU can subsequently be revealed by specific antibodies.

3.1. IVTT

1. Perform IVTT experiments in a 25 µL reaction mixture composed of 1 µg plasmid DNA, 0.5 µL TNT buffer, 0.5 µL RNasin, 0.5 µL TNT amino acid mix without methionine, 12.5 µL TNT reticulocyte lysate, 0.5 µL TNT T7 polymerase, 2 µL [^{35}S]methionine, and, if desired, microsomal membranes (see **Notes 1 and 2**). All steps of the IVTT procedure, including SDS-PAGE, must be carried out under appropriate radio safety conditions to prevent exposure to or contamination with [^{35}S]methionine. All radioactive waste must be collected and disposed of properly.

2. The amount of microsomal membranes to be added to the IVTT reaction must be titrated for each protein and batch of membranes, as the IVTT reaction can be inhibited by the

microsomal preparation. Start with a range between 0.5 and $3.0 \mu L$.

3. Perform the IVTT reaction with the TNT T7 polymerase at 30°C for 90 min.

4. To examine posttranslational membrane association, stop the IVTT reaction by adding puromycin to 1.25 mM, incubate at 30°C for 10 min, and subsequently add microsomal membranes.

5. IVTT reactions can be analyzed directly by autoradiography. Membrane association of the in vitro translated proteins can be studied by membrane sedimentation or flotation. Furthermore, the strength of membrane association can be defined by extraction experiments, as detailed below.

6. For membrane sedimentation, adjust the volume of the IVTT reaction to 50 μL with NTE buffer. For membrane flotation analysis, the IVTT reaction is adjusted with 200 μL 75% Nycodenz in 2× PBS and 175 μL water. For membrane extraction experiments, the IVTT reaction is centrifuged in a benchtop centrifuge at 16,000g (e.g., at 13,000 rpm in an Eppendorf 5415 benchtop centrifuge, Eppendorf AG, Hamburg, Germany) at 4°C for 10 min, and the pellet containing the microsomal membranes is resuspended in 150 μL of NTE buffer.

7. IVTT reactions are analyzed directly or after membrane sedimentation, membrane extraction, or membrane flotation. Samples are separated as usual by SDS-PAGE. Gels are fixed by incubation in fixation buffer 1 for 30 min, followed by fixation buffer 2 for 5 min. Gels are placed on filter paper and dried in a vacuum gel-drying device at 80°C for 2 h. Autoradiography can be detected by direct application of an x-ray film. Alternatively, protein detection can be performed by a phosphorimager device (e.g., Fuji BAS 1500) that allows rapid quantification of the protein amount (avoid saturated pixels to ensure a linear range).

3.2. Hypotonic Cell Lysis

1. Approximately 5×10^7 cells expressing the protein of interest are required. Place cell culture dishes on ice, rinse twice with cold PBS, harvest with a rubber policeman (a hand-held flexible natural-rubber scraper attached to a rod), and transfer to a 15 mL Falcon tube. Unless otherwise instructed, perform the subsequent steps at 4°C.

2. Centrifuge at 300g (e.g., at 1500 rpm in an Heraeus Multifuge 3 S-R, Kendro laboratory products, Germany) for 5 min.

3. Discard supernatant and resuspend the pellet in five times the estimated pellet volume of cold hypotonic lysis buffer. Centrifuge immediately at 300g for 5 min.

4. Resuspend the cell pellet in three times the estimated pellet volume of cold hypotonic lysis buffer and let the cells swell on ice for 10 min.

5. Homogenize the samples by 15–20 strokes in a Dounce homogenizer.

6. Centrifuge samples at 1000g (e.g., at 3,300 rpm in an Eppendorf 5415 benchtop centrifuge, Eppendorf AG, Hamburg, Germany) to remove the nuclei and cell debris.

7. For flotation analyses, adjust postnuclear lysates 1:1 with 75% Nycodenz in 2× PBS to produce a total volume of 400 μL and perform membrane flotation as indicated below.

8. For membrane extraction experiments, adjust postnuclear lysates to 1× PBS and centrifuge them at 100,000g (e.g., 40,000 rpm, Beckman Optima TLX ultracentrifuge, TLS 55 rotor, Beckman Coulter Inc., Fullerton, CA) for 40 min. Redissolve the membrane pellet in 150 μL NTE buffer and subject it to membrane extraction as indicated below (*see* **Note 8**).

9. Analyze equal amounts of protein from the pellet fraction and the supernatant after the 100,000 g centrifugation step by western blot to determine the membrane bound protein.

3.3. Membrane Extraction

1. Derive the samples from the IVTT reaction or hypotonic cell lysis as indicated above.

2. Mix 25 μL of each sample with 25 μL NTE buffer, 25 μL 2× high-salt extraction buffer, 25 μL 2× alkaline extraction buffer, 25 μL 2× urea extraction buffer, or 25 μL 2× membrane dissolving buffer. Apply extraction conditions on ice for 30 min (*see* **Notes 8 and 9**).

3. Analyze samples by either membrane sedimentation or membrane flotation. Samples can be directly subjected to membrane sedimentation. For membrane flotation experiments, adjust samples to 37.5% Nycodenz in PBS in a volume of 400 μL.

3.4. Membrane Sedimentation

1. Samples derived from IVTT or membrane extraction procedures can be analyzed (*see* **Notes 10 and 11**).

2. Centrifuge samples at 16,000g (e.g., at 13,000 rpm in an Eppendorf 5415 benchtop centrifuge, Eppendorf AG, Hamburg, Germany) at 4°C for 15 min. Collect the supernatant and resuspend the membrane pellet in a volume of NTE buffer equal to the volume of the supernatant.

3. Analyze equal volumes of the supernatant and pellet fractions by western blot or autoradiography. The membrane-bound protein is found in the pellet fraction (**Fig. 15.2**).

3.5. Membrane Flotation

1. Samples derived from IVTT, hypotonic cell lysis or membrane extraction can be analyzed by membrane flotation.

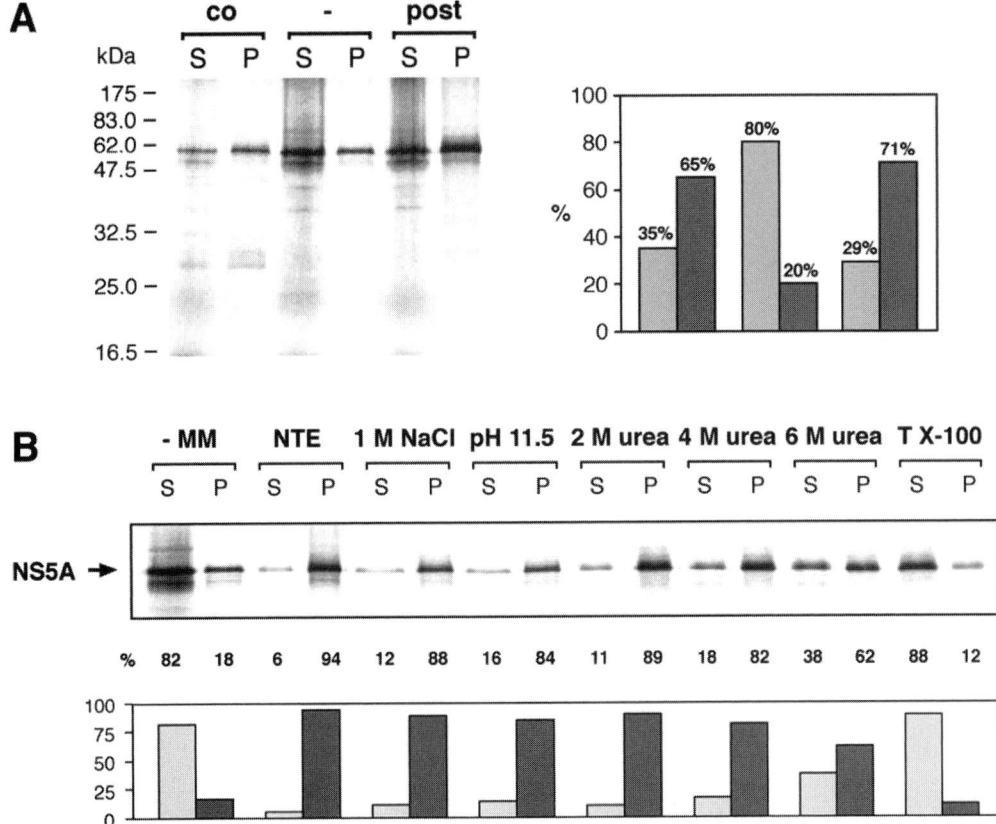

Fig. 15.2. Characterization of the membrane association of NS5A analyzed by IVTT, membrane sedimentation, and membrane extraction. IVTT reactions of NS5A were performed in the presence (co) or absence of microsomal membranes (post and −) as described under "Methods." (**A**) Membrane sedimentation analyses were performed to define the fraction of membrane-bound protein. Equal volumes of supernatant (S) and pellet fractions (P) were separated by 12% SDS-PAGE. [^{35}S]methionine-labeled translation products were detected by autoradiography. Quantification was performed with a Fuji BAS1000 PhosphorImager and the Fuji MacBAS version 2.4 software. Light bars represent supernatant, and dark bars represent pellet fractions. NS5A efficiently binds to membranes under both, co- and posttranslational conditions. (**B**) For membrane extraction experiments, microsomal membranes were posttranslationally added to IVTT reactions of NS5A. After sedimentation of microsomal membranes, the membrane extraction procedure was performed as described under "Methods." NS5A behaves as an integral membrane protein, as membrane extraction occurs exclusively after disruption of membranes by Triton-X100. Reprinted with permission from **ref.** 8.

2. Place each sample, adjusted to 400 µL 37.5% Nycodenz in PBS, at the bottom of a 1.4 mL, thick-walled ultracentrifugation tube (e.g., 11 × 34 mm PC Tube, Beckman Instruments, Spinco Division, Palo Alto, CA).

3. Overlay it carefully with 900 µL 37% Nycodenz and 100 µL 5% Nycodenz (*see* **Notes 3 and 12**).

4. Centrifuge it at 100,000g at 4°C for 20 h (e.g., 40,000 rpm, Beckman Optima TLX ultracentrifuge, TLS 55 rotor, Beckman Coulter, Fullerton, CA).

5. The membrane layer should be visible in the upper one-third of the gradient. Carefully collect fractions (e.g., 8 fractions of 175 μL) from the top to the bottom.

6. Separate the fractions by SDS-PAGE and analyze them by autoradiography or immunoblot. Membrane-associated proteins are found in the upper, low-density fractions, and soluble proteins and aggregates are found in the lower, high-density fractions (**Fig. 15.3**).

3.6. Indirect Immunofluorescence

1. Grow cells on sterile glass coverslips to a confluency of approximately 50–70%.

2. Rinse the cell monolayer twice with PBS.

3. Fix the cells with 2% PFA in PBS at 20°C for 40 min (*see* **Note 4**). PFA is toxic, so the fixation procedure must be done under an appropriate fume hood.

4. Discard the fixation solution in an appropriate hazardous-waste container.

5. Rinse cells three times with PBS.

6. Permeabilize cells with 0.05% saponin in PBS at 20°C for 20 min (*see* **Note 5**).

7. Remove permeabilization solution, add appropriately diluted primary antibody in PBS containing 0.05% saponin and 3% bovine serum albumin (*see* **Note 6**), and incubate at 20°C for 60 min.

8. Rinse cells three times with PBS containing 0.05% saponin and once with PBS containing 0.05% saponin and 3% bovine serum albumin.

9. Add appropriately diluted secondary antibody in PBS containing 0.05% saponin and 3% bovine serum albumin and incubate at 20°C for 60 min. Protect specimens from light (*see* **Note 13**).

10. Rinse cells three times with PBS containing 0.05% saponin and once with PBS.

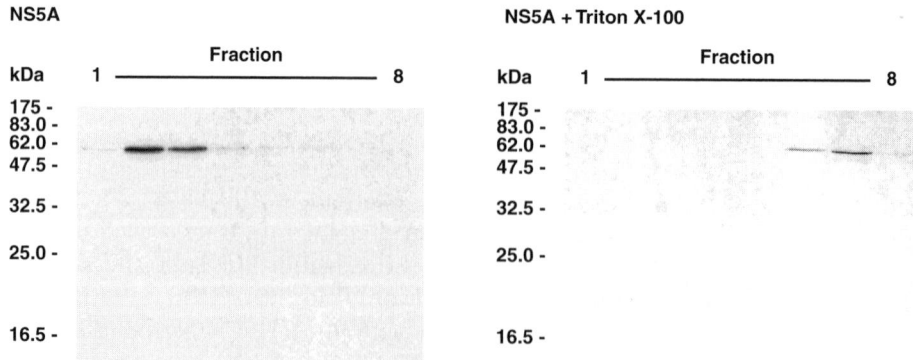

Fig. 15.3. Membrane flotation analysis of HCV NS5A. U-2 OS cells transiently transfected with a NS5A expression construct were analyzed by membrane flotation as described under "Methods." Fractions were collected from the top and analyzed by immunoblot using mAb 11H against NS5A. Treatment of the postnuclear lysate with 1% Trixon-X100 disrupts membrane association of NS5A.

11. Mount the coverslip in antifade reagent on a glass slide and seal with nail enamel.

3.7. Fluorescent In Situ Hybridization of Viral Positive-Strand RNA (See Note 7)

1. Grow cells on sterile glass coverslips to a confluency of approximately 50–70%.
2. Rinse the cell monolayer twice with PBS.
3. Fix the cells with 4% PFA in PBS at 20°C for 10 min (*see* **Note 4**).
4. Fix/permeabilize cells with 4% PFA/0.2% Triton-X100 at 20°C for 10 min.
5. Rinse cells twice with PBS.
6. Incubate cells in 0.5 M NH$_4$Cl at 20°C for 7 min.
7. Rinse cells twice with PBS.
8. Generate fluorochrome-labeled probe (e.g., 5 µL 10× transcription buffer, 5 µL 100 mM NaCl, 1.25 µL RNAse inhibitor, 5 µL of each 10 mM ATP, CTP, GTP, 6.5 mM UTP, 1.75 µL 10 mM FITC (or Texas Red)-UTP, approximates 500 ng linearized template, 2.5 µL T7 or SP6 polymerase (final concentration 1 U/µL), add 50 µL dH$_2$O, and incubate at 37°C for 2 h; subsequently add 5 µL DNAse (final concentration 1 U/µL) at 37°C for 15 min and stop the reaction with 5 µL 100 mM EDTA).
9. Remove unincorporated nucleotides by spinning over a Biospin column 6 at 1200g (e.g., 3500 rpm in an Eppendorf 5415 benchtop centrifuge) for 4 min before performing ethanol precipitation. Resuspend the probe in 50 µL dH$_2$O.
10. Hydrolyze the probe by adding 5.5 µL NaHCO$_3$/Na$_2$CO$_3$ and incubate it at 37°C in a water bath for an appropriate time (to generate approximately 100 nt fragments, use the following formula: t[min] = (transcript length[kb] − desired probe length[kb])/(0.11[kb/min] × transcript length [kb] × desired probe length[kb]).
11. Add 6.2 µL 1 M NaOAc, pH 6.0, and 3.1 µL acetic acid (1:10 diluted); precipitate probe with 190 µL ethanol and resuspend it in 30 µL of dH$_2$O.
12. To remove small fragments, spin over a Biospin 30 column at 1200g (e.g., 3500 rpm in an Eppendorf 5415 benchtop centrifuge) for 4 min.
13. To determine the optimal working dilution of the probe, prepare 1:2 dilution series in hybridization buffer and incubate cells with diluted probe at 40°C overnight (*see* **Note 13**).
14. Wash the cells four times in 0.1 × SSC at 20°C for 10 min and mount the coverslips.

3.8. BrU Labeling of Nascent Viral RNA (See Note 7)

1. Grow cells on sterile glass coverslips to a confluency of approximately 50–70%.

2. To block cellular RNA synthesis, add actinomycin D to the culture medium to a final concentration of 5 μg/mL and incubate at 37°C for 30 min.

3. Transfect BrUTP using Lipofectin (e.g., for each well of a 12-well tissue culture plate dilute 4 μL Lipofectin with 16.85 μL H$_2$O and incubate at 20°C for 10 min. Add

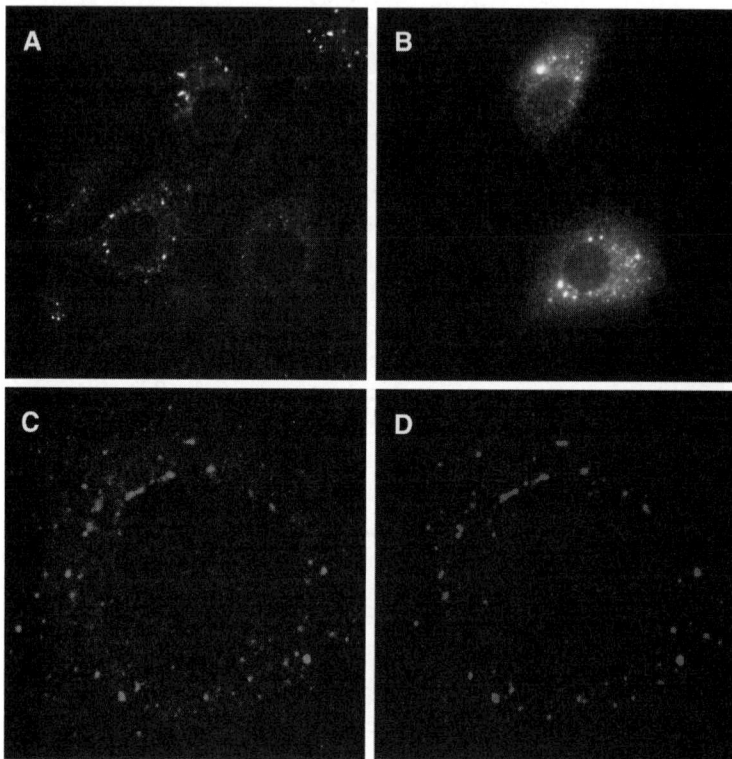

Fig. 15.4. Detection of HCV Proteins, Positive-strand HCV RNA, and nascent HCV RNA by immunofluorescence, fluorescent in situ hybridization, and metabolic labeling. (**A**) Human hepatoma cells harboring a subgenomic HCV replicon were analyzed by indirect immunofluorescence microscopy using mAb 11H directed against NS5A. NS5A was found predominantly in the cytoplasm as brightly fluorescing dots and in a reticular staining pattern. Very similar staining patterns can be observed with monoclonal antibodies (mAbs) directed against NS3 (when expressed together with NS4A), NS4A, NS4B, and NS5B (data not shown). (**B**) HCV positive-strand RNA was detected by FISH with a directly FITC-labeled riboprobe of negative polarity. Like the viral proteins, HCV positive-strand RNA accumulates in dot-like structures in the cytoplasm. (**C, D**) For detection of nascent HCV RNA, Huh-7 cells harboring a subgenomic HCV replicon were metabolically labeled with 5-bromouridine 5′-triphosphate in the presence of actinomycin D and then subjected to double-labeling confocal laser scanning microscopy. (C) NS5A is detected with the polyclonal antiserum WU144 and a FITC-labeled secondary antibody; (D) newly synthesized, bromouridine-labeled viral RNA is visualized with a mAb against bromodeoxyuridine and a Texas-Red-conjugated secondary antibody. Both signals for NS5A and nascent RNA clearly colocalize to the same cytoplasmic dots. The observation that NS5A colocalizes with newly synthesized viral RNA demonstrates that these dot-like structures represent the site of viral RNA synthesis and, therefore, the HCV replication complex.

2.85 µL of the BrUTP stock solution to the diluted Lipo-fectin and incubate at 20°C for 20 min. Subsequently, add 90 µL DMEM and 1.55 µL actinomycin D stock solution to the BrUTP–Lipofectin mixture).

4. Wash the cells with DMEM containing 5 µg/mL actinomycin D, add the BrUTP–Lipofectin mixture, and incubate at 37°C for 3 h.

5. Rinse the cells twice with PBS.

6. Proceed to indirect immunofluorescence as detailed above. Dilute primary and secondary antibodies in PBS with 1 IU/µL RNAse inhibitor. For permeabilization, Triton-X100 is preferred to saponin, and no BSA is added to buffers to avoid contamination with RNAses. To minimize unspecific antibody binding, a quenching step with 1 mL 0.5 M ammonium chloride can be included after fixation (**Fig. 15.4**).

4. Notes

1. Before membrane sedimentation experiments, clear TNT rabbit reticulocyte lysate of traces of membranes by centrifugation at 16,000g (e.g., at 13,000 rpm in an Eppendorf 5415 benchtop centrifuge, Eppendorf AG, Hamburg, Germany) at 4°C for 15 min.

2. Membrane association studies in the IVTT setting, including the investigation of the membrane topology and posttranslational mechanisms of membrane association as well as glycosylation studies, depend on the quality of the microsomal membrane preparation. Therefore, each batch of microsomes should be tested with well-characterized control proteins.

3. The different Nycodenz solutions for the membrane flotation should be prepared freshly, and a test gradient should demonstrate the accuracy of the gradient before each individual experiment.

4. PFA solution should be prepared freshly. For convenience, 8% stock solutions can be prepared and stored at −20°C. If the solubility is problematic, PFA can be dissolved in 60–70°C warm water by addition of sodium hydroxide before adjustment to 1× PBS. Before use, adjust pH to 7.4.

5. Alternatively, perform permeabilization with 0.2–0.3% Triton-X100.

6. Filter blocking buffer before use to remove remnants of undissolved BSA that could result in granular background staining.

7. Perform the BrU labeling and FISH procedure under RNAse-free conditions to prevent degradation of the nascent RNA. To this end, prepare all reagents with RNAse-free

water in plastic bottles, and filter-sterilize reagents. All steps of the BrU detection should be carried out on ice.

8. If membrane extractions are performed from cell lysates, increase the volumes of the extraction procedure if necessary as a result of high cell concentrations, which require larger volumes of hypotonic lysis buffer.

9. Membrane extraction with urea usually is performed at a final concentration of 4 M, but further concentrations, for example, 2 M and 8 M, can be included in the extraction procedure.

10. Protein aggregation can be problematic for the interpretation of membrane sedimentation experiments in the IVTT setting. To control for the amount of sedimented protein resulting from aggregation, include for each investigated protein an IVTT reaction without addition of microsomal membranes.

11. To analyze hydrophobic proteins in the IVTT setting and to minimize aggregation, low concentrations of digitonin (e.g., 0.01%) can be added to the reaction. Digitonin must be prepared freshly and a careful titration conducted to determine the concentration that reduces aggregation but leaves the membranes unaffected. Digitonin is toxic and must be handled accordingly.

12. In principle, other gradients in different volumes are feasible, but this approach using rather low volumes for the gradient yields high protein concentrations in each fraction and thus avoids additional protein concentration procedures. Furthermore, small amounts of material can be analyzed.

13. Fluorescent labeled antibodies and RNA probes are susceptible to bleaching. To obtain an optimal fluorescence signal, carry out all steps after the addition of labeled antibodies and RNA probes in the dark.

References

1. Mackenzie, J. (2005) Wrapping things up about virus RNA replication. *Traffic* **6**, 967–977.

2. Salonen, A., Ahola, T., and Kaariainen, L. (2005) Viral RNA replication in association with cellular membranes. *Curr. Top. Microbiol. Immunol.* **285**, 139–173.

3. Lyle, J. M., Bullitt, E., Bienz, K., and Kirkegaard, K. (2002) Visualization and functional analysis of RNA-dependent RNA polymerase lattices. *Science* **296**, 2218–2222.

4. Schwartz, M., Chen, J., Janda, M., Sullivan, M., den Boon, J., and Ahlquist, P. (2002) A positive-strand RNA virus replication complex parallels form and function of retrovirus capsids. *Mol. Cell.* **9**, 505–514.

5. Gosert, R., Egger, D., Lohmann, V., Bartenschlager, R., Blum, H. E., Bienz, K., et al. (2003) Identification of the hepatitis C virus RNA replication complex in Huh-7 cells harboring subgenomic replicons. *J. Virol.* **77**, 5487–5492.

6. Egger, D., Wölk, B., Gosert, R., Bianchi, L., Blum, H. E., Moradpour, D., et al. (2002) Expression of hepatitis C virus proteins induces distinct membrane alterations including a candidate viral replication complex. *J. Virol.* **76**, 5974–5984.

7. Wölk, B., Sansonno, D., Krausslich, H. G., Dammacco, F., Rice, C. M., Blum, H. E., et al. (2000) Subcellular localization, stability, and trans-cleavage competence of the

hepatitis C virus NS3-NS4A complex expressed in tetracycline-regulated cell lines. *J. Virol.* **74,** 2293–2304.

8. Brass, V., Bieck, E., Montserret, R., Wölk, B., Hellings, J. A., Blum, H. E., et al. (2002) An amino-terminal amphipathic alpha-helix mediates membrane association of the hepatitis C virus nonstructural protein 5A. *J. Biol. Chem.* **277,** 8130–8139.

9. Penin, F., Brass, V., Appel, N., Ramboarina, S., Montserret, R., Ficheux, D., et al. (2004) Structure and function of the membrane anchor domain of hepatitis C virus nonstructural protein 5A. *J. Biol. Chem.* **279,** 40835–40843.

10. Moradpour, D., Brass, V., Bieck, E., Friebe, P., Gosert, R., Blum, H. E., et al. (2004) Membrane association of the RNA-dependent RNA polymerase is essential for hepatitis C virus RNA replication. *J. Virol.* **78,** 13278–13284.

11. Schmidt-Mende, J., Bieck, E., Hugle, T., Penin, F., Rice, C. M., Blum, H. E., et al. (2001) Determinants for membrane association of the hepatitis C virus RNA-dependent RNA polymerase. *J. Biol. Chem.* 276, 44052–44063.

12. Moradpour, D., Penin, F., and Rice, C. M. (2007) Replication of Hepatitis C Virus. *Nat. Rev. Microbiol.* **5,** 453–463.

13. Evans, M. J., Rice, C. M., and Goff, S. P. (2004) Phosphorylation of hepatitis C virus nonstructural protein 5A modulates its protein interactions and viral RNA replication. *Proc. Natl. Acad. Sci. USA* **101,** 13038–13043.

14. Wang, C., Gale, M., Jr., Keller, B. C., Huang, H., Brown, M. S., Goldstein, J. L., et al. (2005) Identification of FBL2 as a geranylgeranylated cellular protein required for hepatitis C virus RNA replication. *Mol. Cell* **18,** 425–434.

15. Watashi, K., Ishii, N., Hijikata, M., Inoue, D., Murata, T., Miyanari, Y., et al. (2005) Cyclophilin B is a functional regulator of hepatitis C virus RNA polymerase. *Mol. Cell* **19,** 111–122.

16. Moradpour, D., Brass, V., and Penin, F. (2005) Function follows form: the structure of the N-terminal domain of HCV NS5A. *Hepatology* **42,** 732–735.

17. Tellinghuisen, T. L., Marcotrigiano, J., and Rice, C. M. (2005) Structure of the zinc-binding domain of an essential component of the hepatitis C virus replicase. *Nature* **435,** 374–379.

18. Egger, D., Gosert, R., and Bienz, K. (2002) Role of cellular structures in viral RNA replication, in *Molecular Biology of Picornaviruses* (Semler, B. and Wimmer, E., eds.), ASM Press, Washington, D.C., pp. 247–253.

19. Miller, D. J. and Ahlquist, P. (2002) Flock house virus RNA polymerase is a transmembrane protein with amino-terminal sequences sufficient for mitochondrial localization and membrane insertion. *J. Virol.* **76,** 9856–9867.

Chapter 16

Regulation of Interferon Regulatory Factor 3-Dependent Innate Immunity by the HCV NS3/4A Protease

Kui Li

Abstract

Interferon regulatory factor 3 (IRF-3) is a ubiquitously expressed latent cellular transcription factor that plays a pivotal role in control of innate, type I interferon (IFN) antiviral responses. After viral infections, IRF-3 is activated by specific C-terminal phosphorylation, which induces its dimerization and nuclear translocation, whereupon IRF-3 activates the transcription of type I IFNs and a number of other antiviral effector genes. Many viruses have evolved strategies that antagonize signaling mechanisms leading to IRF-3 activation. Recent studies have shown that hepatitis C virus blocks IRF-3 activation and subsequent IFN induction by cleaving critical cellular substrates within the intracellular antiviral signaling pathways upstream of IRF-3 with its major protease, NS3/4A.

Key words: Interferon regulatory factor 3, hepatitis C virus, Sendai virus, innate immunity, native PAGE, phosphorylation, dimer, nuclear translocation, immunofluorescence, confocal microscopy.

1. Introduction

Cellular responses to viral infections are coordinated by a complex regulatory network that triggers the activation of a number of different cellular transcription factors, interferon regulatory factors (IRFs), nuclear factor-kappa B, and ATF/c-Jun, upon sensing of viral products by cellular pathogen-recognition receptors. Once activated, these transcription factors coordinately regulate the expression of type I interferons (IFN-α and -β) and a number of cytokines/chemokines that contribute to the establishment of an antiviral state, which limits viral replication and spread *(1–3)*. Interferon regulatory factor 3 (IRF-3) is a constitutively expressed, latent transcription factor that plays a central role in

Hengli Tang (ed.), *Hepatitis C: Methods and Protocols, Second Edition, vol. 510*
© 2009 Humana Press, a part of Springer Science+Business Media
DOI 10.1007/978-1-59745-394-3_16 Springerprotocols.com

type I IFN responses. Its C-terminal region contains a cluster of serine/threonine residues that are known to be phosphorylated during viral infections. These include the Ser385, Ser386, Ser396, Ser398, Ser402, Thr404, and Ser405 sites, which are involved in activation of IRF-3. Virus-induced C-terminal phosphorylation is thought to change the conformation of IRF-3, resulting in its dimerization, translocation to and retention in the nucleus, where it associates with transcriptional coactivators, CBP/p300, and activates type I IFN synthesis (4,5). During coevolution with their hosts, many viruses have acquired mechanisms that specifically perturb the signaling mechanisms leading to IRF-3 activation. A prime example is the hepatitis C virus (HCV), which uses its NS3/4A serine protease to cleave critical adaptor molecules in the innate intracellular antiviral signaling pathways, MAVS/IPS-1/Cardif/VISA and TRIF/TICAM-1, thereby disrupting virus-induced IRF-3 activation (6–11).

Here, we describe detailed experimental procedures for determining the activation status of IRF-3. As Ser396 is a minimal phosphoacceptor site required for in vivo activation IRF-3 in response to virus and double-stranded RNA, virus-activated IRF-3 phosphorylation can be detected by immunoblot analysis of IRF-3 with an antibody that specifically recognizes phosphorylated IRF-3 at residue Ser396 (12). IRF-3 phosphorylation can also be monitored by immunodetection of the slow-migrating, hyperphosphorylated IRF-3 species separated on sodium dodecyl sulfate polyacrylamide gel (SDS-PAGE) with an antibody against full-length IRF-3. Virus-induced dimerization of IRF-3 is determined by separation of total cell lysates under native conditions followed by immunoblot detection of both IRF-3 monomer and dimer (13). Immunofluorescence staining and subsequent confocal microscopy allow confirmation that activated IRF-3 has been translocated to the nucleus. Finally, promoter-based reporter-gene assay is used to determine the activation of IRF-3 target promoters.

2. Materials

2.1. Cell-Culture Reagents, Plasmids, and Virus Stocks

1. Dulbecco's modified Eagle's medium (DMEM), heat-inactivated fetal bovine serum (FBS), OPTi-MEM I, Dulbecco's phosphate-buffered saline (DPBS) without calcium and magnesium, Trypsin-EDTA.
2. Experimental reporter plasmids (firefly luciferase reporter gene driven by IRF-3–dependent promoters): IFN-β-Luc (a gift from Rongtuan Lin), ISG56-Luc (a gift from Michael Gale), or ISRE-Luc (Stratagene). All reporter plasmids are

prepared with endotoxin-free procedures (e.g., Qiagen End-ofree Maxiprep kit, *see* **Note 1**).

3. Internal control reporter vector: pCMVβGal (Clontech, expresses β-galactosidase under the constitutive CMV promoter), prepared with endotoxin-free procedures (*see* **Note 2**).

4. DNA transfection reagents: choose the reagent that gives best transfection efficiency in the cell lines to be used. We use TransIT-LT1 (Mirus Bio) for Hepa1-6 cells.

5. Sendai virus (SeV), Cantell Strain (Charles River Laboratories): This virus stock comes with 100 mL SeV grown in allantoic fluid (minimal titer 2000 HAU/mL) shipped on dry ice. Upon receiving and thawing on ice, spin down the debris, divide the virus stocks (supernatant) into aliquots, and store at −80°C.

2.2. Cell Lysis for Western Blotting

1. Cell lysis buffer: 25 mM Tris-HCl, pH 7.5, 150 mM NaCl, 1% Triton X-100.

2. Protease inhibitor cocktail (Sigma): Make small aliquots and store at −20°C. Add fresh to cell lysis buffer (1:100 dilution).

3. Okadaic acid (Calbiochem): a specific inhibitor of serine/threonine protein phosphatase 1 and 2A. Dissolve okadaic acid in dimethylsulfoxide or ethanol at 1 mg/mL and store in $10 - \mu$L aliquots at −20°C. Add fresh to cell lysis buffer to a final concentration of $1\,\mu$g/mL.

2.3. SDS-PAGE

1. 30% acrylamide/bis solution (37.5:1, with 2.67% C): Weigh 29.2 g acrylamide 0.8 g N′N′-bis-methylene-acrylamide; add ddH_2O to 100 ml. Filter and store at 4°C in the dark (up to 30 days). Alternatively, use the premixed 30% acrylamide/bis solutions (37.5:1) available from many commercial sources. Caution: unpolymerized acrylamide/bis is a neurotoxin, and direct contact with skin should be avoided.

2. Separating gel buffer: 1.5 M Tris-HCl, pH 8.8. Weigh 27.23 g Tris base, add approx 80 ml ddH_2O to dissolve, and adjust pH to 8.8 with 6 N HCl. Add ddH_2O to 150 mL and store at 4°C.

3. Stacking gel buffer: 0.5 M Tris-HCl, pH 6.8. Weigh 6 g Tris base, add approx 60 mL ddH_2O to dissolve, and adjust pH to 6.8 with 6 N HCl. Add ddH_2O to 100 mL and store at 4°C.

4. 10% SDS: Dissolve 10 g SDS in ddH_2O and adjust to 100 mL, store at room temperature (RT).

5. Isopropanol.

6. 10% ammonium persulfate (APS): Dissolve 0.1 g APS in 1 mL ddH_2O. Store at 4°C for approx. 1 month.

7. N,N,N′,N′-tetramethylethylenediamine (TEMED): store at RT.

8. 4× SDS sample buffer (10 mL): Mix 2.5 mL 1 M Tris-HCl, pH 6.8, 4 mL glycerol, 0.8 g SDS, 2 mL 2-mercaptoethanol, 0.5 mg bromophenol blue. Adjust to 10 mL with ddH$_2$O and store in aliquots at −20°C.

9. 10× SDS running buffer: To 30 g Tris base, 144 g glycine, 10 g SDS, add ddH$_2$O to 1 liter, and store at 4°C.

2.4. Native PAGE

1. 4× native sample buffer (10 mL): Mix 5 mL 0.5 M Tris-HCl, pH 6.8, 4 mL glycerol, 0.4 g sodium deoxycholate (DOC, Sigma), 0.5 mg bromophenol blue. Add ddH$_2$O to 10 mL, and store in aliquots at −20°C.

2. 10% DOC: dissolve 10 g DOC in 100 mL ddH$_2$O. The mixture may need stirring and slight heating to go into solution. Store at RT.

3. 10× native running buffer: 30 g Tris base, 144 g glycine. Add ddH$_2$O to 1 liter, and store at 4°C. Immediately before use, add DOC to 1% for the upper chamber buffer (cathode side).

2.5. Western Blotting of IRF-3

1. Transfer buffer: 3.03 g Tris base, 14.4 g glycine. Add ddH$_2$O to 800 mL, then add 200 mL methanol, mix well, and store at RT.

2. Nitrocellulose or PVDF membranes. We use Hybond^{TM} ECL membrane (Amersham/GE Health).

3. PBS: prepare 10× stock: 40 g NaCl, 7.2 g Na$_2$HPO$_4$, 1.2 g KH$_2$PO$_4$, 1 g KCl. Add ddH$_2$O to 500 mL and autoclave. Store at RT.

4. PBS-T: Add Tween-20 to 0.1% in 1× PBS.

5. Ponceau S solution: For reversibly staining protein bands on nitrocellulose or PVDF membranes. Prepare 0.1% (w/v) Ponceau S in 1% acetic acid; store at RT. Can be reused.

6. Blocking buffer: 3% nonfat dry milk in PBS.

7. Antibody dilution buffer: 3% nonfat dry milk in PBS or 5% BSA in PBS (*see* **Note 3**).

8. Rabbit anti-IRF-3 antibody (FL-425, from Santa Cruz); rabbit anti-phosphoSer396 IRF-3 (a gift from John Hiscott, *see* **Note 4**).

9. Secondary antibody: Goat anti-rabbit IgG-HRP (Southern Biotech).

10. Enhanced chemiluminescent reagents: Amersham/GE Health or Pierce.

11. X-ray film.

2.6. Stripping and Reprobing

1. Stripping buffer: 62.5 mM Tris-HCl, pH 6.8, 2% SDS, 100 mM 2-mercaptoethanol. For example, to prepare 500 mL stripping buffer, mix 62.5 mL 0.5 M Tris-HCl, pH 6.8, 100 mL 10% SDS, and 334 mL ddH$_2$O. Immediately before use, add 3.5 mL 2-mercaptoethanol (14.4 M stock)

and warm up to the desired stripping temperature (usually $50 - 70°C$).

2. Wash buffer: PBS-T.

3. Primary antibodies: monoclonal antibodies to HCV NS3 (Vector Labs) and β-actin (Sigma).

4. Secondary antibody: goat anti-mouse IgG-HRP (Southern Biotech).

2.7. Immuno-fluorescence Staining and Confocal Microscopy for IRF-3 Subcellular Localization

1. Four-well chamber slides.

2. Wash buffer: PBS.

3. Fixation buffer: 4% (w/v) paraformaldehyde in PBS. Prepare the fixation buffer in a fume hood. It requires stirring and heating to dissolve. We suggest that fixation buffer be prepared fresh each time, but we have found that the solution is usually good for two to three weeks when stored at RT.

4. Permeabilization buffer: 0.2% Triton X-100 in PBS.

5. Antibody dilution buffer: 3% BSA in PBS.

6. Secondary antibody: goat anti-rabbit IgG-FITC conjugate (Southern Biotech).

7. Nuclear stain: $4',6$-diamidino-2-phenylindole (DAPI). Prepare stock solution at $1 \, mg/mL$ in ddH$_2$O and store in small aliquots at $-20°C$, protected from light. Use $1 \, μg/mL$ working concentration (diluted in PBS) for counterstaining cellular nuclei.

8. VECTASHIELD Mounting Medium (Vector Labs). Store refrigerated in the dark.

9. Clear nail polish.

2.8. Reporter Gene Assay of IRF-3-Dependent Promoter Activity

1. Reporter lysis buffer (RLB, $5\times$) (Promega) can be stored at $-20°C$ or RT. Prepare fresh $1\times$ RLB by adding 4 volumes of ddH$_2$O each time before cell lysis. This buffer is compatible with both firefly luciferase and β-galactosidase assays.

2. Luciferase Assay System, 10-Pack (Promega, stored at $-20°C$). This kit comes with 10 vials of lyophilized luciferase assay substrate, and $10\times 10 \, ml$ of luciferase assay buffer. For reconstitution of the luciferase substrate, add $10 \, mL$ of the luciferase assay buffer per vial of lyophilized substrate. Once reconstituted, the substrates should be divided into aliquots and stored at $-80°C$. For frequent use, aliquots can be stored at $-20°C$ for up to a month.

3. Luminometer: Turner Designs 20/20 or other brands.

4. Recombinant β-galactosidase (from *E. coli*): Sigma or Roche.

5. $2\times$ assay buffer for β-galactosidase assay ($50 \, mL$): mix $20.5 \, mL \, 0.4 \, M \, Na_2HPO_4$, $4.5 \, mL \, 0.4 \, M \, NaH_2PO_4$, $0.5 \, mL$ $0.2 \, M \, MgCl_2$, $0.35 \, mL$ 2-mercaptoethanol, $66.5 \, mg$ of 2-nitrophenyl β-D-galactopyranoside (ONPG, Sigma), and

24.15 mL of ddH$_2$O. The 2× assay buffer is stored at −20°C in 10 mL aliquots.

6. Stop solution for β-galactosidase assay: 1 M sodium carbonate (Na$_2$CO$_3$). Store at either RT or −20°C.

3. Methods

3.1. Cell Culture and Sendai Virus Infection (as an Inducer for IRF-3 Activation)

1. Split subconfluent cells that do and do not harbor replicating genome-length HCV RNAs (14), or with and without expression of HCV NS3/4A (15), into six-well plates at a density of 2 × 10^5 cells per well in complete medium and culture them overnight.

2. Immediately before virus infection on the following day, prepare enough SeV inoculum by diluting the virus stock to 100 HAU/mL in prewarmed DMEM. Remove the cell-culture medium and wash cells in 2 mL of prewarmed DMEM once. Carefully remove all of the DMEM wash and load cells with either 1 mL prewarmed DMEM (mock) or SeV inoculum.

3. After 1 h incubation at 37°C, carefully aspirate the DMEM or SeV inoculum completely and wash cells once with prewarmed DMEM (*see* **Notes 5 and 6**). Then refeed cells with complete culture medium and culture them for a further 15 h before cell lysis.

3.2. Whole Cell Lysate Preparation for PAGE and Western Blotting

1. Sixteen hours after infection or mock infection with SeV, carefully aspirate culture medium from cells (*see* **Note 6**), wash cells three times with ice-cold PBS, and remove as much of the last wash as possible (but do not let cells dry out!).

2. Add 100 μL of cell lysis buffer supplemented with fresh protease inhibitor cocktails and okadaic acid (as described in **Section 2.2**). Place the cell plate on ice with gentle rocking, which promotes an even covering of the cells with lysis buffer.

3. After 10 min of incubation on ice, scrape lysed cells off the plate and transfer the individual samples to prechilled microcentrifuge tubes. Centrifuge the samples at 13,000g at 4°C for 10 min and recover the supernatants (whole cell lysates).

4. Proceed to protein quantification using the desired methods and fractionation on SDS- or native PAGE (see below), or immediately freeze cell lysates and store at −80°C until analysis.

3.3. SDS-PAGE

The instructions assume the use of the Mini-PROTEAN® II Electrophoresis Cell from Bio-Rad (16):

1. Assemble the clean glass plate sandwiches using 1.5 mm thick combs.

2. Prepare 10 ml of 7.5% separating gel monomer solution by mixing 4.85 mL ddH$_2$O, 2.5 mL separating gel buffer, 100 μL 10% SDS, 2.5 mL acrylamide/bis (30% stock), 50 μL 10% APS, and 5 μL TEMED. Pour the gel, leaving space for a stacking gel, and immediately overlay the monomer solution with isopropanol. Let the gel polymerize (which usually occurs after 30 to 45 min).

3. Pour off the isopropanol overlay and rinse the top of the gel twice with ddH$_2$O.

4. Prepare 5 mL of 4% stacking gel monomer solution by mixing 3.05 mL ddH$_2$O, 1.25 mL stacking gel buffer, 50 μL 10% SDS, 0.67 mL 30% acrylamide/bis stock, 25 μL 10% APS, and 5 μL TEMED. Quickly apply the stacking gel solution on top of the polymerized separating gel. Insert the comb. Let the stacking gel polymerize (which should occur within 20 to 30 min).

5. Prepare 1× SDS running buffer by diluting the 10× stock with ddH$_2$O. Prepare the samples to be loaded by mixing 50 μg of protein per sample with 1/3 volume of the 4× SDS sample buffer. Similarly, prepare the MW marker (high-range rainbow marker, Amersham) with the SDS sample buffer. Heat the microcentrifuge tubes containing the samples/MW marker for 10 min at 95°C and return them to RT.

6. Remove the comb from the polymerized stacking gel, assemble the gel-running apparatus, and immerse the gel in 1× SDS running buffer. Carefully flush the wells to remove unpolymerized gel if necessary. Briefly spin down the heated samples and load them onto the gel. Put the lid on and run the gel with a constant voltage between 160 and 200 V. Stop running until the bromophenol blue dye goes to the very bottom of the gel (which usually takes about 45 min to 1 h, depending on the voltage used).

3.4. Native PAGE for Separation of IRF-3 Dimers from Monomers (13)

1. Pour a 7.5% native separating gel by the process described in **Section 3.3** with the following exceptions: do not add SDS to the gel monomer solution, and use no stacking gel (*see* **Note 7**).

2. Fill the upper tank (cathode side) with 1× native running buffer freshly supplemented with 1% DOC, and fill the lower tank (anode side) with regular 1× native running buffer.

3. Prerun the gel at 40 mA for 30 min.

4. Mix 10–20 μg protein per sample with 1/3 volume of 4× native PAGE buffer. Do *not* heat the samples.

5. Load samples onto native gel, and run at a constant current (25 mA) for 80 min.

3.5. Western Blotting of IRF-3

1. Upon completion of SDS-PAGE (**Section 3.3**) or native PAGE (**Section 3.4**), disconnect the power supply and carefully disassemble the gel unit. Discard the stacking gel (SDS-PAGE) and keep the separating gel for electrophoretic transfer with Bio-Rad's Mini Trans-Blot Module.

2. Prewet the two fiber pads, two pieces of filter paper (slightly larger than the gel), and the membrane in prechilled transfer buffer. Assemble the "sandwich" as follows, ensuring that no bubbles are trapped between the layers: cathode side, fiber pad, filter paper, gel, membrane, filter paper, fiber pad, anode side. Insert the sandwich into the Trans-Blot Module, immerse it in the prechilled transfer buffer, and place it, along with a cold pack, in the transfer tank. Put the lid on and transfer the proteins on gel to the membrane at a constant voltage (100 V) for 1 h (*see* **Note 8**).

3. Once the transfer is completed, disassemble the transfer module. The MW marker should be clearly visible on the membrane.

4. Optional step: Stain the membrane with Ponceau S solution for 1 min, and rinse it with ddH$_2$O to view the protein bands. Take pictures if necessary. Then destain the membrane by rinsing it in PBS.

5. Block the membrane in >20 mL of blocking buffer at RT for 1 h.

6. Briefly rinse the membrane with PBS.

7. Incubate the membrane with the primary antibody (rabbit anti-IRF-3 at 1:500 in blocking buffer or rabbit anti-phosphoSer396-IRF-3 at 1:10000 in 5% BSA) at 4°C overnight (*see* **Notes 3 and 9**).

8. Wash the membrane in PBS three times, at 5, 10, and 5 min.

9. Incubate the membrane with goat anti-rabbit IgG-HRP conjugate (Southern Biotech, 1:10000) in blocking buffer at RT for 1 h.

10. Wash the membrane in PBS-T three times, at 5, 10, and 5 min.

11. Wash the membrane once in PBS for 5 min (*see* **Note 10**).

12. Incubate the membrane for 1 min with ECL Western Blotting Detection Reagents (prepare enough to cover the entire membrane); drain reagents and wrap the membrane in plastic wrap before exposing it to X-ray film. Perform various exposures for different time periods to obtain the desired signal intensity of target bands. Examples of the western blotting results for IRF-3 detection are shown in **Fig. 16.1** (*see* **Note 11**).

3.6. Stripping and Reprobing

1. Wash the membrane 2× 10 min in PBS-T. At this time prewarm enough stripping buffer to the desired stripping temperature (usually 50 − 70°C).

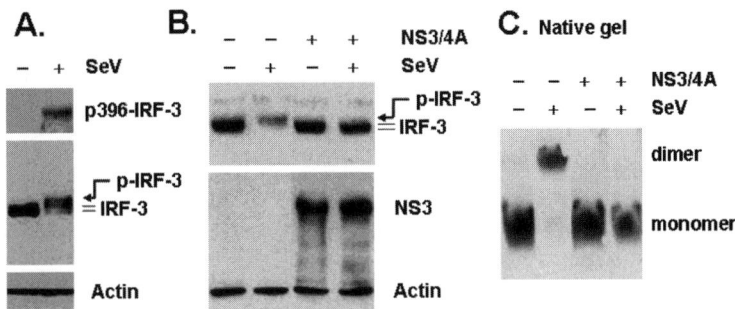

Fig. 16.1. Virus-induced hyperphosphorylation and dimerization of IRF-3 blocked by expression of HCV NS3/4A. **A.** HeLa cells were mock infected or infected with SeV for 16 h before cell lysis and SDS-PAGE, followed by immunoblot analysis with antibodies specific for phospho-Ser396 IRF-3 (*top panel*), total IRF-3 (*middle panel*), and actin (*bottom panel*). *The arrow* and *hatch marks* in the *middle panel* denote hyperphosphorylated (virus-activated, p-IRF-3) and inactive IRF-3 isoforms, respectively. **B** and **C.** Conditional NS3/4A expression was repressed or induced in osteosarcoma cells, which were subsequently mock-infected or challenged with SeV for 16 h. Equal amounts of cell lysates were separated by SDS-PAGE (B) or native PAGE (C), followed by immunoblot analysis of total IRF-3 (C and *top panel* in B) or NS3 and actin (*bottom panel* in B).

2. Strip the membrane in stripping buffer at 50 − 70°C for 15–30 min with occasional agitation. The temperature and time adequate for stripping must be determined empirically for individual antibodies.

3. Wash the membrane 2 × 10 min in PBS-T. Then reblock the membrane in blocking buffer for 1 h at RT.

4. The membrane is then ready for reprobing with a mixture of monoclonal antibodies to NS3 (at a dilution of 2 μg/mL) and β-actin (at a dilution of 1:5000) in PBS plus 1% milk or 1% BSA (*see* **Note 12**) at 4°C overnight. After PBS washes, incubate the membrane with goat anti-mouse IgG-HRP (Southern Biotech, 1:10000) in blocking buffer for 1 h, wash it again, and subject it to ECL detection, by procedures similar to those described in 3.5. A sample result is shown in **Fig. 16.1B**, bottom panel.

3.7. Immuno-fluorescence Staining and Confocal Microscopy for IRF-3 Subcellular Localization

1. Plate Huh-7 2-3 cells harboring autonomous replicating, genome-length HCV RNAs and their IFN-cured counterparts, 2-3c cells, in four-well chamber slides at a density of 2×10⁴ cells/well. On the following day, either mock-infect or infect cells with 25 HAU of SeV (250 μL SeV at 100 HAU/mL) in DMEM for 1 h. Then carefully remove the virus inocula and refeed the cells with complete medium and culture them for an additional 15 h (*see* **Notes 5 and 6**).

2. Aspirate the medium from the cells and wash the cells once in PBS. Do not let the cells dry (*see* **Note 13**)!

3. Aspirate the PBS wash and fix the cells in 4% paraformaldehyde at RT for 30 min.

4. Wash the cells three times with PBS, and add permeabilization buffer for 10 min at RT. Do not incubate longer than 15 min.

5. Wash the cells in PBS three times, and carefully remove the last wash.

6. Block nonspecific binding by incubating the cells with antibody dilution buffer (3% BSA in PBS) for 1 h at RT (*see* **Note 14**).

7. Remove the blocking solution and incubate the cells with rabbit anti-IRF-3 (1:100) diluted in antibody dilution buffer for 1 h at RT.

8. Remove primary antibody and wash the cells with PBS, three times.

9. Incubate the cells with goat anti-rabbit FITC (Southern Biotech) at 1:200 in antibody dilution buffer at RT for 45 min or at 37°C for 30 min in a humidified chamber.

10. Wash the cells in PBS, three times.

11. Counterstain the nuclei with DAPI (at 1 μg/mL) diluted in PBS and incubate them at RT for 5 min (*see* **Note 15**).

12. Wash the cells in PBS, three times. Shake excess liquid from the slide.

13. Carefully remove the plastic wells and gasket, add one drop of mounting medium, and put the cover slip on. Seal the slide with clear nail polish, and view it under a Zeiss 510 META confocal microscope. For each field, capture two images, one each at the excitation wavelengths of FITC (IRF-3) and DAPI (nuclei). Sample images of IRF-3 immunofluorescence and nuclei are shown in **Fig. 16.2**. If the sealed slides are not to be immediately viewed, they can be wrapped in aluminum foil and stored at 4°C for a week without significant loss of fluroscent signal.

3.8. Reporter Gene Assay of IRF-3-Dependent Promoter Activity

3.8.1. DNA Transfection and Virus Infection

1. Plate 1.5×10^4 cells/well (or a cell number that allows cells to reach 50–70% confluency on the following day) in 48-well plates in 300 μL of complete growth medium without antibiotics and let cells grow overnight. Prepare three wells of cells for each gene of interest (e.g., empty vector or NS3/4A), for mock and SeV infection in triplicate later.

2. On day 2, transfect the cells in each well with the following DNAs, using TransIT-LT1 reagent diluted in Opti-MEM I at a ratio of 1 μg: 3 μL: 50 μL (DNA:TransIT-LT1:Opti-MEM I): 50 ng of IFN-β-Luc (or ISG56-Luc or ISRE-Luc), 50 ng pCMVβGal, and 100 ng of plasmid DNA encoding the gene of interest (empty vector or NS3/4A). Prepare the master transfection mix for each gene of interest (a total of six wells)

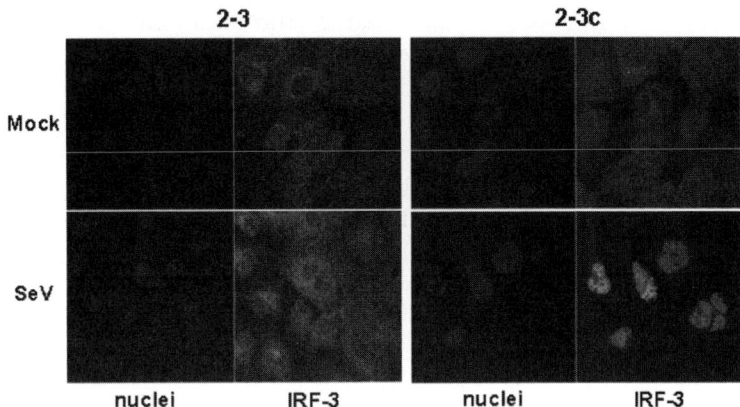

Fig. 16.2. Virus-induced IRF-3 nuclear translocation blocked in human hepatoma Huh-7 cells harboring autonomous replicating, genome-length HCV RNAs (2–3 cells). Confocal microscopy of IRF-3 subcellular localization in 2–3 cells (*left panel sets*) or the interferon-cured 2–3c cells (*right panel sets*), after either mock infection (*top panel sets*) or infection with SeV (*bottom panel sets*). IRF-3 was detected by indirect immunofluorescence (*right panels*); nuclei were visualized by DAPI staining (*left panels*).

to ensure pipetting accuracy. Load aliquots of the transfection complex onto the cells in complete growth medium.

3. At 24 h after transfection, either mock-infect or infect triplicate wells of cells expressing each gene of interest with 100 HAU/mL of SeV for 16 h, by a procedure similar to that described in **Section 3.1**, except that the volume of SeV or mock inoculum per well of a 48-well plate format is 125 μL.

3.8.2. Cell Lysate Preparation for Both Luciferase and β-galactosidase Assays

1. Carefully aspirate the culture medium and wash cells once with 1× PBS, being careful not to dislodge attached cells. Remove as much of the PBS as possible (*see* **Note 16**). Add enough 1× RLB (e.g., 80 μL per well of a 48-well plate) and rock the plate to ensure complete coverage of the cells by RLB. Perform a single freeze-thaw cycle to ensure complete lysis (*see* **Note 17**).

2. Pipette up and down to get lysed cells completely off of the plate. If 12-well or larger plates are used, cells can be conveniently scraped off with a scraper. Transfer the cell lysate into a microcentrifuge tube.

3. Vortex the microcentrifuge tube briefly, then spin it at 13,000g for 30 s at RT. Then transfer the supernatant to a new tube for reporter activity assays.

3.8.3. Luciferase Assay (The Standard Procedure with a Manual Luminometer) (17)

1. Thaw enough reconstituted luciferase assay substrate and warm it to RT. Dispense 100 μL of the substrate into luminometer tubes, one tube per sample.

2. Program the luminometer to perform a 2-s measurement delay followed by a 12-s measurement read for luciferase activity.

3. Add $10\,\mu L$ of cell lysate to a luminometer tube containing the substrate. Mix by pipetting two to three times or vortex briefly (*see* **Note 18**).

4. Place the tube in the luminometer and initiate reading. Record the reading on paper if the luminometer is not connected to a printer.

3.8.4. β-Galactosidase Assay (The Standard Assay Procedure with a Colorimetric Method) (18)

1. Thaw the $2\times$ assay buffer and warm it to RT. Prepare enough $1\times$ RLB for dilution of samples.

2. Pipette 10 to $150\,\mu L$ aliquots of cell lysate into labeled microcentrifuge tubes. The exact volume depends on the efficiency of the transfection and must be optimized (*see* **Note 19**). Adjust to $150\,\mu L$ by adding $1\times$ RLB. Also prepare 5 tubes for the reactions using known amount of standards (0, 1, 3, 6, and 10 milliunits of β-galactosidase; adjust to $150\,\mu L$ by adding $1\times$ RLB).

3. Add $150\,\mu L$ of $2\times$ assay buffer to each of the microcentrifuge tubes.

4. Mix all samples by vortexing briefly. Incubate the reactions in a 37°C water bath for up to 30 min or until a faint yellow color has developed. Approximately 5 min before the end of the reactions, turn on the spectrophotometer to warm it up.

5. Stop the reactions by adding $500\,\mu L$ of 1 M Na_2CO_3. Mix by briefly vortexing the tubes and read the absorbance immediately at 420 nm.

6. Enter the data into an Excel spreadsheet. Calculate the β-galactosidase milliunits per microliter of each sample on the basis of the standard curve constructed from the β-galactosidase standards.

3.8.5. Data Calculation and Presentation

Calculate the relative luciferase activity in each sample by dividing the firefly luciferase units by the β-galactosidase milliunits. Present the data in a bar graph using the average $+/-$ standard deviation of the relative luciferase activity of each transfection group (expressing the gene of interest or a control vector) from triplicate wells and per treatment (*see* **Fig. 16.3** for a sample illustration). Alternatively, calculate the fold change caused by the treatment (SeV infection) in each transfection group by dividing the relative luciferase activities of cells challenged with SeV by those of the mock-infected cells and present the fold change in a bar graph. Conduct Student's t test (between two groups) or Analysis of Variance (ANOVA) (among three or more groups) to determine the statistical differences among different experiment groups.

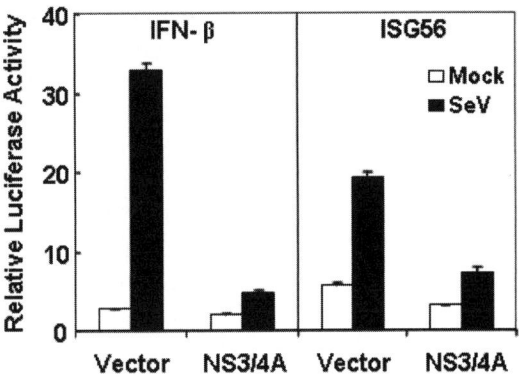

Fig. 16.3. Luciferase reporter assays in Hepa1-6 cells showing significant inhibition of viral activation of IRF-3–dependent promoters, IFN-β (*left*) or ISG56 (*right*) by ectopic expression of HCV NS3/4A. Cells were cotransfected with IFN-β-Luc (*left*) or ISG56-Luc (*right*) and pCMVβgal and a control vector or a vector encoding HCV NS3/4A. Twenty-four h later, cells were either mock-infected (*open bars*) or infected with SeV at 100 HAU/mL for 16 h (*solid bars*) before cell lysis for both luciferase and β-galactosidase assays. Bars show relative luciferase activity normalized to β-galactosidase activity, that is, promoter activity.

4. Notes

1. The quality of the DNA is very important for efficient transfection. Poor-quality DNA due to, for example, lipopolysaccharide (LPS) contamination, may cause cell death and culminate in irreproducible results.

2. An internal control reporter vector is used in combination with the experimental reporter plasmid to cotransfect cells and normalize transfection efficiencies. This step is necessary for interpretation of results, as overexpression of some gene products may affect transfection efficiency or cause changes in the growth features of cells. Besides β-galactosidase, renilla luciferase (under the control of CMV or TK promoters, available from Promega) is another widely used internal control for reporter gene assays. In this case, a dual-luciferase assay (Promega) is used to measure both firefly and renilla luciferase activities in a given sample.

3. For phosphospecific antibody, 5% BSA is preferred as the primary antibody dilution buffer. Try to avoid using nonfat dry milk, as it will mask some epitopes.

4. Phospho-Ser396 IRF-3 antibody is now commercially available from Upstate and Cell Signaling Tech.

5. Alternatively, after the 1 h absorption of SeV (or mock infection in DMEM), add an equal volume of DMEM containing 20% FBS without removing the SeV inoculum/DMEM. This

step generates results similar to those of the standard procedure with removal of inoculum.

6. Keep in mind that the culture medium containing SeV should be treated and disposed of properly as a biohazardous material, even though SeV is not a human virus.

7. Alternatively, use the 7.5% Tris-HCl Ready Gel from Bio-Rad as a substitute for the homemade nondenaturing gel, as it yields similar results.

8. Alternatively, the transfer can be conducted at 80 V for 2 h or at 30 V overnight. Larger or smaller proteins may take longer or shorter times to transfer.

9. For detection of total IRF-3 with the regular IRF-3 antibody, the primary antibody incubation can also be conducted at RT for 1 h, but when phosphospecific antibodies (in this case, phospho-Ser396 IRF-3) are used, overnight incubation at 4°C is important.

10. The final wash is conducted with PBS without Tween-20, as the latter is known sometimes to cause some background in ECL detection.

11. In unstimulated cells, IRF-3 exists in two species (see hatch marks in **Fig. 16.1A and B**), nonphosphorylated and basal-phosphorylated. Virus infection or dsRNA stimulation triggers the formation of hyperphosphorylated forms of IRF-3 that migrate more slowly in SDS-PAGE (arrows in **Fig. 16.1A and B**). Virus infection also triggers the proteasomal degradation of IRF-3, as demonstrated by a decrease in the protein abundance of IRF-3 (**Fig. 16.1B**), although cell type-, and virus-specific differences seem to be associated with this decrease *(4, 19)*

12. For western detection with monoclonal antibodies, 1% milk or BSA is usually enough for blocking nonspecificity *(4, 19)* during primary antibody incubation. Use of less blocking reagent produces stronger signal detection.

13. Make sure cells do not dry out during the washing step before fixation, as drying will cause high background immunofluorescence.

14. For most monoclonal antibodies, this step can be omitted or reduced to 30 min.

15. Steps 11 and 12 can be omitted if a mounting medium with DAPI (available from Vector Labs) is used in Step 13.

16. For cells that detach easily (e.g., HEK293 cells), the wash step can be omitted as long as the culture medium is removed as completely as possible.

17. If the assay is not to be done immediately, the plate can be kept frozen at −20°C for approx 2 weeks.

18. Repeat the assay using less cell lysate (adjust to 10 μL using 1× RLB) if the reading is above the linear range of the luminometer.

19. This optimum can be easily estimated by the following quick method. Prepare 3 microcentrifuge tubes, each containing 100 μL of 2× assay buffer. Add 1 milliunit of β-galactosidase standard to one of the three and 12 milliunits to a second; add an estimated volume of a sample to the third tube. Adjust to equal volume by adding 1× RLB. Incubate the three tubes in a 37°C water bath and view the yellow color development from time to time. If the sample tube develops color between the two standards, the dilution used is satisfactory. Otherwise, adjust the sample quantity and dilution as required.

Acknowledgments

The author would like to thank Drs. Stanley Lemon, Darius Moradpour, John Hiscott, Rongtuan Lin, and Michael Gale for providing reagents and Mardelle Susman and Charlette Estevanes for help with the manuscript. This work was supported by grants from the National Institutes of Health and the American Liver Foundation.

References

1. Akira, S., Uematsu, S., and Takeuchi, O. (2006) Pathogen recognition and innate immunity. *Cell* **124**, 783–801.
2. Sen, G. C. (2001) Viruses and interferons. *Annu. Rev. Microbiol.* **55**, 255–281.
3. Servant, M. J., Grandvaux, N., and Hiscott, J. (2002) Multiple signaling pathways leading to the activation of interferon regulatory factor 3. *Biochem. Pharmacol.* **64**, 985–992.
4. Lin, R., Heylbroeck, C., Pitha, P. M., and Hiscott, J. (1998) Virus-dependent phosphorylation of the IRF-3 transcription factor regulates nuclear translocation, transactivation potential, and proteasome-mediated degradation. *Mol. Cell Biol.* **18**, 2986–2996.
5. Yoneyama, M., Suhara, W., Fukuhara, Y., Fukuda, M., Nishida, E., and Fujita, T. (1998) Direct triggering of the type I interferon system by virus infection: activation of a transcription factor complex containing IRF-3 and CBP/p300. *EMBO J.* **17**, 1087–1095.
6. Foy, E., Li, K., Wang, C., Sumpter, R., Ikeda, M., Lemon, S. M., et al. (2003) Regulation of interferon regulatory factor-3 by the hepatitis C virus serine protease. *Science* **300**, 1145–1148.
7. Li, K., Foy, E., Ferreon, J. C., Nakamura, M., Ferreon, A. C. M., Ikeda, M., et al. (2005) Immune evasion by hepatitis C virus NS3/4A protease-mediated cleavage of the Toll-like receptor 3 adaptor protein TRIF. *Proc. Natl. Acad. Sci. USA* **102**, 2992–2997.
8. Li, X. D., Sun, L., Seth, R. B., Pineda, G., and Chen, Z. J. (2005) Hepatitis C virus protease NS3/4A cleaves mitochondrial antiviral signaling protein off the mitochondria to evade innate immunity. *Proc. Natl. Acad. Sci. USA* **102**, 17717–17722.
9. Lin, R., Lacoste, J., Nakhaei, P., Sun, Q., Yang, L., Paz, S., et al. (2006) Dissociation of a MAVS/IPS-1/VISA/Cardif-IKKepsilon molecular complex from the mitochondrial outer membrane by hepatitis C virus NS3-4A proteolytic cleavage. *J. Virol.* **80**, 6072–6083.
10. Loo, Y. M., Owen, D. M., Li, K., Erickson, A. K., Johnson, C. L., Fish, P. M., et al. (2006) Viral and therapeutic control of IFN-beta promoter stimulator 1 during hepatitis C virus infection. *Proc. Natl. Acad. Sci. USA* **103**, 6001–6006.
11. Meylan, E., Curran, J., Hofmann, K., Moradpour, D., Binder, M., Bartenschlager, R., et al. (2005) Cardif is an adaptor protein in the RIG-I antiviral pathway and is targeted by hepatitis C virus. *Nature* **437**, 1167–1172.

12. Servant, M. J., Grandvaux, N., tenOever, B. R., Duguay, D., Lin, R. T., and Hiscott, J. (2003) Identification of the minimal phosphoacceptor site required for in vivo activation of interferon regulatory factor 3 in response to virus and double-stranded RNA. *J. Biol. Chem.* **278**, 9441–9447.

13. Iwamura, T., Yoneyama, M., Yamaguchi, K., Suhara, W., Mori, W., Shiota, K., et al. (2001) Induction of IRF-3/-7 kinase and NF-kappaB in response to double-stranded RNA and virus infection: common and unique pathways. *Genes Cells* **6**, 375–388.

14. Scholle, F., Li, K., Bodola, F., Ikeda, M., Luxon, B. A., Lemon, S. M. (2004) Virus-host cell interactions during hepatitis C virus RNA replication: impact of polyprotein expression on the cellular transcriptome and cell cycle association with viral RNA synthesis. *J. Virol.* **78**, 1513–1524.

15. Wolk, B., Sansonno, D., Krausslich, H. G., Dammacco, F., Rice, C. M., Blum, H. E., et al. (2000) Subcellular localization, stability, and trans-cleavage competence of the hepatitis C virus NS3-NS4A complex expressed in tetracycline-regulated cell lines. *J. Virol.* **74**, 2293–2304.

16. Bio-Rad Laboratories, Mini-PROTEAN® II Electrophoresis Cell Instruction Manual.

17. Promega Technical Bulletin #281

18. Promega Technical Bulletin #097

19. Collins, S. E., Noyce, R. S., and Mossman, K. L. (2004) Innate cellular response to virus particle entry requires IRF3 but not virus replication. *J. Virol.* **78**, 1706–1717.

Chapter 17

Selection and Characterization of Drug-Resistant HCV Replicons In Vitro with a Flow Cytometry-Based Assay

Jason M. Robotham, Heather B. Nelson, and Hengli Tang

Abstract

Because HCV RNA-dependent RNA polymerase is error-prone and the viral RNA has a high turnover rate, the genetic diversity of HCV is very high both in vitro and in vivo. The mutation rate in long-term replicon cultures approaches 3.0×10^{-3} base substitutions/site/year in this in vitro replication model. A direct consequence of the high mutation rate is the rapid emergence of drug-resistant variants, both in cell culture and in patients. Selectable replicons have been used extensively to isolate and characterize drug-resistant HCV genomes in vitro. Typically, replicon cells are plated at a low density and then subjected to a double selection by G418 and escalating dosages of a compound of choice. Here we describe an alternative screening assay that takes advantage of an HCV replicon that is amenable to live-cell sorting with a suitable flow cytometer. We also present a strategy for determining the relative contribution to the resistance by viral genome and host cells. We use selection and characterization of Cyclosporine A (CsA)-resistant replicons as a example to present the protocols, but this method can easily be adapted for the selection of replicon cells resistant to other chemical compounds as long as the compound does not fluoresce at the same wavelength as the fluorescent reporter protein in the replicon.

Key words: Flow cytometry, live-cell sorting, electroporation, drug resistance.

1. Introduction

The quasispecies nature of HCV in patients is mirrored by the emergence of numerous mutations across the replicon genome in long-term replicon cultures in vitro. Many of these mutations increase replication, whereas others contribute to drug resistance (1–3). The replicon system has been used to evaluate the resistance profiles of a variety of compounds under consideration for development as anti-HCV drugs (4–10). When a viral protein

Hengli Tang (ed.), *Hepatitis C: Methods and Protocols, Second Edition, vol. 510*
© 2009 Humana Press, a part of Springer Science+Business Media
DOI 10.1007/978-1-59745-394-3_17 Springerprotocols.com

is the target of a compound, mutations in the coding region of the target gene typically arise that confer resistance. For example, many well-defined mutations have been mapped to the binding sites of the inhibitors acting on either the viral protease NS3 or the polymerase NS5B (4–10). For compounds that act on cellular targets to inhibit HCV (Interferon, Ribovirin, CsA, etc.), the mechanism of resistance is less understood and can be attributed to changes in the host cells, mutations in the viral genome, or both (11–16).

Stable replicon cells can be obtained by antibiotic (i.e., with Geneticin, G418) selection as a result of the incorporation of an antibiotic-resistance gene (neomycin phosphotransferase gene [NPT] in most cases) into the first cistron of the replicon (see a detailed overview by Lohmann in **Chapter 11**). In fact, the efficiency of the replicon RNA in forming G418-resistant colonies has been widely used a measure of replication capacity. Similarly, double selection with G418 and antiviral compounds is a common approach to isolation of drug-resistant replicons in vitro. To facilitate high-throughput analysis of antivirals, others have modified the replicon system to incorporate various additional reporter genes, including luciferase, GFP, and SEAP (17, 18). We recently adopted a GFP replicon (17) and developed a flow cytometry-based assay for measuring HCV replication that is simple, fast, and unbiased against cell growth arrest (19). Here we present methods for isolating and characterizing Cyclosporine A-resistant replicons using this assay.

2. Materials

2.1. GFP-Based Sorting and Selection of GS5 Cells

1. Growth media: Dulbecco's modified Eagle's medium (DMEM) supplemented with 10% fetal bovine serum, 1×concentration of nonessential amino acids, and penicillin/streptomycin.
2. Trypsin EDTA: 0.05% trypsin, 0.53 mM EDTA (Mediatech, Herndon, VA).
3. Cell sorting buffer: 1 × PBS + 5% fetal bovine serum (Cell Generation, Fort Collins, CO).
4. BD FACS Aria (BD Biosciences, San Jose, CA).

2.2. Cyclosporine A-Resistant Cell Selection and GFP-Based Cell Sorting

1. 1 mg/mL cyclosporine A: Dissolve 1 mg CsA in 1 ml 100% EtOH. Store at −20°C (Alexis, San Diego, CA).
2. 500 μg/mL G418 sulfate: Add 1 mL stock G418 50 mg/mL to 100 mL complete DMEM (Mediatech, Herndon, VA). Store at 4°C.

2.3. Extraction *of Resistant Cell RNA*	1. TRIzol® Reagent (Invitrogen, Carlsbad, CA) 2. 70% ethanol.
2.4. Identification *of Viral RNA as CsA Resistance Determinant*	1. Gene Pulser Xcell Electroporator (Bio-Rad, Hercules, CA). 2. Electroporation cuvettes, 4 mm (VWR Scientific Products, West Chester, PA). 3. OPTI-MEM Reduced Serum Media (Invitrogen, Carlsbad, CA). 4. 10 units/μL Interferon-α, IFN αA/D: Add 1000 units to 100 μL of sterile 1 × PBS (Sigma, St. Louis, MO). Store at −80°C.
2.5. RT-PCR *of Resistant Replicon Genome*	1. Superscript III First-Strand Synthesis System for RT-PCR (Invitrogen, Carlsbad, CA) 2. Custom Primers (IDT DNA, Coralville, IA) 3. Platinum Taq DNA Polymerase High Fidelity (Invitrogen, Carlsbad, CA).
2.6. TOPO TA Cloning *and Sequencing of Resistant Replicon Genome*	1. QIAquick Gel Extraction Kit (Qiagen Sciences, Valencia, CA). 2. Taq DNA polymerase (Invitrogen, Carlsbad, CA). 3. 100 mM dNTP Set (Fisher Scientific, Suwanee, GA). 4. TOPO TA Cloning Kit (Invitrogen, Carlsbad, CA). 5. TOPO XL PCR Cloning Kit (Invitrogen, Carlsbad, CA). 6. QIAprep Spin Miniprep Kit (Qiagen Sciences, Valencia, CA).
2.7. Site-Directed *Mutagenesis of the HCV NS5B Gene*	1. pCR-2.1-TOPO vector (Invitrogen, Carlsbad, CA). 2. Restriction enzymes: *Xho*I, *Eco*RV, *Cla*I, *Dpn*I (Invitrogen, Carlsbad, CA). 3. T4 DNA ligase (Invitrogen, Carlsbad, CA). 4. Stratagene QuikChange® Kit (Stratagene, La Jolla, CA). 5. *E. coli* competent cells
2.8. In Vitro *Transcription of Mutant Replicon RNA*	1. MEGAscript T7 Kit (Ambion, Austin, TX). 2. Chloroform/isoamyl alcohol, 24:1 (Acros Organics, NJ). 3. Phenol, saturated, pH 6.6/7.9 (Fisher Scientific, Suwanee, GA). 4. Phenol, saturated, pH 4.3 (Fisher Scientific, Suwanee, GA). 5. Formaldehyde solution, 37% w:w (Fisher Scientific, Suwanee, GA). 6. 10 × MOPS buffer. Weigh 8.37 g of MOPS (Sigma) and 2.177 g sodium acetate and dissolve them in 160 mL of nuclease-free H_2O. Add 4 mL of 500 mM EDTA and bring to 200 mL with nuclease-free H_2O. Filter through a 0.2 μm filter unit and protect from light. Store at 4°C. Discard when color turns from clear to yellow. 7. Nuclease-free H_2O.

3. Methods

3.1. GFP-Based Sorting and Selection of GS5 Cells

GS5 cells are derived from a genotype 1b replicon that has a GFP gene inserted into the coding region of NS5A *(17, 19)*. In these replicon cells, the NS5A-GFP fusion protein exhibits properties of both the fluorescent protein and the HCV NS5A, as it supports HCV replication and renders the cells harboring the hybrid replicon fluorescent under a fluorescent microscope and flow cytometry analysis. Live-cell sorting can be performed on these cells to select for cells that express high levels of NS5A-GFP.

1. Culture cells under nonconfluency conditions, expanding when necessary, until approximately 9×10^7 (*see* **Note 1**) cells are obtained.
2. Remove medium, wash with PBS, and trypsinize cells using 0.05% Trypsin.
3. Resuspend cells in growth medium, place them in sterile centrifuge tubes, and pellet cells at 290 g for 10 min at 4°C.
4. Remove medium, taking care not to disturb the cell pellet.
5. Resuspend all collected cells in 5 mL FACS buffer and proceed to cell sorting as quickly as possible.
6. Subject cells to live-cell sorting under sterile conditions and recover the 5% of cells with the highest GFP intensity in growth medium.
7. Seed an appropriate-sized flask with recovered cells, with 20% FBS-supplemented medium. After 24 h, remove medium, wash cells with PBS, and add 10% FBS-supplemented medium with 500 µg/mL G418.
8. Steps 1–7 can be repeated with subsequent cell populations until a consistently strong fluorescent signal in over 90% of the cells is achieved. In this protocol, "GS4 cells" (GFP-Sort #4), are the I/5A-GFP cell line that has undergone this process four times.

3.2. Cyclosporine A-Resistant Cell Selection and GFP-Based Cell Sorting

1. Seed 10 cm dishes with GS4 cells at a density of 5×10^6 per dish, with 10% FBS-supplemented medium without 500 µg/mL G418.
2. Add CsA directly to the medium to a final concentration of 1 µg/mL.
3. Maintain cells under these conditions for 3 days.
4. Aspirate medium, wash with PBS, and add fresh medium supplemented with 500 µg/mL G418 and 1 µg/mL CsA.
5. Maintain cells under these conditions for 3 weeks. Change medium as needed but keep 500 µg/mL G418 and 1 µg/mL CsA in the medium (*see* **Note 2**).

6. Identify colonies at the end of the selection process, which is marked by the absence of any dead cells floating in the medium.

7. Aspirate medium, wash with PBS, and trypsinize cells. Combine colonies and use them to seed a new, appropriate-sized flask (*see* **Note 3**).

8. Expand these cells (designated CsA-R cells in this example) until approximately 4×10^7 cells are obtained (*see* **Note 4**).

9. Treat the cells with $2 \mu g/mL$ of CsA for 3 days. In parallel, similarly treat a small sample of the control cells (e.g., GS4).

10. Subject the above treated cells to flow cytometry analysis. The percentage of GFP+ cells remaining among the control cells after this treatment is typically below 2%, whereas that in the selected-cell pool approaches 10–15%.

11. Recover the top 5–10% of GFP-positive cells from the selected-cells pool under sterile live-cell sorting conditions.

12. Expand the recovered cells (designated CsA-RS1) in the presence of $2 \mu g/mL$ CsA.

13. Steps 8–12 can be repeated with CsA-RS1 cells and higher concentration of CsA. In this example, we obtained CsA-RS2 by treating CsA-RS1 cells with $4 \mu g/mL$ of CsA.

14. Alternatively, single cell clones of CsA-RS1 cells could be obtained.

3.3. Identification of Viral RNA as CsA Resistance Determinant

For compounds that inhibit HCV by unknown mechanisms (e.g., hit compounds that are the result of replicon-based screening), it is important to determine whether resistance to a particular compound is conferred by the host cells or the viral RNA before embarking on sequencing efforts. Here we describe a strategy for distinguishing the contributions by the cell or the replicon RNA, taking advantage of two well-established techniques used in HCV replicon research. First, de novo replication can be established by introduction of total cellular RNA of an existing replicon, which contains the HCV RNA, into "naïve" hepatoma cells (Huh-7 and its derivatives) by electroporation. This procedure allowed us to separate the HCV RNA of resistant cells from their original host cells and transfer them into a naïve host cell background. Second, prolonged treatment with IFN-α can completely eliminate the viral RNA from established replicon cells and "cure" the cells of HCV replication. This feature permitted the removal of HCV RNA from the resistant cells and left the host cells ready to accept "naïve" replicon RNA. The sensitivity of our CsA-resistant cells to IFN-α treatment made this approach possible. A schematic representation of the experimental design is shown in **Fig. 17.1**.

3.3.1. Production of Interferon-α Cured Cells

1. Seed a $25 \, cm^2$ flask with 5×10^5 replicon cells and add Interferon-α (IFN) to a final concentration of 50 units/mL (*see* **Note 5**).

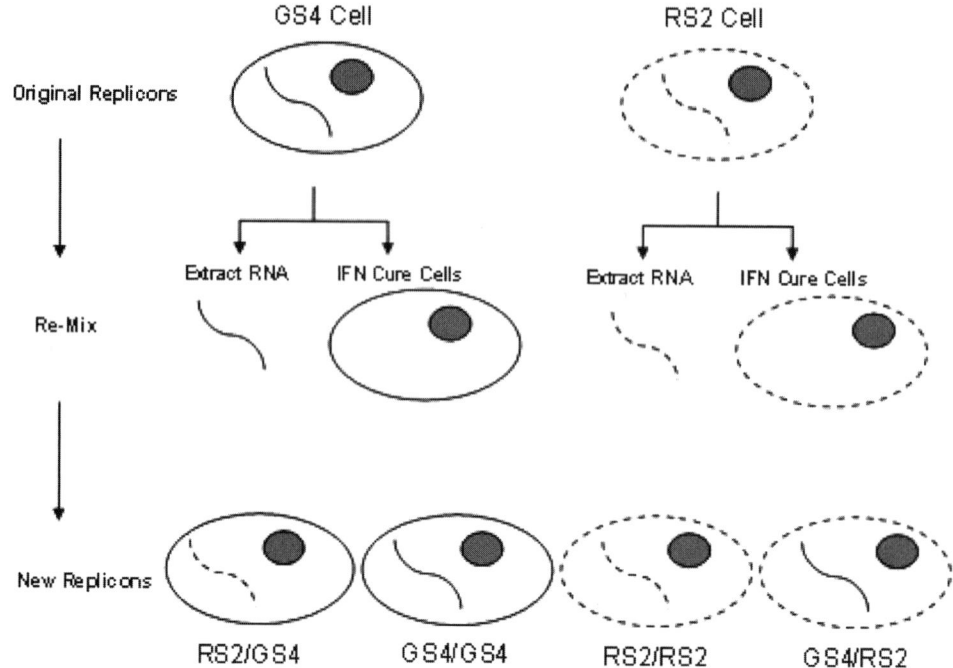

Fig. 17.1. Remix experiments designed to determine the contribution to CsA resistance by the host cell and the replicon RNA. Testing the resistant profiles of the new replicons can reveal the relative contributions by the cell and the virus.

2. Culture cells under these conditions for 4 days.

3. Trypsinize the cells, use 25% of the cells to seed a new 25 cm² flask, and add IFN to a final concentration of 50 units/mL.

4. Repeat step 3 every 4 days for 4 weeks to obtain the cured cells.

3.3.2. Testing of Cured Cells for the Absence of HCV Replicon

1. Seed a 25 cm² flask with 3×10^5 of the cured cells and add 500 µg/mL G418 to the medium.

2. Monitor the cells in the test flask for 2–3 weeks. Complete cell death under G418 treatment indicates that the replicon has been cured (*see* **Note 6**).

3.3.3. Isolation of CsA-Resistant RNA (Protocol for a 75 cm² Cell-Culture Vessel)

1. Aspirate medium, wash with $1 \times$ PBS and add 7.5 mL of Trizol solution.

2. Resuspend cells in Trizol and add to 1.5 mL centrifuge tubes (1 mL each).

3. Add 200 µL of chloroform to each tube.

4. Shake tubes vigorously for 15 s.

5. Incubate at room temperature for 3 min.

6. Centrifuge at 12,000g for 15 min at 4°C.

7. Transfer the clear, upper phase only to fresh tubes and discard the remainder.

8. Add 500 µL of isopropanol to each tube. Mix by inversion 5–10 times.

9. Incubate at room temperature for 10 min.

10. Centrifuge at 12,000g for 10 min at 4°C.

11. Aspirate supernatant, taking care not to disturb the pellet.

12. Add 1 mL of 75% ethanol to each tube and vortex briefly.

13. Centrifuge at 7500g for 5 min at 4°C.

14. Aspirate or pipette away the ethanol, taking care not to disturb the pellet.

15. Allow the ethanol to dry by leaving the cap of the tube open for approximately 5 min.

16. Add 20 μL of RNase-free water to each tube. The sample should be pipetted up and down several times to ensure that the RNA pellet has dissolved completely.

17. Store RNA at −80°C.

3.3.4. Electroporations of Total Cellular RNA Containing Replicon

1. Seed T-75 flasks with the appropriate amounts of the IFN-cured cells 3–4 days before the electroporation experiment. Cells should be confluent at the time of harvest. A confluent T-75 flask typically contains 1–1.2×10^7 cells and is sufficient for 2–3 electroporations.

2. Trypsinize the cells, resuspend in 10% FBS-supplemented medium, place in sterile centrifuge tubes, take a small sample for counting, and pellet the rest at 290 g for 10 min at 4°C.

3. Resuspend cells in OPTI-MEM at a density of 1×10^7 cells/mL.

4. Load 400 μL of cells in OPTI-MEM into an electroporation cuvette with a 0.4 cm gap width and add 10 μg of the total cellular RNA isolated in **Section 3.3.3**.

5. Perform electroporations with the conditions of 270 V, 950 uF, and $\infty\Omega$.

6. Plate electroporated cells onto a 10 cm dish in 20% FBS-supplemented medium as quickly as possible.

7. Twenty-four h after electroporation, aspirate medium, wash with PBS, and supply cells with 10% FBS-supplemented medium and 500 μg/mL G418.

8. Change medium frequently (every 2–3 d) to remove dead cells.

9. Once cells have stopped dying and colonies have formed, collect the cells from the plate and trypsinize and expand the new replicon cell line.

3.4. RT-PCR of Resistant Replicon RNA

1. Amplify the HCV replicon genome spanning the HCV IRES, NS3, NS4A, NS4B, NS5A, and NS5B genes by RT-PCR using RNA extracted from CsA-resistant replicon cells (shown diagrammatically in **Fig. 17.2**, left). Carry out reverse transcription using the following reaction mixtures and conditions (*see* **Note 7**):
5 μg total RNA
10 mM dNTP mix (1 μL)

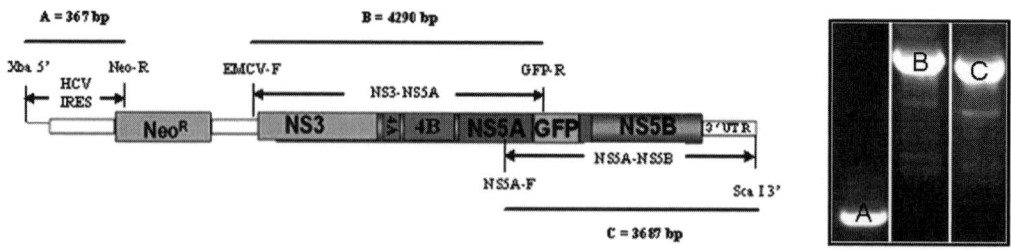

Fig. 17.2. RT-PCR and sequencing of the CsA-resistant replicon RNA. (**A**) Diagram of GS5 (I/5A-GFP) replicon. Primers Xba 5′ and Neo-R were used to generate fragment A, which covers the entire HCV 5′ NTR; primers EMCV-F and GFP-R were used to generate fragment B, which covers the coding regions of NS3, NS4A/B, and NS5A; primers NS5A-F and *Sca*I 3′ were used to generate fragment C, which covers parts of NS5A, GFP, and NS5B and the entire 3′ NTR. (**B**) Electrophoresis gel image of the fragments, which were then directly cloned into TA vectors for sequencing.

Gene-specific reverse primer ($2\,\mu M$)

Nuclease-free water to final volume of $10\,\mu L$

Incubate in a thermocycler for 10 min at 65°C, cool on ice for 1 min, and then combine each sample with $10\,\mu L$ cDNA-synthesis mix formulated as follows:

$10 \times$ RT buffer ($2\,\mu L$)

25 mM $MgCl_2$ ($4\,\mu L$)

100 mM DTT ($2\,\mu L$)

RNase OUT [$40\,U/\mu L$] ($1\,\mu L$)

Superscript III RT [$200\,U/\mu L$]($1\,\mu L$)

Incubate in the thermocycler for 50 min at 50°C and then 5 min at 85°C. Add $1\,\mu L$ of RNase H to each reaction and incubate for 20 min at 37°C.

2. After cDNA synthesis, perform subsequent rounds of PCR amplification using the following PCR amplification mixtures and conditions.

cDNA sample from step 1 ($2\,\mu L$)

$10\times$ Hi-Fi buffer ($5\,\mu L$)

10 mM dNTP mix ($1\,\mu L$)

50 mM $MgSO_4$ ($1.5\,\mu L$)

$10\,\mu M$ forward primer $10\,\mu M$ reverse primer

Platinum Taq ($0.5\,\mu L$)

DEPC-treated water to final volume of $50\,\mu L$

Once the PCR amplification mixture is set up, carry out amplification as described below:

One cycle of a 94°C denaturing step for 1 min 30 cycles of following:

94°C denaturing step for 1 min

50°C annealing step for 45 s (*see* **Note 2**)

72°C extension step for 1 min/kb of the insert

One cycle of a final extension at 72°C for 10 min

The nucleotide sequence of primers used in the first-strand cDNA synthesis and subsequent PCR reactions are summarized below in **Table 17.1**.

Table 17.1
Primers used for CsA-resistant HCV cDNA synthesis and PCR amplification

Reverse primer*	Nucleotide sequence	Nucleotide position**	cDNA fragment***
Neo-R	CTTTGAGGTTTAGGATTCGTGCTC	344–367	A
GFP-R	CTTGCCGTAGGTGGCATCGC	6026–6045	B
*Sca*I 3′	TGATCTGCAGAGAGGCCAGTATC	8747–8775	C
Forward primer	Nucleotide sequence	Nucleotide position	Amplicon (bp)***
Xba 5′	GCCAGCCCCCGATTGGGG	1–18	A (367)
EMCV-F	CGAACCACGGGGACGTGG	1756–1773	B (4290)
NS5A-F	TTCACAGAAGTGGATGGGGTGC	5089–5110	C (3687)

*The reverse primers shown above were used for cDNA synthesis as well as the proceeding PCR reactions.
**Nucleotide position is based on the sequence of the I/5A replicon plasmid, provided by Dr. Charles Rice.
***cDNA fragments and PCR amplicons are labeled according to **Fig. 17.2**.

3.5. TOPO TA Cloning and Sequencing of Resistant Replicon DNA

1. Determine the purity of the PCR products 5 μL of the PCR reactions by loading them on ethidium-bromide-containing 0.8–1.5% agarose gels, depending on the expected amplicon size, and visualize them on a UV light box. A sample gel is shown in **Fig. 17.2**, right.

2. After gel analysis, load the remaining 45 μL of the PCR reactions onto a new gel and excise and purify the desired bands using the QIAquick Gel Extraction kit (*see* **Note 8**).

3. Use the purified fragments in TOPO TA cloning reactions (*see* **Note 9**). Use the pCR-2.1-TOPO vector for cloning the 367-bp fragment corresponding to the HCV IRES and the pCR-XL-TOPO vector for the two larger 4390- and 3687-bp fragments. In all cases, carry out the cloning and subsequent transformation reactions as described below. Prepare the following TOPO TA cloning mixture:
   PCR product (4 μL) (*see* **Note 10**)
   Salt solution (1 μL)
   Appropriate TOPO vector (1 μL)
   Mix samples gently and incubate them for 30 min at room temperature (*see* **Note 11**).
   Add 2 μL of the TOPO TA cloning reaction from above to a vial of One Shot Chemically Competent *E. coli* cells (e.g., DH5α or TOPO10) and mix gently by inverting tube.
   Incubate on ice for 30 min.
   Heat-shock the cells for 30 s at 42°C.
   Transfer tubes to ice and add 250 μL S.O.C. medium.
   Shake for 1 h at 37°C.

During this time prewarm agar plates containing the appropriate antibiotics (e.g., 50 μg/mL kanamycin or 50 μg/mL ampicillin) to 37°C and spread 40 μL each of 100 mM IPTG and 40 mg/mL X-gal onto each plate. Add 10 μL of the transformation media to the prepared agar plates and incubate overnight at 37°C.

4. Pick white colonies the next day and use them to inoculate 3 mL of Circlegrow medium supplemented with the appropriate antibiotics for an overnight culture and subsequent plasmid preparation.

5. Subject plasmids made from white colonies to DNA sequencing. For the smaller fragment A, sequencing with the TOPO 2.1 specific M13F or M13R primer is sufficient to yield the complete sequence. For the larger fragments B and C, successive rounds of sequencing are needed to yield the complete sequences. The portion of the 3687-bp fragment C corresponding to the inserted GFP gene was not sequenced. Primers used in all sequencing reactions are shown in **Table 17.2**.

3.6. Site-Directed Mutagenesis of the HCV NS5B Gene

1. Use the Con1=SG-Neo_(I).new replicon (Rep 1b) as the backbone for generating site-specific CsA-resistant mutations that reflect those identified through the cloning and sequencing efforts.

2. Use the restriction enzymes *Xho*I and *Eco*RV to digest the Rep 1b plasmid and pCR-2.1-TOPO vector (*see* **Note 12**).

Table 17.2
Primers used for sequencing TOPO TA cloned HCV DNA fragments

Forward primer	Nucleotide sequence	Nucleotide position	Insert in TOPO clone (bp)	HCV nucleotides sequenced
M13F	GTAAAACGACGGCCAG	*See* TOPO 2.1	A (367)	1–367
M13F	CTGGCCGTCGTTTTAC	*See* TOPO XL	B (4290)	1801–2583
1b-2506-F	CCCTAGGTTTCGGGGCG	2507–2523	B (4290)	2584–3493
1B-1651-F	TCCCAGACTAAGCAGGC	3445–3461	B (4290)	3494–4142
M13R	CAGGAAACAGCTATGA	*See* TOPO XL	B (4290)	4143–5110
M13R	CAGGAAACAGCTATGA	*See* TOPO XL	C (3687)	5111–5929
GFP-F	GCTGCCCGACAACCACTACC	6504–6523	C (3687)	6526–7553
1b-7515	ACAGGCCATAAGGTCGC	7515–7531	C (3687)	7554–8541

*Note that the missing portion of the 3687-bp fragment (5930–6525) pertains to a portion of the GFP gene.

3. Gel purify and ligate overnight at 16°C the *Xho*I/*Eco*RV-digested 2672-bp fragment (corresponding to a large portion of the Rep 1b NS5B gene) from Rep 1b and the similarly digested ~ 3.9 − kb pCR-2.1-TOPO vector, using high-concentration T4 DNA ligase in a reaction as described below:

   25 ng digested pCR-2.1-TOPO vector DNA (See **Note 13**)

   75 ng digested Rep 1b NS5B fragment

   2 μL 5× ligation buffer

   1 μL T4 DNA Ligase (5 U/μL)

   Nuclease-free water to 10 μL

4. The next day, use the entire 10 μL ligation mixture to transform *E. coli* cells as described in **Section 3.5**.

5. Pick white colonies the next day and use them to inoculate 3 mL of Circlegrow medium for an overnight culture and subsequent plasmid preparation.

6. Sequence plasmids using M13F and M13R primers to confirm the presence of the Rep 1b NS5B fragment in the pCR-2.1-TOPO vector background (Rep 1b NS5B/TOPO) and use them subsequently in site-directed mutagenesis reactions.

7. Using custom primers designed to introduce a single-site amino acid change and the Stratagene QuikChange® Kit, introduce site-specific CsA-resistant mutations into the Rep 1b NS5B/TOPO plasmid. The reaction mixture for the mutagenesis PCR is described below:

   50 ng Rep 1b NS5B/TOPO

   10 *Pfu* buffer (5 μL)

   10 mM dNTP mixture (2.4 μL)

   10 μM forward primer (1.5 μL)

   10 μM reverse primer (1.5 μL)

   Pfu DNA polymerase (1 μL)

   Nuclease-free water to 50 μL

   Once the mutagenesis mixture is set up, carry out amplification as follows:

   1 cycle of a 94°C denaturing step for 1 min

   16 cycles of the following:

      94°C denaturing step for 1 min

      55°C annealing step for 1 min

      65°C extension step for 13 min (~2 min/kb plasmid)

   1 μL *Dpn*I restriction enzyme was added to the PCR reaction mixture, mixed gently with a pipette, and spun down briefly at top speed before being allowed to incubate at 37°C for 2 h to digest the nonmutated parental DNA.

8. Use 4 μL of the reaction mixture to transform competent XL1-blue cells, plate 100 μL of the transformation mixture onto prewarmed agar plates containing 50 μg/mL ampicillin, and allow to grow overnight at 37°C.

9. The next day use colonies to inoculate 3 mL of Circle-grow medium supplemented with $50 \mu g/mL$ ampicillin for overnight culture and subsequent plasmid preparation.

10. Sequence plasmids with M13F, M13R, or Rep 1b NS5B gene-specific primers to confirm the presence of mutations before proceeding to the next cloning step.

11. Digest the Rep 1b NS5B/TOPO plasmids carrying the desired mutations with *Xho*I and *Cla*I, and gel purify and ligate the resulting 2510-bp fragment back into a similarly digested and gel purified 8803-bp Rep 1b vector using T4 DNA ligase.

12. Transform *E. coli* cells, prepare plasmids, and sequence them to confirm the presence of the desired mutations, as described in **Section 3.5**.

3.7. In Vitro Transcription of Mutant Replicon RNA

1. Linearize $3 \mu g$ of each mutated plasmid by digesting them overnight with either *Sca*I (Rep1b, H77) or *Xba*I (pSGR2a, pJFH-1, pFL-J6/JFH-1) at 37°C.

2. Run samples on a 0.7% agarose gel and view them on a UV box (Fisher Scientific) to confirm complete vector linearization.

3. Purify linearized vector DNA from the digest mixture by phenol:chloroform extraction and subsequent isopropanol precipitation as follows: Bring digestion mixture to $135 \mu L$ with RNase-free water. Add $15 \mu L$ ammonium acetate stop solution and invert the tube five times. Add $150 \mu L$ 1:1 saturated phenol (pH 6.6/7.9):chloroform mixture and invert the tube five times. Centrifuge the tube 14,000g for 5 min at 4°C. Remove the supernatant and place it in a clean 1.5 mL centrifuge tube containing $150 \mu L$ chloroform; invert the tube five times. Spin the tube at 14,000g for 5 min at 4°C. Remove the supernatant and place it in a clean 1.5 mL centrifuge tube. Add an equal volume of isopropanol, invert the tube five times, and incubate at −20°C for 1 h to overnight. Pellet the DNA at 14,000g for 5 min at 4°C, gently aspirate off the supernatant, wash the pellet with 70% ethanol, and resuspend it in $20 \mu L$ nuclease-free water (*see* **Note 14**).

4. Using the MEGAscript T7 Kit, use the purified linear template DNA for in vitro transcription reactions (*see* **Note 15**): Combine $1 \mu g$ purified linear template DNA; $2 \mu L$ 10× reaction buffer; $2 \mu L$ each of ATP, CTP, GTP, and UTP ribonucleotides; and $2 \mu L$ enzyme mix. Incubate the in vitro transcription mixture for 4 h to overnight at 37°C. Add $1 \mu L$ TURBO DNase, mix gently, and incubate for 15 min 37°C.

5. Purify in vitro transcribed RNA in the same manner as linearized DNA (*see* above), substituting pH 4.3 saturated phenol at the phenol:chloroform extraction step.

6. Determine the integrity of the in vitro transcribed RNA samples on 1% agarose gels, prepared and run as follows: Dissolve 300 mg agarose in an autoclaved flask with 26.1 ml DEPC-treated water and 3 ml 10 × MOPS. Cool the solution to 65°C, move it to a fume hood, and add 600 μL 37% formaldehyde solution and 1 μL ethidium bromide. When the gel solidifies, placed it in an electrophoresis chamber containing 1 × MOPS (diluted in DEPC-treated water) and allow it to equilibrate for 30 min. Mix 1 μg of each RNA sample with an equal volume of RNA gel-loading buffer, heat it to 65°C for 10 min and cooled it on ice for 5 min. Load samples, run them for several hours at 70 V (10 V/cm distance between anode and cathode of running chamber), and view them on a UV box before proceeding to the electroporation step.

4. Notes

1. The number of cells required for sorting depends on the percentage you wish to collect from the population, as well as the efficiency of cell recovery. We have found that from 9×10^7 cells, or about 9 near-confluent 75 cm² flasks, recovering the top 5% of cells, we can collect between 1 and 2 million viable cells.

2. During the beginning of the selection process, cells will die rapidly and medium should be changed every 2–3 days. As the selection progresses, fewer cells will die, colonies will start to form, and the medium can be changed less frequently, normally once every 7 d.

3. The size of the flask depends on the number of colonies obtained. A T-25 is appropriate for 50–100 colonies. Alternatively, a single cell clone can be isolated at this point, but we recommend pooling the cells and performing additional selection with flow cytometry to obtain cells with higher levels of resistance.

4. As this population of cells is expanded, a portion of the cells should be frozen to preserve the cell line before the next round of treatment and sorting is done.

5. A small cell-culture flask (such as a 25 cm² flask) is used to reduce the amount of IFN required. Once the cells have been cured, the population can be expanded to produce more cells.

6. More sensitive methods to detect HCV RNA using real-time RT-PCR can also be used to verify the absence of HCV replicon in the cured cells, but we found that the sensitivity to G418 is a reliable indicator.

7. For RT-PCR reactions in which the desired cDNA was large (over 2 kb), performing the cDNA synthesis and PCR reactions separately appeared to help in producing full-length products, but when the desired fragment was small, as in the case of fragment A (367 bp) the Superscript III One-Step RT-PCR Platinum Taq Hi-Fi kit (Cat no. 12574-030: Invitrogen, Carlsbad, CA) proved to be suitable. Note also that, when RT is performed through the 3'X portion of the replicon, using a reverse transcriptase with proofreading capabilities to ensure accurate reading of the template RNA by the enzyme is beneficial. We found Accuscript High Fidelity RT (Cat no. 600089-51: Stratagene, La Jolla, CA) suitable for this purpose.

8. TOPO TA cloning should not be carried out unless the PCR product clearly represents the major (and preferably only) band in the reaction mixture. If additional bands of equal or higher intensity can be visualized, the RT-PCR should be repeated.

9. Gel purification of the PCR product is recommended but not always necessary. TOPO TA cloning can be performed directly from the reaction mixture if the target band is the only product in the mixture, as shown in **Fig. 17.2**. Alternatively, gels can be run, bands excised and melted at 65°C, and 4 μL of the resulting sample used for TOPO TA cloning.

10. Should problems arise in TOPO TA cloning of the PCR product or purified fragment, addition of 3'A-overhangs may increase the efficiency. After amplification with a proofreading polymerase, 1 unit of Taq polymerase (Invitrogen, Carlsbad, CA) can be added directly to the PCR reaction mixture and incubated for 10 min at 72°C.

11. Incubation times as short as 5 min may be sufficient for TOPO TA cloning, but when larger fragments of DNA (> 1 kb) are used, increasing the incubation to 30 min can increase efficiency.

12. Because the efficiency of the QuikChange® mutagenesis reaction decreases when larger plasmids are used as templates, smaller constructs were created where only the region of the gene chosen for mutagenesis was cloned into the pCR-2.1-TOPO vector. The restriction enzymes used to generate these target fragments will depend on the genotype-specific restriction sites and the molecular cloning site of the pCR-2.1 TOPO vector; following the protocol given in **Section 3.9** for other genotypes may therefore not suffice. For example, when NS5B mutant clones were created on the JFH-1 full-length plasmid (pJFH-1), the NS5B gene was cut with *Xho*I, made blunt by Klenow fragment of DNA polymerase (Cat no. 18012-021: Invitrogen, Carlsbad, CA), digested again

with *Bam*HI (Cat no. 15201-023: Invitrogen, Carlsbad, CA), and ligated into the blunt *Eco*RV and corresponding *Bam*HI site of the pCR-2.1 TOPO vector. Once mutagenesis was performed on this plasmid, a region of the gene containing the desired mutated DNA (*Xba*I to *Rsr*II) was digested with *Rsr*II and *Xba*I and ligated into the similarly digested full-length JFH-1_pUC plasmid.

13. For ligation reactions, 100 ng total DNA with a 3:1 molar ratio of insert:vector should be used whenever possible.

14. Often no pellet could be seen after linearized DNA was centrifuged. The purification has not necessarily failed, however, as a pellet of highly purified DNA will typically not be seen at this point. Resuspension should still be performed in nuclease-free water.

15. RNA degradation frequently occurs as a result of RNase contamination, so we recommend cleaning the bench, pipettes, and tubes with a decontaminating spray (e.g., RNase AWAY (Molecular BioProducts, San Diego, CA).

Acknowledgments

We thank Drs. Charles Rice and Takaji Wakita for I/5A-GFP replicon and full-length clone of JFH-1, respectively. Ruth Didier assisted us in flow cytometry, and the molecular cloning core facility helped with subcloning and sequencing of HCV replicon fragment. H.T. is supported by the American Heart Association and the James Esther King Biomedical Research Program.

References

1. Blight, K. J., Kolykhalov, A. A., and Rice, C. M. (2000) Efficient initiation of HCV RNA replication in cell culture. *Science* **290**, 1972–1974.

2. Kato, N., Nakamura, F., Dansako, H., Namba, K., Abe, K., Nozaki, A., et al. (2005) Genetic variation and dynamics of hepatitis C virus replicons in long-term cell culture. *J. Gen. Virol.* **86**, 645–656.

3. Lohmann, V., Korner, F., Koch, J., Herian, U., Theilmann, L., and Bartenschlager, R. (1999) Replication of subgenomic hepatitis C virus RNAs in a hepatoma cell line. *Science* **285**, 110–113.

4. Kukolj, G., McGibbon, G. A., McKercher, G., Marquis, M., Lefebvre, S., Thauvette, L., et al. (2005) Binding site characterization and resistance to a class of non-nucleoside inhibitors of the hepatitis C virus NS5B polymerase. *J. Biol. Chem.* **280**, 39260–39267.

5. Le Pogam, S., Jiang, W. R, Leveque, V., Rajyaguru, S., Ma, H., Kang, H., et al. (2006) *In vitro* selected Con1 subgenomic replicons resistant to 2′-C-Methyl-Cytidine or to R1479 show lack of cross resistance. *Virology* **351**, 349–359.

6. Le Pogam, S., Kang, H., Harris, S. F., Leveque, V., Giannetti, A. M., Ali, S., et al. (2006) Selection and characterization of replicon variants dually resistant to thumb- and palm-binding nonnucleoside polymerase inhibitors of the hepatitis C virus. *J. Virol.* **80**, 6146–6154.

7. Lin, C., Lin, K., Luong, Y. P., Rao, B. G., Wei, Y. Y., Brennan, D. L., et al. (2004) *In vitro* resistance studies of hepatitis C virus serine protease inhibitors, VX-950 and BILN 2061: structural analysis indicates different resistance mechanisms. *J. Biol. Chem.* **279**, 17508–17514.

8. Lu, L., Pilot-Matias, T. J., Stewart, K. D., Randolph, J. T., Pithawalla, R., He, W., et al. (2004) Mutations conferring resistance to a potent hepatitis C virus serine protease inhibitor in vitro. *Antimicrob. Agents Chemother.* **48,** 2260–2266.

9. Migliaccio, G., Tomassini, J. E., Carroll, S. S., Tomei, L., Altamura, S., Bhat, B., et al. (2003) Characterization of resistance to non-obligate chain-terminating ribonucleoside analogs that inhibit hepatitis C virus replication in vitro. *J. Biol. Chem.* **278,** 49164–49170.

10. Mo, H., Lu, L., Pilot-Matias, T., Pithawalla, R., Mondal, R., Masse, S., et al. (2005) Mutations conferring resistance to a hepatitis C virus (HCV) RNA-dependent RNA polymerase inhibitor alone or in combination with an HCV serine protease inhibitor *in vitro. Antimicrob. Agents Chemother.* **49,** 4305–4314.

11. Naka, K., Takemoto, K., Abe, K., Dansako, H., Ikeda, M., Shimotohno, K., et al. (2005) Interferon resistance of hepatitis C virus replicon-harbouring cells is caused by functional disruption of type I interferon receptors. *J. Gen. Virol.* **86,** 2787–2792.

12. Pfeiffer, J. K. and Kirkegaard, K. (2005) Ribavirin resistance in hepatitis C virus replicon-containing cell lines conferred by changes in the cell line or mutations in the replicon RNA. *J. Virol.* **79,** 2346–2355.

13. Sumpter, R., Jr., Wang, C., Foy, E., Loo, Y. M., and Gale, M., Jr. (2004) Viral evolution and interferon resistance of hepatitis C virus RNA replication in a cell culture model. *J. Virol.* **78,** 11591–11604.

14. Young, K. C., Lindsay, K. L., Lee, K. J., Liu, W. C., He, J. W., Milstein, S. L., et al. (2003) Identification of a ribavirin-resistant NS5B mutation of hepatitis C virus during ribavirin monotherapy. *Hepatology* **38,** 869–878.

15. Zhu, H., Nelson, D. R, Crawford, J. M., and Liu, C. (2005) Defective Jak-Stat activation in hepatoma cells is associated with hepatitis C viral IFN-alpha resistance. *J. Interferon Cytokine Res.* **25,** 528–539.

16. Robida, J. M., Nelson, H. B., Liu, Z., and Tang, H. (2007) Characterization of hepatitis C virus subgenomic replicon resistance to cyclosporine *in vitro. J. Virol.* **81,** 5829–5840.

17. Moradpour, D., Evans, M. J., Gosert, R., Yuan, Z., Blum, H. E., Goff, S. P., et al. (2004) Insertion of green fluorescent protein into nonstructural protein 5A allows direct visualization of functional hepatitis C virus replication complexes. *J. Virol.* **78,** 7400–7409.

18. Bourne, N., Pyles, R. B., Yi, M., Veselenak, R. L., Davis, M. M., and Lemon, S. M. (2004) Screening for hepatitis C virus antiviral activity with a cell-based secreted alkaline phosphatase reporter replicon system. *Antiviral Res.* **67,** 76–82.

19. Nelson, H. B. and Tang, H. (2006) Effect of cell growth on hepatitis C virus (HCV) replication and a mechanism of cell confluence-based inhibition of HCV RNA and protein expression. *J. Virol.* **80,** 1181–1190.

Chapter 18

High-Throughput Screening of HCV RNA Replication Inhibitors by Means of a Reporter Replicon System

Weidong Hao and Rohit Duggal

Abstract

Efforts to find effective treatment for hepatitis C virus (HCV) have been hampered by the lack of a robust in vitro infectious tissue-culture system for this virus. A subgenomic replicon system was first developed in 1999 and has since been extensively optimized to accommodate the need for conveniently measuring HCV replication in vitro and widely adopted in HCV drug-discovery efforts. Here we describe the adaptation of a modified replicon system for a high-throughput screening (HTS) in anti-HCV drug discovery. In this system, the antiviral activity and cytotoxicity of any experimental compound are measured from a single well. This duplex measurement greatly increases the efficiency of the HTS while lowering the cost. The usefulness of this approach has been supported by the recent discovery of many new lead compounds from our HTS efforts in the past two years.

Key words: Hepatitis C virus, replicon, HTS, luciferase, antiviral activity, cytotoxicity.

1. Introduction

Hepatitis C virus (HCV), which belongs to the genus *Hepacivirus* within the family Flaviviridae, infects more than 170 million people worldwide *(1)*. More than 4 million people are infected in the United States alone *(2)*. The majority of those infected with HCV become chronic carriers, and most have relatively mild disease with slow progression. HCV infection is a major cause of cirrhosis, end-stage liver disease, and liver cancer *(3)*.

The current standard-of-care regimen for patients infected with HCV includes pegylated alpha interferon (IFN) plus ribavirin *(4)*, but a significant percentage of HCV infections do not

Hengli Tang (ed.), *Hepatitis C: Methods and Protocols, Second Edition, vol. 510*
© 2009 Humana Press, a part of Springer Science+Business Media
DOI 10.1007/978-1-59745-394-3_18 Springerprotocols.com

respond to this combination. Clearly, more specific and effective treatments are urgently needed.

The development of the HCV replicon system in 1999 was a major milestone in basic HCV research and drug development (5). The original replicon system was constructed by replacement of genes from the HCV genome that are not essential for HCV RNA replication, the structural genes (Core, E1, E2), p7 and NS2, with a genetic cassette encoding an antibiotic-resistance gene and the internal ribosomal entry site (IRES) from encephalomyocarditis virus (EMCV). The result was the formation of a dicistronic selectable subgenomic HCV replicon. The first-generation replicons, either the selectable version or the reporter and selectable genes-containing (reporter–selectable) version, were not conducive to high-throughput screening because of the high degree of labor involved in quantitative RT-PCR in the former and the low signal-to-noise ratio of the latter. Here, we describe the generation of a novel reporter replicon and a cell line capable of simultaneously identifying inhibitors of HCV replication and nonspecific cytotoxic compounds (6). The development of a robust, high-throughput HCV replicon assay in which the effect of inhibitors can be monitored for antiviral activity and cytotoxicity should greatly facilitate HCV drug discovery (6).

2. Materials

2.1. Cell Culture

1. Dulbecco's modified Eagle's medium (DMEM) (Invitrogen, Carlsbad, CA) supplemented with 10% fetal bovine serum (Invitrogen), 100 μM nonessential amino acids (Invitrogen), 50 units/mL of penicillin-streptomycin (Invitrogen), 1 mM L-glutamine (Invitrogen).
2. Blasticidin (VWR, West Chester, PA), final concentration: 6 μg/mL; hygromycin (Invitrogen), final concentration: 50 μg/mL.
3. 384-well clear-bottom black plates (VWR), 384-deep-well block (VWR).

2.2. Inhibitors

1. Interferon alpha (IFN, Sigma, St. Louis, MO) and 5,6-dichloro-1-beta-D-ribofuranosyl-benzimidazole (DRB) are obtained from Sigma. IFN is diluted in cell-culture medium. DRB is dissolved to 10 mM in 100% dimethylsulfoxide (DMSO, Sigma), stored in single-use aliquots at −20°C.
2. Compound AG-021541 (6) is a nonnucleoside HCV NS5B polymerase inhibitor synthesized by Pfizer Global Research & Development (La Jolla, CA). It is dissolved to 1 mM in 100% DMSO and stored in single-use aliquots at −20°C.

2.3. Replicon Constructs and Cell Lines

A dicistronic dual reporter (DDR) HCV replicon cell line *(6)* is used for the high-throughput screen (HTS). This cell line contains the dicistronic reporter-selectable HCV replicon that includes a humanized renilla luciferase (hRLuc) gene separated from the selectable marker, NeoR, by a short peptide cleavage site (FMDV2A^{pro}) (**Fig. 18.1**). The adaptive mutations E1202G, T1280I, and S2197P *(7)* are also introduced to enhance replicative capability. The hRLuc signal is used to monitor the antiviral activity of the experimental compounds. For simultaneous monitoring of the cytotoxicity of the experimental compounds, the firefly luciferase (FLuc) reporter gene is introduced into the host cell genome of the dicistronic single reporter (DSR)-containing replicon Huh-7 cells, generating the DDR cell line.

2.4. Chemiluminescence Detection Reagent

Custom dual-glo II reagent purchased from Promega (Madison, WI): This reagent is very similar to the dual-glo reagent manufactured by Promega *(8)*. Basically, the firefly luciferase signal is detected first by addition of the luciferase glo substrate, and the residual firefly luciferase signal is quenched by the addition of the renilla luciferase reagent prepared in Stop-n-Glo buffer.

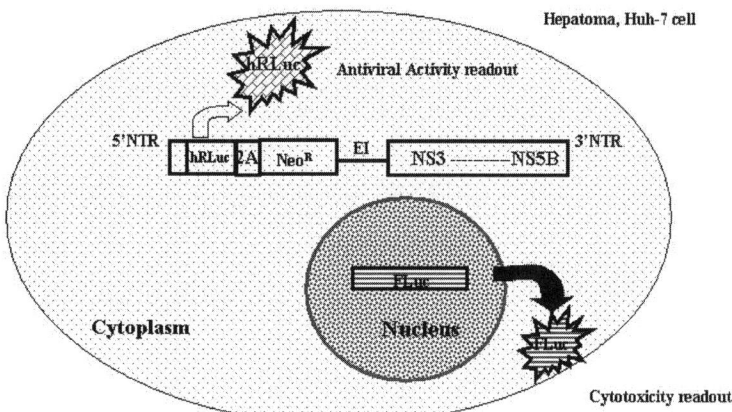

Fig. 18.1. Illustration of the replicon-containing cell line (DDR) used in the high-throughput screen described here. The reporter-selectable dicistronic HCV replicon contains two internal ribosomal entry sites (IRESs). The HCV IRES (HI) drives the translation of the first cistron: the amino terminus of core (36 nt), the humanized renilla luciferase gene (hRLuc), the foot-and-mouth disease virus protease 2A cleave site, and the neomycin phosphotransferase (Neo^R) gene. The EMCV IRES (EI) drives the translation of the second citron: nonstructural (NS) proteins 3–5B. Three adaptive changes in NS3 and NS5A (E1202G, T1280I, and S2197P) *(7)* have been engineered into the replicon. The hRLuc signal is used to monitor the antiviral activity of the experimental compounds. The firefly luciferase (FLuc) reporter gene has been introduced into the host cell genome of the Huh-7 cells to generate the DDR cell line and is used for measuring cytotoxicity. Reprinted with permission from *(6)*.

2.5. Automation Equipment

1. Multidrop: purchased from Thermo Fisher Scientific, Waltham, MA.
2. MiniGene Line Washer: purchased from E&K Scientific, Campbell, CA.
3. PlateMate Plus: purchased from Matrix Technologies, Hudson, NH.
4. Microbeta Jet Trilux with 12 detectors: purchased from Perkin Elmer, Wellesley, MA.
5. Titan Stacker: purchased from Titertek, Huntsville, AL.
6. PlateCrane: purchased from Hudson Control, Springfield, NJ.

3. Methods

The typical assay cycle with the DDR replicon cell line is 4 days. On day 1, 1.6×10^3 cells are seeded into each well of a 384-well plate. The seeded plate is then incubated for 24 h so that the cell number in each well will be doubled. The only reason for this doubling step is that it reduces the number of cell flasks that must be carried to feed the ongoing screening effort.

On day 2, the main activities include the dilution of experimental compounds, addition of controls (AG-021541 and DRB) in the predetermined positions on the compound plates, and whole-plate transfer of compounds to the cell plates prepared on day 1.

On day 4, after the assay plates have been incubated at 37°C under 5% CO_2 for 72 h, the antiviral and cytotoxic effects of the experimental compound in each well are measured according to the formula below.

$$\left(1 - \frac{\text{(Luc signal of compound well)}}{\text{(Mean Luc signal of DMSO wells)}}\right) \times 100$$

A pilot screen must be run to determine the robustness, or Z' factor, of the assay. The Z' factor can be calculated according to the following formula (9):

$$Z' = 1 - \frac{(3\sigma_{\max} + 3\sigma_{\min})}{(\mu_{\max} - \mu_{\min})}$$

where "max" is the maximum signal derived from wells containing DMSO (1%) alone, and "min" is the minimum signal for the antiviral (hRLuc) and cytotoxicity (FLuc) endpoints derived from wells that contained EC_{95} and CC_{95} levels of AG-021541 and DRB, respectively. σ is the standard deviation and μ is the mean

(6). The DDR cell line demonstrates low signal variability within the HTS format as indicated by a calculated Z' value of 0.8 (6).

3.1. Growing and Preparing DDR Replicon Cells

1. Maintain DDR replicon cells in DMEM supplemented with 10% FBS, 1 mM nonessential amino acids, 1 mM pen-strep, and 1 mM L-glutamine. The final concentrations of blasticidin and hygromycin are 6 μg/mL and 50 μg/mL, respectively. Pass cells every 3 days or 1 day before assay set-up. Split cells 1:3 or 1:4.
2. On the day of assay set-up, trypsinize and collect cells from 2 to 3 confluent T-225 flasks and centrifuge them at 200g for 5 min. Discard the supernatant. Resuspend the cell pellet in complete DMEM without selection antibiotics (blasticidin and hygromycin). Count cells and dilute them to 3.2×10^4 cells/mL in complete DMEM without selection antibiotics.
3. Seed each well of 384-well clear-bottom black plates (VWR) with 50 μL of cell suspension, using the Multidrop 384, to produce a final concentration of 1.6×10^3 cells/well (see **Note 1**).
4. Incubate the seeded plates for 24 h at 37°C, under 5% CO_2.

3.2. Preparing Compound Control Master Plate

1. Place two different controls on each assay plate, the antiviral (AV) control and the cytotoxicity (CYT) control. AG-021541 is used as the AV control at a final concentration of 10 μM. DRB is used as the CYT control at a final concentration of 100 μM.
2. Freshly prepare 1 mM AG-021541 and 10 mM DRB in 100% DMSO on the day of the assay set-up or use single aliquots stored at −80°C.
3. Estimate the amount (100 μL for 25 plates) of 1 mM AG-021541 or 10 mM DRB needed for the assay, add it to the predetermined position on a 384-deep-well block (VWR), and place this control-filled block on the PlateMate Plus.

3.3. Transferring Controls to Compound Plates

1. After the compound plates are thawed at room temperature, centrifuge them at 800g for 5 min to ensure that no residual compound remains at the top or on the sides of the well.
2. Scan the barcode on each plate, remove the lids from the compound plates, and label the plates sequentially.
3. Place the compound plates on PlateMate Plus.
4. Using the PlateMate Plus, add the controls in the 384-deep-well block from **Section 3.2** to the predetermined control wells in each compound plate.

3.4. Diluting Compounds

Add 45 μL of DMEM (no antibiotic selection, with 10% FBS, nonessential amino acids, pen-strep, and L-glutamine) to compound plates containing $3\,\mu L$ 1 mM compound (100% DMSO) using the Multidrop with titan stacker to produce a final concentration of the compound in each well of $62.5\,\mu M$.

3.5. Transferring Compounds to Cell Plates

1. Remove the 384-well clear-bottom black plates seeded with replicon-containing cells from the CO_2 incubator the day before; remove the lids from these plates.
2. Place these cell plates in "right stacker b" of the PlateMate Plus.
3. Place the plates with diluted compounds from **Section 3.4** in "left stacker a."
4. Mark the cell plates according to the numbering of the compound plates.
5. Run the compound transfer program to transfer $10\,\mu L$ from each well of the compound plate to the corresponding well of the cell plate, for a final compound concentration in each well of $10\,\mu M$ (*see* **Note 2**).
6. When the transfer is finished, replace the lids on the cell plates and incubate them at 37°C under 5% CO_2 for 72 h (*see* **Note 3**).

3.6. Detecting Luciferase Reporter Signals

1. Make dual-glo II substrate. Thaw the luciferase buffer and Stop-n-Glo buffer in a 37°C water bath. Mix one bottle of buffer with one bottle of substrate for both firefly luciferase and renilla luciferase to make final working reagent. The final working reagent can be prepared one day before the assay development and refrigerated until needed (do not subject it to more than two freeze–thaw cycles).
2. Three days after addition of the compound to the cell plates and incubation at 37°C under 5% CO_2, develop the assay by aspirating all but $10\mu L$ of the medium from the cell plates, using a Minigene 12-head line washer. Add $20\mu L$ of the luciferase reagent to each well of the 384-well plate, using the Multidrop with titan stacker. Incubate the plates at room temperature for 10 min.
3. Load all of the assay plates into the stacker of the Hudson PlateCrane, which is used for executing the reading of all plates in combination with the Microbeta Jet with 12 detectors (*see* **Note 4**).
4. After the firefly luciferase signal has been read, add $20\,\mu L$ of the Stop-n-Glo reagent to each well of the 384-well plate, again using the Multidrop with titan stacker. Then record the renilla luciferase signal of each well using the Microbeta Jet with 12 detectors (*see* **Note 5**).

3.7. Disposal of Plates

Proper disposal of the assay plates is very important because they contain compounds with unknown properties, organic solvents, and the remains of an immortalized human cell line. Remove the liquid from each well to a carboy, using the Minigene 12-well aspirator. Dispose of the dry plates as biological waste.

4. Notes

1. The replicon assay can also be modified to 96-well format. In this case, seed each well with 1×10^4 cells if the procedure outlined here is followed. If compounds are added on the same day as cell seeding, seed each well with 2×10^4 cells.

2. The compound transfer can also be carried out with the Evolution P3 by Perkin Elmer. The aspiration of the medium from the 384-well plates after the 72-h incubation can also be performed with a Tecan 96/384-well aspirator.

3. A visual check for fungal contamination on day 4 is important because no fungicide is used in the medium. Once fungus is spotted, do not open the plate; just wrap it in a plastic bag and dispose it of accordingly.

4. The luciferase signal can also be detected by the Microbeta Jet Trilux with six detectors, which has two injectors for use of flash-type luciferase reagents (*see* **Note 6**).

5. Hits are those compounds that show $\geq 60\%$ AV and $< 40\%$ CYT. Confirmation assays are carried out in the same way as the primary assay described above, except that each compound is tested in duplicate. The percentage inhibition results of AV and CYT are averaged. Confirmed hits are defined as compounds with $\geq 50\%$ AV and $< 40\%$ CYT that demonstrate no renilla luciferase enzymatic inhibition. In a reporter assay like the one we describe here, the reduction of the reporter signal could be due to the reduced translation of the reporter gene as a result of genuine AV activity (for hRLuc) or cytotoxicity (for both FLuc and hRLuc). On the other hand, the reduction could also be due to enzymatic inhibition by the experimental compounds. The FLuc enzymatic inhibition assay is unnecessary, as either a FLuc-specific enzymatic inhibitor ($\geq 40\%$ CYT, $< 50\%$ AV) or a cytotoxic compound ($\geq 40\%$ CYT, $\geq 50\%$ AV) will not be counted as hit. RLuc enzymatic inhibition must clearly be ruled out, however, for compounds that are not cytotoxic yet show RLuc inhibition ($\geq 50\%$ AV, $< 40\%$ CYT). The hRLuc inhibition assay is set up in the same way as the regular assay, except that no experimental compounds are added at the beginning. The assay plates are incubated for 69 h before addition of experimental compounds when the reduction of hRLuc signals are used to identify RLuc enzymatic inhibitors.

6. When the assay is set up in a 96-well plate, either flash-type or glo-type luciferase substrates can be used. The flash-type

luciferase substrate has a higher dynamic range and a very short half-life. In contrast, the glo-type substrate has a half-life of about 2 h but only about 1/100 of the signal strength of the flash-type substrate.

Acknowledgments

The authors would like to thank Eiann Sha and Tihua Chen for developing and validating the experiment procedures. The authors would also like to acknowledge Stephanie Shi for critical reading of the manuscript. The replicon construct used in the HTS were licensed from Apath LLC (St. Louis, MO) and ReBLikon GmbH (Mainz, Germany).

References

1. Fanning, L. J. (2005) Anti-viral therapies for hepatitis C virus infection: current options and evolving candidate drugs. *Lett. Drug Des. Discov.* **2**, 150–161.
2. Centers for Disease Control and Prevention (2005) *Centers for Disease Control and Prevention Viral Hepatitis C fact sheet.*
3. Hugle, T. and Cerny, A. (2003) Current therapy and new molecular approaches to antiviral treatment and prevention of hepatitis C. *Rev. Med. Virol.* **13**, 361–371.
4. Lake-Bakaar, G. (2003) Current and future therapy for chronic hepatitis C virus liver disease. *Curr. Drug Targets Infect. Disord.* **3**, 247–253.
5. Lohmann, V., Korner, F., Koch, J., Herian, U., Theilmann, L., and Bartenschlager, R. (1999) Replication of subgenomic hepatitis C virus RNAs in a hepatoma cell line. *Science* **285**, 110–113.
6. Hao, W., Herlihy, K. J., Zhang, N. J., Fuhrman, S. A., Doan, C., Patick, A. K., et al. (2007) Development of a novel dicistronic reporter-selectable hepatitis C virus replicon suitable for high-throughput inhibitor screening. *Antimicrob. Agents Chemother.* **51**, 95–102.
7. Lohmann, V., Korner, F., Dobierzewska, A., and Bartenschlager, R. (2001) Mutations in hepatitis C virus RNAs conferring cell culture adaptation. *J. Virol.* **75**, 1437–449.
8. Promega Manual TM058. Available at http://www.promega.com/tbs/tm058/tm058.pdf.
9. Zhang, J. H., Chung, T. D., and Oldenburg, K. R. (1999) A simple statistical parameter for use in evaluation and validation of high-throughput screening assays. *J. Biomol. Screen.* **4**, 67–73.

Chapter 19

Inhibition of HCV Replication by Small Interfering RNA

Ratna B. Ray and Tatsuo Kanda

Abstract

Small interfering RNAs (siRNAs) and short hairpin RNAs (shRNAs) have been reported to suppress gene expression significantly. HCV seems a suitable candidate for targets of siRNAs, as HCV is a positive single-strand RNA virus and replicates in the cytoplasm. Efficient inhibition by siRNAs requires access to target RNAs, which usually possess secondary structure. We have shown that shRNAs suppressing the HCV internal ribosomal entry site (IRES) can inhibit different HCV genotypes grown in cell culture and replicon replication, suggesting the potential of siRNA as an additional therapeutic option against HCV infection.

Key words: siRNA, short hairpin RNA, HCVcc, HCV replicon, reporter assay, immortalized human hepatocytes.

1. Introduction

HCV is a major cause of cirrhosis and hepatocellular carcinoma. PEG-interferon alone or together with ribavirin is currently the only therapy for HCV infection, but a significant number of HCV-infected individuals do not respond to this treatment. Development of new therapeutic options against HCV is therefore a matter of urgency. MicroRNA (miRNA) and small interfering RNA (siRNA) are small RNAs of 18–25 nucleotides that play important roles in the regulation of gene expression. Although, both miRNA and siRNA use the RNA-induced silencing complex (RISC) for gene silencing, their mechanisms of inhibiting protein synthesis differ. The siRNAs shut down gene expression at the posttranscriptional level through mRNA degradation (1–3). In mammals, exposure to double-stranded RNAs greater

Hengli Tang (ed.), *Hepatitis C: Methods and Protocols, Second Edition, vol. 510*
© 2009 Humana Press, a part of Springer Science+Business Media
DOI 10.1007/978-1-59745-394-3_19 Springerprotocols.com

than 30 bp in length induces a generalized antiviral interferon response that globally represses mRNA translation *(4, 5)*, but introduction of siRNA into mammalian cells leads to mRNA degradation with exquisite sequence specificity without activating an interferon response. siRNA is therefore a promising vehicle for induction of intracellular immunity. Unlike classical antisense techniques, siRNA taps into existing gene-silencing pathways. Resistance of particular RNAs to RNAi-mediated degradation has also been observed in cases where accessibility of the target sequence was restricted. For viral RNAs, resistance can also be related to intracellular location and/or nucleocapsid association of genomic RNA molecules *(6)*. RNAi effectors can be delivered to cells in two different ways: (a) chemically synthesized siRNAs can be delivered as a drug or (b), in a gene-therapy approach, DNA-encoding shRNA-expression cassettes are delivered into cells, which are then processed into active siRNAs by the host cell *(7)*. Delivery of DNA expression vectors is possible either by integration into the genome or in self-replicating episomal form, which could allow for constitutive expression of the shRNA cassette. shRNAs transfected into cells are initially transcribed in the nucleus and are thought to be exported into the cytoplasm with the aid of Exportin 5, such as miRNAs *(8)*. The loops of shRNA are trimmed by Dicer in the cytoplasm to become siRNA complexes for mRNA degradation, although translational repression might be involved in gene silencing by vector-based shRNA *(9)*. Therefore, RNA silencing provides a new platform that may effectively treat HCV infection in addition to traditional antiviral therapies.

We have shown that siRNA targeted at NS5A knocks down NS5A expression and impairs NS5A-mediated IL-8 activation *(10)*. Other investigators have also reported the inhibition of HCV replication using Core, NS3, or NS5B sequences as targets *(11–15)*. Therefore, RNAi can inhibit HCV replication, and we have demonstrated the proof of concept, but HCV is error-prone in replication and produces mutated progeny molecules or quasispecies.

Some of these natural mutations help viruses to escape immune surveillance or reduce inhibition by antiviral drugs, and they may prevent recognition by siRNAs. To overcome these obstacles, we must attack multiple sites of viral RNA sequences that are conserved and normally invariant across HCV strains or to attach several viral sequences simultaneously. By the methods given here, we have shown that siRNAs aimed at the HCV genomes efficiently act upon replicon systems and cell-culture-grown HCV, inhibiting virus genome replication, suggesting a potential clinical application of this novel approach *(10, 16)*.

2. Materials

2.1. Cell Culture and Lysis

1. Dulbecco's modified Eagle's medium (DMEM) (Cambrex, Walkersville, MD) containing 10% fetal bovine serum (FBS, Gibco/Invitrogen, Carlsbad, CA), 200 U/mL of penicillin G, and 200 µg/mL of streptomycin (Cambrex).
2. Small-airway cell basal medium (SABM, Cambrex) supplemented with growth factors and antimicrobials containing 5% chemical denatured serum (Invitrogen) *(17)*.
3. Solution of trypsin (0.05%) and 0.53 mM EDTA (Cellgro Mediatech, Herndon, VA).
4. G418 active ingredient (800 µg/mL; geneticin, Gibco/Invitrogen).
5. Hygromycin B (Sigma, Saint Louis, MO).
6. Cell Scrapers (BD Falcon, Bedford, MA).
7. HEBS and Phospate-buffered saline (PBS, Cambrex, Walkersville, MD).

2.2. shRNA-Expression Vectors and Electroporation

1. pRNAT-H1.3/Hygro (GenScript Corporation, Piscataway, NJ) *(16)*.
2. Oligo DNA-encoding shRNA sequences.
3. *Bam*HI, *Hin*dIII, calf intestinal alkaline phosphatase (CIAP), and T4 DNA ligase (Invitrogen).
4. T4 polynucleotide kinase (New England Biolabs, Ipswich, MA). 5. GenePulser Xcell (Bio-Rad, Hercules, CA), and Gene pulser cuvette, 0.4 cm.

2.3. Transfection and Reporter Assay

1. 35 mm dishes (Corning Incorporated, Corning, NY).
2. Lipofectamine (Invitrogen, San Diego, CA).
3. The reporter lysis buffer (Promega, Madison, WI), luciferase assay system (Promega), and a luminometer Optocomp II (MGM Instruments, Hamden, CT).
4. The Belly Dancer shaker (Stovall Life Science, Greensboro, NC).

2.4. RNA Extraction, Reverse Transcription-Polymerase Chain Reaction (RT-PCR) and Real-Time RT-PCR

1. Purescript RNA isolation kit (Gentra Systems, Minneapolis, MN).
2. Superscript one-step RT-PCR with platinum Taq kit (Invitrogen).
3. PCR primers: HCV 5′NTR sense primer (5′-TCTGCG GAACCGGTGAGTA-3′) and antisense primer (5′-TCAGG CAGTACCACAAGGC-3′) and GAPDH sense primer (5′-GCCATCAATGACCCCTTCAT-3′), and antisense primer (5′-TCTCGCTCCTGGAAGATGGT-3′) *(18)*.
4. DNA Thermal Cycler 480 (Perkin Elmer, Wellesley, MA).

5. Agarose gel electrophoresis buffer, $1 \times$ TBE buffer: 10.8 g Trizma base (Invitrogen), 5.5 g boric acid, electrophoresis grade (Fisher Scientific, Houston, TX), 0.7 g EDTA (Sigma), distilled water added up to 1000 mL. Store at room temperature.

6. Agarose and ethidium bromide (Sigma, Saint Louis, MO).

7. TaqMan RT reagents, ABI PRISM 7700, and ABI PRISM 96-well optical reaction plate (Applied Biosystems, Foster City, CA).

8. SYBR green PCR master mix (Applied Biosystems).

2.5. Western Blotting for HCV NS5A and Actin

1. $2 \times$ SDS sample buffer (10 mL): 4.6 mL 10% sodium dodecyl sulfate (SDS, electrophoresis grade, Invitrogen), 0.4 mL double-distilled H_2O, 2.0 mL glycerol, 1.0 mL 2-mercaptomethanol, 2.0 mL 0.25 M Tris–HCl, pH 6.8, pinch of bromophenol blue.

2. Running buffer ($10\times$): 30.2 g Tris–HCl, 188 g glycine, 10 g SDS, distilled water to bring the volume to 1000 mL.

3. 30% acrylamide/bis solution, 29:1 (BIO-RAD), and N, N, N', N'-tetramethyl-ethylenediamine (TEMED, Invitrogen).

4. Ammonium persulfate (Invitrogen/Gibco): prepare 10% solution in water and immediately freeze in 10 mL aliquots at 4°C.

5. Prosieve Color Protein Markers (Cambrex Rockland, East Rutherford, NJ).

6. 10% polyacrylamide gel transferred onto a nitrocellulose membrane (Bio-Rad). Ready Gel Blotting Sandwiches (0.45 μm nitrocellulose with filter papers 7×8.5 cm).

7. Transfer buffer: 4.5 g Tris–HCl, 22.2 g glycine, 300 mL methanol, 1200 mL distilled water.

8. Blocking buffer: 5% fat-free milk (instant nonfat dry milk, Nestle, Vevey, Switzerland), $1 \times$ TBST (10 mL 1 M Tris–HCl, pH 8, 8.76 g NaCl, 0.5 mL Tween-20, distilled water to bring volume to 1000 mL).

9. Monoclonal antibody to NS5A (Biodesign International Saco, ME), actin (Santa Cruz Biotechnology, Santa Cruz, CA).

10. ECL detection kit (Amersham Pharmacia, Piscataway, NJ) and image analyzer to quantify the density of the protein bands using Image Quant software (Amersham Molecular Dynamics, Sunnyvale, CA).

11. $10\times$ Re-blot plus strong (Chemicon International, Temecula, CA).

2.6. Colony-Formation Assay of Replicon-Harboring Cells

1. Phosphate buffered saline without calcium or magnesium (Cambrex).

2. 3.7% formaldehyde (Sigma).

3. 1% crystal violet (Gram Crystal Violet Primary Stain, DIFCO, Kansas City, MO).

2.7. Immuno-fluorescence Study

1. T-Octylphenoxypolyethoxyethanol (Triton-X-100)(Sigma).
2. Albumin Bovine Fraction V, fatty-acid poor, lyophilized (BSA, Calbiochem, La Jolla, CA).
3. NS4-specific FITC-conjugated monoclonal antibody (Biodesign International) and TO-PRO3-iodide (Invitrogen).
4. Prolong Gold antifade reagent (Invitrogen/Molecular Probes).

2.8. Treatment of Cells Harboring HCV Genome

1. Sonic Dismembrator Model 500 (Fisher Scientific).
2. Hoefer SE-250 gel system (Hoefer Scientific Instrument, San Francisco, CA).
3. Bio-Rad tank system for western blotting.
4. 10% gel: Mix 4.0 mL double-distilled H_2O with 3.3 mL 30% acrylamide/bis solution, 29:1 (Bio-Rad), 2.5 mL 1.5 M Tris–HCl, pH 8.8, 0.1 mL 10% SDS, 0.1 mL ammonium persulfate solution, and 4 μL TEMED
5. 5% stacking gel: Mix 1.4 mL double-distilled H_2O with 0.33 mL 30% acrylamide/bis solution, 29:1, 0.25 mL 1.5 M Tris–HCl, pH 6.8, 0.02 mL 10% SDS, 0.02 mL ammonium persulfate solution, and 2 μL TEMED.
6. Secondary antibody: Goat anti-mouse HRP; 1 mg/mL (Bio-Rad).
7. Re-blot plus strong (Chemicon International, Billerica, MA).
8. Enhanced chemiluminescent ECL detection kit (Amersham Pharmacia).
9. Image Quant software for Windows NT version 5.2 (Amersham Molecular Dynamics, Sunnyvale, CA).

---

3. Methods

3.1. Making of shRNA-Expression Vectors

1. Search the GenBank database for unique sequences within HCV to exclude identity with known cellular genes. Make forward and reverse oligonucleotides according to manufactures' instructions.
2. Digest pRNAT-H1.31 Hygro with *Hin*dIII at 37°C for 2 h and with *Bam*HI at 37°C for 2 h, then treatment with CIAP at 37°C for 90 min.
3. Elute the band after running on a 0.7% agarose gel.
4. Perform the PNK reaction for each oligonucleotide strand at 37°C for 30 min.
5. Anneal both strands of oligonucleotides. Heat to 90°C for 10 min and cool slowly to room temperature.
6. Ligate the vector and insert oligonucleotides at 16°C overnight.
7. Conduct chemical transformation of *E. coli* DH5-alpha and extraction of vectors according to standard methods.

3.2. Reporter Assay

1. Use the reporter plasmids to screen the plasmids expressing shRNA (*see* **Note 1**).
2. Passage human hepatoma cell line, Huh-7 cells, when approaching ~ 70% confluency with trypsin/EDTA to provide new maintenance cultures on T-25 tissue culture flask and experimental cultures on 35 mm dishes. A 1:15 split will provide experimental cultures whose numbers will be approximately 1×10^5 cells per dish (Corning Incorporated, Corning, NY) at transfection 24 h after splitting cells.
3. Transfect plasmid expressing shRNA (0.4 μg each) into Huh-7 cells with pSV40-HCV IRES-luc (0.6 μg) using Lipofectamine.
4. 72 h after transfection, wash the cells once and then add 200 μL reporter lysis buffer and incubate them at room temperature for 20 min with tilting. Using a scraper, collect the cell lysates. Determine the luciferase assay activities of Fluc with a luminometer Optocomp II following the manufacturer's protocol. Normalize the activities with respect to the protein concentration of the cell lysates (*see* **Note 2**).

3.3. Treatment of Cells Harboring HCV Genome

1. Maintain Huh-7 cells *(19)* harboring subgenomic replicon (SR) or full-length (FL) replicon *(20, 21)* of HCV genotype 1b in 10% FCS, DMEM with 800 μg/mL G418. Introduce vector encoding shRNA into these cell lines using Lipofectamine as described above. After splitting the cells into 35 mm plates, use 10% FCS, DMEM without 800 μg/mL G418. Cell-culture grown HCV (genotypes 1a and 2a) can also be used *(18)*. Harvest cells at 48 h or 72 h after transfection for RNA and protein analysis.
2. RNA Extraction: Remove medium from plates by aspiration and wash the cells with ice-cold PBS. Immediately add 350 μL of cell lysis buffer and isolate total cellular RNA using a Purescript RNA isolation kit. Resuspend RNA with ~ 50 μL RNA hydration solution. Keep RNA samples at −70°C.
3. To analyze the development of escape mutants, analyze the corresponding region of the HCV replicon using a Superscript one-step RT-PCR with platinum Taq kit. PCR primers for HCV 5′NTR and GAPDH are described above. Run in a DNA Thermal Cycler 480 as follows: 33 min at 45°C; 2 min at 94°C; then 28 to 50 cycles of at 94°C for 30 sec, 57°C for 1 min, and 72°C for 1 min.
4. Subject PCR products to electrophoresis on a 1.8% agarose gel. Measure the ratio of HCV 5′NTR/GAPDH from three independent experiments using Scion Image.

3.3.1. Real-Time PCR

1. Detect HCV-specific RNA by real-time PCR as the increase in fluorescence of SYBR Green I on an ABI PRISM 7700. Use house-keeping gene GAPDH as a control for normalization. Perform each real-time PCR assay in triplicate.
2. For one tube for reverse transcription (RT), use 5 μL 50 ng/mL random hexamers and 5 μL RNA sample.
3. Heat to 80°C for 5 min and quick chill. Add master mix and perform RT as follows: 25°C for 10 min, 48°C for 30 min, and 95°C for 5 min.
4. For one tube of real-time PCR reaction, combine 25 μL 2× SYBR mix, forward primer to a final concentration of 50 to 900 nM, reverse primer to a final concentration of 50 to 900 nM, 5 μL (1 to 100 ng) template, 20 μL and water to bring to a total volume of 50 μL, the perform PCR as follows: 95°C for 10 min, 40 cycles of 95°C for 15 s and 60°C for 1 min, then 95°C for 20 min (dissociation curve).

3.3.2. Preparation of Samples for Western Blotting

1. Remove the medium by aspiration and wash the cells with ice-cold PBS.
2. Immediately add 50–100 μL of 2× SDS sample buffer and scrape the material into the appropriate labeled tube.
3. Close the tubes, then sonicate them for 20 s three times in Sonic Dismembrator Model 500, and boil them for 5 min.
4. Centrifuge them at $300 \times g$ for 10 min, after which they are ready for separation by SDS-PAGE.

3.3.3. Sodium Dodecyl Sulfate-Polyacrylamide Gel Electrophoresis (SDS-PAGE)

These instructions assume the use of a Hoefer SE-250 gel system. Glass plates for the gels must be scrubbed clean with a water and 70% ethanol solution after use and rinsed with distilled water.

1. Prepare a 1.5 mm thick, 10% SDS polyacrylamide gel by mixing 4.0 mL double-distilled H_2O with 3.3 mL 30% acrylamide/bis solution, 29:1, 2.5 mL 1.5 M Tris–HCl (pH 8.8), 0.1 mL 10% SDS, 0.1 mL 10% ammonium persulfate solution, and 4 μL TEMED .
2. Pour the gel, leaving space for a stacking gel, and overlay it with distilled water. The gel should polymerize in about 1 hr.
3. Pour off the water and rinse the top of the gel twice with water.
4. Prepare 5% stacking gel by mixing 1.4 mL double-distilled H_2O with 0.33 mL 30% acrylamide/bis solution, 29:1, 0.25 mL 1.5 M Tris–HCl (pH 6.8), 0.02 mL 10% SDS, 0.02 mL ammonium persulfate solution, and 2 μL TEMED.
5. Pour the gel and insert the 10-well comb. The stacking gel should polymerize within ~ 1 hr.
6. Prepare the running buffer by diluting 100 mL of the 10× running buffer with 900 mL of water in a graduated cylinder.

7. Once the stacking gel has set, carefully remove the comb and wash the wells with running buffer.

8. Add the running buffer to the upper and lower chambers of the gel unit and load up to $10\,\mu L$ of each sample into a well. Include one well for prestained molecular weight markers (for examples, Prosieve color protein markers, Cambrex).

9. Complete the assembly of the gel unit and connect to a power supply (for example, Power-Pac Basic, BIO-RAD). Gel is run during the day ($\sim 4\,h$) at 85 V.

3.3.4. Western Blotting for HCV Antigen

Transfer the samples that have been separated by SDS-PAGE to a $0.45\,\mu m$ nitrocellulose membrane electrophoretically, using the Bio-Rad tank system.

1. Prepare a tray of transfer buffer that is large enough to accommodate a transfer cassette with two sheets of filter paper and a sheet of nitrocellulose.

2. Disconnect the gel unit from the power supply and disassemble.

3. Remove and discard the stacking gel.

4. Lay nitrocellulose membrane over top of separating gel.

5. Place the cassette in the transfer tank such that the nitrocellulose membrane is between the gel and the anode, according to the manufacturer's protocol. Orientation is very important; it prevents the proteins from being lost from the gel into the buffer rather than transferred to nitrocellulose.

6. Place the lid on the tank and activate the power supply. Transfers can be accomplished at 21 constant volts overnight.

7. Once the transfer is complete, remove cassette out from the tank and carefully disassemble it, removing the top sponge and sheets of white paper. Correct orientation is validated by colored marker. Excess nitrocellulose is trimmed using a razor blade and discarded.

8. Incubate the nitrocellulose in 50 mL blocking buffer for 2 h at room temperature in tray with gentle swirling using the Belly Dancer shaker.

9. Discard the blocking buffer and rinse the membrane quickly before adding a 1:1000 dilution of the NS5A antibody (1 mg/mL) in TBST and leaving for 1 h at room temperature in a 100 mm dish on Belly Dancer.

10. Remove the primary antibody and wash the membrane five times for 5 min each with $\sim 50\,mL$ TBST.

11. Prepare the appropriate secondary antibody freshly for each experiment, at 1:7000-fold dilution in blocking buffer, add it to the membrane, incubate for 45 min at room temperature.

12. Discard the secondary antibody and wash the membrane 6 times for 5 min each with TBST.

13. After the final wash, mix well 1 mL aliquots of each portion of the ECL reagent and then immediately add it to the blot for 1 min, ensuring even coverage.

14. Remove the blot from the ECL reagents, blot with Kim-Wipes, wrap it with plastic film, and then place it on an X-ray film into a cassette of suitable size. Suitable exposure time for the film is $\sim$ 60 min.

3.3.5. Stripping and Reprobing Blots for Actin

Once a satisfactory exposure for the result of NS5A has been obtained, strip the membrane of that signal and reprobe it with actin antibody, which provides a loading control that confirms equal recovery of the samples throughout the experiment.

1. Incubate the blot for 20 min in 1× re-blot plus strong (Chemicon International).

2. Once the blot is stripped, washed it in blocking buffer twice, for 10 min each.

3. Reprobe the membrane with anti-actin (1:1000 in TBST) with washes, secondary antibody, and ECL detection as described above.

4. Visualize proteins using an enhanced chemiluminescent ECL detection kit, scan them by image analyzer and measure the density of the protein bands with Image Quant software.

3.4. Colony-Formation Assay

Transfect vector encoding shRNA into cells harboring SR or FL cell lines by electroporation.

1. Detach subconfluent cells by trypsin treatment, collect them by centrifugation (500g, 10min), and wash them twice in RNase-free PBS.

2. Mix 3.0 µg plasmid vectors with 0.4 mL (2.0 × 10^6) of cells in a 4 mm gap-width cuvette (Bio-Rad). Pulse samples using a Bio-Rad GenePulser X cell (electroporation conditions: 270 V and 950 µF).

3. Leave pulsed cells to recover for 10 min at room temperature.

4. Seed pulsed cells into two 100 mm culture dishes.

5. Twenty-four hours later and every 3–4 days during selection, replace the medium with fresh DMEM (10% FCS with PC/SM) supplemented with 800 µg of G418 for 3 weeks in SR-harboring cells or FL-harboring cells.

6. Use one dish from each transfection for the colony-formation assay.

7. After washing the cells twice, fix G418-resistant colonies with 3.7% formaldehyde and stain them with 1% crystal violet.

8. In the other dish, pool and expand the G418-resistant colonies and use them for RNA and/or protein analyses.

9. For selection of stable cells, use 20–100 µg/mL hygromycin.

**3.5. Immuno-
fluorescence Study**

1. Passage IHH cells when approaching ∼ 70% confluency, using trypsin/EDTA, when maintaining cultures on T-25 tissue culture flask.

2. Electroporate IHH with 2 μg shRNA-expression vector or 2 μg control-shRNA-expression vector and put each of them into a 35 mm collagen-coated plate (*see* **Note 3**). Two days after transfection, cell confluency is 60–80%, and cell-culture grown HCV (HCVcc) infects both cells.

3. After washing them with HEBS twice, add 4500 viruses in 0.7 mL SABM to each 35 mm plates, tilt the plates every 30 min, and 8 h later add 1.3 mL 5% CDS SABM.

4. For intracellular immunofluorescence, at day 3 after infection, wash infected hepatocytes with PBS twice (*see* **Note 4**) and fix them with 3.7% formaldehyde for 30 min.

5. Wash the cells with PBS three times, permeabilize them with 0.2% Triton-X-100 for 5 min (*see* **Note 5**), and wash them with PBS three more times.

6. Block them with 3% BSA in PBS for 1 h, wash them with PBS once, and incubated them at room temperature for 1 h with a NS4-specific FITC-conjugated monoclonal antibody (1:150 in 3% BSA with PBS).

7. Wash them with PBS once.

8. Perform nuclear staining with TO-PRO3-iodide (1:10,000 in PBS) for 5 min.

9. Wash them with PBS four times.

10. Mount them with antifade and keep them at 4°C until observation with confocal microscopy (Bio-Rad 1024), and count focus-forming units/mL *(18)*.

---

4. Notes

1. For screening shRNA against HCV internal ribosomal entry site (IRES), we used pSV40-HCV IRES-luc *(22)* (kindly provided by Professor Martin Kruger), which encodes in a bicistronic fashion the *Renilla reniformis* luciferase (Rluc), the entire HCV core gene under translational control of HCV 5′ noncoding region (5′NTR), and the firefly luciferase (Fluc) followed by 3′ noncoding region of HCV *(20)*.

2. Because siRNA can degrade messenger RNA, suppression of Fluc activity is correlated with that of Rluc activity, indicating that sequences located downstream and upstream of the target site are degraded. If translation is suppressed, suppression of Fluc activity is not correlated with that of Rluc activity.

3. IHH cells can grow in collagen-coated plates.

4. Because cells are easily detached, take extreme care to wash gently.

5. Permeabilization takes exactly 5 min.

Acknowledgments

The authors would like to thank Drs. M. Kruger, R. Bartenschlager, T. Wakita, and C. M. Rice for providing reagents and Robert Steele for critical reading. This work was supported by research grant AI45144 from the National Institutes of Health.

References

1. Elbashir, S. M., Harborth, J., Lendeckel, W., Yalcin, A., Weber, K., and Tuschl, T. (2001) Duplexes of 21-nucleotide RNAs mediate RNA interference in cultured mammalian cells. *Nature* **411**, 494–498.

2. Liu, J., Valencia-Sanchez, M. A., Hannon, G. J., and Parker, R. (2005) MicroRNA-dependent localization of targeted mRNAs to mammalian P-bodies. *Nat. Cell Biol.* **7**, 719–723.

3. McCaffrey, A. P., Meuse, L., Pham, T. T., Conklin, D. S., Hannon, G. J., and Kay, M. A. (2002) RNA interference in adult mice. *Nature* **418**, 38–39.

4. Kumar, M. and Carmichael, G. G. (1998) Antisense RNA: function and fate of duplex RNA in cells of higher eukaryotes. *Microbiol. Mol. Biol. Rev.* **62**, 1415–1434.

5. Stark, G. R., Kerr, I. M., Williams, B. R., Silverman, R. H., and Schreiber, R. D. (1998) How cells respond to interferons. *Annu. Rev. Biochem.* **67**, 227–264.

6. Uprichard, S. L., Boyd, B., Althage, A., and Chisari, F. V. (2005) Clearance of hepatitis B virus from the liver of transgenic mice by short hairpin RNAs. *Proc. Natl. Acad. Sci. USA* **102**, 773–778.

7. Brummelkamp, T. R., Bernards, R., and Agami, R. (2002) A system for stable expression of short interfering RNAs in mammalian cells. *Science* **296**, 550–553.

8. Barik, S. (2006) RNAi in moderation. *Nat. Biotechnol.* **24**, 796–797.

9. Petersen, C. P., Bordeleau, M. E., Pelletier, J., and Sharp, P. A. (2006) Short RNAs repress translation after initiation in mammalian cells. *Mol. Cell* **21**, 533–542.

10. Sen, A., Steele, R., Ghosh, A. K., Basu, A., Ray, R., and Ray, R. B. (2003) Inhibition of hepatitis C virus protein expression by RNA interference. *Virus Res.* **96**, 27–35.

11. Kapadia, S. B., Brideau-Andersen, A, and Chisari, F. V. (2003) Interference of hepatitis C virus RNA replication by short interfering RNAs. *Proc. Natl. Acad. Sci. USA* **100**, 2014–2018.

12. Randall, G., Grakoui, A., and Rice, C. M. (2003) Clearance of replicating hepatitis C virus replicon RNAs in cell culture by small interfering RNAs. *Proc. Natl. Acad. Sci. USA* **100**, 235–240.

13. Seo, M. Y., Abrignani, S., Houghton, M., and Han, J. H. (2003) Small interfering RNA-mediated inhibition of hepatitis C virus replication in the human hepatoma cell line Huh-7. *J. Virol.* **77**, 810–812.

14. Wilson, J. A., Jayasena, S., Khvorova, A., Sabatinos, S., Rodrigue-Gervais, I. G., Arya, S., et al. (2003) RNA interference blocks gene expression and RNA synthesis from hepatitis C replicons propagated in human liver cells. *Proc. Natl. Acad. Sci. USA* **100**, 2783–2788.

15. Wilson, J. A. and Richardson, C. D. (2005) Hepatitis C virus replicons escape RNA interference induced by a short interfering RNA directed against the NS5b coding region. *J. Virol.* **79**, 7050–7058.

16. Kanda, T., Steele, R., Ray, R., and Ray, R. B. (2007) Small interfering RNA targeted to hepatitis C virus 5′nontranslated region exerts potent antiviral effect. *J. Virol.* **81**, 669–679.

17. Ray, R. B., Meyer, K., and Ray, R. (2000) Hepatitis C virus core protein promotes immortalization of primary human hepatocytes. *Virology* **271**, 197–204.

18. Kanda, T., Basu, A., Steele, R., Wakita, T., Ryerse, J. S., Ray, R., and Ray, R. B. (2006) Generation of infectious hepatitis C virus in immortalized human hepatocytes. *J. Virol.* **80**, 4633–4639.

19. Blight, K. J., McKeating, J. A., and Rice, C. M. (2002) Highly permissive cell lines for subgenomic and genomic hepatitis C virus RNA replication. *J. Virol.* **76,** 13001–13014.

20. Lohmann, V., Korner, F., Koch, J., Herian, U., Theilmann, L., and Bartenschlager, R. (1999) Replication of subgenomic hepatitis C virus RNAs in a hepatoma cell lines. *Science* **285,** 110–113.

21. Pietschmann, T., Lohmann, V., Kaul, A., Krieger, N., Rinck, G., Rutter, G., et al. (2002) Persistent and transient replication of full-length hepatitis C virus genomes in cell culture. *J. Virol.* **76,** 4008–4021.

22. Korf, M., Jarczak, D., Berger, C., Manns, M. P., and Kruger, M. (2005) Inhibition of hepatitis C virus translation and subgenomic replication by siRNAs directed against highly conserved HCV sequence and cellular HCV cofactors. *J. Hepatol.* **43,** 225–234.

Part IV

HCV Entry and Infection

Chapter 20

Cellular Receptors and HCV Entry

Mike Flint and Donna M. Tscherne

Abstract

After attachment to specific receptors on the surfaces of target cells, hepatitis C virus (HCV) particles are thought to be internalized to endosomes, where low pH induces fusion between the viral and cellular membranes, delivering the HCV genome into the cytoplasm. Here, we describe methods to study the early events in HCV infection; the interactions with cellular receptors and the mechanism of entry.

Key words: HCV E2 glycoprotein, CD81, pH-dependent, entry.

1. Introduction

HCV encodes two envelope glycoproteins, E1 and E2, believed to be type I integral membrane proteins that interact with cellular receptors and mediate viral entry. Until recently, the absence of cell-culture systems permitting robust HCV replication necessitated the use of surrogate assays for the study of HCV entry. One such assay uses a truncated, soluble form of the E2 glycoprotein (sE2); removal of the C-terminal transmembrane domain of E2, which mediates localization to the endoplasmic reticulum (ER), allows the secretion of sE2 from expressing cells. sE2 can bind to human cells and was used to identify interactions with several cell-surface molecules, including CD81, scavenger receptor class B type I, and dendritic cell-specific intercellular adhesion molecule-3–grabbing nonintegrin [reviewed in **refs.** *(1, 2)*]. Originally identified simply as a binding partner for sE2, CD81 was subsequently confirmed to have an essential role in entry of both HCV pseudoparticles (HCVpp) and cell-culture–propagated HCV (HCVcc).

Hengli Tang (ed.), *Hepatitis C: Methods and Protocols, Second Edition, vol. 510*
© 2009 Humana Press, a part of Springer Science+Business Media
DOI 10.1007/978-1-59745-394-3_20 Springerprotocols.com

Here, we describe a protocol for expressing sE2 protein suitable for studying the interaction with CD81 and other cellular receptors. It was used to express the genotype 1a, strain H, E2 protein with an N-terminal tissue plasminogen activator (tpa) signal sequence, by transient expression in a human embryonic kidney (HEK 293) cell line. sE2 was harvested from the media of the expressing cells and bound to CD81 (either as a recombinant protein or on the surfaces of cells) and was used to characterize the interaction between E2 and CD81 (3–6).

The C-terminal transmembrane domain of E2 is predicted to span amino acids 718–746 in the polyprotein. Consequently, truncation of E2 to amino acid positions 660, 661, or 715 results in loss of membrane association. To direct translocation into the ER and entry into the secretory transport pathway, a signal sequence should be fused to sE2. The endogenous signal sequence from the C-terminus of the upstream E1 protein can be used for this purpose. Although the exact boundary of the E1 signal sequence is unclear, addition of amino acids 364–383 from E1 to the N-terminus of sE2 directs translocation. Alternatively, an N-terminal signal sequence from a heterologous protein, such as tpa, directs efficient secretion. To facilitate detection, an epitope tag such as a myc tag, can be added to the C-terminus of the sE2 sequence. The presence of such a tag does not prevent the interaction between sE2 and CD81 (4, 7).

The HCV glycoproteins tend to form disulfide-linked aggregates that are thought to result from a misfolding pathway. Such aggregates are present in preparations of sE2 but do not interact with CD81 (5, 8). The extent of aggregation can be assessed by separation by reducing and nonreducing sodium dodecyl sulfate polyacrylamide gel electrophoresis (SDS-PAGE) followed by immunoblotting for E2. On a nonreducing gel, the aggregated sE2 appears as high molecular mass material that barely enters the gel. In contrast, secreted, nonaggregated sE2 (truncated to residue 661) migrates as a diffuse smear centered around 60 kDa. Diverse HCV sequences from multiple genotypes have been shown to require CD81 for entry when the full-length E1 and E2 proteins were incorporated into HCVpp (9). When expressed as sE2 proteins, however, different E2 sequences differ in their ability to bind to CD81 (10, 11). Lack of CD81 binding may be related to the tendency of the sequence to form disulfide-linked aggregates, or to fold into another nonnative conformation, when the sequence is truncated and expressed in the absence of E1.

In the protocol described here, sE2 is harvested from the medium of transfected HEK 293 cells and can be used to bind to recombinant CD81 in an ELISA protocol (*see* **Section 3.3** and **Fig. 20.1A**) or to the surface of cells and detected by flow cytometry (*see* **Section 3.4** and **Fig. 20.1B**). The large-scale

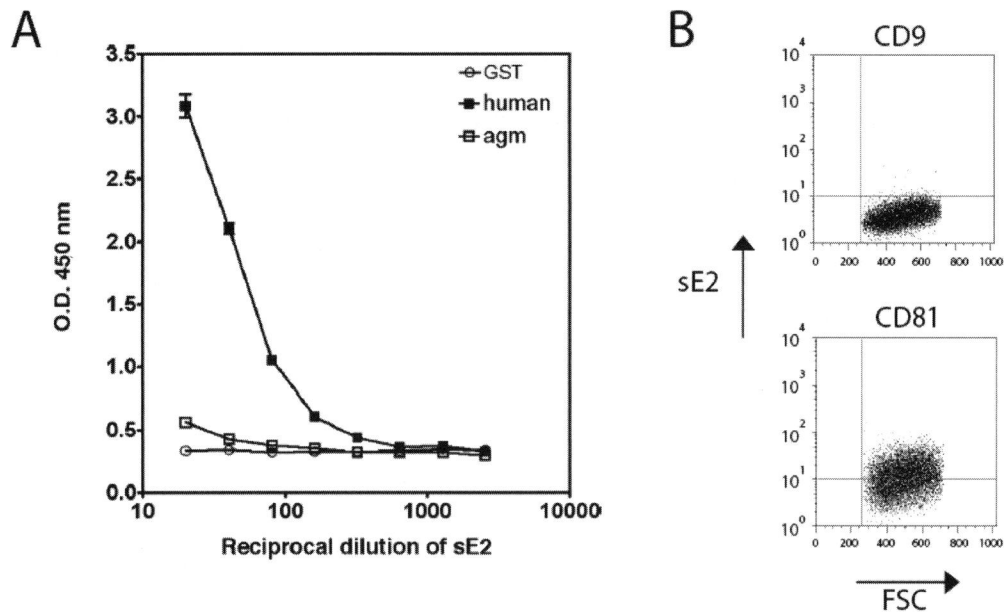

Fig. 20.1. **A.** Binding of sE2 to GST-CD81-LEL fusion proteins. A serial dilution of sE2 protein was allowed to bind to microtiter plates coated with either GST-human CD81-LEL, GST-African green monkey (agm) CD81-LEL, or GST protein. After washing, bound sE2 was detected with rat anti-E2 MAbs and anti-rat-HRP as described under **Section 3.3**. Each dilution was tested in triplicate, and the mean absorbance is shown with standard deviation bars (not visible for many points). O.D., optical density. agm, African green monkey. **B.** Binding of sE2 to cells. HepG2 cells transduced to express CD9 (as a nonbinding control) or CD81 were incubated with sE2 protein, and the cell surface–bound sE2 was detected with rat anti-E2 MAbs and anti-rat Alexa 488 as described under **Section 3.4**. FSC, forward scatter.

expression of sE2, as well as extensive purification methods, have been described by others (8, 12). If a known standard is available, sE2 can be quantified by comparison to the standard by ELISA in which *Galanthus nivalis* lectin is used to capture sE2 (4).

CD81 is a member of the tetraspanin superfamily, possessing short intracellular N- and C- termini, four transmembrane domains, and a small and a large extracellular loop (SEL and LEL) (13, 14). The LEL sequence (residues 115 to 202 of the human CD81 amino acid sequence) is critical for HCV entry (3, 15) and recombinant LEL alone is sufficient for interaction with sE2 (16). We describe the expression of glutathione S-transferase (GST)-CD81-LEL fusion proteins in *Escherichia coli* that can be purified by affinity chromatography on glutathione-sepharose. These recombinant proteins can then be used to study the interaction with sE2 (*see* **Section 3.3**) or to block infection by HCVpp or HCVcc. Similar fusion proteins, but with maltose binding protein or thioredoxin as the affinity-purification tag, have also been used successfully (7, 16).

The CD81-LEL is stabilized by disulfide bonds (17), the integrity of which are important for the interaction with sE2 *(6, 18)*. These bonds in the recombinant protein can be promoted by expression in an *E. coli* host that favors their formation. Rosetta-gami *E. coli* has been successfully used to express CD81 proteins with correctly formed disulfide bonds *(3)*. This strain bears the thioredoxin reductase (*trxB*) mutation that allows the formation of disulfide bonds in the *E. coli* cytoplasm (selectable on kanamycin). Bond formation is further enhanced by an additional mutation in the glutathione reductase (*gor*) gene (selectable on tetracycline). In addition, the Rosetta-gami strain carries the chloramphenicol-resistant plasmid pRARE2, which supplies tRNAs for six codons that are rarely used in *E. coli*, two of which are used in the human CD81-LEL sequence. To confirm the formation of disulfide bonds within GST-CD81-LEL, the protein can be separated by reducing and nonreducing SDS-PAGE and immunoblotted with an anti-CD81 monoclonal antibody (MAb) whose reactivity depends on the integrity of the LEL disulfide bonds. Several antibodies with this specificity have been reported; for example, 1.3.3.22 is a mouse MAb that recognizes human CD81 (available from Santa Cruz Biotechnology, Santa Cruz, CA, sc-7637). In **Section 3.3** we describe the use of GST-CD81-LEL to detect interactions with sE2.

After the attachment of viral particles to cellular receptors, enveloped viruses can mediate fusion between viral and cellular membranes by one of two strategies. For viruses such as HIV, fusion occurs at the plasma membrane, through a pH-independent mechanism triggered by interactions between the viral envelope proteins and cellular receptors. Other viruses, such as influenza, use a pH-dependent mechanism; virus particles are internalized in endosomes, where acid pH induces the conformational changes in the viral envelope proteins that mediate membrane fusion. HCV entry is pH-dependent *(19–21)* and is therefore sensitive to inhibitors of endosomal acidification such as bafilomycin A1 and concanamycin A. Here, we describe protocols by which these inhibitors can be used to probe the mechanism of fusion (**Section 3.5**). The timing of drug addition can be varied for investigation of the kinetics of viral entry; the drug loses effectiveness over time. These drugs are known to be cytotoxic at high concentrations, so controls of well-characterized pH-dependent and -independent viruses must be included (e.g., Sindbis virus and Herpes simplex virus-1, respectively). Enveloped viruses that are internalized by endocytosis are generally very sensitive to inactivation by low pH treatment, but although HCV undergoes pH-dependent entry, HCVcc is not irreversibly inactivated by acid pH *(20)*. In **Section 3.6** we describe a protocol by which the acid sensitivity of HCVcc particles can be tested.

2. Materials

2.1. Expression of sE2

1. Plasmid encoding sE2 protein under control of a eukaryotic promoter and the corresponding empty vector plasmid for generation of a negative control protein preparation.
2. HEK 293 cells cultured in 100 mm dishes.
3. Opti-MEM I Reduced Serum Medium (Invitrogen, Carlsbad, CA).
4. Lipofectamine 2000 transfection reagent (Invitrogen).
5. Fetal bovine serum (FBS). Heat at 56°C for 30 min; store at −20°C.
6. 1 M HEPES stock solution (Sigma, St. Louis, MO).
7. Dulbecco's modified Eagle's medium (DMEM; Invitrogen) supplemented with 2% FBS and 15 mM HEPES.
8. Amicon Centriplus YM-30 Centrifugal Filter Devices (Millipore, Billerica, MA).

2.2. Expression of GST-CD81-LEL

1. Rosetta-gami *E. coli* (Novagen, Madison, WI) freshly transformed with GST-CD81-LEL expression plasmid *(3)*.
2. Kanamycin: 10 mg/mL in water. Filter-sterilize and store in small aliquots at −20°C.
3. Tetracycline: 5 mg/mL in ethanol. Tetracycline is light sensitive; store it at −20°C.
4. Chloramphenicol: 34 mg/mL in ethanol. Store it at −20°C.
5. Ampicillin: 100 mg/mL in water. Filter-sterilize and store in small aliquots at −20°C.
6. Luria-Bertani (LB) medium: to 950 mL of deionized water add 10 g tryptone, 5 g yeast extract, 10 g NaCl. Stir until the solutes have dissolved, then adjust to pH 7.0 with 5 N NaOH. Adjust the volume of the solution to 1 liter with deionized water and sterilize it by autoclaving for 20 min at 15 psi on liquid cycle.
7. Isopropyl-β-D-1-galactopyranoside (IPTG): 1 M stock. Filter-sterilize and store in small aliquots at −20°C.
8. Lysis buffer: 50 mM HEPES, pH 7.5, 150 mM KCl, 10% glycerol. Sterilize the buffer by autoclaving for 20 min at 15 psi on liquid cycle and store at 4°C.
9. GSTrap HP affinity column (GE Healthcare, Piscataway, NJ).
10. Elution Buffer: 100 mM Tris-HCl, pH 8.0, 150 mM KCl, 5% glycerol, 10 mM reduced glutathione.
11. 10% SDS-PAGE gels.
12. Coomassie blue stain: 40% methanol, 10% acetic acid, 0.05% Coomassie brilliant blue R-250 (Sigma, St. Louis, MO).
13. Bio-Rad protein assay kit (Bio-Rad, Hercules, CA).

2.3. ELISA for sE2 Binding to GST-CD81-LEL

1. Clear, flat-bottomed, 96-well microtiter plates: Nunc-Immuno Maxisorp (Nalgene Nunc International, Rochester, NY) or equivalent.
2. Phosphate-buffered saline (PBS; Invitrogen).
3. Bovine serum albumin, fraction V (BSA; Sigma).
4. Blocking buffer: 1 × PBS supplemented with 5% (w/v) BSA.
5. Antibody diluent: 1 × PBS, 5% (w/v) BSA, 20% (v/v) sheep serum, 0.05% (v/v) Tween-20. Store the diluent in aliquots at −20°C. Thaw and equilibrate it to room temperature before use.
6. sE2 protein (*see* **Section 3.1**).
7. GST-CD81-LEL protein (*see* **Section 3.2**).
8. Detection antibodies: rat anti-E2 MAbs 6/1a, 7/59, 7/16 *(4)*, each diluted 1:10 in antibody diluent.
9. Horseradish peroxidase (HRP)-conjugated donkey anti-rat IgG (H + L) antibody (Jackson ImmunoResearch Laboratories, West Grove, PA) or equivalent, diluted 1:2000 in antibody diluent.
10. Tetramethylbenzidine (TMB) substrate (Sigma). Store the substrate at 4°C protected from light and equilibrate it to room temperature before use.
11. Stop solution (0.5 N sulfuric acid). Store at room temperature.
12. Microtiter plate reader with 450-nm filter.

2.4. Binding of sE2 to Cells

1. Dulbecco's phosphate-buffered saline (DPBS) without calcium or magnesium (Invitrogen).
2. EDTA, 0.5 M in DPBS (Sigma).
3. P/B/A: 1 × DPBS, 2% (w/v) BSA, 0.02% (w/v) sodium azide. Store solution at 4°C.
4. Round-bottomed test tubes, 5 ml, 12 × 75 mm (Becton Dickinson, Franklin Lakes, NJ).
5. sE2 preparation (*see* **Section 3.1**) diluted 1:2 in P/B/A.
6. Detection antibodies: rat anti-E2 MAbs 6/1a, 7/59, 7/16 *(4)*, each diluted 1:10 in P/B/A.
7. Alexa fluor 488, donkey anti-rat IgG (Invitrogen) or equivalent diluted in P/B/A to a final concentration of 2 µg/mL.
8. FACSfix: 2% paraformaldehyde in PBS. To dissolve paraformaldehyde, heat the solution to 70°C in a fume-hood for approx 1 h. Cool it to room temperature before adjusting the pH to 7.2. Store it at 4°C protected from light. This solution is stable for at least 1 month, as long as a pH of approx 7.2 is maintained. Check the pH periodically and discard the solution if it has become acidic.

2.5. Treatment with Chemical Inhibitors of Endosomal Acidification

1. Target cells (e.g., Huh-7.5 cells) and appropriate cell-culture media.
2. Concanamycin A (Sigma). For the stock solution, dissolve concanamycin A in ethanol to a final concentration of 25 μM. Working concentration is 25–50 nM.
3. Bafilomycin A1 (Sigma). For the stock solution, dissolve bafilomycin A1 in dimethylsulfoxide to a final concentration of 50 μM. Working concentration is 50–100 nM.
4. HCVcc reporter virus: e.g., FL-J6/JFH-5′C19Rluc2AUbi *(20)* or other HCVcc stock.
5. 1 M HEPES stock solution (Sigma, St. Louis, MO).
6. Dulbecco's phosphate-buffered saline (DPBS) without calcium or magnesium (Invitrogen).

2.6. Acid Sensitivity Assay

1. HCVcc reporter virus: e.g., FL-J6/JFH-5′C19Rluc2AUbi *(20)* or other HCVcc stock.
2. Citric acid buffer, pH 5.0 or pH 7.0: 15 mM citric acid, pH 4.1 or 7.0, 150 mM NaCl. Adjust pH to 4.1 or 7.0 using HCl or NaOH. A 1 to 5 dilution of cell-culture medium containing 200 mM HEPES into citric acid buffer pH 4.1 will give a final pH of 5.0 ± 0.1.
3. Dulbecco's modified Eagle's medium (DMEM; Invitrogen).
4. Neutralization buffer: DMEM supplemented with 2 M HEPES.
5. Target cells (e.g., Huh-7.5 cells) and appropriate cell-culture medium.

3. Methods

3.1. Expression of sE2

1. On the day before transfection, seed 100 mm cell-culture dishes with HEK 293 cells, such that they will be approx 90% confluent the following day (*see* **Note 1**).
2. On the day of transfection, remove the medium and gently rinse the cells with Opti-MEM I.
3. Transfect the cells with the sE2-expression plasmid and Lipofectamine 2000. Also perform a transfection with the empty vector plasmid, to generate a negative control preparation.
4. Culture the cells for 4–6 h at 37°C in the presence of transfection complexes.
5. Carefully remove the transfection mix and replace it with DMEM supplemented with 2% FBS and 15 mM HEPES (*see* **Note 2**).
6. Culture the cells for 48–72 h at 37°C.

7. Harvest the cell-culture medium. Clarify the medium by centrifuging at 3000 g for 15 min at 4°C.

8. To concentrate the sE2 preparation, transfer the supernatant into the reservoir of a Centriplus-30 concentrator. Centrifuge at 3000 g for 60 min at 4°C or until the retentate volume has diminished 5- to 10-fold.

9. Harvest the retentate from the Centriplus concentrator using the inverted spin method. Divide into small aliquots and store at −80°C.

3.2. Expression of GST-CD81-LEL

1. Using sterile technique, inoculate a single colony of freshly transformed Rosetta-gami *E. coli* into LB supplemented with 15 μg/mL kanamycin, 12.5 μg/mL tetracycline, 34 μg/mL chloramphenicol, and 100 μg/mL ampicillin.

2. Incubate this starter culture overnight in an orbital shaker at 30°C and 250 rpm.

3. Dilute the starter culture 1 to 100 into fresh LB supplemented with antibiotics as in step 1 (*see* **Note 3**).

4. Incubate the culture in an orbital shaker at 30°C and 250 rpm until the optical density at 600 nm reaches 0.75.

5. Chill the cultures at 4°C for 30–60 min.

6. Induce expression of the GST-CD81-LEL protein by adding IPTG to a final concentration of 100 μM. Incubate the culture overnight in an orbital shaker at 18°C and 250 rpm.

7. Collect the cells by centrifugation at 6000g for 10 min at 4°C.

8. Resuspend the cell pellets in ice-cold lysis buffer (20 mL/L of culture).

9. Lyse bacteria with three passes through an Avestin air emulsifier (*see* **Note 4**).

10. Clarify the lysates by centrifugation at 25,000 g for 30 min at 4°C.

11. Apply the clarified lysate to a GSTrap HP affinity column (*see* **Note 5**).

12. Wash the column with five column volumes of lysis buffer.

13. Elute the protein with five column volumes of elution buffer.

14. Collect fractions of the eluate and analyze them for protein content using 10% SDS-PAGE and Coomassie blue staining.

15. Pool the peak fractions and determine the protein concentration using the Bio-Rad protein assay (*see* **Notes 6, 7**). Store in aliquots at −80°C.

3.3. ELISA for sE2 Binding to GST-CD81-LEL

1. Dilute the GST-CD81-LEL protein to a final concentration of 5 μg/mL in PBS and add 50 μL to each well of the microtiter plate. To determine background levels of binding, coat the wells with GST alone.

2. Wrap the plate in plastic wrap and incubate it overnight at 4°C.

3. Wash the plate with PBS three times.
4. To block nonspecific binding sites, add $100\,\mu L$ of blocking buffer to each well and incubate for 1 h at room temperature.
5. Prepare a dilution series of sE2 in antibody diluent. Prepare a similar dilution series using the supernatant from the empty vector transfection as a negative control.
6. Wash the plate with PBS.
7. Add the diluted sE2 to the appropriate wells.
8. Wrap the plate in plastic wrap and incubate it overnight at 4°C to allow the sE2 to bind to the GST-CD81-LEL.
9. Remove unbound sE2 by washing the plate with PBS three times.
10. Add $50\,\mu L$ of detection antibodies in antibody diluent to each well (*see* **Note 8**).
11. Incubate the plate for 1 h at room temperature.
12. Wash the plate with PBS three times.
13. Add $50\,\mu L$ of HRP-conjugated antibody in antibody diluent to each well.
14. Incubate the plate for 1 h at room temperature.
15. Wash the plate with PBS three times.
16. Add $50\,\mu L$ of TMB substrate to each well.
17. Incubate the plate at room temperature for a maximum of 20 min.
18. Halt the reaction by adding $50\,\mu L$ of stop solution to each well.
19. Measure absorbance at 450 nm.

3.4. Binding of sE2 to Cells

1. Remove the medium from the monolayers of cells. Wash cells with DPBS supplemented with 2 mM EDTA and discard the wash.
2. Add sufficient DPBS supplemented with 2 mM EDTA to cover the monolayer and incubate at 37°C for 5 min (*see* **Note 9**).
3. Gently rinse the cells to dissociate them from the plastic.
4. Transfer the cells to a conical tube and centrifuge them at 300 g for 8 min.
5. Discard the supernatant and resuspend the cell pellet in P/B/A.
6. Determine the viable cell count by trypan blue exclusion.
7. For each reaction, transfer 1×10^6 cells to a 5 mL round-bottom tube.
8. Centrifuge the tubes at 300 g for 8 min and resuspend the cell pellet in $100\,\mu L$ of sE2 in P/B/A.
9. Incubate the tubes at room temperature for 1 h (*see* **Note 10**).

10. Wash the cells by resuspending them in 3 mL of P/B/A, then centrifuge the suspension at 300 g for 8 min.
11. Resuspend the cell pellet in 100 µL of rat detection antibodies in P/B/A.
12. Incubate at room temperature for 30 min.
13. Wash the cells by resuspending them in 3 mL of P/B/A, then centrifuge the suspension at 300 g for 8 min.
14. Resuspend the cell pellet in 100 µL of anti-rat-Alexa 488 in P/B/A.
15. Incubate at room temperature for 30 min.
16. Wash the cells by resuspending them in 3 mL of P/B/A, then centrifuge the suspension at 300 g for 8 min.
17. Wash the cells by resuspending them in 3 mL of PBS, then centrifuge the suspension at 300 g for 8 min.
18. Resuspend the cell pellet in 500 µL of FACSfix.
19. Analyze the suspension by flow cytometry.

3.5. Treatment with Chemical Inhibitors of Endosomal Acidification

1. Seed a 12-well culture plate with target cells (*see* **Note 11**) to achieve approx 70% confluence (approx 5×10^4 cells/well for Huh-7.5) and incubate the dish at 37°C, under 5% CO_2, until the cells have attached.
2. Remove the medium and add cell-culture medium supplemented with inhibitor (concanamycin A or bafilomycin A1) or vehicle (ethanol or DMSO, respectively).
3. Incubate the cells for 1 h at 37°C under 5% CO_2.
4. During this time, prepare the virus inoculum. Add HEPES buffer to the HCVcc reporter virus to a final concentration of 200 mM (*see* **Note 12**). Add bafilomycin A1, concanamycin A, or vehicle controls as appropriate. Mix the inocula and chill them on ice.
5. Remove the media from cells and add the inoculum (approx 100–150 µL/well).
6. Incubate the cells at 4°C to allow virus to bind (*see* **Note 13**).
7. Wash the cells with ice-cold DPBS.
8. Add cell-culture medium supplemented with bafilomycin A1, concanamycin A, or vehicle controls as appropriate (0.5 mL/well).
9. Incubate the cultures at 37°C, under 5% CO_2.
10. Measure HCVcc infection using an appropriate readout (e.g. luciferase activity, HCV RNA levels) within 24 h after infection (*see* **Note 14**).

3.6. Acid Sensitivity Assay

1. In a 15 mL conical tube, dilute the HCVcc reporter virus 1 to 5 in citric acid buffer, either pH 4.1 for the acid-treated sample or pH 7 as a control (*see* **Notes 15, 16**).
2. Vortex the tube gently.

3. Incubate it in a water bath at 37°C for 10 min.

4. Add 1/10th volume of neutralization buffer and vortex gently.

5. To ensure that the pH is completely neutralized, dilute the reporter virus 5- to 10-fold in DMEM supplemented with 200 mM HEPES.

6. Infect target cells with the inoculum and incubate them at 37°C.

7. Measure HCVcc infection 24 h after infection using an appropriate readout (e.g. luciferase activity, HCV RNA levels).

4. Notes

1. For the highest transfection efficiency and lowest cytotoxicity, we recommend optimizing the transfection conditions by varying cell density as well as DNA and lipofectamine 2000 concentrations. In our hands, optimal results are obtained when the cell monolayer is approx 90% confluent at the time of transfection, and the DNA:lipofectamine ratio is 1 μg : 1 μL. When seeding the cells for transfection, do not include antibiotics in the medium, as these may decrease the transfection efficiency. In addition to the cationic lipid-based transfection described here, we have successfully generated sE2 using a calcium-phosphate transfection protocol.

2. The low concentration of FBS (2%) reduces the growth rate of the cells. HEPES is added to the medium to minimize pH changes upon confluency. Even so, the cell monolayer will be confluent and the medium turning yellow-orange at the end of the 48- to 72-h culture period.

3. To ensure adequate aeration of the culture, use a flask with a volume at least four times that of the culture. Baffled flasks can be used to increase aeration further.

4. Other methods of bacterial cell lysis, such as sonication, can be employed.

5. Lysates can be applied to the GSTrap HP affinity column by fast protein liquid chromatography (FPLC). Alternatively, for a fast, small-scale preparation, the lysates, as well as subsequent wash and elution buffers, can be applied to a GSTrap FF column with a syringe.

6. The glutathione in the elution buffer interferes with Lowry or BCA-based protein-determination assays. Instead, use an assay based on the Bradford dye-binding procedure, such as the Bio-Rad protein assay indicated.

7. The expected yield for GST-CD81-LEL is approx 10 mg/L of culture.

8. Antibodies for detection of sE2 while it is bound to CD81 should be chosen carefully, as the epitopes recognized by some anti-E2 MAbs are obscured in the sE2-CD81 complex. Here, three rat MAbs that recognize the complex are used in combination to increase the sensitivity of detection. If the sE2 includes a C-terminal tag, an antitag antibody can be used. An anti-CD81 or anti-GST antibody can be used in place of the sE2 to confirm binding of the GST-CD81-LEL protein to the plate.

9. EDTA (in PBS without calcium or magnesium) is used to remove cells from tissue-culture vessels, as trypsin will cause proteolysis of cell-surface molecules and may alter the binding of sE2.

10. Unusually for cell staining, binding of sE2 to cells is performed at room temperature, as binding is reduced when the cells are incubated on ice. Sodium azide in the P/B/A buffer helps to reduce internalization of the cell-bound sE2 by inhibiting metabolic activity.

11. When a poorly adhering cell line such as 293T is used, culture dishes should be coated with a poly-L-lysine solution, which prevents cells from detaching during the assay.

12. HEPES buffer is necessary to maintain the pH of the media while cells are incubated at 4°C outside of a CO_2 incubator.

13. Incubation at 4°C permits the virus particles to bind to, but not enter, the cells. Entry is initiated upon transfer to 37°C. For an unsynchronized infection, pretreat the cells with the drug, omit the 4°C incubation, and infect cells in the presence of the drug at 37°C.

14. Significant toxicity has been noted when the drug is left in the medium for more than 24–36 h. If longer incubations are required, treat cells for approx 12 h, remove media containing the drug, and replace it with fresh media.

15. If the virus stock is in cell-culture medium, the starting pH of the citric acid buffer that will yield the desired final pH must be determined empirically. This determination is less important when purified virus in PBS is used.

16. Because HCVcc is relatively resistant to the acid treatment, a positive control such as Sindbis virus, which is highly sensitive to low pH, is suggested.

Acknowledgments

We are indebted to Jane McKeating and Charlie Rice with whose support and in whose laboratories these protocols were devised.

References

1. Cocquerel, L., Voisset, C., and Dubuisson, J. (2006) Hepatitis C virus entry: potential receptors and their biological functions. *J. Gen. Virol.* **87**, 1075–1084.

2. Bartosch, B., and Cosset, F. L. (2006) Cell entry of hepatitis C virus. *Virology* **348**, 1–12.

3. Flint, M., von Hahn, T., Zhang, J., Farquhar, M., Jones, C. T., Balfe, P., et al. (2006) Diverse CD81 proteins support hepatitis C virus infection. *J. Virol.* **80**, 11331–11342.

4. Flint, M., Maidens, C., Loomis-Price, L. D., Shotton, C., Dubuisson, J., Monk, P., et al. (1999) Characterization of hepatitis C virus E2 glycoprotein interaction with a putative cellular receptor, CD81. *J. Virol.* **73**, 6235–6244.

5. Flint, M., Dubuisson, J., Maidens, C., Harrop, R., Guile, G. R., Borrow, P., et al. (2000) Functional characterization of intracellular and secreted forms of a truncated hepatitis C virus E2 glycoprotein. *J. Virol.* **74**, 702–709.

6. Higginbottom, A., Quinn, E. R., Kuo, C. C., Flint, M., Wilson, L. H., Bianchi, E., et al. (2000) Identification of amino acid residues in CD81 critical for interaction with hepatitis C virus envelope glycoprotein E2. *J. Virol.* **74**, 3642–3649.

7. Drummer, H. E., Wilson, K. A., and Poumbourios, P. (2002) Identification of the hepatitis C virus E2 glycoprotein binding site on the large extracellular loop of CD81. *J. Virol.* **76**, 11143–11147.

8. Heile, J. M., Fong, Y. L., Rosa, D., Berger, K., Saletti, G., Campagnoli, S., et al. (2000) Evaluation of hepatitis C virus glycoprotein E2 for vaccine design: an endoplasmic reticulum-retained recombinant protein is superior to secreted recombinant protein and DNA-based vaccine candidates. *J. Virol.* **74**, 6885–6892.

9. McKeating, J. A., Zhang, L. Q., Logvinoff, C., Flint, M., Zhang, J., Yu, J., et al. (2004) Diverse hepatitis C virus glycoproteins mediate viral infection in a CD81-dependent manner. *J. Virol.* **78**, 8496–8505.

10. Patel, A. H., Wood, J., Penin, F., Dubuisson, J., and McKeating, J. A. (2000) Construction and characterization of chimeric hepatitis C virus E2 glycoproteins: analysis of regions critical for glycoprotein aggregation and CD81 binding. *J. Gen. Virol.* **81**, 2873–2883.

11. Roccasecca, R., Ansuini, H., Vitelli, A., Meola, A., Scarselli, E., Acali, S., et al. (2003) Binding of the hepatitis C virus E2 glycoprotein to CD81 is strain specific and is modulated by a complex interplay between hypervariable regions 1 and 2. *J. Virol.* **77**, 1856–1867.

12. Rosa, D., Campagnoli, S., Moretto, C., Guenzi, E., Cousens, L., Chin, M., et al. (1996) A quantitative test to estimate neutralizing antibodies to the hepatitis C virus: cytofluorimetric assessment of envelope glycoprotein 2 binding to target cells. *Proc. Natl. Acad. Sci. USA* **93**, 1759–1763.

13. Levy, S. and Shoham, T. (2005) The tetraspanin web modulates immune-signalling complexes. *Nat. Rev. Immunol.* **5**, 136–148.

14. Levy, S., Todd, S. C., and Maecker, H. T. (1998) CD81 (TAPA-1): a molecule involved in signal transduction and cell adhesion in the immune system. *Annu. Rev. Immunol.* **16**, 89–109.

15. Zhang, J., Randall, G., Higginbottom, A., Monk, P., Rice, C. M., and McKeating, J. A. (2004) CD81 is required for hepatitis C virus glycoprotein-mediated viral infection. *J. Virol.* **78**, 1448–1455.

16. Pileri, P., Uematsu, Y., Campagnoli, S., Galli, G., Falugi, F., Petracca, R., et al. (1998) Binding of hepatitis C virus to CD81. *Science* **282**, 938–941.

17. Kitadokoro, K., Bordo, D., Galli, G., Petracca, R., Falugi, F., Abrignani, S., et al. (2001) CD81 extracellular domain 3D structure: insight into the tetraspanin superfamily structural motifs. *EMBO J.* **20**, 12–18.

18. Petracca, R., Falugi, F., Galli, G., Norais, N., Rosa, D., Campagnoli, S., et al. (2000) Structure-function analysis of hepatitis C virus envelope-CD81 binding. *J. Virol.* **74**, 4824–4830.

19. Hsu, M., Zhang, J., Flint, M., Logvinoff, C., Cheng-Mayer, C., Rice, C. M., et al. (2003) Hepatitis C virus glycoproteins mediate pH-dependent cell entry of pseudotyped retroviral particles. *Proc. Natl. Acad. Sci. USA* **100**, 7271–7276.

20. Tscherne, D. M., Jones, C. T., Evans, M. J., Lindenbach, B. D., McKeating, J. A., and Rice, C. M. (2006) Time- and temperature-dependent activation of hepatitis C virus for low-pH-triggered entry. *J. Virol.* **80**, 1734–1741.

21. Koutsoudakis, G., Kaul, A., Steinmann, E., Kallis, S., Lohmann, V., Pietschmann, T., and Bartenschlager, R. (2006) Characterization of the early steps of hepatitis C virus infection by using luciferase reporter viruses. *J. Virol.* **80**, 5308–5320.

Chapter 21

Studying HCV Cell Entry with HCV Pseudoparticles (HCVpp)

Birke Bartosch and François-Loïc Cosset

Abstract

HCV infection leads in 50 to 80% of cases to chronic hepatitis, liver cirrhosis, or hepatocellular carcinoma. Interferons and the nucleoside analog ribavirin form the basis for treatment but are not sufficiently effective and have numerous side effects. Although about 300 million people worldwide are estimated to be infected, the characterization of HCV biology and associated pathologies and development of new therapeutics have been slow. Systems that support HCV replication and particle formation in vitro have emerged only over the last few years, over 15 years after the discovery of the virus. The available infection models have remained limited to chimpanzee *(1)* and immunodeficient mice carrying engrafted human liver cells *(2)*. HCV pseudoparticles (HCVpp) were the first in vitro infection system to become available for investigation of entry and neutralization of this major human pathogen. HCVpp are formed by incorporation of the full-length hepatitis C virus glycoproteins E1 and E2 onto lenti- or retroviral core particles. HCVpp have been validated by many research groups, closely mimic the functionality of the wild-type virus in terms of cell entry and neutralization, and have even been used to isolate the recent HCV receptor Claudin-1. HCVpp are a useful model system not only because of the functional conservation of the envelope glycoproteins with those of the wild-type virus, but also because the retro- or lentiviral vectors used to form them offer of a number of significant technical advantages.

Key words: Hepatitis C virus, glycoprotein, retro/lentiviral vector, cell entry, receptor, neutralization, lipoprotein.

1. Introduction

1.1. The HCV Pseudoparticle System

The description of HCV pseudoparticles (HCVpp) as the first surrogate HCV infection system described the incorporation of functional, unmodified HCV glycoproteins (GP) into heterologous retro- and lentiviral core particles. Initially, the concept to incorporate intracellular, ER membrane-localized HCV GPs onto retroviral cores was disputed, because retroviral and lentiviral

Hengli Tang (ed.), *Hepatitis C: Methods and Protocols, Second Edition, vol. 510*
© 2009 Humana Press, a part of Springer Science+Business Media
DOI 10.1007/978-1-59745-394-3_21 Springerprotocols.com

core proteins were known to assemble and recruit viral GPs at the cell surface. This paradox was resolved by two observations: First, upon overexpression of the HCV GPs, a small population of the GPs leaves the ER and traffics through endosomal compartments toward the cell surface *(3–5)*. Second, studies on retroviral assembly have shown that the assembly of retro- and lentiviral capsid proteins does not necessarily take place at the cell surface but that virions can also assemble intracellularly in endosomal compartments *(6)*. In 293T producer cells, HCVpp assembly has now been confirmed to take place intracellularly, possibly in multivesicular bodies *(7)*. These and other studies confirmed furthermore that the presence of intact transmembrane domains of the GPs is required for functionality of the resulting E1E2 complex *(8)*. HCVpp are produced by transfection of human 293T cells with expression vectors encoding the full-length E1E2 polyprotein, retroviral or lentiviral core proteins, and a packaging-competent retro- or lentiviral genome carrying a marker gene *(3)*. Transfected 293T cells secrete assembled viruses into the supernatant, which can be harvested and used to infect naïve target cells. Infection of target cells can be monitored by assessment of expression of the viral marker gene. Production of HCVpp is relatively efficient, producing an average of 5×10^5 infectious units per mL supernatant, and can be performed at a convenient safety level.

1.2. Conserved Functions of the HCV Glycoproteins in the HCVpp Context

HCVpp allow investigation of all functions mediated by the HCV GPs, comprising cell entry and neutralization. HCVpp have been shown to mimic closely the entry and serological properties of wild-type HCV [*see* **ref.** *(9)* and references therein]. The use of HCVpp has contributed to analysis of the roles of the HCV coreceptors CD81 *(10)*, scavenger receptor 1 *(11)*, low-density lipoprotein receptor *(12)*, and a number of capture receptors, as well as to the identification and analysis of the recently described coreceptor Claudin-1 *(13)*. Cell entry properties and HCV receptor use of the recently developed recombinant, cell-culture-produced HCV particles are remarkably consistent with those described for HCVpp *(14–16)*. Similarly, a significant number of antibodies and sera directed against cellular receptors or the HCV glycoproteins themselves have neutralizing activities on HCVpp similar to those on cell-culture-produced HCV particles *(14, 16)*. Finally HCVpp displaying GP sequences of all major genotypes have been described *(17–19)*.

1.3. HCVpp, A Powerful Genetic System

The use of retroviral vectors has become popular because of the ability of retroviruses to integrate themselves into the host-cell genome and maintain persistent gene expression even over the long term. Using HCVpp-containing vectors encoding nonviral marker genes for infection does not lead to productive viral replication, because the structural genes of the virus are not transferred into the newly infected cell. The advantage of this abortive infection cycle is that it allows tracking and measurement of single infection events. The measurement of single infection events can be advantageous for quantification of neutralization and for comparison of viral glycoprotein mutants in cell entry and/or neutralization. Another advantage of lentiviral vectors, in particular, is that they can infect cells at both mitotic and postmitotic stages of the cell cycle, thus allowing study of nondividing cells and tissues. Furthermore, incorporation of ectopic sequences, marker genes, and even proteins into the core of oncoretro- and lentiviral pseudoparticles is easily achievable. In contrast, modulation of the assembly of the wild-type virus is more restrictive, because assembly is closely linked to viral replication. Both of these events require very specific cell lines and culture conditions. In contrast, assembly of HCV GPs onto retroviral core particles is a very flexible process that leaves a wide choice of producer cells and target cells.

2. Materials

2.1. Cell Lines and Culture

1. Among the most suitable producer cell lines are 293T human embryo kidney cells (CRL-1573), but other cells which are efficiently transfected and allow high expression levels of ectopic constructs are also suitable.
2. Target cells include mainly liver-derived cell lines. Among the most commonly used are Huh-7 human hepatocellular carcinoma (20), Hep3B human hepatocellular carcinoma (HB-8064), HepG2 human hepatocellular carcinoma (HB-8065) overexpressing CD81 (21,22), and PLC/PRF/5 human hepatoma (CRL-8024). With the recent discovery of Claudin-1 as an HCV coreceptor (13), SW13 human adrenocortical carcinoma (CCL-105) and 293(T) cells can be constructed to become permissive as well. Cells should be grown according to American Type Culture Collection recommendations.
3. Dulbecco's modified Eagle's medium (DMEM, #41966-029, GIBCO/Invitrogen) with 0.11 g/L sodium pyridoxine or RPMI (#21875, GIBCO/Invitrogen). Both media must be supplemented with 10% fetal calf serum (#10270-106, GIBCO/Invitrogen) and 100 µg/L penicillin/streptomycin (#15140-122, GIBCO/Invitrogen). All cells can be passed by washing of (semi)confluent plates with Dulbecco's phosphate-buffered saline (#10270-106, GIBCO/Invitrogen) after resuspension of cells in trypsin (#25300-054, GIBCO/Invitrogen).

2.2. Consumables

1. Ø10 cm sterile petri dishes (#430167) from Corning and 6-, 12-, 24-, or 48-well Costar plates with flat bottoms (Corning).
2. 10- or 20 mL syringes.
3. 0.45 – μm filters (Millipore).

2.3. Plasmids

1. HCV E1E2 Expression Constructs. The standard expression plasmid for the HCV E1E2 precursor encodes the last 60 residues of HCV core and the full-length coding sequence of the E1 and E2 glycoproteins, terminating in a stop codon. Coexpression of plasmids encoding E1 only and E2 only with appropriate signal peptide sequences also leads to efficient assembly of HCV pseudoparticles (see **Fig. 21.1A**). Presence of a strong promoter is important to achievement of high expression levels, and we have so far mainly worked with CMV promoter–driven plasmids (see **Fig. 21.1B**).

2. Packaging and marker transfer vector constructs: Retro- as well as lentiviral vector systems can be used to assemble HCVpp efficiently *(3)*. Among the best-characterized retoviral vector systems is that of murine leukemia virus (MLV). A CMV-Gag-Pol MLV packaging construct, encoding the MLV gag and pol genes, and the MLV-GFP plasmid, encoding an MLV-based proviral transfer vector containing a CMV-GFP internal transcriptional unit, have been described previously [see **Fig. 21.1B** and **ref.** *(3)*]. A commonly used lentiviral vector system is based on the human immunodeficiency virus 1. The HIV-1 Gag-Pol packaging construct pCMV8.2 contains all HIV-1 genes except the envelope glycoprotein precursor and nef. Presence of the HIV accessory genes is required when nondividing cells are infected. The HPPT-EF1-GFP HIV-1 proviral transfer vector contains the EF1-α internal promoter driving, in this case, an eGFP marker gene *(23)*. The use of minimal HIV vector systems such as the one based on pCMV8.91, which expresses, besides gag and pol, only the auxiliary genes tat and rev, offers other advantages. By simple co-transfection of a plasmid encoding a chimeric version of the accessory gene vpr fused to lacatmase or gfp, one can label the viral core particles (*see* **Note 4**). Fusion proteins between Vpr and beta-lactamase have indeed been shown to be incorporated into forming particles *(24)*, and their enzymatic activity can for example be used as a readout signal in HCVpp fusion assays *(25)*. Besides HIV-based vector systems, simian immunodeficiency virus–derived vectors can be used with equal efficiently to package the HCV glycoproteins *(26)*.

3. Viral glycoprotein expression constructs: A significant number of viral glycoproteins have been reported to be incorporated and to form infectious pseudoparticles. These can

A

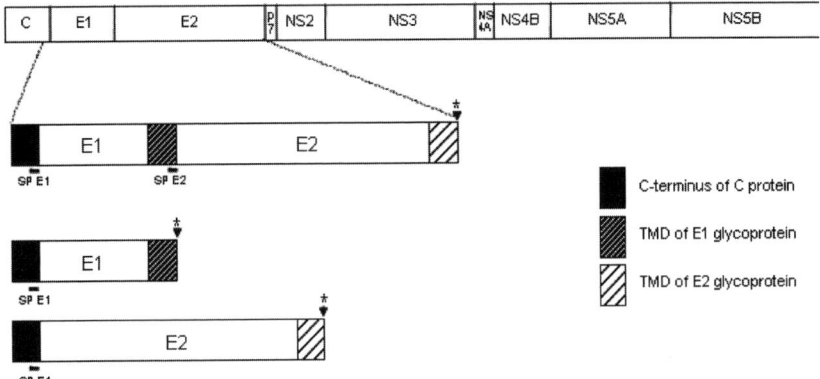

B

Expression constructs for :

Retroviral core proteins

GFP transfer vector

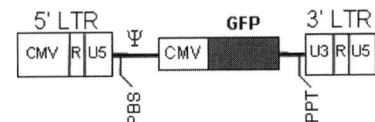

Envelope glycoproteins

Fig. 21.1. HCV E1E2 and retroviral expression constructs. (**A**) A cDNA derived from the HCV polyprotein gene was used to express the E1E2 glycoproteins and the COOH terminus of the C protein, which provides the signal peptide for E1 (SP E1). The position of stop codons (asterisk) inserted in the expression constructs to terminate translation of the proteins is shown. The transmembrane domain (TMD) of E1 provides the signal peptide (SP E2) for the E2 glycoprotein. (**B**) The expression constructs encoding the different components required to assemble infectious pseudoparticles are shown. The shaded boxes represent the viral genes and the marker gene (GFP) transferred to the infected cells. The open boxes show the cis-acting sequences. LTR, long terminal repeat, comprising U3, R and U5 sequences; CMV, cytomegalovirus immediate-early promoter; PBS, primer binding site; ψ, packaging sequence; PPT, poly-purine track; polyA, polyadenylation site.

be useful as control pseudoparticles, because of either their specific functional properties or their thorough characterization. The phCMV-G envelope-expression vector encodes the vesicular stomatitis virus G protein, which displays very high titers and has a ubiquitous tropism. It is therefore often used as a control in host-range studies *(3)*. phCMV-RD114 encodes the feline endogenous virus RD114 glycoprotein, which is complement resistant and is therefore most useful as a control in neutralization studies *(27)*. phCMV-HA/NA vectors, encoding the hemagglutinin and neuraminidase of fowl plague virus is a glycoprotein complex whose fusion properties are very well characterized *(25, 28)*.

2.4. Reagents for Transfection and Infection

1. Transfection: CalPhos Mammalian transfection kit (# K2051-1) from Clontech, or a home made version of the same kit. The solutions of the kit comprise H_2O, $2 \times HBS$, and $2\,M$ $CaCl_2$ and must be filter sterilized. The $2 \times HBS$ buffer consists of $280\,mM$ NaCl, $10\,mM$ KCl, $1.5\,mM$ $Na_2HPO_4 - 2H_2O$, $12\,mM$ D-glucose, and $20\,mM$ HEPES, final pH adjusted to 7.5 with NaOH. All solutions can be stored at $4°C$ or $-20°C$.

2. Infection: The only reagent required for infection is polybrene (hexa dimethrine bromide, # H9268, Sigma) as a filter-sterilized stock solution of $8\,mg/mL$ in Dulbecco's phosphate-buffered saline. This solution can be stored at $4°C$ for use and at $-20°C$ for long-term storage.

3. Methods

3.1. Outline: Time Line

A minimum of 6 d are necessary for production and testing of HCVpp in an infection reaction (see **Fig. 21.2**).

1. Day 1: In the morning of day 1, seed producer cells for transfection of the vectors the next day.

2. Day 2: Use a calcium-phospate method for tranfection on day 2. We perform the transfection normally in the afternoon or evening of day 2 and leave the transfection mixture on the cells overnight.

3. Day 3: In the morning of day 3, replace the transfection medium on the cells with the production medium of choice. If one has transfected a vector encoding a gfp marker, one should already be able to detect gfp expression in a high proportion of the cells under the UV microscope. Also on day 3, seed the target cells for infection in the desired well format.

4. Day 4: On day 4, harvest and filter the supernatant of the producer cells containing the pseudoparticles. It can be used

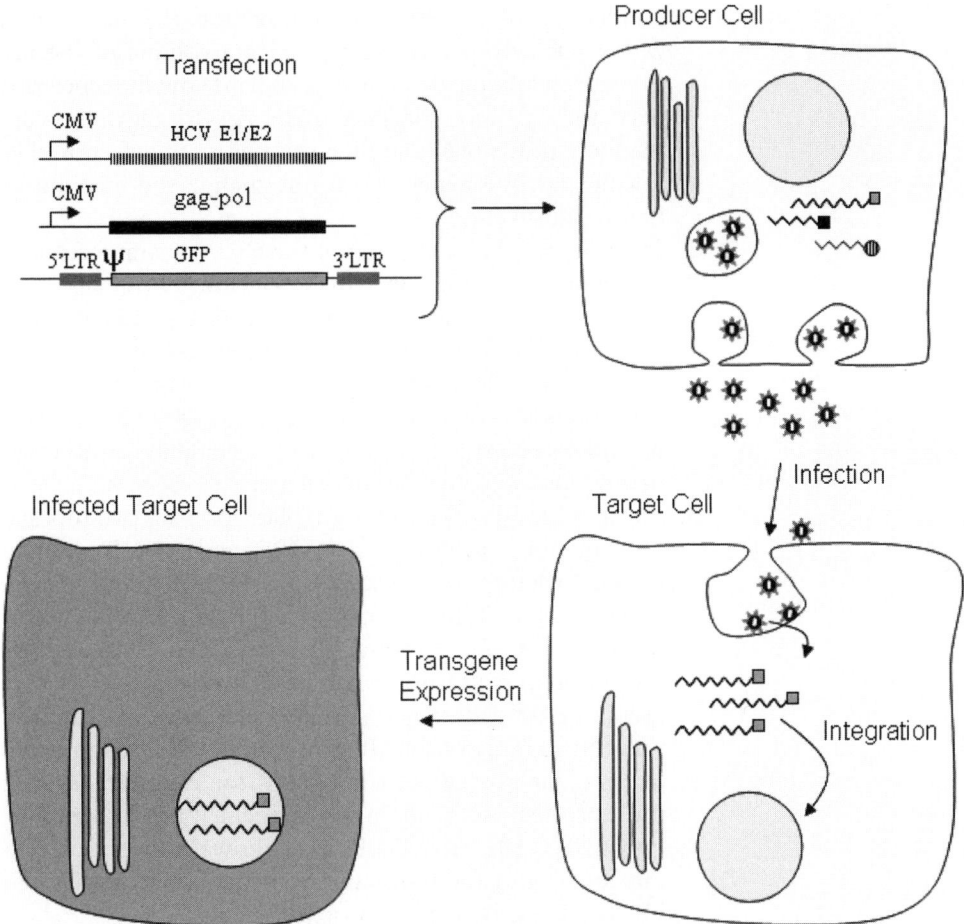

Fig. 21.2. Scheme of HCVpp production and infection. Producer cells are transfected with CMV-Gag-Pol, GFP transfer vector and HCV GP expression constructs. Successfully transfected producer cells assemble HCVpp intracellularly and secrete them into the supernatant. The HCVpp-containing supernatant can be harvested. The supernatant can then either be used unmodified or conditioned (centrifugation, drug treatement etc.) to infect target cells. HCVpp attach to the target cells, become endocytosed and fuse with the endocytic membranes to release the retroviral core containing the gfp transfer vector into the cytoplasm. The gfp transfer vector is then reverse transcribed and integrated into the host-cell genome. 24 to 96 hours after infection, transgene expression can be analysed.

directly for infection reactions. Once the infection is carried out, incubate the infected target cells for 48 to 72 h.

5. Days 6–8: Analyze the infection reactions.

3.2. Production of HCVpp

The quality of the producer cells is of major importance for HCVpp production (*see* **Note 1**). Besides 293T cells, we have successfully used a range of other cell lines for HCVpp production, but because of their lower transfection efficiencies, the resulting HCVpp titers are usually lower than those with 293T producer cells.

1. Depending of the amount of HCVpp needed, seed 293T cells into Ø10 cm plates or six-well plates. To obtain a cell

confluency of ca. 30% for transfection the next day, add 2.1×10^6 cells per Ø10 cm plate in a volume of 7–8 mL or 3×10^5 cells in a volume of 2–2.5 mL medium per 6 well. Seed the cells not too late in the day to allow the cells to readhere to the plates and recover before transfection the following day. We generally do not change the medium again before transfection.

2. In the afternoon or evening of the next day, tranfect the cells. We generally use a home made calcium-phosphate method for transfection and the protocol below is adapted to this kit (*see* **Note 2**). Transfection efficiencies of over 80% are required to produce HCVpp supernatants with good infectious titers. Besides the quality of the cells, well-purified vector preparations are necessary, but standard column chromatography of the plasmids is generally sufficient.

3. For transfection of one Ø10 cm plate, prepare two tubes: tube A, containing 450 µL of 2 × HBS, and tube B, containing the viral vectors and expression plasmids in a final volume of 395 µL H_2O. Of the vectors, we use 8 µg of marker gene vector, 8 µg of gagpol expression plasmid, and 4 µg of the HCV glycoprotein expression plasmid. For production of control pseudoparticles displaying other envelopes, one may have to vary the quantity of the GP expression plasmid slightly. Vesicular stomatitis virus G expression, for example, is toxic to the cells because of its high fusiogenicity; we therefore use only 1.8 µg plasmid. **Table 21.1** shows an overview of plasmid quantities for transfection adequate to produce either HCVpp or control pseudoparticles.

4. After mixing the DNA and H_2O thoroughly in tube B, add 55 µL of 2 M $CaCl_2$ dropwise while holding tube B on a vortexer to ensure immediate mixing.

Table 21.1
Plasmid quantities (in µg) for transfection of a Ø10 cm plate

Pseudoparticles	gagpol	Marker gene vector	GP expression plasmid	
HCVpp	8	8	phCMV-E1E2	4.0
VSV-Gpp	8	8	phCMV-G	1.8
RD114pp	8	8	phCMV-RD114	2.0
HA/NApp	8	8	phCMV-HA&NA	2 × 2.0
noEnvpp	8	8	–	–

5. Then take the $DNA/H_2O/CaCl_2$ mixture up with a pipette and add it dropwise, again on the vortexer, to tube A, containing $2\times$ HBS.

6. No more than 20 min later, apply the resulting transfection mixture dropwise to the producer cells.

7. Because of its toxicity, do not leave the transfection mixture on the cells for more than 16 h. Replace it early the next morning with ca. 7 mL of the culture medium of choice. Take care in replacing the medium, because 293T cells are easily detached. At this point, when transfecting with a marker gene that can be visualized such as a fluorescent protein, a high proportion of transfected cells should already be detectable.

8. In choosing the culture medium, consider whether it should be supplemented with bovine or human serum factors or other lipids. Particles produced in human serum have high specific infectivity (see **Fig. 21.3**) but are in turn less sensitive to neutralization with many mono- and polyclonal antibodies *(27,29,30)*. The choice of culture medium is therefore an important factor and is discussed in more detail in **Chapter 32**.

9. Twenty-four h after the medium change on, i.e., ca. 40 h after transfection, harvest the HCVpp-containing supernatant from the producer cells.

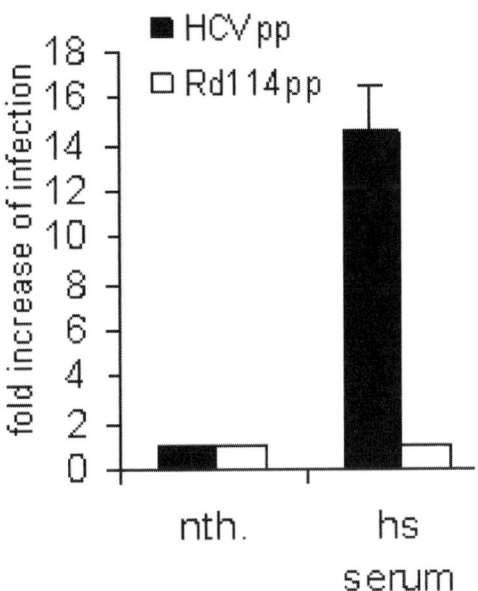

Fig. 21.3. The specific infectivity of HCVpp is enhanced by human serum. Infection assays on Huh-7 cells using HCVpp of genotype 1a and RD114pp as control. Virions were produced in cell-culture media containing no serum or 2% human serum. Results show the fold increase of infection determined by calculating the ratio between average infectious titres determined in the presence or absence of serum (n=5).

10. If desired, verify the transfection efficiency of the producer cells, which should be above 80%. Verification is easy if, e.g., gfp was transfected. For that purpose, the 293T producer cells can be trypsinized and analyzed with a Fluorescence Activated Cell Sorter (FACS). Note that transfection efficiency should be assessed with producer cells that have been transfected with only gagpol and marker gene vector constructs, without a GP expression construct. Indeed, in the presence of viral glycoproteins, infectious virions that have been formed early after transfection can lead to transfer and amplification of the gfp signal into cells that were originally not transfected. In contrast, assembly of retroviral core proteins can occur in the absence of viral glycoproteins but results in the secretion of core particles with viral marker genomes incorporated that are noninfectious and cannot disseminate the gfp transgene in the culture.

3.3. Infection with HCVpp

1. Harvest HCVpp supernatant with syringes and filter it through $45 - \mu m$ filters to ensure removal of producer cells and cellular debris.
2. Either use the supernatant for infection immediately or store it at 4°C for some days. If necessary, HCVpp supernatants can be frozen and stored in liquid nitrogen.
3. Seed target cells the day before infection. In the case of Huh-7 cells, use 5.5×10^3, 4×10^4, 8×10^4, or 2×10^5 per 96, 24, 12, or 6 wells, respectively. Infection reactions with HCVpp are generally performed in the presence of the cationic liposome polybrene, which is added to the reaction to a final concentration of $4 \mu g/mL$ (*see* **Note 3**). The total volume in which the infection is performed is important because it influences how much virus can interact with the target cells. Good ratios of total infection volumes to HCVpp input volumes for different well sizes are listed in **Table 21.2**. For control of the background of infection, the

Table 21.2
Infection conditions

Well format	Number of target cells	Total infection volume (μL)	HCVpp input volume (μL)
96	5.5×10^3	50	5–40
24	4×10^4	300	10–100
12	8×10^4	600	50–300

use of nonenveloped retroviral particles, generated by transfection of viral gagpol and marker vectors only, gives most accurate results. Infection reactions are usually allowed to proceed for 4–6 h or even overnight at 37°C, before the viral supernatant is replaced with the culture medium required by the target cell type and cells are incubated until harvested for analysis of infection.

3.4. Analysis

3.4.1. Analysis of Transgene Expression

Analysis of transgene expression is based on detection of the genome that the virus delivers and integrates into the genome of a given target cell type. The process of viral cell entry, retroviral reverse transcription, proviral integration up to expression of the viral marker gene takes at least 24 h, so detection of viral infection by transgene expression is possible only 1.5 to 2 d after infection and is not suitable for investigation of the early steps of infection. The earliest time at which transgene expression can be analyzed also depends on the strength of the promoter driving transgene expression in a given target cell type as well as the multiplicity of infection.

1. Depending on the type of marker gene incorporated into HCVpp, process the cells differently. Among the most commonly used transgenes are fluorescent marker genes like GFP and derivates.

2. Prepare gfp-expressing cells for FACS analysis by simple trypsinization. If cells are processed immediately, leave them in 1:10 diluted trypsin during FACS analysis. If safety issues require it, cells can also be fixed after trypsinization.

3. Calculate the infectious titers, expressed as transducing or infectious units (TU or IU) per milliliter, from the total number of target cells present at the time of infection and the percentage of GFP-positive cells.

4. Other commonly used viral marker genes include luciferase enzymes, which require lysis of target cells for subsequent analysis. For detection of firefly and/or renilla luciferases, use commercial kits from e.g. Promega according to the manufacturer's recommendations. Because of the need for target cell lysis, luciferase analysis does not allow direct measurement of single infection events but is based instead on the total number of copies of luciferase enzyme present in an infected cell lysate.

5. When drugs that affect target cell survival and quality are used, use of luciferase as a reporter can become problematic, and target cell survival must be controlled. In this case, use dual reporter luciferase detection systems, in which one measures viral infection with firefly luciferase vectors on target cells expressing renilla luciferase. Luciferase detection is very

sensitive, has a low background, and can be useful for work with HCVpp displaying low titers.

6. Another widespread set of transgenes are selection markers such as puromycin, hygromicin, blasticidin, neomycin, and other resistance genes. Like luciferase, these markers have the disadvantage that one cannot directly follow transfection efficiencies during production. Furthermore, the appropriate drugs can be added to infected target cells only once the expression levels of the selection markers are sufficient to allow cell survival. Selection markers have proven very useful for expression cloning (13, 31) and are similarly useful for work done at the detection limit of infection. To determine infectious titers using selection markers, simply count the colonies that have survived the selection process and extrapolate this number to 1 mL of virus input.

7. Finally, a number of other transgene markers have been described. Amongst them are the enzymes β-galactosidase (32) and the cell surface targeted gene Thy1 (33). An important factor for determination of infectious titers, which applies to any transgene, is to choose dilutions of the viral supernatant in the right range. Only values within the linear range should be considered.

3.4.2. Analysis of Cell Entry with Virions Carrying Protein Markers

Pseudoparticles can be modified and tagged for investigation of very early infection events. Besides incorporation of fluorescent probes, an assay that can be performed with any given virus, retroviral pseudoparticles offer the distinct advantage of incorporating enzymatic proteins into the viral core, in the form of fusion proteins with viral components (25).

1. The best-characterized approaches are based on the HIV-specific accessory protein Vpr (24). Beta-lactamase-Vpr chimeric proteins (BlaM-Vpr) can easily be incorporated into HIV-1 virions (see **Note 4**), and their subsequent delivery into the cytoplasm of target cells or the lumen of artificial lipidic vesicles is a result of virion fusion. Detect this transfer of BlaM-Vpr by enzymatic cleavage of CCF2 (CCF2-FA; free acid form; Invitrogen Life Technologies), the fluorescent substrate of beta-lactamase (BlaM), loaded into the target cells or the lumen of liposomal vesicles (24, 25). Beta-lactamase cleaves CCF2 into a non-fluorescent fluorescein-derived moiety and a fluorescent coumarin-based moiety. Therefore, visualize cleavage by monitoring the increase of coumarin fluorescence kinetically.

2. Besides BlaM-Vpr tagged viruses, other methods of pseudoparticle detection have been described, based on retroviral or lentiviral gag sequences fused to fluorescent proteins (34), but these approaches seem to be less sensitive.

4. Notes

1. The quality of the producer cells is of major importance for HCVpp production. Indeed, incorporation of the HCV glyco-proteins onto retroviral cores occurs less efficiently than that of other retroviral glycoproteins, so healthy, exponentially growing 293T cells are absolutely essential for efficient production. Besides 293T cells, we have used a range of other cell lines successfully for HCVpp production, but because of lower transfection efficiencies, resulting HCVpp titers are usually lower those for 293T producer cells.

2. We generally use a home made calcium-phosphate method for transfection for financial reasons. Other transfection kits, e.g. lipofectamine-based kits, can also be used with similar success, but cell density and timing of transfection must be adapted.

3. Polybrene is thought to favor virus-cell interactions by reducing the electrostatic repulsion between viral and cellular membranes (35). Indeed, the use of polybrene augments infection efficiency with HCVpp between 2- and 5-fold, but in the presence of polybrene, neutralization efficiencies can be lower. The use of polybrene must therefore be adapted to individual need.

4. For Vpr-Bla incorporation into HIV-based pseudoparticles, a minimal HIV vector system must be used to prevent competition for incorporation between wild-type and the Bla-fusion forms of Vpr. Incorporation is achieved simply by coexpression of 4 µg of the Vpr-Bla expression construct in addition to the 8 µg of HIV minimal gagpol, HIV marker vector, and 4 µg of viral glycoprotein expression construct.

Acknowledgments

We are grateful to our co-workers and colleagues for encouragement and advice. Our work is supported by INSERM, ENS Lyon, and CNRS and by grants from the Ligue Nationale Contre le Cancer, the European Union (LSHB-CT-2004-005246), and the Agence Nationale de Recherches sur le SIDA et les Hépatites Virales (ANRS).

References

1. Lanford, R. E. and Bigger, C. (2002) Advances in model systems for hepatitis C virus research. *Virology* **293**, 1–9.

2. Mercer, D. F., Schiller, D. E., Elliott, J. F., Douglas, D. N., Hao, C., Rinfret, A., et al. (2001) Hepatitis C virus replication in mice with chimeric human livers. *Nat. Med.* **7**, 927–933.

3. Bartosch, B., Dubuisson, J., and Cosset, F. L. (2003) Infectious hepatitis C virus pseudoparticles containing functional E1-E2 envelope protein complexes. *J. Exp. Med.* **197**, 633–642.

4. Drummer, H. E., Maerz, A., and Poumbourios, P. (2003) Cell surface expression of functional hepatitis C virus

E1 and E2 glycoproteins. *FEBS Lett.* **546**, 385–390.

5. Hsu, M., Zhang, J., Flint, M., Logvinoff, C., Cheng-Mayer, C., Rice, C. M., et al. (2003) Hepatitis C virus glycoproteins mediate pH-dependent cell entry of pseudotyped retroviral particles. *Proc. Natl. Acad. Sci. USA* **100**, 7271–7276.

6. Sherer, N. M., Lehmann, M. J., Jimenez-Soto, L. F., Ingmundson, A., Horner, S. M., Cicchetti, G., et al. (2003) Visualization of retroviral replication in living cells reveals budding into multivesicular bodies. *Traffic* **4**, 785–801.

7. Sandrin, V., Boulanger, P., Penin, F., Granier, C., Cosset, F. L., and Bartosch, B. (2005) Assembly of functional hepatitis C virus glycoproteins on infectious pseudoparticles occurs intracellularly and requires concomitant incorporation of E1 and E2 glycoproteins. *J. Gen. Virol.* **86**, 3189–3199.

8. Buonocore, L., Blight, K. J., Rice, C. M., and Rose, J. K. (2002) Characterization of vesicular stomatitis virus recombinants that express and incorporate high levels of hepatitis C virus glycoproteins. *J. Virol.* **76**, 6865–6872.

9. Bartosch, B., and Cosset, F.-L. (2006) Cell entry of hepatitis C virus. *Virology* **348**, 1–12.

10. Pileri, P., Uematsu, Y., Campagnoli, S., Galli, G., Falugi, F., Petracca, R., et al. (1998) Binding of hepatitis C virus to CD81. *Science* **282**, 938–941.

11. Scarselli, E., Ansuini, H., Cerino, R., Roccasecca, R. M., Acali, S., Filocamo, G., et al. (2002) The human scavenger receptor class B type I is a novel candidate receptor for the hepatitis C virus. *EMBO J.* **21**, 5017–5025.

12. Agnello, V., Abel, G., Elfahal, M., Knight, G. B., and Zhang, Q. X. (1999) Hepatitis C virus and other Flaviviridae viruses enter cells via low density lipoprotein receptor. *Proc. Natl. Acad. Sci. USA* **96**, 12766–12771.

13. Evans, M. J., von Hahn, T., Tscherne, D. M., Syder, A. J., Panis, M., Wolk, B., et al. (2007) Claudin-1 is a hepatitis C virus co-receptor required for a late step in entry. *Nature* **446**, 801–805.

14. Wakita, T., Pietschmann, T., Kato, T., Date, T., Miyamoto, M., Zhao, Z., et al. (2005) Production of infectious hepatitis C virus in tissue culture from a cloned viral genome. *Nat. Med.* **11**, 791–796.

15. Zhong, J., Gastaminza, P., Cheng, G., Kapadia, S., Kato, T., Burton, D. R., et al. (2005) Robust hepatitis C virus infection *in vitro*. *Proc. Natl. Acad. Sci. USA* **102**, 9294–9299.

16. Lindenbach, B. D., Evans, M. J., Syder, A. J., Wolk, B., Tellinghuisen, T. L., Liu, C. C., et al. (2005) Complete replication of hepatitis C virus in cell culture. *Science* **309**, 623–626.

17. Meunier, J. C., Engle, R. E., Faulk, K., Zhao, M., Bartosch, B., Alter, H., et al. (2005) Evidence for cross-genotype neutralization of hepatitis C virus pseudo-particles and enhancement of infectivity by apolipoprotein C1. *Proc. Natl. Acad. Sci. USA* **102**, 4560–4565.

18. Lavillette, D., Tarr, A. W., Voisset, C., Donot, P., Bartosch, B., Bain, C., et al. (2005) Characterization of host-range and cell entry properties of the major genotypes and subtypes of hepatitis C virus. *Hepatology* **41**, 265–274.

19. McKeating, J. A., Zhang, L. Q., Logvinoff, C., Flint, M., Zhang, J., Yu, J., et al. (2004) Diverse hepatitis C virus glycoproteins mediate viral infection in a CD81-dependent manner. *J. Virol.* **78**, 8496–8505.

20. Nakabayashi, H., Taketa, K., Yamane, T., Oda, M., and Sato, J. (1985) Hormone control of α-fetoprotein secretion in human hepatoma cell lines proliferating in chemically defined medium. *Cancer Res.* **45**, 6379–6383.

21. Zhang, J., Randall, G., Higginbottom, A., Monk, P., Rice, C. M., and McKeating, J. A. (2004) CD81 is required for hepatitis C virus glycoprotein-mediated viral infection. *J. Virol.* **78**, 1448–1455.

22. Bartosch, B., Vitelli, A., Granier, C., Goujon, C., Dubuisson, J., Pascale, S., et al. (2003) *J. Biol. Chem.* **278**, 41624–41630.

23. Bovia, F., Salmon, P., Matthes, T., Kvell, K., Nguyen, T. H., Werner-Favre, C., et al. (2003) Efficient transduction of primary human B lymphocytes and nondividing myeloma B cells with HIV-1-derived lentiviral vectors. *Blood* **101**, 1727–1733.

24. Cavrois, M., De Noronha, C., and Greene, W. C. (2002) A sensitive and specific enzyme-based assay detecting HIV-1 virion fusion in primary T lymphocytes. *Nat. Biotechnol.* **20**, 1151–1154.

25. Lavillette, D., Bartosch, B., Nourrisson, D., Verney, G., Cosset, F. L., Penin, F., et al. (2006) Hepatitis C virus glycoproteins mediate low pH–dependent membrane fusion with liposomes. *J. Biol. Chem.* **281**, 3909–3917.

26. Sandrin, V. and Cosset, F. L. (2006) Intracellular versus cell surface assembly of retroviral pseudotypes is determined by the cellular localization of the viral glycoprotein, its capacity to interact with Gag, and the expression of the Nef protein. *J. Biol. Chem.* **281**, 528–542.

27. Bartosch, B., Verney, G., Dreux, M., Donot, P., Morice, Y., Penin, F., et al. (2005) An interplay between hypervariable region 1 of the hepatitis C virus E2 glycoprotein, the scavenger receptor BI, and high-density lipoprotein promotes both enhancement of infection

and protection against neutralizing antibodies. *J. Virol.* **79**, 8217–8229.

28. Szecsi, J., Drury, R., Josserand, V., Grange, M. P., Boson, B., Hartl, I., et al. (2006) Targeted retroviral vectors displaying a cleavage site–engineered hemagglutinin (HA) through HA-protease interactions. *Mol. Ther.* **14**, 735–744.

29. Dreux, M., Pietschmann, T., Granier, C., Voisset, C., Ricard-Blum, S., Mangeot, P. E., et al. (2006) High density lipoprotein inhibits hepatitis C virus–neutralizing antibodies by stimulating cell entry via activation of the scavenger receptor BI. *J. Biol. Chem.* **281**, 18285–18295.

30. Voisset, C., Op de Beeck, A., Horellou, P., Dreux, M., Gustot, T., Duverlie, G., et al. (2006) High-density lipoproteins reduce the neutralizing effect of hepatitis C virus (HCV)-infected patient antibodies by promoting HCV entry. *J. Gen. Virol.* **87**, 2577–2581.

31. Tailor, C. S., Nouri, A., Lee, C. G., Kozak, C., and Kabat, D. (1999) Cloning and characterization of a cell surface receptor for xenotropic and polytropic murine leukemia viruses. *Proc. Natl. Acad. Sci. USA* **96**, 927–932.

32. Bagnis, C., Chabannon, C., and Mannoni, P. (1999) β-galactosidase marker genes to tag and track human hematopoietic cells. *Cancer Gene Ther.* **6**, 3–13.

33. Lemoine, F. M., Mesel-Lemoine, M., Cherai, M., Gallot, G., Vie, H., Leclercq, V., et al. (2004) Efficient transduction and selection of human T-lymphocytes with bicistronic Thy1/HSV1-TK retroviral vector produced by a human packaging cell line. *J. Gene Med.* **6**, 374–386.

34. Andrawiss, M., Takeuchi, Y., Hewlett, L., and Collins, M. (2003) Murine leukemia virus particle assembly quantitated by fluorescence microscopy: role of Gag-Gag interactions and membrane association. *J. Virol.* **77**, 11651–11660.

35. Porter, C. D., Lukacs, K. V., Box, G., Takeuchi, Y., and Collins, M. K. (1998) Cationic liposomes enhance the rate of transduction by a recombinant retroviral vector *in vitro* and in vivo. *J. Virol.* **72**, 4832–4840.

Chapter 22

Screening of Small-Molecule Compounds as Inhibitors of HCV Entry

Jian-Ping Yang, Demin Zhou, and Flossie Wong-Staal

Abstract

The hepatitis C virus (HCV) has infected some 170 million people worldwide, and is expected to pose a significant medical problem for the foreseeable future. No vaccine is presently available, and the current antiviral therapies (pegylated interferon-α and ribavirin) are characterized by limited efficacy, high costs, and substantial side effects. Initiation of infection requires attachment of the HCV virus to the cell surface followed by viral entry and represents a critical determinant of tissue tropism and pathogenesis. Small molecules that inhibit the virus at the stage of viral entry, for example, by blocking the interactions between viral envelope glycoprotein and cellular receptor or coreceptor or by inhibiting the viral fusion process, would serve as attractive antiviral drugs. Recent development of HCV pseudoparticles (HCVpp), displaying unmodified and functional HCV glycoprotein on the surface of retroviral core particles, has greatly facilitated studies of HCV entry and provides an essential tool for the identification and characterization of molecules that block HCV entry. We have adapted the HCVpp infection assay with HCVpp harboring a luciferase reporter to a 96-well format and screened a small-molecule compound library to identify inhibitors of HCV entry. Such active viral entry inhibitors have the potential to be first-in-class antiviral drugs that can be incorporated into combinations of multiple drugs with different targets for the treatment of chronic HCV infection.

Key words: HCV entry, HCVpp, VSV-Gpp, compound screening, inhibitor, luciferase assay.

1. Introduction

The development of antiviral drugs for HCV has been focused largely on inhibition of virus replication. NS3, a protease, and NS5B, an RNA-dependent RNA polymerase, have been the major targets for HCV drug development *(1, 2)*, but because of HCV's genomic plasticity, viral resistance is a looming issue, as it has been in HIV therapy. Identification of small-molecule compounds

Hengli Tang (ed.), *Hepatitis C: Methods and Protocols, Second Edition, vol. 510*
© 2009 Humana Press, a part of Springer Science+Business Media
DOI 10.1007/978-1-59745-394-3_22 Springerprotocols.com

that inhibit other steps in HCV replication, such as virus binding, entry, and postentry events, will greatly expand our antiviral repertoire. Combination chemotherapy with different classes of inhibitors including entry inhibitors may result in synergistic inhibition of viral infection *(3)*.

The mechanism by which HCV enters target cells has not been fully elucidated. HCV envelope proteins E1 and E2 are type I integral transmembrane proteins with extensively glycosylated ectodomains and are required for virus particle formation and infection of host cells. Several surrogate systems such as recombinant HCV envelope glycoproteins *(4)*, HCV-like particles *(5)*, and HCV pseudoparticles (HCVpp) *(6, 7)* have been developed

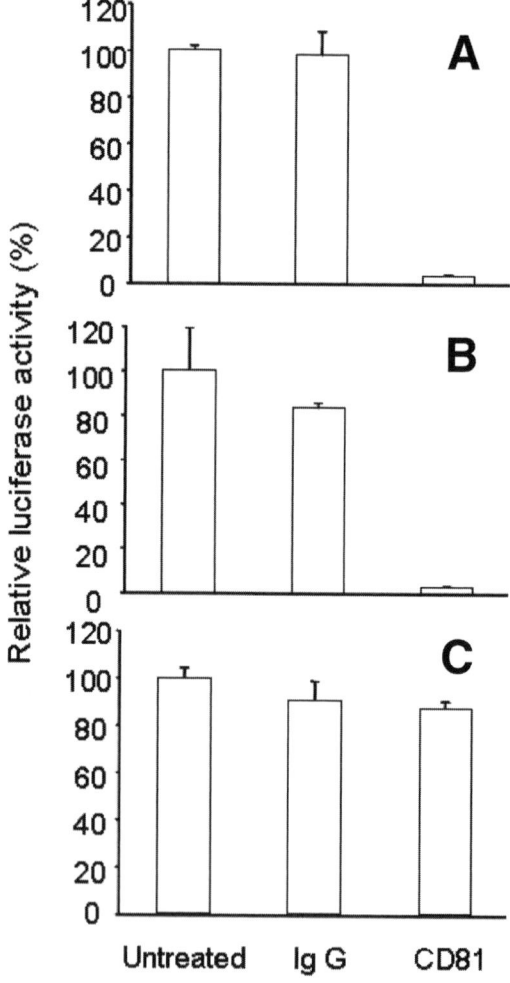

Fig. 22.1. Validation of HCVpp. Huh-7 cells were preincubated with or without 10 μg/mL of CD81 antibody or a control IgG for 30 min at room temperature and infected with (**A**) HCVpp-1a, (**B**) HCVpp-2b, or (**C**) VSV-Gpp (1:100 dilution) in triplicates. Infection was quantified by luciferase assay 72 h after infection. Results were expressed as the percentage of inhibition compared to the cells without treatment.

and used to study viral attachment, entry, and infection. More recently, cell-cultured infectious HCV (HCVcc) has been developed *(8–10)*, but its use for study of HCV entry is confounded by multiple steps and multiple cycles of infection and, further, pertains so far to only one genotype (genotype 2a).

HCVpp is considered the most biologically relevant reporter system for the study of HCV entry. It closely mimics the entry and serological properties of native HCV infection, such as the tropism for primary human hepatocytes and hepatocyte cell lines, pH dependence of the infection process, and neutralization by patient sera as well as monoclonal antibodies (MAbs) specific for E2 *(6,7,11–13)*. The involvement of human CD81 in HCV entry was confirmed in this surrogate system *(14)*.

HCVpp consists of unmodified HCV envelope glycoproteins assembled onto retroviral core particles carrying a reporter gene such as luciferase, green fluorescence protein, or antibiotic resistance genes. Use of HCVpp harboring a luciferase reporter permits easy detection of productive viral entry by the very sensitive luciferase assay. Similar pseudovirus particles bearing other viral envelope proteins (e.g., VSV-G protein) can be used as controls for events irrelevant to entry.

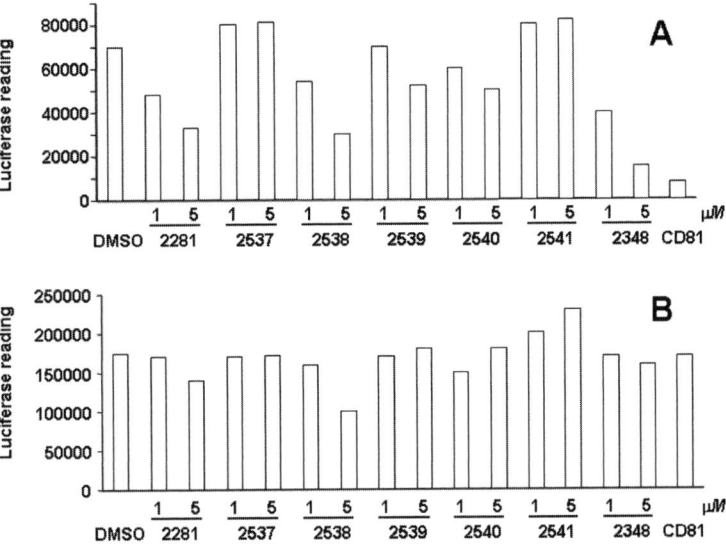

Fig. 22.2. Compound screen based on HCVpp/VSV-Gpp entry assay. (**A**) Efficacy test for blocking HCVpp entry. Two concentrations, 1 μM and 5 μM, were tested for each compound. DMSO is a negative control for monitoring the maximum HCVpp entry in the presence of 1% DMSO. CD81 is a positive control to reflect the maximum blocking of HCVpp entry. In this assay, the compounds 2281, 2348, and 2538 showed inhibition of HCVpp infection. (**B**) Specificity test using VSV-Gpp entry assay. VSV-Gpp and HCVpp are identical except in their envelope proteins, and thus any chemicals that are active in both HCVpp and VSV-G assays, like 2538, are likely to inhibit a step other than entry. Such compounds will be excluded. Those chemicals, like 2281 and 2348, that are active only in the HCVpp entry assay will be selected as potential hits.

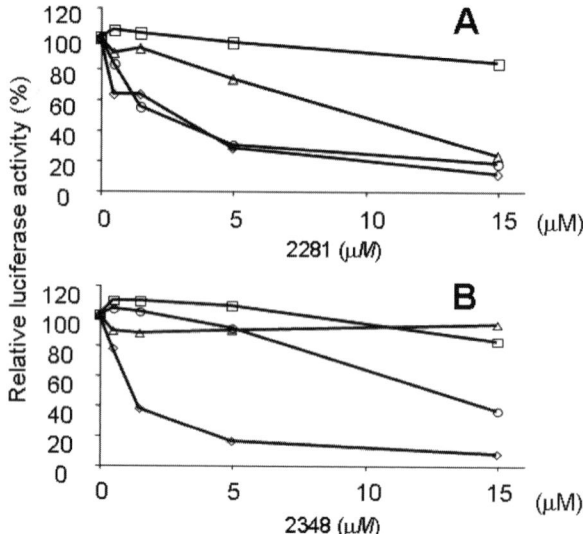

Fig. 22.3. Characterization of chemical hits 2281 (**A**) and 2348 (**B**) in entry assays using HCVpp-1a (*diamonds*), HCVpp-2b (*circles*), VSV-G control (*triangles*), and toxicity assay using Alamar Blue (*squares*). Each compound was tested in five concentrations: 0, 0.5, 1.5, 5, and 15 μM. 2281 shows good and specific activity against both HCVpp-1a and 2b at low concentration (<5 μM); 2348 blocks HCVpp-1a but shows less activity against HCVpp-2b and VSV-G.

Here we describe our experience with HCVpp and VSV-Gpp in screening small-molecule compounds for inhibition of HCV entry. The specificity of the vector system was tested with antibodies to CD81 (**Fig. 22.1**). Several compounds were identified that inhibited HCVpp infection but not VSV-G infection (**Fig. 22.2**). We further evaluated the compounds' dose dependence, specificity (e.g., activity on HCVpp harboring envelope proteins of other HCV genotypes), and cellular toxicity (Alamar Blue assay) (**Fig. 22.3**). Finally, the compounds were tested in the HCVcc system (data not shown). We believe the relatively high-throughput HCVpp system can be used to discover compounds that selectively inhibit HCV entry.

2. Materials

2.1. Viral Production

1. pNL4-3 HIV proviral DNA (AIDS Reagent Program, NIH, Bethesda, MD).
2. pcDNA-E1E2 1a and 2b [cloned from total RNA prepared from HCV-infected plasma as described previously (15)].
3. pcDNA-VSV-G (a plasmid encoding the vesicular stomatitis virus (VSV) glycoprotein G protein).
4. pAdVantage (Promega, Madison, WI).

5. 293T cells (American Type Culture Collection, Manassas, VA).

6. 293T cell-culture medium: Dulbecco's modified Eagle's medium (DMEM) (Gibco/BRL, Bethesda, MD) supplemented with 10% heat-inactivated fetal bovine serum (FBS; HyClone, Ogden, UT), 2 mM L-glutamine, 0.1 mM MEM nonessential amino acids.

7. Medium for viral production and infection: DMEM, supplemented with 3% FBS, 1 mM MEM sodium pyruvate, 2 mM L-glutamine, 0.1 mM MEM nonessential amino acids.

8. TransIT-LT1 transfection reagent (Mirus, Madison, WI).

9. Opti-MEM medium (Invitrogen).

10. Solution of trypsin (0.25%) from Gibco/BRL.

11. CD81 antibody (BD Biosciences).

2.2. Viral Infection Assay

1. Huh-7 cell line.

2. Polybrene: Prepare a 10-mg/mL stock solution in deionized, sterile water. Filter-sterilize it and dispense 1 ml aliquots into sterile microcentrifuge tubes. Store at −20°C for long-term storage (Sigma).

2.3. Luciferase Assay

1. Costar TC-treated solid white 96-well tissue-culture plate with flat bottom (Corning Incorporated, NY, USA).

2. Bright-Glo™ Luciferase Assay System (Promega).

2.4. Alamar Blue Assay

1. 96-well Microtest tissue-culture plate, flat bottom with low-evaporation lid (Becton Dickinson Labware, Franklin Lakes, NJ).

2. Alamar Blue reagent (Biosource).

3. Methods

3.1. HCVpp Production

Here, we outline the optimal procedure for producing HCVpp and VSV-Gpp harboring a luciferase reporter, including plasmid transfection, viral supernatant harvesting, and HCVpp quality evaluation.

3.1.1. Transfection of 293T Cells with HIV Proviral DNA, pcDNA-E1E2, or pcDNA-VSV-G and pAdVAtage

1. Plate 5×10^5 293T cells/well in 2 mL of 10% DMEM culture medium in a 6-well plate and culture overnight at 37°C under 5% CO_2, humidified. They should be approximately 80% confluent by the next day.

2. In two sterile plastic tubes, prepare the following mixtures for each transfection sample:

a. Add 5 μL of Mirus TransIT-LT1 directly into 100 μL of Opi-MEM Medium per well; mix gently by pipetting up and down.

b. Add $1\,\mu g$/well HIV-Luc, $0.7\,\mu g$/well pcDNA-E1E2, and $0.3\,\mu g$/well pAdVantage to $100\,\mu L$ of Opti-MEM Medium per well and mix. To generating VSV-Gpp, add $1\,\mu g$/well HIV-Luc and $0.7\mu g$/well pcDNA-VSV-G to $100\,\mu L$ of Opti-MEM Medium per well and mix. DNA is suspended in sterile water at a concentration of $0.2 - 3.0\,\mu g/\mu L$ (*see* **Note 1**).

3. Slowly mix "a" and "b" together with gentle vortexing.

4. Incubate the DNA/LT1 complex at room temperature for 15–30 min to precipitate the DNA.

5. Gently resuspend any DNA-lipid complexes by pipetting the suspension up and down; add the DNA suspension to 293T cells in 2 ml of culture medium in a drop-wise fashion, swirling gently to prevent the cells from being lifted from the plate and to distribute the DNA suspension evenly (*see* **Notes 2 and 3**).

6. Return the tissue-culture plates to the 37°C incubator.

7. After 4-6 h of incubation, remove the medium from the plates and replace it with 2 mL/well of 3% FBS DMEM supplemented with 1 mM sodium pyruvate that has been pre-warmed to 37°C.

8. Incubate the cells at 37°C under 5% CO_2 for 48 h.

3.1.2. Harvest of Viral Supernatants

1. Harvest virus-containing supernatants of the transfected 293T cells at the 48-h time point.

2. Refeed cells with 2 mL/well of 3% FBS DMEM with 1 mM sodium pyruvate and incubate them at 37°C under 5% CO_2 overnight. Harvest the supernatant at the 72-h time point.

3. Filter the viral supernatants through membranes of $0.45\,\mu m$ pore size to remove the cell debris (*see* **Note 4**).

4. Store aliquots of the viral supernatants of desired volumes at -80°C (*see* **Note 5**).

3.1.3. Evaluation of the Quality of HCVpp

1. Seed a 96-well plate with Huh-7 cells at 1×10^5 cells/well and incubate it at 37°C under 5% CO_2 overnight (*see* **Notes 6 and 7**).

2. Dilute CD81 antibody and control IgG to $10\,\mu g$/mL in 3% FBS DMEM culture medium (*see* **Note 8**).

3. Remove the culture medium from the Huh-7 cells.

4. Add $50\,\mu L$ of CD81, IgG, or 3% FBS DMEM culture medium to the cells and incubate them at room temperature for 30 min.

5. Add $50\,\mu L$ of HCVpp or VGV-Gpp (1:100 dilution) with $8\,\mu g$/mL polybrene per well to the cells.

6. Incubate at 37°C under 5% CO_2 for 72 h.

7. Perform the luciferase assay (*see* **Section 3.3**). A sample result is shown in **Fig. 22.1** (*see* **Notes 9, 10, and 11**).

**3.2. Screening
Small-Molecule
Compounds**

*3.2.1. Compound
Preparation*

1. Dilute chemical compounds in DMSO to make $1500\,\mu M$ and $500\,\mu M$ stock solutions in clear 96-well plates.
2. Further dilute stock compounds to working concentrations as follows:
   a. Add $12\,\mu L$ of $1500\,\mu M$ stock solution to $588\,\mu L$ of 3% FBS DMEM culture medium to make a $30\,\mu M$ compound solution containing 2% DMSO.
   b. Add $12\,\mu L$ of $500\,\mu M$ stock solution to $588\,\mu L$ of 3% FBS DMEM culture medium to make a $10\,\mu M$ compound solution containing 2% DMSO.
   c. Add $3.6\,\mu L$ of $500\,\mu M$ stock solution and $8.4\,\mu L$ of DMSO to $588\,\mu L$ of 3% FBS DMEM culture medium to make a $3\,\mu M$ compound solution containing 2% DMSO.
   d. Add $2.4\,\mu L$ of $500\,\mu M$ stock solution and $9.6\,\mu L$ of DMSO to $588\,\mu L$ of 3% FBS DMEM culture medium to make a $2\,\mu M$ compound solution containing 2% DMSO.
   e. Add $1.2\,\mu L$ of $500\,\mu M$ stock solution and $10.8\,\mu L$ of DMSO to $588\,\mu L$ of 3% FBS DMEM culture medium to make a $1\,\mu M$ compound solution containing 2% DMSO.
3. As a negative control, add $12\,\mu L$ of DMSO to $588\,\mu L$ of 3% FBS DMEM culture medium to make a control cell-culture medium containing 2% DMSO.

3.2.2. Seeding the Wells

1. Count Huh-7 cells and make a suspension of 1×10^5 cells/mL in 10% FBS DMEM culture medium.
2. Divide the cell suspension into aliquots of $100\,\mu L$ (1×10^4 cells) per well in two white 96-well tissue-culture plates for the luciferase assay and one clear plate for the cell-toxicity assay.
3. Incubate the cells at 37°C under 5% CO_2 overnight.

*3.2.3. HCVpp and
VSV-Gpp Infection Assays*

1. Flip plates over to empty wells of culture medium and tap them on a bench over paper (Kimwipes) to empty wells completely.
2. Add $50\,\mu L$ of diluted compound to each well.
3. Incubate plates at room temperature for 30 min.
4. During the incubation, thaw the viral stock and dilute the appropriate amount of virus in 3% FBS complete medium. We usually dilute the VSV-Gpp stock 1:100 because of its high infectivity.
5. Add polybrene to viral supernatants to yield $8\,\mu g/mL$.
6. Add $50\,\mu L$ of HCVpp or VSV-Gpp per well to the cells.
7. Incubate the cells at 37°C under 5% CO_2 for 4 h.
8. Remove the medium containing virus and compound, and add $100\,\mu L$ of fresh culture medium, and incubate the cells at 37°C under 5% CO_2 for 72 h.

3.3. Luciferase Assay

1. Make Bright-Glo reagent by reconsititution of Bright-Glo Substrate with room temperature Bright-Glo buffer; avoid exposing it to light.
2. Add 100 μL of Bright-Glo reagent per well to cells.
3. Incubate the plate at room temperature for 5 min in dark.
4. Read the plates on a Veritase Luminometer.

3.4. Cell-Toxicity Assay

3.4.1. Treatment the Huh-7 Cells with Compounds

1. Add 50 μL of diluted compound per well to Huh-7 cells in a clear 96-well plate.
2. Incubate the plate at room temperature for 30 min.
3. Add 50 μL of 30% FBS DMEM/well over the 50 μL of compound in each well.
4. Incubate the plate at 37°C under 5% CO_2 for 72 h.

3.4.2. Alamar Blue Assay

1. Dilute 10× stock Alamar Blue reagent 1:5 in PBS (see **Note 12**).
2. Add 100 μL of 2× Alamar Blue reagent per well over the cells.
3. Incubate the plate at 37°C under 5% CO_2 for 3–4 h.
4. Read plates in Cytofluor plate reader (emission 590 nm and excitation 530 nm).

4. Notes

1. We observed that cotransfection of pAdVantage increased the viral titer at least 10-fold, but when the VSV-Gpp vector is packaged, including pAdVantage in the transfection is not recommended, because high expression of VSV-G is toxic to cells. We recommend making HCVpp and VSV-Gpp at different times to avoid cross contamination.
2. Transfection efficiency is the key to production of high-titer virus. Early-passage 293T stocks should be used; cells should be passed at high density and thoroughly trypsinized to prevent clumping; the cell density should be about 80% confluent at the time of transfection.
3. 293T cells are weakly adherent, especially after transfection, so all medium changes should be performed with extreme care.
4. The filter used for viral purification should be cellulose acetate polysulfonic (low protein binding), not nitrocellulose. Nitrocellulose binds proteins in the retrovirus membrane and destroys the virus.
5. Viral supernatants can generally be collected 48 h and 72 h after transfection if many cells are still attached to the plate and look healthy at these times. Minimal differences in viral titer are observed for harvests at these two times.

6. To avoid the edge effect, we generally exclude the outer wells of 96-well plates from screening. To prevent the plates from drying out, add 200 μL of PBS to each outer well to humidify plate.

7. We observed that Huh-7 cells seeded in 3% FBS DMEM medium are infected by psuedoviruses more efficiently than that in 10% FBS DMEM medium.

8. We noticed that antibodies with sodium azide affected the infection of both HCVpp and VSV-Gpp because of cell toxicity. We recommend dialyzing the antibodies to remove sodium azide before use.

9. This protocol for viral production is optimized in a six-well plate format. The cotransfection experiment can be scaled up if a large volume of virus is desired.

10. Viral vector is stable for a week at 4°C. Frozen at −80°C, the virus can be stable for at least 6 months, but the titer will be decreased twofold by each freeze-thaw cycle. Any viral supernatant that is not used immediately upon harvesting should be frozen in aliquots at −80°C.

11. Luciferase activity is temperature dependent. The most convenient and effective method for thawing or temperature equilibrating cold reagent is to place it in a water bath at room temperature. Do not use a water bath above 25°C.

12. Upon thawing Alamar Blue reagent, warm it to 37°C and mix it well to assure complete redissolution. Store the Alamar Blue reagent away from light to prevent changes in its absorbance properties.

Acknowledgments

The authors would like to thank Jing Zhang and Maureen Ibanez for technical assistance.

References

1. Perni, R. B. (2000) NS3.4A protease as a target for interfering with hepatitis C virus replication. *Drug News Perspect.* **13,** 69–77.

2. De Francesco, R., Tomei, L., Altamura, S., Summa, V., and Migliaccio, G. (2003) Approaching a new era for hepatitis C virus therapy: inhibitors of the NS3-4A serine protease and the NS5B RNA-dependent RNA polymerase. *Antiviral Res.* **58,** 1–16.

3. Pawlotsky, J. M. (2006) Therapy of hepatitis C: from empiricism to eradication. *Hepatology* **43,** S207–S220.

4. Pileri, P., Uematsu, Y., Campagnoli, S., Galli, G., Falugi, F., Petracca, R., et al. (1998) Binding of hepatitis C virus to CD81. *Science* **282,** 938–941.

5. Baumert, T. F., Ito, S., Wong, D. T., and Liang, T. J. (1998) Hepatitis C virus structural proteins assemble into viruslike particles in insect cells. *J. Virol.* **72,** 3827–3836.

6. Bartosch, B., Dubuisson, J., and Cosset, F. L. (2003) Infectious hepatitis C virus pseudoparticles containing functional E1-E2 envelope protein complexes. *J. Exp. Med.* **197,** 633–642.

7. Hsu, M., Zhang, J., Flint, M., Logvinoff, C., Cheng-Mayer, C., Rice, C. M., et al. (2003) Hepatitis C virus glycoproteins mediate

pH-dependent cell entry of pseudotyped retroviral particles. *Proc. Natl. Acad. Sci. USA* **100**, 7271–7276.

8. Lindenbach, B. D., Evans, M. J., Syder, A. J., Wolk, B., Tellinghuisen, T. L., Liu, C. C., et al. (2005) Complete replication of hepatitis C virus in cell culture. *Science* **309**, 623–626.

9. Wakita, T., Pietschmann, T., Kato, T., Date, T., Miyamoto, M., Zhao, Z., et al. (2005) Production of infectious hepatitis C virus in tissue culture from a cloned viral genome. *Nat. Med.* **11**, 791–796.

10. Zhong, J., Gastaminza, P., Cheng, G., Kapadia, S., Kato, T., Burton, D. R., et al. (2005) Robust hepatitis C virus infection in vitro. *Proc. Natl. Acad. Sci. USA* **102**, 9294–9299.

11. Bartosch, B., Bukh, J., Meunier, J. C., Granier, C., Engle, R. E., Blackwelder, W. C., et al. (2003) In vitro assay for neutralizing antibody to hepatitis C virus: evidence for broadly conserved neutralization epitopes. *Proc. Natl. Acad. Sci. USA* **100**, 14199–14204.

12. Meunier, J. C., Engle, R. E., Faulk, K., Zhao, M., Bartosch, B., Alter, H., et al. (2005) Evidence for cross-genotype neutralization of hepatitis C virus pseudo-particles and enhancement of infectivity by apolipoprotein C1. *Proc. Natl. Acad. Sci. USA* **102**, 4560–4565.

13. Logvinoff, C., Major, M. E., Oldach, D., Heyward, S., Talal, A., Balfe, P., et al. (2004) Neutralizing antibody response during acute and chronic hepatitis C virus infection. *Proc. Natl. Acad. Sci. USA* **101**, 10149–10154.

14. Cormier, E. G., Tsamis, F., Kajumo, F., Durso, R. J., Gardner, J. P., and Dragic, T. (2004) CD81 is an entry coreceptor for hepatitis C virus. *Proc. Natl. Acad. Sci. USA* **101**, 7270–7274.

15. McKeating, J. A., Zhang, L. Q., Logvinoff, C., Flint, M., Zhang, J., Yu, J., et al. (2004) Diverse hepatitis C virus glycoproteins mediate viral infection in a CD81-dependent manner. *J. Virol.* **78**, 8496–8505.

Chapter 23

Isolation of JFH-1 Strain and Development of an HCV Infection System

Takaji Wakita

Abstract

Detailed analysis of hepatitis C virus (HCV) has been hampered by the lack of an appropriate viral culture system and small animal models of infection. My group and others have recently reported the production of infectious virus after full-length HCV RNA transfection into Huh-7 cells. This system depends primarily on isolation of a JFH-1 strain from a patient with fulminant hepatitis. The JFH-1 strain belongs to genotype 2a and has high colony-formation efficiency when tested with a subgenomic replicon system. Here, I describe various protocols for isolation of the JFH-1 strain and construction of the HCV infection system. The HCV infection system contributes to our understanding of HCV virology and may permit development of novel antiviral strategies.

Key words: Fulminant hepatitis, JFH-1, patient sera, hepatocytes, nested RT-PCR, virus particles.

1. Introduction

To date, propagation of HCV in cultured cells has been difficult (1) for a number of reason, including low replication capacity of the virus and its tropism for highly differentiated hepatocytes. Inoculation of patient sera or plasma into cultured cells results in only a limited level of HCV replication, as determined by nested RT-PCR. This problem hindered the efforts of a number of HCV researchers, but in 1999, Lohmann et al. (2) were the first to report efficient replication of an HCV subgenomic replicon, in which an HCV structural region was replaced with a neomycin-resistance gene. After transfection of replicon RNA into Huh-7 hepatocellular carcinoma cells, followed by several weeks of G418 selection culture, replicons were established, and robust replicon

Hengli Tang (ed.), *Hepatitis C: Methods and Protocols, Second Edition, vol. 510*
© 2009 Humana Press, a part of Springer Science+Business Media
DOI 10.1007/978-1-59745-394-3_23 Springerprotocols.com

RNA replication was observed in these cells. Adaptive mutations were found in most replicon genomes that increased virus replication at different levels, and some combinations of these adaptive mutations were observed to increase replication strongly *(3–5)*. Genomic replicons containing a structural region with adaptive mutations in a nonstructural region demonstrated efficient replication in transfected Huh-7 cells *(6–8)*, but viral particles were not produced from these genomic replicons. Furthermore, a full-length viral RNA genome with adaptive mutations synthesized in vitro was not infectious in chimpanzees, unlike the wild-type genome *(9)*. These results suggest that adaptive mutations enhance the replication capacity of the HCV RNA genome in cultured cells at the expense of efficient viral particle formation in cultured cells and in vivo.

The JFH-1 strain was isolated from a 32-year-old male patient *(10)*. He was admitted with acute liver failure and had serum aspartate aminotransferase (AST) and alanine aminotransferase (ALT) concentrations of 9160 IU/L and 6970 IU/L, respectively. The minimum prothrombin time was 16%. Stage II encephalopathy developed 5 days after admission, after which he was diagnosed with fulminant hepatitis. HCV RNA was detected by reverse transcription polymerase chain reaction (RT-PCR) with sera obtained during the acute phase. Anti-HCV antibody was also tested for but not detected on admission (by second-generation enzyme-linked immunosorbent assay, Ortho Diagnostics, Tokyo, Japan). All viral markers indicating exposure to other hepatitis viruses were negative. After admission, the patient's liver function and clinical condition improved with conservative treatment. Anti-HCV antibody became positive 6 weeks after admission. These findings suggest that his fulminant hepatitis was in fact due to HCV infection. The infectious strain of HCV was analyzed in 12 sets of nested RT-PCR, as well as 5′ RACE and 3′ RACE RT-PCR, which covered the entire HCV genome. All of the PCR products were cloned and sequenced. Five clones of each PCR fragment were sequenced, and the consensus sequence was determined. According to sequence analysis, the JFH-1 strain belongs to genotype 2a, and its sequence deviates slightly from other genotype 2a clones isolated from patients with chronic hepatitis (10).

Subgenomic replicon and full-length constructs were assembled with cloned PCR fragments *(11–13)*. The colony-formation efficiency of the JFH-1 replicon was much greater than that of the Con1 replicon with adaptive mutations. Furthermore, transient transfection of replicon RNA into Huh-7 cells resulted in autonomous RNA replication, as determined by northern-blot analysis *(11, 14)*. Importantly, adaptive mutations were not necessary for efficient JFH-1 replicon replication in Huh-7 cells. In addition, the JFH-1 replicon produced colonies in several other cell lines, including HepG2, IMY-N9, HeLa, and 293 human cells, as well as mouse NIH3T3 fibroblast cells and mouse AML12, MMHD3, and MMH1-1 hepatocytes *(12, 15, 16)*. On the basis of this analysis, JFH-1 demonstrates markedly greater replication efficiency than other reported HCV clones. So that full advantage could be taken of this characteristic of the JFH-1 clone, full-length JFH-1 was examined with regard to replication in Huh-7 cells after transfection of synthesized RNA *(17)*. Transfected full-length RNA replicated efficiently in Huh-7 cells, and surprisingly, infectious virus particles were secreted into the culture medium. Culture supernatant was harvested from the transfected cells, cleared by centrifugation and filtration, and used to inoculate naïve Huh-7 cells. After inoculation, several infected cells were identified by immunostaining with HCV-specific antibodies. In standard Huh-7 cells, the infection efficiency was less than 0.5%, but this infection was specific because an antibody against CD81, as well as anti-E2 antibody, inhibited infection (17 and unpublished data). Furthermore, the virus particles secreted into the culture supernatant were infectious in chimpanzees *(17)*. These results strongly suggest that the secreted virus particles were authentic HCV.

Although the infection efficiency with original Huh-7 cells maintained in our laboratory was quite low, efficient viral infection was achieved with cured cells, such as Huh-7.5 and Huh-7.5.1 *(18, 19)*. These cell lines were produced by interferon treatment of subgenomic replicons. Huh-7.5 cells have a defective point mutation in RIG-I, resulting in defective intracellular interferon signaling against HCV RNA replication *(20)*. Transfection of the JFH-1 genome or inoculation of Huh-7.5 and Huh-7.5.1 cells with infectious JFH-1 virus thus produces robust replication and HCV virus infection *(18, 21)*. We therefore tested the permissiveness of JFH-1 replication in these cell lines by transient transfection of a subgenomic JFH-1 replicon. We found that JFH-1 RNA replication was not greater in these cell lines than in original Huh-7 cells (Wakita, unpublished data), but we observed a stable increase in the cell-surface expression of CD81 in these cell lines over that in original Huh-7 cells. Of note, Koutsoudakis et al. *(22)* have reported that the level of CD81 cell-surface expression is a key determinant of HCV infection. In fact, we have observed that the Huh-7 cells used in the initial infection assay were in fact a mixture of cell clones with varying levels of CD81 expression and infectivity *(23)*. Interestingly, we have isolated several Huh-7 subclones without cell-surface CD81 expression. Among these subclones, some support highly efficient subgenomic replicon replication. We therefore transfected a CD81 expression vector into these cell clones to produce stable cells with a high level of ectopic CD81 expression. These cells supported a greater degree of infectivity of the JFH-1 virus than did original Huh-7 cells or the cured cell lines *(23)*. Therefore achievement of a high level of infectivity depends on cell-surface expression of CD81, as well as the replicon replication efficiency of Huh-7 cells. Huh-7 cells were first isolated more than 20 y ago *(24)* and were distributed worldwide. The phenotype of Huh-7 cells, including permissiveness for JFH-1 virus infectivity, may differ in subclones maintained in different laboratories.

Interestingly, CD81 may play an important role after virus binding to the cell surface *(25, 26)*. Furthermore, SR-BI and SR-BII are thought to play roles in early infection, but the role of SR-BI/II receptors in HCV infection remains unclear *(27, 28)*. Heparan sulfate and heparinase reduce cell-surface HCV binding, and anti-E2 and E1 antibodies also block HCV infection in cultured cells. These results suggest that HCV first binds to the cell surface by means of a heparan sulfate proteoglycan at low affinity, after which it may be transferred to high-affinity receptors, such as CD81, which may facilitate virus internalization *(25, 26)* but CD81 may also be involved in initial virus attachment to the cell surface. HCV has also been reported to enter target cells by clathrin-mediated endocytosis, followed by fusion within an acidic endosomal compartment *(29)*. Recently, Evans et al. *(30)* reported that claudin-1, a tight-junction component, is a coreceptor for HCV infection. 293T cells are both CD81 and SR-B1 positive but not permissive for HCV infection. 293T gained permissiveness for HCV infection when claudin-1 was ectopically expressed *(30)*. Further studies will be necessary to elucidate how these molecules are cooperatively involved in the process of virus entry. These observations may also explain the tissue tropism of HCV.

Human hepatocytes (immortalized by HCV core protein [IHH]) were also used to develop a permissive cell line for HCV infection. A similar degree of virus particle production was observed upon transfection of RNA from genotype 1a (H77) and genotype 2a (JFH-1) into IHH, so IHH may support HCV genome replication and virus assembly *(31)*. These results suggest that a number of host factors are involved in the virus-host interaction and thus determine the permissiveness of a host cell for HCV infection. Furthermore, a regulatory link may exist between innate antiviral and inflammatory cellular responses to viral infection. HCV infection triggers dsRNA signaling pathways that induce CXCL-8 (IL-8) through transcriptional activation and mRNA stabilization *(32)*. Proinflammatory cytokines induce indoleamine-2,3-dioxygenase (IDO), which is an important mediator of peripheral immune tolerance. Huh-7 cells supporting HCV replication express higher levels of IDO mRNA than do noninfected cells when stimulated with IFN-gamma or when cocultured with activated T cells *(33)*. Proinflammatory cytokines induced by HCV infection may therefore play an important role in the pathogenesis of HCV infection and escape from host immune responses.

Development of an efficient therapy that eliminates infected HCV from chronic carriers is important. To date, interferon and ribavirin have been used in clinics, with limited efficacy. NS3 protease inhibitors, as well as NS5B polymerase inhibitors and other drugs, are undergoing clinical trials but have not yet been approved for therapeutic use [reviewed in **ref.** *(34)*]. The HCV infection system described here, using a JFH-1 clone, may provide a good method for screening new antiviral agents. Furthermore, stable JFH-1 cDNA-transfected cell lines capable of producing infectious virus may be suitable for screening antiviral agents *(35, 36)*. Further understanding of the HCV life cycle remains important because each step provides a potential target for control of HCV infection and replication. HCV-infected cell systems enable us to characterize the subcellular localization of HCV structural proteins in the context of an infectious cycle. Interestingly, Rouillé et al. *(37)* have reported colocalization of core and NS3 proteins in infected cells, which may suggest that interaction of structural and nonstructural proteins is important for infectious virus formation. On the other hand, Shirakura et al. *(38)* have identified ubiquitin ligase E6AP as an HCV core–binding protein. E6AP has been observed to bind to the core protein and promote its degradation. Exogenous expression of E6AP has also been found to decrease intracellular core protein levels, as well as supernatant HCV infectivity titers, in HCV JFH-1–infected Huh-7 cells. Furthermore, knockdown of endogenous E6AP by RNA interference has been observed to increase intracellular core protein levels and supernatant HCV infectivity titers in HCV JFH-1–infected cells. These studies suggest several novel targets for control of HCV infection.

A great deal of research has also been focused on the development of an HCV vaccine [reviewed in **ref.** *(39)*]. Efforts to produce an HCV vaccine have been met with skepticism because the presence of neutralizing antibody after recovery from HCV infection is difficult to demonstrate, but the JFH-1 infection system, as well as a pseudotype retrovirus carrying HCV envelope proteins, have been used to demonstrate that most chronic HCV carriers develop neutralizing antibodies in coexistence with virus particles within their circulation *(17, 40)*. Determining how HCV evades host immune surveillance remains important, as does determining how protective immunity against HCV infection develops; both will contribute to development of efficient vaccines and immunotherapies.

The development of an HCV infection system using the JFH-1 strain will aid our virological understanding of this important virus. A genotype 1a strain, H77S containing 5 adaptive mutations, has been reported to produce infectious virus after synthesized RNA transfection into Huh-7 cells, albeit with limited efficiency *(41)*. Clearly, therefore, JFH-1 is not the only HCV strain that can be propagated in cultured cells. Further study will be necessary to develop other genotypic infectious HCV in cell culture. Understanding why JFH-1 is the only strain that replicates efficiently without adaptive mutations in cultured cells will also be important. Further mechanistic analysis of evaluation of the mechanics of JFH-1 genomic replication will help further these goals.

2. Materials

2.1. Cell Culture

1. Huh-7 cells (24), which can be purchased from Cell Bank, RIKEN BioResource Center (Cat. no. RCB1366).
2. Dulbecco's modified Eagle's medium (DMEM) (high glucose; Sigma-Aldrich Japan K.K., Tokyo, Japan) supplemented with 10% fetal bovine serum (an appropriate lot for Huh-7 cells and derivatives), 0.1 mM MEM nonessential amino acids solution (Invitrogen, Carlsbad, CA), 2 mM L-glutamine, 10 mM HEPES, pH 7.4, 1 mM sodium pyruvate, 100 U/mL penicillin, 100 μg/mL streptomycin.

2.2. RNA Extraction and cDNA Synthesis

1. TRIzol and TRIzol LS (Invitrogen).
2. Nuclease-free water (Ambion, Austin, TX).
3. Random hexamer (TAKARA Bio, Kyoto, Japan).
4. RNase inhibitor (TAKARA Bio).
5. Moloney murine leukemia virus reverse transcriptase (Superscript II, Invitrogen).

2.3. RT-PCR for Isolation of HCV cDNA

1. TaKaRa LA Taq (TAKARA Bio).
2. Primers (**Table 23.1**).

2.4. 5′ RACE RT-PCR

1. 5′ RACE System, Version 2.0 (Invitrogen, cat no. 18374-058).
2. TaKaRa Ex Taq (TAKARA Bio).
3. Primers (**Table 23.1** and included in the 5′ RACE System).

2.5. 3′ RACE RT-PCR

1. Poly(A) Polymerase (TaKaRa).
2. 5× Poly(A) Polymerase buffer (200 mM Tris-HCl, pH 8.0, 50 mM $MgCl_2$, 12.5 mM $MnCl_2$, 1.25 M NaCl, 2.5 mg/mL bovine serum albumin).
3. Nuclease-free water
4. RNase inhibitor.
5. TRIzol.
6. Superscript II.

Table 23.1
Primers used in cloning the JFH-1 strain

Name		Sequence (5′ > 3′)
1st PCR	44S	5′-CTG TGA GGA ACT ACT GTC TT-3′
	1323R	5′-GGC GAC CAG TTC ATC ATC AT-3′
2nd PCR	44S	
	486R	5′-GTC GTG CGC ACA CCC AAC CT-3′
2nd PCR	317S	5′-GGG AGG TCT CGT AGA CCG TG-3′
	849R	5′-GGT AGG TTC CCT GTT GCA TA-3′
2nd PCR	617S	5′-TGG GCA GGA TGG CTC CTG TC-3′
	1323R	
1st PCR	1050S	5′-GGT GTT GGG TGC CAG TCT C-3′
	2445R	5′-TCC ACG ATG TTC TGG TGA AG-3′
2nd PCR	1141S	5′-TGT CCG CCA CCT TCT GCT-3′
	2367R	5′-CAT TCC GTG GTA GAG TGC A-3′
1st PCR	2099S	5′-ACG GAC TGT TTT AGG AAG CA-3′
	3568R	5′-TGT TCC GAG GAA GGA CTG AG-3′
2nd PCR	2285S	5′-AAC TTC ACT CGT GGG GAT CG-3′
	3509R	5′-TCC TGT CAC GCC CCG TCA-3′
1st PCR	3425S	5′-CTT CTC GCC CCC ATC ACT G-3′
	4706R	5′-TTG CAG TCG ATC ACG GAG TC-3′
2nd PCR	3471S	5′-TGG GCG CCA TAG TGG TGA G-3′
	4665R	5′-TCG GTG GCG ACG ACC AC-3′
1st PCR	4547S	5′-AAG TGT GAC GAG CTC GCG G-3′
	5970R	5′-TTC TCG CCA GAC ATG ATC TT-3′
2nd PCR	4547S	5′-AAG TGT GAC GAG CTC GCG G-3′
	5970R	5′-TTC TCG CCA GAC ATG ATC TT-3′
1st PCR	5714S	5′-GCT TCC ATG ATG GCA TTC AG-3′
	7220R	5′-TGT AAT CAG GCC GTG CCC A-3′
2nd PCR	5883S	5′-TGG GTA AGG TGC TGG TGG A-3′
	7003R	5′-GTG GTG CAG GTG GCT CGC A-3′
1st PCR	6537S	5′-TCA ATT GTT ACA CGG AGG GC-3′
	8091R	5′-TTT TTG GCC ATG ATG GTT GTA-3′
2nd PCR	6950S	5′-GAG CTC CTC AGT GAG CCA G-3′

(Continued)

Table 23.1
(Continued)

	8035R	5′-CCA CAC GGA CTT GAT GTG GT-3′
1st PCR	7848S	5′-ACG CCC ATT ATG ACT CAG TC-3′
	8892R	5′-AGC CAT GAA TTG ATA GGG GA-3′
2nd PCR	7952S	5′-TCT GCA AGA TCC AAG TAT GG-3′
	8892R	5′-AGC CAT GAA TTG ATA GGG GA-3′
1st PCR	8337S	5′-TTT CGT ATG ATA CCC GAT GCT T-3′
	9330R	5′-GCG CCG ACG GTG AAC CAA CT-3′
2nd PCR	8680S	5′-CTT CAC GGA GGC CAT GAC CA-3′
	9283R	5′-CAA TGG AGT GAG TTT GAG CTT-3′
1st PCR	9095S	5′-TAC TCT CAC CAC GAA CTG AC-3′
	3X-75R	5′-TAC GGC ACT CTC TGC AGT CA-3′
2nd PCR	9231S	5′-GCC GAT ATC TCT TCA ATT GG-3′
	3X-54R	5′-GCG GCT CAC GGA CCT TTC AC-3′
5′RACE		
cDNA synthesis	444R	5′-TAT ACT CCG CCA ACG ATC TG-3′
1st PCR	408R	5′-TTA ACG TCT TCT GGG CGA CG-3′
2nd PCR	258R	5′-ACT CGG CTA GCA GTC TTG CG-3′
3′RACE		
	3X-10S	5′-ATC TTA GCC CTA GTC ACG GC-3′
	CACT35	5′-CAC TTT TTT TTT TTT TTT TTT TTT TTT TTT TTT TTT TT-3′

7. Primers (**Table 23.1**).
8. Ribonuclease H (TaKaRa).
9. AmpliTaq Gold DNA polymerase (Applied Biosystems Japan, Tokyo, Japan).

2.6. Cloning of PCR Products

1. QIAquick Gel Extraction Kit (QIAGEN K.K., Tokyo, Japan).
2. pGEM-T EASY vector (Promega Corp., Madison, WI).
3. Big Dye Terminator Mix and an automated DNA sequencer (Applied Biosystems Japan).

2.7. HCV Clone and Plasmid Construction

1. pGEM-T easy vector.
2. Restriction enzymes.

2.8. Plasmid DNA Preparation

1. DH5α competent cells.
2. Luria-Bertani medium.
3. Solution 1: 25 mM Tris-HCl, pH 8.0, 10 mM EDTA, 50 mM glucose; solution 2: 0.2 N NaOH, 1% sodium dodecyl sulfate (SDS); solution 3: 3 M KOAc, 11.5% glacial acetic acid, for plasmid DNA preparation
4. TE buffer: 10 mM Tris-HCl, pH 8.0, 1 mM EDTA, pH 8.0.
5. 10 mg/mL ethidium bromide solution (Nippon Gene, Tokyo, Japan).
6. Cesium chloride (Iwai Chemicals, Tokyo, Japan, molecular biology grade).
7. Opti-Seal polyallomer centrifuge tube (Beckmann, Palo Alto, CA).
8. TLN100 rotor (Beckmann).
9. Optima TLX Ultracentrifuge (Beckmann).
10. Isopropanol saturated with NaCl.

2.9. RNA Synthesis

1. *Xba*I
2. Mung bean nuclease (New England Biolabs, Beverly, MA).
3. MEGAscript™ T7 kit (Ambion, Austin, TX).

2.10. RNA Transfection

1. Opti-MEM I™ reduced-serum medium (Invitrogen).
2. Cytomix buffer *(42)*.
3. Electroporation cuvette (Precision Universal Cuvettes, Thermo Hybrid, Middlesex, UK).
4. Gene Pulser II™ apparatus (Bio-Rad, Hercules, CA).

2.11. Northern Blot Analysis

1. To prepare DEPC-treated water, add 0.1% of diethylpyrocarbonate (DEPC) to distilled water and shake well. Incubate the solution at 37°C for 2 h, and then autoclave it before use.
2. TRIzol.
3. 37% formaldehyde.
4. Sample buffer: 0.4× 3-(N-morpholino) propanesulfonic acid (MOPS), 6.7% formaldehyde (*see* **Note 1**), 50% formamide.
5. 0.4× MOPS, prepared from 10× MOPS buffer stock: 0.2 M MOPS, pH 7.0, 50 mM sodium acetate, 10 mM EDTA, pH 8.0.
6. Gel-loading buffer: 1% SDS, 50% glycerol, 0.05% bromophenol blue).
7. To prepare 100 mL of 1% denaturing agarose gel, melt 1 g of SeaKem GTG Agarose (Cambrex Bio Science Rockland, Inc., Rockland, ME) completely in 50 mL of DEPC-treated water, then add 10 mL of prewarmed 10× MOPS buffer and 18 mL of prewarmed 37% formaldehyde in a fume hood.
8. 20× SSC: 3 M NaCl, 0.3 M sodium citrate, pH 7.0.
9. Hybond-N+(GE Healthcare, Piscataway, NJ).
10. Stratalinker™ UV crosslinker (Stratagene, La Jolla, CA).

11. Megaprime™ DNA labeling system (GE Healthcare).
12. S-300HR (GE Healthcare).
13. Rapid-Hyb™ buffer (GE Healthcare).

2.12. Quantification of HCV Core Protein

1. HCV core protein immunoassay (Ortho-Clinical Diagnostic K. K., Tokyo, Japan).
2. Lysis buffer: 10 mM Tris-HCl, pH 7.4, 0.5% NP40, 0.15 M NaCl, 1 mM EDTA, pH 8.0, 0.1% SDS.

2.13. Quantification of HCV RNA by Real-Time RT-PCR

1. TaqMan EZ RT-PCR Core Reagents kit (Applied Biosystems Japan).
2. 7500 Real-Time PCR System (Applied Biosystems Japan).
3. MEGAscript™ T7 kit (Ambion, Austin, TX).
4. RNA standard dilution buffer (10 mM DTT, 2000 U/mL RNase inhibitor, 0.2 mg/mL transfer RNA).

2.14. Infection of Cells with Secreted HCV and Determination of Infectivity

1. A disk filter with a 0.45 μm pore size (Millipore, Bedford, MA).
2. Amicon Ultra-15 (100,000 MWCO; Millipore).
3. poly-D-lysine-coated 96-well plates (Corning, New York, NY).
4. Blocking buffer: PBS(-) containing 1% bovine serum albumin and 2.5 mM EDTA.
5. Anticore antibody (e.g., C7-50) solution.
6. AlexaFluor 488-conjugated anti-mouse IgG (Molecular Probes, Eugene, OR).

---

3. Methods

3.1. Patients

A 32-year-old man was admitted with acute liver failure to Jikei University Hospital (Daisan). He was diagnosed with fulminant hepatitis, and HCV RNA was detected by RT-PCR in sera during the acute phase of his illness. Serum HCV RNA was quantified with an Amplicor Monitor HCV test (Roche Diagnostic Systems, NJ). The titer was 10^5 copies/mL at admission. Anti-HCV antibody was negative at admission. All viral markers of the other hepatitis viruses, anti-HAV antibodies (IgG and IgM), and HBV markers (HBsAg, anti-HBs, HBeAg, anti-HBe, anti-HBc, and HBV DNA), as well as GB virus-C/hepatitis G virus RNA, were negative. Analysis of antibodies to Epstein Barr virus and cytomegalovirus revealed a past history of infection.

3.2. RNA Extraction and cDNA Synthesis

1. Extract total RNA from 250 μL of patient serum using the acid-guanidinium-isothiocyanate-phenol-chloroform method (TRIzol LS), in accordance with the manufacturer's instructions (*see* **Note 2**).

2. Resuspend the RNA pellet in 20 μL of nuclease-free water.

3. Transfer 10 μL of the RNA sample to a new tube, and add 8 μL of 5× 1st strand buffer: 10 μL of 2 mM dNTP mixture, 2 μL of 0.1 M DTT, 2 μL of 25 μm random hexamer, 5 μL of nuclease-free water.

4. Mix it well and incubate it at 90°C for 3 min, then transfer the tube onto ice.

5. Spin it down briefly, and add 1 μL of RNase inhibitor and 2 μL of Superscript II reverse transcriptase.

6. Spin it down briefly and incubate at 42°C for 1 h, and then at 70°C for 15 min to terminate the reaction.

7. The cDNA sample can be stored at −70°C until use.

3.3. RT-PCR for Isolation of HCV cDNA

1. Prepare primer sets for the 1st and 2nd PCR to amplify the entire HCV genome as shown in **Table 23.1**.

2. Transfer 1 μL of the cDNA to a PCR tube and add 5 μL of 10× LA Taq buffer, 50 pmol of appropriate sense and antisense primers, 4 μL of 2.5 mM dNTP mix, 2.5 U of TaKaRa LA Taq, and distilled water to bring total volume to 50 μL.

3. Conduct 40 cycles of PCR, each of denaturing at 95°C for 30 s, annealing at 60°C for 30 s, and extension at 72°C for 1 min.

4. Amplified products can be separated by agarose gel electrophoresis

3.4. 5′ RACE RT-PCR

For determination of the terminal 5′ end sequence, cDNA can be synthesized with a 5′UTR primer (antisense), tailed with terminal deoxynucleotidyl transferase and a dCTP homopolymer, and then amplified by PCR (5′ RACE System for Rapid Amplification of cDNA Ends Version 2.0)

1. Transfer 5 μL of the RNA sample to a new tube, add 2.5 μL of 1 μm antisense primer (444R, **Table 23.1**), and mix.

2. Incubate the mixture at 70°C for 10 min and then transfer onto ice.

3. Add 2.5 μL of 10× Ex Taq buffer, 2.5 μL of 25 mM MgCl$_2$, 1 μL of 10 mM dNTP mix, 1.2 μL of 0.1 M DTT, 9.3 μL of nuclease-free water, and 1 μL of Superscript II.

4. Mix gently and incubate the mixture at 42°C for 1 h, then at 70°C for 15 min.

5. Spin it down briefly and maintain it at 37°C.

6. Add 1 μL of RNase Mix and incubate at 37°C for 30 min.

7. Purify cDNA using a RACE DNA Purification Spin Cartridge according to the manufacturer's instructions.

8. Transfer 10 μL of purified cDNA into a new tube and add 6.5 μL of nuclease-free water, 5 μL of 5× tailing buffer, and 2 mM dCTP.

9. Incubate the mixture at 94°C for 3 min, and then transfer onto ice.

10. Add 1 μL of recombinant Terminal deoxynucleotidyl Transferase (TdT), and mix gently.

11. Incubate the mixture at 37°C for 10 min, and then at 65°C for 10 min.

12. Perform the 1st PCR of dC-tailed cDNA using the Abridged Anchor Primer and 408 R primer (**Table 23.1**). Prepare the PCR mixture as follows:

dC-tailed cDNA		5 μL
primer	10 pM	2 μL
Abridged Anchor Primer		2 μL
dNTP mix	10 mM	1 μL
10× Ex Taq buffer		5 μL
MgCl$_2$	25 mM	3 μL
Distilled water		31.5 μL
TaKaRa Ex Taq	5 U/μL	0.5 μL

13. Conduct 35 cycles of PCR, each of denaturing at 94°C for 30 s, annealing at 55°C for 30 s, and extension at 72°C for 1 min.

14. Prepare the 2nd PCR mixture as follows:

1st PCR product		1 μL
primer	10 pM	2 μL
Abridged Universal Anchor Primer		1 μL
dNTP mix	10 mM	1 μL
10× Ex Taq buffer		5 μL
MgCl$_2$	25 mM	3 μL
Distilled water		33.5 μL
TaKaRa Ex Taq	5 U/μL	0.5 μL

15. Conduct 35 cycles of PCR, each of denaturing at 94°C for 30 s, annealing at 55°C for 30 s, and extension at 72°C for 1 min.

16. Separate amplified products by 3% agarose gel electrophoresis.

3.5. 3′ RACE RT-PCR   For determination of the terminal 3′ end sequence, extracted RNA can be polyadenylated with Poly(A) Polymerase (Takara Biochemicals), converted to cDNA with a 38-mer oligonucleotide containing (T)$_{35}$, and amplified with a 3′UTR primer and a primer used for reverse transcription (*see* **Note 3**).

1. Transfer 5 μL of the RNA sample to a new tube and incubate it at 90°C for 3 min and then transfer it onto ice.
2. Spin it down briefly and add 10 μL of 5× Poly(A) Polymerase buffer, 5 μL of 10 mM DTT, 1 μL of RNase inhibitor, 5 μL of 10 mM ATP, and 23 μL of nuclease-free water, followed by 1 μL of Poly(A) Polymerase, and mix gently.
3. Incubate the mixture at 37°C for 60 min.
4. Extract poly(A)-tailed RNA using TRIzol reagent according to the manufacturer's instructions.
5. Resuspend the RNA pellet in 10 μL of nuclease-free water.
6. Add 8 μL of 5× 1st strand buffer, 10 μL of 2 mM dNTP mixture, 2 μL of 0.1 M DTT, 5 μL of 10 μm CACT35, and 2 μL of nuclease-free water.
7. Mix well and incubate the mixture at 90°C for 3 min, then transfer the tube onto ice.
8. Spin it down briefly and add 1 μL of RNase inhibitor and 2 μL of Superscript II.
9. Mix well and incubate the tube at 50°C for 1 h, then at 72°C for 15 min to terminate the reaction.
10. Add 12 U of Ribonuclease H and incubate at 37°C for 20 min, then at 95°C for 3 min to terminate the reaction.
11. The cDNA sample can be stored at −70°C until use.
12. Conduct PCR of poly(A)-tailed cDNA using 3X-10S and CACT35 primers. Prepare the PCR mixture as follows:

poly(A)-tailed cDNA		2 μL
3X-10S primer	10 pM	1 μL
CACT35 Primer	10 pM	1 μL
dNTP mix	10 mM	2 μL
10× PCR Gold Buffer		2 μL
Distilled water		11.75 μL
AmpliTaq Gold DNA polymerase	0.05 U/μL	0.25 μL

13. Incubate the PCR mixture at 95°C for 7 min before PCR cycling, and perform PCR using 70 cycles, each of denaturing at 95°C for 30 s and annealing and extension at 60°C for 1 min.
14. Amplified products can be separated by 3% agarose gel electrophoresis

3.6. Cloning of PCR Products

1. Amplified products can be separated by agarose gel electrophoresis.
2. Excise the DNA fragment containing the PCR product from the agarose gel under UV light.

3. Purify the DNA from the excised agarose gel using the QIAquick Gel Extraction Kit (QIAGEN) according to the manufacturer's instructions.
4. Purified PCR products can be cloned into a pGEM-T EASY vector (Promega).
5. Sequence the cloned DNA using Big Dye Terminator Mix and an automated DNA sequencer (Applied Biosystems Japan). In our laboratory, at least five clones for each RT-PCR fragment were sequenced and a consensus sequence for JFH-1 determined (accession number: AB047639).

3.7. HCV Clone and Plasmid Construction

1. To isolate the JFH-1 strain as it was isolated in our laboratory, amplify 14 fragments of HCV cDNA, covering the entire genome, by RT-PCR, then purify and clone all amplified products into a number of pGEM-T easy^{TM} vectors: pGEM1-258, pGEM44-486, pGEM317-849, pGEM617-1323, pGEM1141-2367, pGEM2285-3509, pGEM3471-4665, pGEM4547-5970, pGEM5883-7003, pGEM6950-8035, pGEM7984-8892, pGEM8680-9283, pGEM9231-9634, and pGEM9594-9678 (the number assigned to each clone indicates its position within JFH-1).
2. Assemble a subgenomic replicon construct of JFH-1 [ours is pSGR-JFH1; accession number AB114136; see **ref.** *(11)*] Based on the consensus sequence of JFH-1. Then assemble the 5′ half of the JFH-1 cDNA (nt 1–5970) into plasmid pGEM1-5970 using nine plasmids containing overlapping cDNA. Cut out an *Age*I and *Eco*T22I fragment (nt 154-5293) from pGEM1-5970 and insert it into pSGR-JFH1; the resulting plasmid (pJFH-1) will contain full-length JFH-1 cDNA downstream of the T7 RNA promoter sequence [pJFH-1; **ref.** *(17)*].
3. Then derive two mutant constructs from pJFH-1: pJFH-1/GND, with a point mutation of GDD to GND, abolishing the RNA polymerase activity of NS5B, and pJFH-1/ΔE1-E2, with an in-frame deletion of 351 amino acids spanning most of the E1-to-E2 region *(17)*.

3.8. Plasmid DNA Preparation

1. Transform DH5α competent cells with pJFH-1 and other plasmid DNA.
2. Inoculate 200 mL Luria-Bertani medium containing 100 μg/mL ampicillin with a single colony.
3. Incubate the culture at 37°C overnight with vigorous shaking.
4. Transfer the culture to a centrifuge bottle, and centrifuge it at 4,000g for 15 min.
5. Discard the supernatant, and resuspend the pellet with 12 mL of solution 1.

6. Add 24 mL of solution 2, mix gently by inverting the bottles five to six times, and then incubate them at room temperature for 10 min.

7. Add 18 mL of solution 3, and mix gently by inverting the bottles five to six times, then incubate them on ice for 10 min.

8. Centrifuge the solution at 12,000g for 15 min at 4°C.

9. Transfer the supernatant to a new centrifuge bottle.

10. Add 32 mL isopropanol, mix vigorously, and incubate at room temperature for 10 min.

11. Centrifuge the mixture at 12,000g for 15 min at 4°C.

12. Discard the supernatant and resuspend the pellet with 5.2 mL of TE buffer, then add 6 g of cesium chloride and mix until the CsCl is dissolved. Add 0.14 mL of 10 mg/mL ethidium bromide solution and incubate at room temperature for 10 min with gentle shaking.

13. Centrifuge the mixture at 3,000g for 10 min at room temperature. Transfer the supernatant to two Opti-Seal polyallomer centrifuge tubes.

14. Place caps on the tubes and insert the tubes into a Beckmann TLN100 rotor. Centrifuge them at $360,000 \times$ g for 3 h, or at $140,000 \times$ g overnight, in a Beckmann Optima TLX Ultracentrifuge at 25°C.

15. After centrifugation, remove the caps from the tubes and insert a 2.5 mL syringe attached to a 21-gauge needle into each tube a few millimeters below the band of DNA and then bring it in line with the bottom edge of the band. Draw the plunger slowly until most of the DNA is collected. Remove the needle from the tube and transfer the DNA solution into a new Opti-Seal polyallomer centrifuge tube.

16. Fill up the tube with CsCl/TE solution (1.1 g/mL, wt/vol) and centrifuge again at $360,000 \times$ g for 3 h or at $140,000 \times$ g overnight in a Beckmann TLN100 rotor at 25°C.

17. After this second round of ultracentrifugation, draw up the DNA band again as above and transfer the DNA solution into a new centrifuge tube.

18. Extract the DNA solution with isopropanol saturated with NaCl to remove ethidium bromide. Extraction should be repeated one more time after the orange color of the ethidium bromide disappears.

19. Add three volumes of TE buffer and eight volumes of 99.5% ethanol to the DNA solution. Mix well and let it stand at −70°C for 20 min.

20. Centrifuge at $9,100 \times$ g for 20 min at 4°C.

21. After centrifugation, discard the supernatant and add 70% ethanol.

22. Centrifuge at $9,100 \times$ g for 20 min at 4°C.

23. Discard the supernatant and dry the pellet completely.

24. Resuspend the pellet in TE buffer. Determine the DNA concentration by measuring optical density at OD260 and OD280. Also, double check the concentration of DNA, purity of the supercoiled DNA, and RNA contamination by agarose gel electrophoresis. Adjust the DNA concentration to 2 mg/mL with TE buffer.

25. Sequence the purified pJFH-1 plasmid at each preparation. Some unintended mutations are incorporated through the procedures.

3.9. RNA Synthesis

This protocol of RNA synthesis has also been described elsewhere *(43)*.

1. Digest pJFH-1 and other plasmids with *Xba*I.
2. Treat the *Xba*I-digested plasmid with mung bean nuclease to remove four nucleotides, leaving the correct 3′ end of the HCV cDNA.
3. Purify the digested plasmid DNA using phenol/chloroform extraction.
4. Synthesize HCV RNA using purified plasmid as templates for RNA with a MEGAscript^{TM} T7 kit in accordance with the manufacturer's instructions.
5. Treat synthesized RNA with DNaseI then by acid phenol extraction to remove any remaining template DNA.
6. Determine the RNA concentration by measuring optical density at OD260 and OD280. Check the purity of synthesized RNA by agarose gel electrophoresis. Adjust the RNA concentration to 10 mg/mL with nuclease-free water.

3.10. RNA Transfection

A detailed protocol of RNA transfection has been described previously (44).

1. Trypsinize Huh-7 cells, wash them with Opti-MEM I^{TM} Reduced-Serum Medium, and resuspend them at 7.5×10^6 cells/mL with Cytomix buffer *(42)*.
2. Add 10 μg RNA to 400 μL of cell suspension (3×10^6 cells) and transfer the mixture into an electroporation cuvette.
3. Pulse the cells at 260 V and 950 μF using a Gene Pulser II^{TM} apparatus.
4. Transfer the transfected cells immediately into two 10 cm culture dishes, each containing 8 mL of culture medium.

3.11. Northern Blot Analysis

RNA replication in the transfected or infected Huh-7 cells can be analyzed by northern-blot analysis.

1. Extract cellular RNA from RNA-transfected cells using TRIzol solution according to the manufacturer's instructions.
2. Determine the concentration of the isolated RNA by measuring optical density. Adjust the concentration to 1 mg/mL with nuclease-free water.
3. Prepare a denaturing 1% agarose gel.

4. Transfer $2\,\mu L$ of $2\,mg/mL$ RNA to a new tube and add $8\,\mu L$ of sample buffer, then incubate at 65°C for 15 min and transfer onto ice.

5. Spin the tube down briefly, add $2\,\mu L$ of gel-loading buffer, and apply the sample to a denaturing agarose gel.

6. After electrophoresis, rinse the agarose gel with DEPC-treated water once for 10 min with gentle shaking.

7. For a minigel, incubate in $150\,mL$ of $50\,mM$ NaOH/$10\,mM$ NaCl solution at room temperature for 20 min.

8. Discard the solution and rinse the gel with DEPC-treated water.

9. Incubate in $150\,mL$ of $0.1\,M$ Tris-HCl, pH 7.4, at room temperature for 20 min.

10. Discard the solution and rinse the gel with DEPC-treated water.

11. Incubate in $150\,mL$ of $20\times$ SSC at room temperature for 20 min.

12. Transfer the gel onto a Hybond-N+ membrane at 50–55 cm H_2O for 1 h (see **Note 4**).

13. Immobilize the membrane using a Stratalinker^{TM} UV crosslinker.

14. Purify the DNA fragment containing the *NS3* to *5B* genes of JFH-1.

15. Transfer 25 ng of the purified DNA fragment into a new tube and add $5\,\mu L$ of primer solution from the Megaprime^{TM} DNA labeling system and distilled water to a total volume of $33\,\mu L$.

16. Boil the solution for 5 min and then transfer it onto ice.

17. Add $10\,\mu L$ of labeling buffer and $5\,\mu L$ of $[\alpha-^{32}P]$dCTP, followed by $2\,\mu L$ of enzyme solution, and then incubate the mixture at 37°C for 10 min.

18. Purify the $[\alpha-^{32}P]$dCTP-labeled DNA probe using a S-300HR spin column.

19. Determine the specific activity with a scintillation counter.

20. Incubate the membrane with $0.125\,mL/cm^2$ Rapid-Hyb Buffer^{TM} at 65°C for at least 15 min. During this incubation period, boil the probe for 5 min and then place it on ice for 2 min.

21. Add the boiled probe at 1×10^5 cpm/mL buffer and incubate at 65°C for 2 h with shaking.

22. Wash the hybridized membrane with $2\times$ SSC/0.1% SDS at room temperature for 10 min three times.

23. Wash the membrane with $1\times$ SSC/0.1% SDS at 65°C for 15 min once.

24. Wash the membrane with $0.1\times$ SSC/0.1% SDS at 65°C for 20 min three times.

25. Expose the membrane to X-ray film.

3.12. Quantification of HCV Core Protein

For estimation of levels of HCV core protein in culture supernatant or cell lysate, concentrations of HCV core protein can be determined by HCV core protein immunoassay. Cell pellets can be lysed in lysis buffer.

1. Dilute the HCV core standard solution with standard diluent and prepare 3,600, 1,200, 400, 133, and 44.4 fmol/L solutions. Use standard diluent as a negative control (0 fmol/L).
2. Add 50 μL of pretreatment solution to 100 μL of sample and mix (see **Note 5**).
3. Incubate at 56–60°C for 30 min and then at room temperature for 5 min.
4. Determine the number of eight-well strips needed for the assay. Insert these into the frame for current use.
5. Add 100 μL of reaction buffer to each well.
6. Add 100 μL of each standard solution and pretreated samples to the appropriate wells. Mix them by pipetting.
7. Cover the plate with a plate cover and incubate it for 60 min at room temperature with shaking on a plate mixer.
8. Thoroughly aspirate or decant solution from the wells and discard the liquid. Take care not to scratch the inside of the well. After aspiration, fill the wells with wash solution, being careful not to overflow the wells. Soak for at least 20 s and then aspirate the liquid. Repeat this washing procedure six times.
9. Add 200 μL of labeled-antibody solution to each well.
10. Cover the plate with a plate cover and incubate it for 30 min at room temperature.
11. Thoroughly aspirate or decant solution from the wells and discard the liquid. Wash the wells six times as described above.
12. Add 200 μL of substrate solution to each well. Incubate for 30 min at room temperature in the dark.
13. Add 50 μL of stop solution to each well. Tap the side of the plate gently to mix.
14. Read the absorbance of each well at 492 nm against two negative control wells.
15. Plot the absorbance of each standard against the standard concentration. Draw the best smooth curve through these points to construct the standard curve.
16. Determine HCV core protein concentrations for unknown samples from the standard curve. Samples with concentrations exceeding the highest standard (3,600 fmol/L) should be diluted and reanalyzed. Samples producing less than 100 fmol/L should also be reanalyzed for confirmation of results.
17. Determine the total lysate protein concentrations, for example using the Bradford method, and adjust them to 1 mg/mL with normal serum. HCV core protein concentrations within

cell lysate can be expressed as fmol/g of total protein when divided by total protein concentrations (*see* **Note 6**).

3.13. Quantification of HCV RNA by Real-Time RT-PCR

Copy numbers of HCV RNA in culture supernatant or infected cells can be determined by real-time detection RT-PCR (RTD-PCR), using TaqMan EZ RT-PCR Core Reagents kit and the 7500 Real-Time PCR System.

1. To synthesize standard RNA for quantification, prepare plasmid DNA containing the HCV IRES sequence.
2. Synthesize an HCV RNA standard using an appropriate commercial kit, such as the MEGAscript^{TM} T7 kit. Check the integrity of the synthesized RNA by denaturing agarose gel electrophoresis as described above.
3. Determine RNA concentrations by measuring optical density at OD260 and OD280. Calculate the copy number of synthesized RNA from the concentration and length of RNA. Adjust the standard RNA concentration to 10^{10} copies/μL with nuclease-free water.
4. Prepare primers and the TaqMan probe listed in **Table 23.2**.
5. Dilute HCV RNA standard solution with RNA standard dilution buffer and prepare 10^7, 10^6, 10^5, 10^4, 10^3, 10^2, and 10^1 copies/μLE solutions. Use nuclease-free water as a negative control.
6. Prepare the reaction mixture as follows:

Nuclease-free water	7.1 μL
5× TaqMan EZ buffer	5 μL
10 mM dATP	0.5 μL
10 mM dGTP	0.5 μL
10 mM dCTP	0.5 μL
20 mM dUTP	0.65 μL
10 μm 130S primer	0.5 μL
10 μm 290R primer	0.5 μL
3 μm TaqMan probe	2.5 μL
25 mM Mn(OAc)$_2$	3 μL
AmpErase UNG	0.25 μL
rTth DNA polymerase	1.5 μL
Total	22.5 μL

7. Add 22.5 μL of reaction mixture to each well of the PCR plate.

Table 23.2
Primers and a probe used in RTD-PCR

Name	Sequence (5′ > 3′)
130S	5′- CGG GAG AGC CAT AGT GG-3′
290R	5′- AGT ACC ACA AGG CCT TTC G-3′
probe	5′- (6-Fam) CTG CGG AAC CGG TGA GTA CAC (Tamra) -3′

8. Add 2.5 μL of each RNA sample and standard to the appropriate wells. Use nuclease-free water as a negative control. All samples and controls should be evaluated in at least two wells.

9. Set the PCR plate in a 7500 Real-Time PCR System or an equivalent system.

10. Set the reaction conditions and incubate at 50°C for 2 min, 60°C for 30 min, 95°C for 5 min, followed by 50 cycles at 95°C for 20 s and 62°C for 1 min.

11. Confirm the absence of amplification in the negative control wells.

12. Determine RNA copy numbers in the samples using Sequence Detection Software.

3.14. Infection of Cells with Secreted HCV and Determination of Infectivity

1. Collect culture medium from RNA-transfected or infected cells.

2. Centrifuge the collected culture medium at 3,000g for 20 min at 4°C, and pass the supernatant through a disk filter with a 0.45 μm pore size.

3. Concentrate the collected culture medium using an Amicon Ultra-15 device (100,000 MWCO) if necessary. Add up to 15 mL of culture medium to the upper chamber of the Amicon Ultra-15. Centrifuge at 3,000g for 30 min at 4°C. The concentrated culture medium can be stored at −80°C until use.

4. Seed wells in poly-D-lysine-coated 96-well plates with Huh-7 cells at a density of 1×10^4 cells on the day before virus inoculation.

5. Dilute culture medium containing infectious JFH-1 virus.

6. Discard the culture medium from the plates seeded with Huh-7 cells.

7. Add 100 μL of serially diluted virus solution to at least six wells per dilution, and incubate for 4 h at 37°C.

8. Remove the inoculum and add 100 μL of fresh complete medium, then incubate the inoculated cells for 72 h at 37°C under 5% CO_2.

9. At the end of incubation, remove the culture medium.

10. Fix the cells by dipping the plate into ice-chilled 100% methanol.

11. Incubate it at −20°C for 20 min.
12. Block the cells at room temperature for 1 h with 100 µL per well of blocking buffer, then wash with PBS(−) once.
13. Add 100 µL of anticore antibody solution to each well to make a concentration of 50 µg/mL in blocking buffer, and incubate at room temperature for 1 h.
14. Aspirate the antibody solution and wash the wells with PBS(−) three times.
15. Add 100 µL per well of AlexaFluor 488-conjugated antimouse IgG in blocking buffer, and incubate at room temperature for 1 h.
16. Aspirate the antibody solution and wash the wells with PBS(−) three times.
17. Count the stained cells using fluorescence microscopy.
18. Calculate the infectivity of the inoculum form the focus number and inoculum dilution, which can be expressed as focus-forming units per milliliter (ffu/mL). Alternatively, determine the TCID50 according to the method of Reed and Muench *(44)*.

4. Notes

1. The final concentration of formaldehyde is shown.
2. If the expected amount of recovered RNA is low, transfer RNA or glycogen should be added before isopropanol precipitation to enhance the precipitation efficiency.
3. If the 3′ residue might be A, other 3′ RACE methods should also be used, such as the linker ligation method.
4. The membrane should be rinsed with DEPC-treated water, then 20× SSC.
5. Multiple dilutions of samples may be necessary to produce core protein concentrations within a standard range. Fetal bovine serum can be used as the diluent.
6. Adjusting the total protein concentration may reduce background contamination.

Acknowledgments

This work was partially supported by a grant-in-aid for Scientific Research from the Japan Society for the Promotion of Science, from the Ministry of Health, Labour and Welfare of Japan, and from the Ministry of Education, Culture, Sports, Science and Technology and by the Research on Health Sciences Focusing on Drug Innovation from the Japan Health Sciences Foundation.

References

1. Bartenschlager, R. and Lohmann, V. (2001) Novel cell culture systems for the hepatitis C virus. *Antiviral Res.* **52,** 1–17.
2. Lohmann, V., Korner, F., Koch, J., Herian, U., Theilmann, L., and Bartenschlager, R. (1999) Replication of subgenomic hepatitis C virus RNAs in a hepatoma cell line. *Science* **285,** 110–113.
3. Blight, K. J., Kolykhalov, A. A., and Rice, C. M. (2000) Efficient initiation of HCV RNA replication in cell culture. *Science* **290,** 1972–1974.
4. Lohmann, V., Korner, F., Dobierzewska, A., and Bartenschlager, R. (2001) Mutations in hepatitis C virus RNAs conferring cell culture adaptation. *J. Virol.* **75,** 1437–1449.
5. Krieger, N., Lohmann, V., and Bartenschlager, R. (2001) Enhancement of hepatitis C virus RNA replication by cell culture-adaptive mutations. *J. Virol.* **75,** 4614–4624.
6. Ikeda, M., Yi, M., Li, K., and Lemon, S. M. (2002) Selectable subgenomic and genome-length dicistronic RNAs derived from an infectious molecular clone of the HCV-N strain of hepatitis C virus replicate efficiently in cultured Huh7 cells. *J. Virol.* **76,** 2997–3006.
7. Pietschmann, T., Lohmann, V., Kaul, A., Krieger, N., Rinck, G., Rutter, G., et al. (2002) Persistent and transient replication of full-length hepatitis C virus genomes in cell culture. *J. Virol.* **76,** 4008–4021.
8. Blight, K. J., McKeating, J. A., Marcotrigiano, J., and Rice, C. M. (2003) Efficient replication of hepatitis C virus genotype 1a RNAs in cell culture. *J. Virol.* **77,** 3181–3190.
9. Bukh, J., Pietschmann, T., Lohmann, V., Krieger, N., Faulk, K., Engle, R. E., et al. (2002) Mutations that permit efficient replication of hepatitis C virus RNA in Huh-7 cells prevent productive replication in chimpanzees. *Proc. Natl. Acad. Sci. USA* 99, 14416–14421.
10. Kato, T., Furusaka, A., Miyamoto, M., Date, T., Yasui, K., Hiramoto, J., et al. (2001) Analysis of hepatitis C virus isolated from a fulminant hepatitis patient. *J. Med. Virol.* **64,** 334–339.
11. Kato, T., Date, T., Miyamoto, M., Furusaka, A., Tokushige, K., Mizokami, M., et al. (2003) Efficient replication of the genotype 2a hepatitis C virus subgenomic replicon. *Gastroenterology* **125,** 1808–1817.
12. Uprichard, S. L., Chung, J., Chisari, F. V., and Wakita, T. (2006) Replication of a hepatitis C virus replicon clone in mouse cells. *Virol. J.* **3,** 89.
13. Date, T., Miyamoto, M., Kato, T., Morikawa, K., Murayama, A., Akazawa, D., et al. (2007) An infectious and selectable full-length replicon system with hepatitis C virus JFH-1 strain. *Hepatol. Res.* in press.
14. Kato, T., Date, T., Miyamoto, M., Sugiyama, M., Tanaka, Y., Orito, E., et al. Detection of anti-hepatitis C virus effects of interferon and ribavirin by a sensitive replicon system. *J. Clin. Microbiol.* **43,** 5679–5684.
15. Date, T., Kato, T., Miyamoto, M., Zhao, Z., Yasui, K., Mizokami, M., et al. (2004) Genotype 2a hepatitis C virus subgenomic replicon can replicate in HepG2 and IMY-N9 cells. *J. Biol. Chem.* **279,** 22371–22376.
16. Kato, T., Date, T., Miyamoto, M., Zhao, Z., Mizokami, M., and Wakita, T. (2005) Nonhepatic cell lines HeLa and 293 cells support efficient replication of hepatitis C virus genotype 2a subgenomic replicon. *J. Virol.* **79,** 592–596.
17. Wakita, T., Pietschmann, T., Kato, T., Date, T., Miyamoto, M., Zhao, Z., et al. (2005) Production of infectious hepatitis C virus in tissue culture from a cloned viral genome. *Nat. Med.* **11,** 791–796.
18. Zhong, J., Gastaminza, P., Cheng, G., Kapadia, S., Kato, T., Burton, D. R., et al. (2005) Robust hepatitis C virus infection in vitro. *Proc. Natl. Acad. Sci. USA* **102,** 9294–9299.
19. Blight, K. J., McKeating, J. A., and Rice, C. M. (2002) Highly permissive cell lines for subgenomic and genomic hepatitis C virus RNA replication. *J. Virol.* **76,** 13001–13014.
20. Sumpter, R., Jr., Loo, Y. M., Foy, E., Li, K., Yoneyama, M., Fujita, T., et al. (2005) Regulating intracellular antiviral defense and permissiveness to hepatitis C virus RNA replication through a cellular RNA helicase, RIG-I. *J. Virol.* **79,** 2689–2699.
21. Lindenbach, B. D., Evans, M. J., Syder, A. J., Wolk, B., Tellinghuisen, T. L., Liu, C. C., et al. (2005) Complete replication of hepatitis C virus in cell culture. *Science* **309,** 623–626.
22. Koutsoudakis, G., Herrmann, E., Kallis, S., Bartenschlager, R., and Pietschmann, T. (2007) The level of CD81 cell surface expression is a key determinant for productive entry of hepatitis C virus into host cells. *J. Virol.* **81,** 588–598.
23. Akazawa, D., Date, T., Morikawa, K., Murayama, A., Miyamoto, M., Kaga, M., et al. (2007) CD81 expression is important for heterogeneous HCV permissiveness of Huh7 cell clones. *J. Virol.* [Epub ahead of print].

24. Nakabayashi, H., Taketa, K., Miyano, K., Yamane, T., and Sato, J. (1982) Growth of human hepatoma cells lines with differentiated functions in chemically defined medium. *Cancer Res.* **42,** 3858–3863.

25. Koutsoudakis, G., Kaul, A., Steinmann, E., Kallis, S., Lohmann, V., Pietschmann, T., et al. (2006) Characterization of the early steps of hepatitis C virus infection by using luciferase reporter viruses. *J. Virol.* **80,** 5308–5320.

26. Morikawa, K., Zhao, Z., Date, T., Miyamoto, M., Murayama, A., Akazawa, D., et al. (2007) The roles of CD81 and glycosaminoglycans in the adsorption and uptake of infectious HCV particles. *J. Med. Virol.* in press.

27. Kapadia, S. B., Barth, H., Baumert, T., McKeating, J. A., and Chisari, F. V. (2007) Initiation of hepatitis C virus infection is dependent on cholesterol and cooperativity between CD81 and scavenger receptor B type I. *J. Virol.* **81,** 374–383.

28. Grove, J., Huby, T., Stamataki, Z., Vanwolleghem, T., Meuleman, P., Farquhar, M., et al. (2007) Scavenger receptor BI and BII expression levels modulate Hepatitis C virus infectivity. *J Virol.* [Epub ahead of print].

29. Blanchard, E., Belouzard, S., Goueslain, L., Wakita, T., Dubuisson, J., Wychowski, C., et al. (2006) Hepatitis C virus entry depends on clathrin-mediated endocytosis. *J Virol.* **80,** 6964–6972.

30. Evans, M. J., von Hahn, T., Tscherne, D. M., Syder, A. J., Panis, M., Wolk, B., (2007) Claudin-1 is a hepatitis C virus co-receptor required for a late step in entry. *Nature* [Epub ahead of print].

31. Kanda, T., Basu, A., Steele, R., Wakita, T., Ryerse, JS., Ray, R., et al. (2006) Generation of infectious hepatitis C virus in immortalized human hepatocytes. *J. Virol.* **80,** 4633–4639.

32. Wagoner, J., Austin, M., Green, J., Imaizumi, T., Casola, A., Brasier, A., et al. Regulation of CXCL-8 (interleukin 8) induction by dsRNA signaling pathways during hepatitis C virus infection. *J. Virol.* **81,** 309–318.

33. Larrea, E., Riezu-Boj, J. I., Gil-Guerrero, L., Casares, N., Aldabe, R., Sarobe, P., et al. (2007) Upregulation of indoleamine 2,3 dioxygenase in hepatitis C virus infection. *J. Virol.* [Epub ahead of print].

34. Francesco, R. D., and Migliaccio, G. (2005) Challenges and successes in developing new therapies for hepatitis C. *Nature* **436,** 953–960.

35. Cai, Z., Zhang, C., Chang, K-Y., Jiang, J., Ahn, B-C., Wakita, T., et al. (2005) Robust production of infectious hepatitis C virus (HCV) from stably HCV cDNA-transfected human hepatoma cells. *J. Virol.* **79,** 13963–13973.

36. Kato, T., Matsumura, T., Heller, T., Saito, S., Sapp, R. K., Murthy, K., et al. (2007) Production of infectious hepatitis C virus of various genotypes in cell culture. *J. Virol.* [Epub ahead of print].

37. Rouillé, Y., Helle, F., Delgrange, D., Roingeard, P., Voisset, C., Blanchard, E., et al. (2006) Subcellular localization of hepatitis C virus structural proteins in a cell culture system that efficiently replicates the virus. *J. Virol.* **80,** 2832–2841.

38. Shirakura, M., Murakami, K., Ichimura, T., Suzuki, R., Shimoji, T., Fukuda, K., et al. (2007) The E6AP ubiquitin ligase mediates ubiquitylation and degradation of hepatitis C virus core protein. *J. Virol.* **81,** 1174–1185.

39. Houghton, M. and Abrignani, S. (2005) Prospects for a vaccine against the hepatitis C virus. *Nature* **436,** 961–966.

40. Meunier, J. C., Engle, R. E., Faulk, K., Zhao, M., Bartosch, B., Alter, H., et al. (2005) Evidence for cross-genotype neutralization of hepatitis C virus pseudo-particles and enhancement of infectivity by apolipoprotein C1. *Proc. Natl. Acad. Sci. USA* **102,** 4560–4565.

41. Yi, M., Villanueva, R. A., Thomas, D. L., Wakita, T., and Lemon, S. M. (2006) Production of infectious genotype 1a hepatitis C virus (Hutchinson strain) in cultured human hepatoma cells. *Proc. Natl. Acad. Sci. USA* **103,** 2310–2315.

42. van den Hoff, M. J., Moorman, A. F., and Lamers, W. H. (1992) Electroporation in "intracellular" buffer increases cell survival. *Nucleic Acids Res.* **20,** 2902.

43. Kato, T., Date, T., Murayama, A., Morikawa, K., Akazawa, D., and Wakita, T. (2006) Cell culture and infection system for hepatitis C virus. *Nature Protocols* **1,** 2334–2339.

44. Reed, L. J. and Muench, H. A. (1938) Simple method of estimating fifty per cent endpoints. *Am. J. Hyg.* 27493–27497.

Chapter 24

Measuring HCV Infectivity Produced in Cell Culture and In Vivo

Brett D. Lindenbach

Abstract

Recently described systems for efficiently growing HCV in cell culture provide powerful new tools with which to dissect the life cycle of this important human pathogen. This chapter describes methods for measuring the infectivity of HCV produced in cell culture or recovered from animals experimentally infected with the cell culture–produced virus.

Key words: Hepatitis C virus, HCV, titer, infectivity assay.

1. Introduction

Early attempts at culturing hepatitis C virus (HCV) met with only limited success [reviewed in **ref.** *(1)*]. Although HCV RNA could be maintained in culture for several days, weeks, or even over a year, the amount of replication that occurred in these systems was very low, requiring sensitive reverse-transcription polymerase chain reaction (RT-PCR) and nested RT-PCR for detection. Viral replication therefore could not be measured with traditional virologic techniques in these early cell-culture systems. The development of subgenomic replicons that autonomously replicate in the human hepatoma line Huh-7 *(2)* allowed study of the intracellular aspects of the HCV life cycle in cell culture. Yet full-length HCV genomes containing cell culture–adaptive mutations did not produce infectious virus in culture *(3, 4)* and were severely attenuated in chimpanzees *(5)*. Based on the HCV JFH-1 subgenomic replicon, which exhibits unusually robust replication *(6)*, full-length

Hengli Tang (ed.), *Hepatitis C: Methods and Protocols, Second Edition, vol. 510*
© 2009 Humana Press, a part of Springer Science+Business Media
DOI 10.1007/978-1-59745-394-3_24 Springerprotocols.com

genotype 2a HCV genomes were constructed and shown to replicate and produce infectious virus in cell culture *(7–9)*. Importantly, cell culture–produced virus was shown to be infectious in animals *(8, 10)* and could be recovered in cell culture *(10)*. Yi and colleagues demonstrated that, in addition to the genotype 2a cell-culture systems, a cell culture–adapted HCV strain H77 (genotype 1a) genome can produce infectious virus in cell culture *(11)*. Although the level of infectivity produced by these systems varies depending on the genetic background of the virus and host cell, all of these systems produce sufficient levels of infectivity to allow classic virologic analysis of HCV.

One of the most widely used methods of determining viral infectivity is the plaque assay *(12)*. Although HCV infection causes cytopathic effects in Huh-7 cells *(13)*, these effects generally take several days or weeks to become apparent, and conditions for HCV plaque formation are still at an early stage of develop-

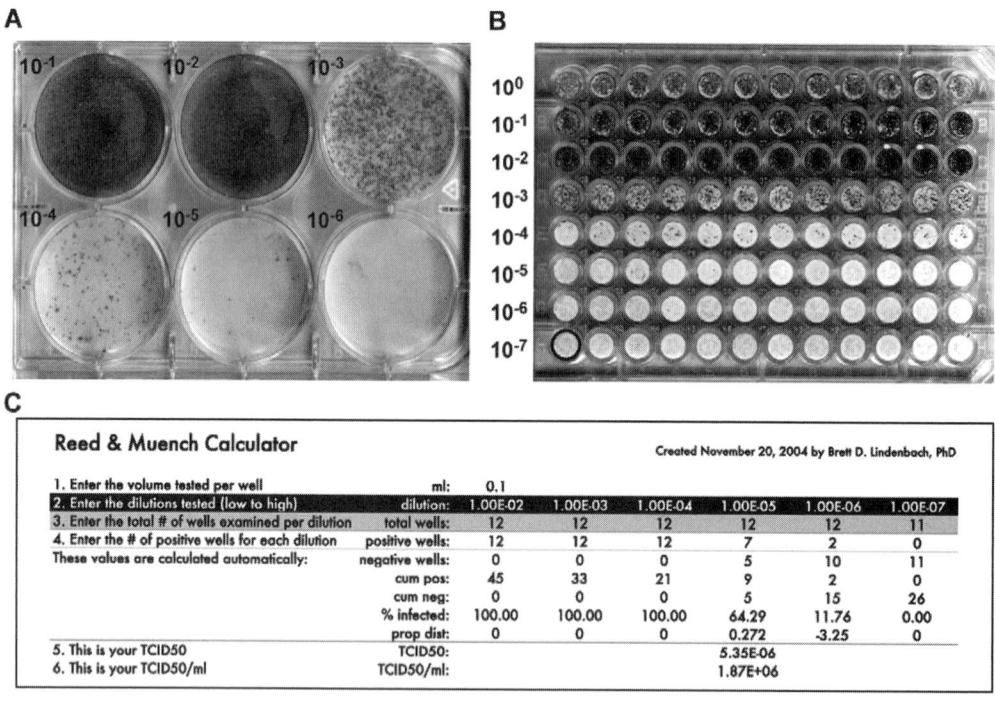

Fig. 24.1. Examples of measuring HCV infectivity. A single stock of cell culture–produced HCV was titered with a focus-forming assay and with the endpoint dilution assay described here. (**A**) In the focus-forming assay, Huh-7.5 cells were infected with 0.2 mL of each virus dilution, overlaid with $1\times$ Dulbecco's modified Eagle's medium containing 5% fetal calf serum and 0.4% LE-agarose, and fixed and immunostained after 4 days. The titer was found to be 1.3×10^6 focus-forming units (ffu)/mL. (**B**) In the endpoint dilution assay, each row of Huh-7.5 cells was infected with 0.1 mL of the indicated virus dilution, and then fixed and immunostained after 4 days. The circled well was intentionally left uninfected as a control for level of background immunostaining. With this method, the titer of this virus stock was determined to be 1.87×10^6 tissue-culture infectious dose, 50% endpoint units (TCID_{50})/mL. (**C**) The sample output of an Excel spreadsheet, available from the author, for calculating HCV titers by the endpoint dilution method. The data correspond to the plate shown in panel (B)

ment *(14)*. An alternative method is to detect HCV-infected cells by means of antibodies specific for the virus. Given that HCV is efficiently released into the cell-culture supernatant *(7, 15)*, accurately measuring the infectivity of the input virus requires that these assays be designed to limit the spread of the virus within the culture. Focus-forming assays therefore require the use of an overlay such as agarose or methylcellulose to avoid overestimation of the viral titer. An example focus-forming assay, made with an agarose overlay, is shown in **Fig. 24.1A**. Counting foci can be difficult, however, for HCV variants that produce minute or irregular foci. We therefore favor the use of the endpoint dilution assay, which is simple, inexpensive, and reproducible and does not require the use of an overlay for accuracy.

2. Materials

2.1. Cell Culture

1. Huh-7 human hepatoma cells or the highly permissive Huh-7.5 subline *(3)*.
2. Complete growth medium: Dulbecco's modified Eagle's medium supplemented with 10% heat-inactivated fetal calf serum and 1 mM nonessential amino acids (all from Invitrogen, Carlsbad, CA).
3. Solution of 0.05% (w:v) trypsin and 1 mM EDTA (Invitrogen).
4. Sterile 70 µm nylon cell strainers (BD Biosciences, Bedford, MA).
5. Poly-L-lysine hydrobromide, molecular weight 30,000–70,000 (catalog # P2636, Sigma, St. Louis, MO) dissolved to 100 µg/mL in Milli-Q water and filter-sterilized (0.2 µm pore size).
6. Filter-sterilized (0.2 µm pore size) Milli-Q water for washing excess poly-L-lysine from plates.
7. (Optional) Heparin, sodium salt (1000 United States Pharmacopeia units/mL).
8. 96-well plates (Corning, Corning, NY).
9. Multichannel pipettor, multichannel aspirator, sterile pipette tips, and sterile pipetting basins.

2.2. Staining

1. Phosphate-buffered saline (PBS): 137 mM NaCl, 2.7 mM KCl, 10 mM sodium phosphate dibasic heptahydrate, and 2 mM potassium phosphate monobasic.
2. 30% H_2O_2.
3. PBS-T: PBS containing 0.1% (v:v) Tween-20.
4. Blocking buffer: PBS-T containing 1% (w:v) bovine serum albumin (Fraction V; Fisher Scientific, Morris Plains, NJ),

0.2% (w:v) dried skim milk, and 0.02% sodium azide. This solution should be filtered ($0.2\,\mu$m pore size) and can be stored at 4°C for several months.

5. Cold methanol (store at −20°C).
6. HCV NS5A-specific monoclonal antibody 9E10 *(7)* or other HCV-reactive antibody.
7. ImmPress anti-mouse horseradish peroxidase secondary detection kit (Vector Labs, Burlingame, CA) (see **Note 1**).
8. Enhanced diaminobenzidine (DAB+) substrate kit (catalog #K3468; Dako USA, Carpinteria, CA) (see **Note 1**).
9. (Optional) Gill 2 hematoxylin (Richard-Allan Scientific, Kalamazoo, MI).

3. Methods

3.1. Growth and Passage of HCV in Cell Culture

1. Maintain Huh-7.5 cells as subconfluent, adherent monolayers in complete growth media at 37°C and 5% CO_2. Split cells 1:3 every third day. To help prevent clumping, pass cells routinely through sterile cell strainers ($70\,\mu$m pore size) during cell passage.
2. Grow HCV in Huh-7.5 cells either after RNA transfection or by passage at a low multiplicity of infection ($\leq 0.1\,\text{TCID}_{50}$/ml) in naïve cells. The kinetics of virus release vary greatly from one HCV isolate to another. For the chimeric genotype 2a FL-J6/JFH and derivatives thereof, peak infectivity is reached in 2–4 days *(7, 16)*.
3. To propagate virus in culture from the serum and plasma of animals infected with cell culture–derived HCV *(10)*, add animal serum directly to complete cell-culture medium and allow infection of Huh-7.5 cells for 4 h. Remove the inoculum, wash the cells with sterile PBS, and add fresh growth medium. To recover infectious virus from animal plasma, add heparin to 1 U/mL before adding complete growth medium to prevent clotting.
4. Harvest HCV by collecting the viral culture media, clarified by centrifugation (2000g for 10 min) and filter-sterilized ($0.2\,\mu$m pore size). Store cell culture–produced HCV for several weeks at 4°C or store aliquot at −80°C.

3.2. Measuring HCV Infectivity by Limiting Dilution

1. The day before the experiment, seed 6.4×10^3 Huh-7 cells/well in a 96-well plate and incubate them at 37°C in 5% CO_2. To help prevent cell loss during fixation and processing, precoat the plate with poly-L-lysine for 10 min,

aspirate the poly-L-lysine, and wash the wells once with sterile water.

2. Under appropriate biosafety containment, prepare serial dilutions (1 or 2 mL each) of the HCV stock to be titered. The range of dilutions to use will depend on the anticipated titer. For example, 10-fold dilutions covering 10^{-1} to 10^{-6} are recommended for measuring HCV grown in cell culture (up to $1 \times 10^{6} TCID_{50}/mL$). A series of two- or threefold dilutions is recommended for measuring relatively low levels of infectivity present in infected animal tissues. Prepare dilutions in complete growth medium and use a small amount of heparin (1 U/mL) in the diluent to prevent clotting if animal plasma is to be tested, but note that this level of heparin decreases cell culture–produced HCV titers by around two- to threefold (10).

3. Gently aspirate the medium from the cells and replace it with the HCV dilutions to be tested. Adding 100 μL to each well in an eight-well column allows testing of two viruses over a range of six dilutions on a single plate, but other arrangements may be more useful for titering multiple viruses over a narrower range of dilutions. In the example provided in **Fig. 24.1B**, a single virus was titered by testing of eight dilutions in replicates of 12. Regardless of the format chosen, include at least one uninfected well on each plate as a negative control.

4. Return the plate to the incubator and incubate for 4–5 days at 37°C in 5% CO_2. During this time the cells will become confluent, and cytopathic effects may be visible at the lowest virus dilutions.

3.3. Fixing

1. Gently aspirate the media and wash the cells with PBS, 100 μL/well.
2. Aspirate the PBS, add 100 μL/well cold (−20°C) methanol, and fix the cells for at least 10 min at −20°C.
3. Remove the methanol and allow the wells to dry fully within the tissue-culture hood.
4. The plates can now be safely worked with on the benchtop.

3.4. Immunostaining

1. Wash the cells twice with PBS, passing the plate under a gentle stream of PBS, rather than using a forced stream from a squirt bottle, which can tear the cell monolayer. For convenience, store PBS near a sink in a 20-L carboy attached to a bottom-draining Bobbitt valve.
2. Quench endogenous peroxidase by filling each well with 100 μL of PBS containing 3% H_2O_2. Incubate for 5 min at room temperature.

3. Wash the plate twice with PBS and fill each well with 40–100 µL of blocking buffer.

4. Place the plate on a rocker and block for 30 min at room temperature or overnight at 4°C.

5. Replace the blocking buffer with PBS-T containing 60 ng/mL anti-NS5A monoclonal antibody 9E10. Incubate with gentle rocking at room temperature for 1 h.

6. Wash twice with PBS and once with PBS-T.

7. Add 40 µL/well ImmPress anti-mouse-HRP, diluted with an equal volume of PBS-T. Incubate with gentle rocking at room temperature for 30 min.

8. Wash twice with PBS and once with PBS-T.

9. Add 40 µL/well freshly prepared DAB+. Incubate at room temperature for 1–10 min, depending on the desired color intensity. Note that shorter incubation times help to minimize background staining.

10. Remove DAB+ solution and dispose it of according to the chemical safety guidelines of your institution. Note that DAB is toxic and a potential carcinogen.

11. Wash cells twice with PBS and once with distilled water.

12. (Optional) Counterstain cell nuclei for a few minutes with hematoxylin and wash them with distilled water.

13. Air-dried plates can be stored indefinitely, but for better cell morphology, fill each well with 100 µL PBS + 0.03% sodium azide, replace the cover, and seal the edges with parafilm.

3.5. Calculate Endpoint Dilution

1. Using a microscope, examine each well and score it as positive if it contains one or more HCV-infected cells or negative if it lacks HCV-infected cells. Examine uninfected wells to determine the level of background staining. Although background is usually very low, nonspecific DAB precipitates will occasionally form, which can be easily discerned from the intense cytoplasmic staining of HCV-infected cells.

2. Calculate the dilution that will infect 50% of the wells using the method of Reed and Muench (17) (see Note 2). An Excel spreadsheet for calculating HCV TCID$_{50}$ and TCID$_{50}$/mL values can be downloaded from http://www.med.yale.edu/micropath/fac_lindenbach.html. An example calculation is shown in Fig. 24.1C. First, for each dilution, calculate the cumulative number of infected and uninfected wells by summing them from the highest to lowest or lowest to highest dilution, respectively. Then, for the dilutions closest to the 50% endpoint, calculate the proportional distance as follows: ((%positive wells above 50%) − 50) divided by ((% positive wells above 50%) − (% positive wells below 50%)). The proportional distance provides an estimate of the fractional dilution, between the two flanking dilutions, in which half of the wells would be

infected. Then calculate the tissue-culture infectious dose 50% endpoint ($TCID_{50}$) on the basis of the dilution whose proportional distance is greater than 0 and equal to or less than 1: $10^{\wedge}((\log_{10} \text{ dilution}) + (\log_{10} \text{ dilution factor} \times \text{proportional distance}))$. To determine the concentration of HCV infectivity, calculate the $TCID_{50}$/mL from the $TCID_{50}$ and the amount of inoculum used per well.

4. Notes

1. We have tested various peroxidase secondary detection reagents and DAB substrates with this assay. Although most of these reagents have produced good results, the Vector Imm-Press HRP and Dako DAB+ provide particularly high signal with low background.

2. The statistical methods of Spearman and Karber *(18, 19)* have also proven to be useful in determining the $TCID_{50}$ in HCV endpoint dilution assays *(16)*.

References

1. Bartenschlager, R., and Lohmann, V. (2000) Replication of hepatitis C virus. *J. Gen. Virol.* **81**, 1631–1648.

2. Lohmann, V., Korner, F., Koch, J. O., Herian, U., Theilmann, L., and Bartenschlager, R. (1999) Replication of subgenomic hepatitis C virus RNAs in a hepatoma cell line. *Science* **285**, 110–113.

3. Blight, K. J., McKeating, J. A., and Rice, C. M. (2002) Highly permissive cell lines for subgenomic and genomic hepatitis C virus RNA replication. *J. Virol.* **76**, 13001–13014.

4. Pietschmann, T., Lohmann, V., Kaul, A., Krieger, N., Rinck, G., Rutter, G., et al. (2002) Persistent and transient replication of full-length hepatitis C virus genomes in cell culture. *J. Virol.* **76**, 4008–4021.

5. Bukh, J., Pietschmann, T., Lohmann, V., Krieger, N., Faulk, K., Engle, R. E., et al. (2002) Mutations that permit efficient replication of hepatitis C virus RNA in Huh-7 cells prevent productive replication in chimpanzees. *Proc. Natl. Acad. Sci. USA* **99**, 14416–14421.

6. Kato, T., Date, T., Miyamoto, M., Furusaka, A., Tokushige, K., Mizokami, M., et al. (2003) Efficient replication of the genotype 2a hepatitis C virus subgenomic replicon. *Gastroenterology* **125**, 1808–1817.

7. Lindenbach, B. D., Evans, M. J., Syder, A. J., Wolk, B., Tellinghuisen, T. L., Liu, C. C., et al. (2005) Complete replication of hepatitis C virus in cell culture. *Science* **309**, 623–626.

8. Wakita, T., Pietschmann, T., Kato, T., Date, T., Miyamoto, M., Zhao, Z., et al. (2005) Production of infectious hepatitis C virus in tissue culture from a cloned viral genome. *Nat. Med.* **11**, 791–796.

9. Zhong, J., Gastaminza, P., Cheng, G., Kapadia, S., Kato, T., Burton, D. R., et al. (2005) Robust hepatitis C virus infection in vitro. *Proc. Natl. Acad. Sci. USA* **102**, 9294–9299.

10. Lindenbach, B. D., Meuleman, P., Ploss, A., Vanwolleghem, T., Syder, A. J., McKeating, J. A., et al. (2006) Cell culture-grown hepatitis C virus is infectious in vivo and can be recultured in vitro. *Proc. Natl. Acad. Sci. USA* **103**, 3805–3809.

11. Yi, M., Villanueva, R. A., Thomas, D. L., Wakita, T., and Lemon, S. M. (2006) Production of infectious genotype 1a hepatitis C virus (Hutchinson strain) in cultured human hepatoma cells. *Proc. Natl. Acad. Sci. USA* **103**, 2310–2315.

12. Condit, R. C. (2007). Principles of Virology. In *Fields Virology*, Knipe, D. M., and Howley, P. M., eds. (Philadelphia, Lippincott-Raven Publishers), pp. 25–57.

13. Zhong, J., Gastaminza, P., Chung, J., Stamataki, Z., Isogawa, M., Cheng, G., et al. (2006) Persistent hepatitis C virus infection in vitro: coevolution of virus and host. *J. Virol.* **80**, 11082–11093.

14. Sekine-Osajima, Y., Sakamoto, N., Mishima, K., Nakagawa, M., Itsui, Y., Tasaka, M.,

et al. (2008) Development of plaque assays for hepatitis C virus-JFH1 strain and isolation of mutants with enhanced cytopathogenicity and replication capacity. *Virology* **371**, 71–85.

15. Gastaminza, P., Kapadia, S. B., and Chisari, F. V. (2006) Differential biophysical properties of infectious intracellular and secreted hepatitis C virus particles. *J. Virol.* **80**, 11074–11081.

16. Pietschmann, T., Kaul, A., Koutsoudakis, G., Shavinskaya, A., Kallis, S., Steinmann, E., et al. (2006) Construction and characterization of infectious intragenotypic and intergenotypic hepatitis C virus chimeras. *Proc. Natl. Acad. Sci. USA* **103**, 7408–7413.

17. Reed, L. J., and Muench, H. (1938) A simple method of estimating fifty percent end points. *Am. J. Hyg.* **27**, 493–497.

18. Karber, G. (1931) Beitrag zur kollektiven Behandlung pharmakologischer Reihenversuche. *Naunyn Schmiedebergs Arch. Exp. Pathol. Pharmakol.* **162**, 480–482.

19. Spearman, C. (1908) The method of right and wrong cases (constant stimuli) without Gauss's formulae. *Br. J. Psychol.* **2**, 227–242.

Chapter 25

Genotype 1a HCV (H77S) Infection System

MinKyung Yi and Stanley M. Lemon

Abstract

HCV is a small RNA virus belonging to the genus *Hepacivirus* within the virus family Flaviviridae. Infection with HCV often leads to chronic liver diseases including chronic hepatitis, liver cirrhosis, and hepatocellular carcinoma. Current therapy, based on the use of interferon-α (IFN-α) in combination with ribavirin, results in limited success, especially in patients infected with the most prevalent genotype 1 viruses. Better therapies are needed, but the inability to propagate HCV in cell culture hampers antiviral drug-discovery efforts. Recently, fully permissive cell-culture systems have been developed that use viral RNA derived from the genotype 2a JFH-1 strain of HCV. Although these systems mark a significant breakthrough for HCV research, the parallel development of a tractable genotype 1a infection system (H77S virus) has provided significant advantages in assessing genotype 1-specific interventions, given the highly heterogeneous nature of HCV. H77S RNA contains five cell culture–adaptive mutations that are placed throughout the nonstructural protein-coding segment of the genome and render the RNA capable of robust replication in human hepatoma (Huh-7) cells. Although significantly less efficient than JFH-1 RNA, H77S RNA produces moderate titers of cell culture–infectious virus when transfected into Huh-7 cells.

Key words: Hepatitis C virus, genotype 1a, infection system, H77S, adaptive mutation.

1. Introduction

HCV is an enveloped, positive-strand RNA virus with a genome approximately 9.6 kb in length (*1*). Its genome encodes one large open reading frame that is co- and posttranslationally processed by cellular and viral proteases into at least 10 different proteins (*2–4*). HCV is the sole member of the genus *Hepacivirus,* within the family Flaviviridae, which also includes flaviviruses and pestiviruses. The genus demonstrates substantial genomic heterogeneity, having six major genotypes with distinct clinical and

Hengli Tang (ed.), *Hepatitis C: Methods and Protocols, Second Edition, vol. 510*
© 2009 Humana Press, a part of Springer Science+Business Media
DOI 10.1007/978-1-59745-394-3_25 Springerprotocols.com

probably antigenic characteristics. Most serious infections with this virus are associated with genotype 1, which is more prevalent and less amenable to interferon-based therapies than other genotypes. Unlike those of many members of this virus family, genome-length RNAs derived from nearly all wild-type HCV (except JFH-1 and HCV-N) are incapable of significant replication in any type of cell culture in the absence of specific adaptive mutations *(5,6)*. This situation has hampered efforts to study this virus, including antiviral drug-discovery research. The genotype 1a Hutchinson strain of HCV (H77) is no exception, and neither H77 virus nor viral RNA is capable of significant levels of replication in cultured cells, but through a tedious, iterative process of adaptation of subgenomic H77 replicons and molecular cloning of the resulting replicon RNAs, we identified a series of five cell culture–adaptive mutations that render the genotype 1a H77 RNA capable of relatively high-level replication in transfected Huh-7 cells *(7)*. We subsequently determined that replication of genome-length H77 RNA containing these adaptive mutations (now referred to as "H77S", see **Fig. 25.1**) results in the production of cell culture–infectious virus that is released into supernatant culture fluids *(8)*.

The discovery that the HCV genotype 2a clone JFH-1 could undergo its entire life cycle in cultured cells has opened a new era in HCV research *(9–11)*. JFH-1 RNA replicates at very high levels in Huh-7 cells without adaptive mutations, as mentioned above *(5)*, but upon transfection of wild-type JFH-1 RNA, relatively low amounts of virus are produced and released into culture supernatants. With continuous passage of cells transfected with JFH-1 RNA, the virus titer increases coincident with the emergence of adaptive mutations in the genome *(12,13)*. For unknown reasons, H77S produces infectious virus at a much lower level than JFH-1. In addition, we have not observed increased virus yields upon passaging of H77S RNA-transfected cells, as occurs with JFH-1 virus *(8,14)*. One potential reason is that H77S RNA replicates less efficiently than JFH-1 RNA *(8)*. A strict comparison of the replication efficiencies of the two RNAs is difficult, as replication of the JFH-1 RNA is considerably more toxic to the cell. Core

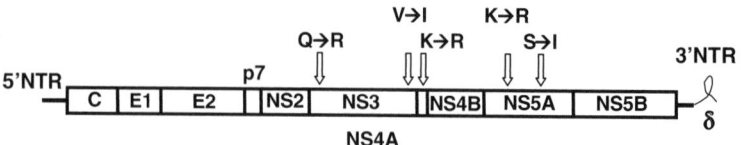

Fig. 25.1. Organization of the H77S insert in pH77S with five tissue-culture-adaptive mutations indicated on the top by arrows. "δ" indicates the hepatitis delta ribozyme sequence introduced downstream of the 3′ terminus of the HCV sequence so as to produce an authentic 3′ end of the RNA after RNA transcription.

protein expression is much greater soon after transfection in JFH-1 than in H77S-transfected cells, suggesting that a substantially higher abundance of the viral RNA, and probably all of the viral proteins as well, is produced in cells transfected with JFH-1 RNA.

Until recently, we have used Huh-7.5 cells to produce H77S virus. This cell line, which Blight et al. *(15)* isolated by curing a Huh-7-derived replicon cell line harboring a genotype 1b Con1 replicon with IFN-α, is deficient in RIG-I signaling *(16)*. Huh-7.5 cells support a higher level of H77S RNA replication and produce a greater infectious virus yield after RNA transfection than do parental Huh-7 cells *(8)*. Lately, however, we have switched to use of a clonal Huh-7 cell line derived from a genome-length H77 replicon cell line that was cured with IFN-α. These cells (8C1 cells) are progeny of our En5-3 Huh-7 subline, which is stably transformed with the gene encoding secreted alkaline phosphatase under control of the human immunodeficiency virus (HIV) long terminal repeat promoter, allowing induction of this phosphatase by HIV tat protein expressed from replicating viral RNA *(17)*. 8C1 cells support efficient H77S RNA replication and show greater infectious virus production than Huh-7.5 cells (Yi and Lemon, unpublished results). These results support the notion that infectious H77S production depends on very efficient RNA replication. Also, H77S replication and virus production depend on conditions that have been suggested to be important for genotype 1b HCV RNA replication *(18)*.

H77S infectivity has been measured as number of foci of infected cells, which are recognized as clusters of cells that are immunostained with HCV-specific antibody after inoculation of naive cells with clarified culture supernatants derived from H77S-transfected cells (**Fig. 25.2**). The number of foci present allows calculation of the starting titer of infectious virus in terms of focus-forming units (FFU). Using chimeric viruses derived from H77S and JFH1, we have demonstrated a strict linear relationship between the dilution of a virus inoculum and the number of

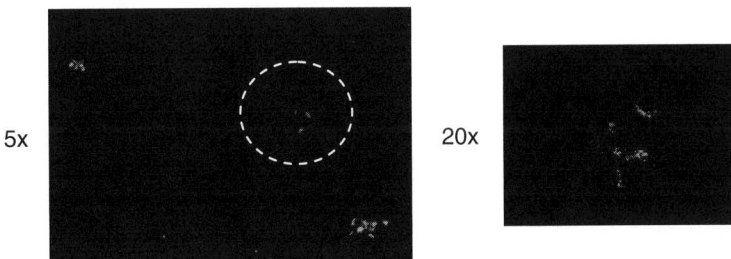

Fig. 25.2. Microscopic images of immunostained cells containing HCV core antigen after infection of Huh-7.5 cells with H77S virus. 5× and 20× indicate the objective magnification. The broken circle shown in the 5× panel on the left indicates a single infection focus. Three foci are visible in this image.

infectious foci identified by immunofluorescent staining (unpublished data). Overlays are therefore not necessary for accurate quantitative assessment of HCV infectivity by this method, so the virus may spread largely in a cell-to-cell fashion in these cultures. To determine the titer of H77S virus, we routinely use a clonal Huh-7-derived cell line, S5C-CD81-3, which has been selected for high CD81 expression. S5C-CD81-3 cells are equivalent to Huh-7.5 cells for permissiveness for H77S, generating similar estimates of FFU, but support a higher level of H77S protein expression than do Huh-7.5 cells, resulting in improved detection of infectious foci. We developed this cell line by first curing an En5-3-derived replicon cell line harboring a genome-length HCV replicon with IFN-α and then transforming it with CD81 cDNA.

Our recent data indicate that the production of infectious virus after transfection of H77S RNA can be improved by the addition of a single mutation within the polyprotein coding region (H77S.2). This mutation was identified during recent experiments in which we characterized compensatory mutations that improve the yield of H77-JFH1 chimeras (14). Combining the use of H77S.2 RNA and the specialized producer (8C1) and indicator (S5C-CD81-3) cell lines described above now allows us to obtain virus yields routinely in the range of 10^3 FFU/mL. With the use of centrifugal concentrators, we have prepared virus preparations with titers as high as 10^4–10^5 FFU/mL.

2. Materials

2.1. DNA Digestion and In Vitro RNA Transcription

1. pH77S (or pH77S.2) and pH77S/aag plasmids (*see* **Note 1**).
2. Restriction enzyme *Xba*I and reaction buffer 2 (New England BioLab).
3. 1% agarose gel with 0.5 μg/mL of ethidium bromide (EtBr). EtBr is a known mutagen and should be handled as a hazardous chemical; wear gloves while handling it.
4. Phenol-chloroform-isoamylalcohol 25:24:1 (vol:vol:vol), molecular biology grade (Sigma). This material is toxic and should be handled with care as a hazardous chemical.
5. Chloroform, molecular biology grade. This material is harmful and should be handled with care as a hazardous chemical.
6. 3 M sodium acetate.
7. Ethanol (molecular biology grade, 100%, 70%).
8. Megascript T7 kit (Ambion).
9. RNeasy Mini kit (Qiagene).

2.2. Electroporation

1. Huh-7-derived cell lines (such as Huh-7.5, 8C1, S5C-CD81-3).
2. PBS (Cellgro, MT21031Cv).

3. Trypsin-EDTA (Gibco, 25200056). Store at 4°C.

4. Electroporation cuvette (0.2 cm gap width, Bio-Rad).

5. Gene Pulser II electroporation apparatus (Bio-Rad).

6. Complete medium: Dulbecco's modified Eagle's medium (DMEM; high-glucose; Gibco, 11965-118) supplemented with 10% fetal bovine serum (Gibco, 16000044), 1× penicillin/streptomycin (Gibco, 15140122), and 5 mM HEPES (Cellgro, MT25060Cl). Store at 4°C.

2.3. Virus Harvest and Titration

1. Polyethersulfone (PES) membrane vacuum filtration system, $0.22\,\mu m$ (TPP).

2. Centricon Plus-20 (UFC2BHK08), Centricon Plus-70 (UFC710008) (Millipore).

3. Eight-well tissue-culture chamber slide (Nunc, Permanox slide).

4. Methanol:acetone 1:1 (vol:vol) solution.

5. Mouse monoclonal anti-core antibody (Affinity Bioreagent, C7-50).

6. Fluorescein isothiocyanate-conjugated goat-anti-mouse IgG antibody (Southern Biotech, 1032-02) or Alexa 488-conjugated goat-anti-mouse IgG antibody (Invitrogen, A11029).

7. Bovine serum albumin (7.5% in DPBS, Sigma).

8. Hoechst 33258, 10 mg/mL (Bisbenzimide H 33258, Sigma).

9. VECTASHIELD® mounting fluid (Vector Labs).

3. Methods

3.1. DNA Digestion and In Vitro Transcription

3.1.1. DNA Digestion

1. Prepare high-quality pH77S and replication-incompetent (negative control) pH77S/aag plasmid DNA (*see* **Note 1**).

2. Digest plasmid DNA with *Xba*I: First, mix $2\,\mu g$ of plasmid DNA, $2.5\,\mu L$ of NEBuffer 2 (10×), and 5 units of *Xba*I, and add nuclease-free water to $25\,\mu L$.

3. Incubate the reaction at 37°C in a water bath for 3 h.

4. Determine the quality of digested DNA by examining a small aliquot of the reaction mixture in a 1% agarose gel containing ethidium bromide ($0.5\,\mu g/mL$).

5. Add TE buffer to the digested plasmid to make $100\,\mu L$ and mix with $100\,\mu L$ of phenol-chloroform-isoamylalcohol, shaking vigorously before centrifuging at 16,000g for 15 min at room temperature.

6. Remove the upper aqueous phase to a new tube and add $100\,\mu L$ of chloroform, and then mix and centrifuge as described above.

7. Remove the upper aqueous phase, place it in a new tube, and add 1/10th volume 3 M sodium acetate and 2.5 volumes 100% ethanol.

8. Chill the solution at –80°C for 15 min or –20°C overnight before centrifugation at 16,000g for 15 min at 4°C.

9. Remove the supernatant, wash the pellet once with 500 μL of 70% ethanol, and centrifuge at 16,000g for 15 min at 4°C.

10. Remove the supernatant and air dry the pellet.

11. Resuspend the pellet with 10 μL of nuclease-free water. The DNA solution can be stored at –20°C until further use. No further treatment of H77S DNA is necessary before transcription (*see* **Note 2**).

3.1.2. In Vitro Transcription and RNA Purification

3.1.2.1. In Vitro Transcription

1. Transcribe HCV RNA using the T7-Megascript kit (Ambion) according to the protocol provided by the manufacturer. Briefly, add 2 μL each of ATP, CTP, GTP, UTP, and 10× reaction buffer and enzyme mix per 1 μg of digested DNA. Add nuclease-free water to make a 20 μL reaction mixture.

2. Incubate the reaction for 4–5 h in a 37°C water bath.

3. Add 1 μL of DNase (included in the kit) and incubate for 15 min in a 37°C water bath.

4. Add 15 μL of stop solution (included in the kit). This solution can be stored at –80°C until further use.

3.1.2.2. RNA Purification

1. Perform RNA purification according to the protocol provided with the RNeasy kit (Qiagene). To purify the RNA, first add nuclease-free water to the RNA solution above to make 100 μL.

2. Add 350 μL of RLT buffer and mix.

3. Add 250 μL of 100% ethanol and mix.

4. Apply the solution to a spin column (a kit component) and centrifuge at 10,000g for 30 s at room temperature.

5. After transferring the column to a new collection tube, wash the column twice with 500 μL RPE buffer.

6. Transfer the column to a new collection tube and centrifuge at 10,000g for 30 s at room temperature.

7. Elute the RNA with 50 μL nuclease-free water and determine the RNA concentration using a spectrometer (*see* **Note 3**). RNA can be stored at –80°C.

3.2. Electroporation

1. Prepare Huh-7-derived cells for electroporation (*see* **Note 4**). Wash the cells twice with PBS and remove from the culture vessel by treatment with trypsin.

2. Wash the cells with ice-cold PBS twice.

3. Resuspend the cells in cold PBS at a density of 1×10^7 cells/mL.

4. Mix 10 μg of RNA with 500 μL of cell suspension and transfer the mix to a 0.2 cm gap-width electroporation cuvette.

5. Place the cuvette in the Gene Pulser II apparatus and pulse twice at 1.4 kV and 25 μF.

6. Transfer the cells to a 25 cm² flask containing 10 mL of medium and incubate at 37°C in a 5% CO_2 environment (*see* **Note 5**).

7. Passage the cells, splitting 1:4 at day 4, and 1:3 every third day thereafter.

8. Transfection efficiency can be determined at day 2 after transfection by seeding of cells in an eight-well tissue-culture-chamber slide at 2×10^4 cells per well at the time of electroporation and detecting HCV antigen by indirect immunofluorescence as described below (see "Methods", **Section 3.3.2**).

3.3. Virus Harvest and Titration

3.3.1. Virus Harvest

1. Collect the cell-culture supernatants at time points later than 48 h after electroporation or 48 h after splitting at successive passages thereafter (*see* **Note 6**).

2. Remove the cell debris by low-speed centrifugation (2000g for 5 min).

3. Filter virus through a 200 μm PES filter if you wish (*see* **Note 7**).

4. To increase the titer of virus, concentrate supernatant using a Centricon Plus-20 (up to 15 mL) or Centricon Plus-70 (up to 50 mL) centrifugal concentrator. Centrifuge at 2000g for 25 min at 4°C (*see* **Note 8**).

3.3.2. Virus Titration

1. Seed an eight-well chamber slide with Huh-7.5 or S5C-500-3 cells at a density of 1×10^5 cells per well 24 h before virus inoculation (*see* **Note 9**).

2. Make serial dilutions of the inoculum in complete medium (2- to 10-fold serial dilutions, depending on the titer of virus).

3. Remove medium from the chamber slide and add 100 μL of the serially diluted inoculum to each well of the chamber slide.

4. Incubate the slide at 37°C in a 5% CO_2 environment.

5. Remove the inoculum at 4 h and add 200 μL of fresh medium (*see* **Note 10**).

6. Incubate slide for 4 days at 37°C in 5% CO_2 (*see* **Note 11**).

7. Aspirate medium from the well, wash cells twice with PBS, and fix the cells with methanol:acetone (1:1) solution for 9 min at room temperature.

8. Remove methanol:acetone solution and wash cells twice with PBS.

9. Dilute primary C7-50 mouse anti-core antibody 1:300 in PBS with 3% BSA, and add 100 μL to each well.

10. Incubate for 2 h at room temperature.

11. After washing twice with PBS, apply 100 μL of FITC- or Alexa 488-conjugated secondary antibody diluted 1:100 in PBS with 3% BSA.

12. Incubate for 1 h at room temperature.

13. Wash cells twice with PBS and apply Hoechst 33258 (diluted 1:1000 in PBS) for 5 min.

14. Wash cells twice with PBS and mount a coverslip on the slide after applying a small amount of VECTASHIELD®.

15. Examine slides under an epifluorescence microscope and count foci of infection, which appear as clusters of antigen-positive cells (typically 2–10 cells per cluster, *see* **Fig. 25.2**). Titers are calculated from the number of foci of infection produced per 1 mL of the starting inoculum (*see* **Note 12**).

4. Notes

1. pH77S/aag contains a substitution destroying the active-site "Gly-Asp-Asp" motif in the NS5B RNA-dependent RNA polymerase coding region of the genome.

2. Because the pH77S plasmid contains a hepatitis delta virus ribozyme sequence immediately downstream of the HCV sequence, self-cleavage of the transcribed RNA produces the exact 3′ HCV terminus (**Fig. 25.1**).

3. The quality of RNA must be checked after electrophoresis on a denaturing agarose gel and ethidium bromide staining to ensure that it is intact and of the correct size.

4. For electroporation, use Huh-7 (or Huh-7-derived subline) cells that are growing in the logarithmic phase and are ~90% confluent.

5. Change the medium every 24 h to stimulate replication of H77S RNA.

6. Huh-7.5 cells secrete low levels of infectious virus for up to 21 days after transfection of H77S RNA *(14)*. Virus can be stored at −80°C if supplemented with 20% serum. Under these conditions, loss of infectivity is minimal after multiple freeze-thaw cycles.

7. Wash the filter first with medium to remove residual chemical on the filter.

8. The medium can be changed to serum-free medium 24 h before virus harvest to facilitate concentration of virus in a centrifugal concentrator.

9. The deduced titer of H77S will vary depending on cell density: Increasing the cell density more than is described does

not increase the titer but does lessen the core antigen signal intensity.

10. The H77S inoculum can be left on the cells for 4–24 h without causing a major difference in the number of FFU.

11. HCV antigen can be detected by day 3 after virus inoculation, but the immunofluorescence signal intensity is significantly stronger at day 4.

12. To assess the infectious titer of an inoculum accurately, use FFU counts generated by dilutions of the inoculum that produce neither too many (overlapping) nor too few (random variation) infectious foci. FFU counts derived from two replicate wells showing between 20 and 60 foci of infection are ideal.

Acknowledgments

We thank Dr. Robert Purcell, National Institutes of Health, for the gift of the H77c clone and Dr. Charles Rice, Rockefeller University, for the gift of Huh-7.5 cells. We also thank Jeremy Yates and Yinghong Ma for their technical assistance. This work was supported in part by National Institute of Allergy and Infectious Diseases grants U19-AI40035, R21-AI063451, and N01-AI25488.

References

1. Lemon, S. M., Walker, C., Alter, M. J., and Yi, M. (2007) Hepatitis C viruses, in *Fields Virology* (Knipe, D., Howley, P., Griffin, D. E., Martin, M. A., Lamb, R. A., Roizman, B., et al, eds.), Lippincott Williams & Wilkins, Philadelphia, pp. 1253–1304.

2. Shimotohno, K., Tanji, Y., Hirowatari, Y., Komoda, Y., Kato, N., and Hijikata, M. (1995) Processing of the hepatitis C virus precursor protein. *J. Hepatol.* **22,** 87–92.

3. Bartenschlager, R. (1999) The NS3/4A proteinase of the hepatitis C virus: unravelling structure and function of an unusual enzyme and a prime target for antiviral therapy. *J. Viral Hepat.* **6,** 165–181.

4. Reed, K. E., Grakoui, A., and Rice, C. M. (1995) Hepatitis C virus-encoded NS2-3 protease: cleavage-site mutagenesis and requirements for bimolecular cleavage. *J. Virol.* **69,** 4127–4136.

5. Kato, T., Date, T., Miyamoto, M., Furusaka, A., Tokushige, K., Mizokami, M., et al. (2003) Efficient replication of the genotype 2a hepatitis C virus subgenomic replicon. *Gastroenterology* **125,** 1808–1817.

6. Ikeda, M., Yi, M., Li, K., and Lemon, S. M. (2002) Selectable subgenomic and genome-length dicistronic RNAs derived from an infectious molecular clone of the HCV-N strain of hepatitis C virus replicate efficiently in cultured Huh7 cells. *J. Virol.* **76,** 2997–3006.

7. Yi, M. and Lemon, S. M. (2004) Adaptive mutations producing efficient replication of genotype 1a hepatitis C virus RNA in normal Huh7 cells. *J. Virol.* **78,** 7904–7915.

8. Yi, M., Villanueva, R. A., Thomas, D. L., Wakita, T., and Lemon, S. M. (2006) Production of infectious genotype 1a hepatitis C virus (Hutchinson strain) in cultured human hepatoma cells. *Proc. Natl. Acad. Sci. USA* **103,** 2310–2315.

9. Zhong, J., Gastaminza, P., Cheng, G., Kapadia, S., Kato, T., Burton, D. R., et al. (2005) Robust hepatitis C virus infection *in vitro. Proc. Natl. Acad. Sci. USA* **102,** 9294–9299.

10. Lindenbach, B. D., Evans, M. J., Syder, A. J., Wolk, B., Tellinghuisen, T. L., Liu, C. C., et al. (2005) Complete replication of hepatitis C virus in cell culture. *Science* **309,** 623–626.

11. Wakita, T., Pietschmann, T., Kato, T., Date, T., Miyamoto, M., Zhao, Z., et al. (2005) Production of infectious hepatitis C virus in tissue culture from a cloned viral genome. *Nat. Med.* **11,** 791–796.

12. Zhong, J., Gastaminza, P., Chung, J., Stamataki, Z., Isogawa, M., Cheng, G., et al. (2006) Persistent hepatitis C virus infection *in vitro:* coevolution of virus and host. *J. Virol.* **80,** 11082–11093.

13. Kaul, A., Woerz, I., Meuleman, P., Leroux-Roels, G., and Bartenschlager, R. (2007) Cell culture adaptation of hepatitis C virus and *in vivo* viability of an adapted variant. *J. Virol.* **81,** 13168–13179.

14. Yi, M., Ma, Y., Yates, J., and Lemon, S. M. (2007) Compensatory mutations in E1, p7, NS2, and NS3 enhance yields of cell culture-infectious intergenotypic chimeric hepatitis C virus. *J. Virol.* **81,** 629–638.

15. Blight, K. J., McKeating, J. A., and Rice, C. M. (2002) Highly permissive cell lines for subgenomic and genomic hepatitis C virus RNA replication. *J. Virol.* **76,** 13001–13014.

16. Sumpter, R., Jr., Loo, M. Y., Foy, E., Li, K., Yoneyama, M., Fujita, T., et al. (2005) Regulating intracellular antiviral defense and permissiveness to hepatitis C virus RNA replication through a cellular RNA helicase, RIG-I. *J. Virol.* **79,** 2689–2699.

17. Yi, M., Bodola, F., and Lemon, S. M. (2002) Subgenomic hepatitis C virus replicons inducing expression of a secreted enzymatic reporter protein. *Virology* **304,** 197–210.

18. Nelson, H. B. and Tang, H. (2006) Effect of cell growth on hepatitis C virus (HCV) replication and a mechanism of cell confluence-based inhibition of HCV RNA and protein expression. *J. Virol.* **80,** 1181–1190.

Chapter 26

Full-Length Infectious HCV Chimeras

Thomas Pietschmann

Abstract

One hallmark of HCV is its pronounced genetic plasticity, caused by error-prone RNA replication, which probably contributes to its remarkable ability to establish chronic infections. On the basis of phylogenetic analyses, HCV variants are classified into six genotypes (GTs), each comprising a variable number of subtypes. Presumably, these genetic differences, which range from 33 to 35% at the nucleotide level among genotypes and from 22 to 25% between subtypes, are reflected by divergent biological properties of the respective isolates. The unprecedented replication efficiency of the JFH1 isolate (a GT2a strain derived from a Japanese patient with fulminant hepatitis) in transfected Huh-7 cells represents a characteristic feature intrinsic to this particular isolate and has very recently made possible the investigation of the complete viral replication cycle in cultured cells. To expand the scope of this novel HCV infection system, several groups have constructed chimeric HCV genomes comprising JFH1-derived replicase proteins and structural proteins from heterologous HCV strains. This chapter describes experimental procedures for evaluation of the properties of infectious full-length HCV chimeras.

Key words: Hepatitis C virus, chimera, reporter virus, reporter assay, Core ELISA, immunofluorescence, replication, infection.

1. Introduction

Analysis of the HCV life cycle was long hampered by poor replication of the virus in cultured cells. This impediment was recently overcome by the development of HCV infection systems based on two genomes with high replication efficiency: the JFH1 isolate and a highly cell culture–adapted variant of the H77 strain (GT1a) *(1–4)*. Although this achievement is important, because of the high degree of HCV variability, limitation to two isolates certainly restricts application of these novel systems.

Hengli Tang (ed.), *Hepatitis C: Methods and Protocols, Second Edition, vol. 510*
© 2009 Humana Press, a part of Springer Science+Business Media
DOI 10.1007/978-1-59745-394-3_26 Springerprotocols.com

To overcome this restraint, despite the generally poor replication of alternative HCV isolates in transfected cells, which prevents their direct use for infection studies in vitro, various groups have explored the possibility of combining JFH1-derived genome segments with portions from heterologous viral strains *(1, 5–8)*. The objective of this strategy is to create chimeras comprising the highly efficient JFH1-derived replication machinery as well as structural proteins from other HCV strains, thus permitting production of infectious viruses from alternative isolates and providing the opportunity for comparative studies of assembly, egress, and infection.

2. Materials

2.1. Cell Culture

1. DMEM complete: Dulbecco's modified Eagle's medium (with 4.5 g/L glucose, without pyruvate) supplemented with 2 mM l-glutamine, nonessential amino acids, 100 U of penicillin per mL, 100 μg of streptomycin per mL (all from Invitrogen, Karlsruhe, Germany), and 10% fetal calf serum (PAA Laboratories, Cölbe, Germany). Store at 4°C for no more than 6 weeks; reheat to 37°C before use.
2. Trypsin/EDTA buffer: 0.05% trypsin (v:v), 0.02% EDTA (v:v) in phosphate-buffered saline (Invitrogen).
3. Cell-culture plates and cell-culture dishes (BD Falcon, Heidelberg, Germany), 0.45 μm filters (Carl Roth, Karlsruhe, Germany).
4. Phosphate buffered saline (PBS): 8 mM Na_2HPO_4, 2 mM NaH_2PO_4, 140 mM NaCl, sterilized by autoclaving.

2.2. In Vitro Transcription

All reagents necessary for our in vitro transcription protocol are listed in **Chapter 11**, "HCV Replicons: Overview and Basic Protocols," **Section 2.2**, "*In vitro transcription, electroporation, and northern hybridization.*"

2.3. Electroporation

All reagents necessary for our electroporation procedure are listed in **Chapter 11**, "HCV Replicons: Overview and Basic Protocols," **Section 2.2**, "*In vitro transcription, electroporation, and northern hybridization.*"

2.4. Firefly-Luciferase Assay

1. Glycyl-glycine stock solution: 0.5 M glycyl-glycine (Sigma-Aldrich, Munich, Germany), pH 7.8 (KOH). Store at room temperature.
2. Luciferin stock solution: 1 mM luciferin (PJK Chemikalien, Kleinbittersdorf, Germany), 25 mM glycyl-glycine. Store in single-use aliquots at −80°C.
3. ATP stock solution: 100 mM ATP, pH 7.8 (KOH) (Sigma-Aldrich). Sterilize by filtration using 0.2 μM pore-size filters. Store in single-use aliquots at −20°C.

4. DTT stock solution: 1 M DTT in 10 mM sodium acetate, pH 5.2. Sterilize by filtration using 0.2 μM pore-size filters. Store in single-use aliquots at −20°C.

5. Luciferase lysis buffer: 25 mM glycyl-glycine, pH 7.8 (KOH), 15 mM $MgSO_4$, 4 mM EGTA, 1% (v:v) Triton-X100. Store at 4°C.

6. Luciferase assay buffer: 15 mM phosphate buffer, pH 7.8, 15 mM $MgSO_4$, 4 mM EGTA. Store at room temperature. Supplement with 2 mM ATP, 1 mM DTT immediately before use.

7. Luciferase substrate buffer: 200 μM luciferin, 25 mM glycyl-glycine, pH 7.8, (KOH). Prepare freshly immediately before use. Keep protected from light.

8. The described reagents and procedures are tailored for a Lumat LB 9507 apparatus (Berthold Technologies, Bad Wildbad, Germany) with measurement of 100 for all sample, added to 360 for all assay buffer, and 200 for all luciferase substrate buffer. Perform measurements for 20 s.

2.5. HCV Core-Specific Immunoassay

1. Components of the Track-C ELISA kit (Ortho Clinical Diagnostics, Neckargemünd, Germany).

2. Lysis buffer: PBS supplemented with 1% Triton-X100 (v:v) and with 1 mM phenylmethylsulfonyl fluoride (PMSF), 40 mg of leupeptin per mL and 0.001 U of aprotinin per mL (all protease inhibitors from Sigma).

2.6. Immunofluorescence Assay

1. PFA buffer: PBS supplemented with 4% paraformaldehyde (w:v). Dissolve 4 g paraformaldehyde (Sigma-Aldrich) in 80 mL warmed water (*caution*: paraformaldeyhde fumes are toxic; use a fume hood to prepare this buffer). Add a few drops of 1 N NaOH to aid dissolution. Supplement with 10 ml 10× PBS and adjust pH to 7.4 with HCl.

2. Staining buffer: PBS supplemented with 5% goat normal serum (v:v; Sigma-Aldrich).

3. 4′,6′-diamidino-2-phenylindole dihydrochloride (DAPI) (Molecular Probes, Karlsruhe, Germany), stored in aliquots at 4°C in the dark.

4. Fluoromount-G (Biozol Diagnostica, Eching, Germany), stored at 4°C.

3. Methods

Those viral proteins that are the major constituents of the virion are designated structural proteins, and evidently these are directly involved in assembly of progeny particles. In many cases, however, additional viral polypeptides take part in morphogenesis and may at least to some extent also be packaged into virions. In the case of HCV, little is known about virus assembly, the proteins involved, or the composition of the virus particles. Although both glycoproteins (E1 and E2) and Core are clearly essential components of the virion and are required for assembly, additional viral factors may also be involved. In line with this, it was recently noted that p7, a small membrane-associated ion channel protein *(9, 10)*, and the NS2 protease are essential for efficient assembly and release of infectious virions *(11, 12)*. Given that proper assembly probably relies on highly specific interactions among all participating viral proteins, construction of chimeric genomes should take into account this interplay and avoid, to the extent possible, introduction of genetic incompatibility between directly consorting viral factors.

To this end, we and others have used a prototypic configuration connecting JFH1 with heterologous HCV strains at the junction between NS2 and NS3 proteins (**Fig. 26.1**) *(1, 5–8)*. In this configuration, on the one hand, all proteins necessary for generating the membrane-bound replicase complex (consisting of NS3 to NS5B proteins) are derived from the JFH1 strain, sustaining highly efficient RNA replication. In addition, the 3′ and 5′ nontranslated regions carrying genetic elements requisite for initiation of translation and RNA replication are derived from this isolate. On the other hand, all structural proteins (Core, E1, E2), and in addition p7 and NS2, which are constituents of a precursor proteins with E2 (E2-p7, E2-p7-NS2) that may directly participate in virus assembly, are derived from the heterologous strain.

Although this approach maintains maximal genetic compatibility at least for all viral determinants known to be essential for RNA replication, one obvious drawback results from the creation of a chimeric NS2–NS3 protease. Depending on the genetic distance between viral strains linked at this junction, both the efficiency and the kinetics of the cleavage at the NS2–NS3 site and/or the structure and function of both proteins may be affected, with possible adverse consequences for RNA replication and/or virion production. We therefore set out to identify an optimized crossover site generally applicable for construction of HCV chimeras by creating a set of intergenotypic chimeric reporter virus constructs between the Con1 (GT1b) and JFH1 (GT2a) strains *(6)*. Guided by a topology model of NS2 *(13)* and a naturally occurring intergenotypic chimera *(14)*, we compared the natural junction [designated C5 junction in our work, *see* **Fig. 26.1** and **ref.** *(6)*] with alternative crossover points located at the C-terminus of E2 (C1 junction), the C-terminus of p7 (C2 junction), a position between the first and second, or the second and third putative transmembrane helices of NS2 (C3 and C4 junctions, respectively), or the prototype crossover at the

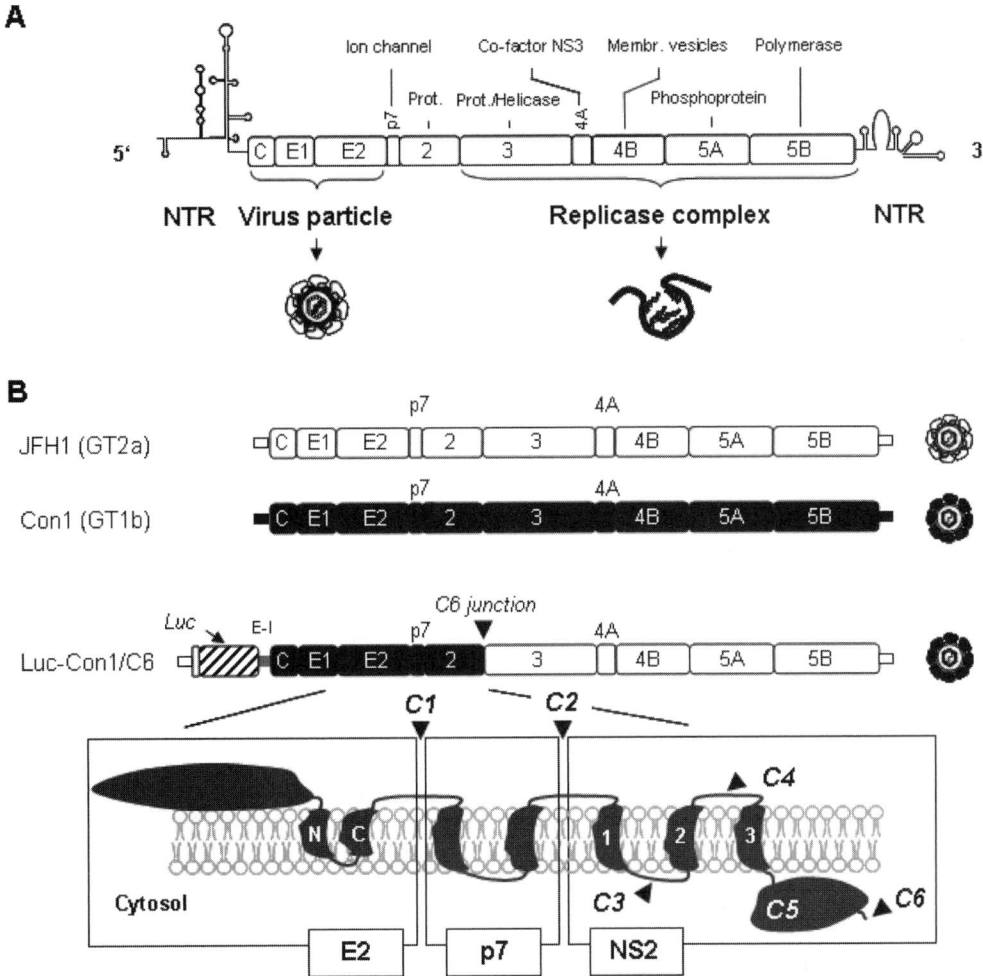

Fig. 26.1. Intergenotypic Con1-JFH1 reporter virus chimeras. (**A**) Schematic representation of the HCV genome with annotation of the individual viral proteins. Replicon studies show that NS3–NS5B proteins and the nontranslated regions of the genome (NTRs) are necessary and sufficient for RNA replication. These proteins, together with the genomic RNA and cellular factors, are believed to assemble the membrane-bound replicase complexes schematically indicated below. Core, E1, and E2 are the minimal structural proteins composing the virus particle. Whether other viral or cellular proteins are constituents of infectious HCV particles is presently unknown. (**B**) Design of chimeric Con1-JFH1 reporter virus genomes and location of various junctions chosen to connect these isolates. JFH1 (*white*) and Con1 (*black*) genomes are given at the top, and the structure of a chimeric reporter virus genome with the C6 junction is depicted below. The luciferase reporter (*striped box*) is expressed by the HCV IRES as fusion protein with the N-terminal 16 amino acids of JFH1 Core, whereas the chimeric polyprotein is translated from the internal ribosome entry site (E-I) of the encephalomyocarditis virus. The topology model of E2, p7, and NS2 at the bottom displays the locations of crossover sites (C1–C6) between Con1 and JFH1 segments.

C-terminus of NS2 (C6). **Table 26.1** lists the exact compositions of the respective constructs.

Because introduction of reporter genes into the HCV genome permits simple and quantitative assessment of RNA replication and infectivity, we employed chimeric reporter virus

Table 26.1
Chimeric Con1-JFH1 HCV genomes

| Chimera | Components of chimeric HCV polyprotein[a] | |
	Con1 (GT1b; AJ238799)	JFH1 (GT2a; AB047639)
Luc-Con1/C1	1–746	751–3034
Luc-Con1/C2	1–809	814–3034
Luc-Con1/C3	1–842	847–3034
Luc-Con1/C4	1–871	876–3034
Luc-Con1/C5	1–949	954–3034
Luc-Con1/C6	1–1026	1031–3034

[a]Amino acid positions of the fused polyprotein fragments are indicated; Gene bank accession numbers of Con1 and JFH1 isolates are given in brackets. Note that JFH1 encodes an E2 protein comprising four more amino acids than does the Con1-derived E2 protein.

genomes for these studies. The results of this mapping analysis identified the C3 junction as the best choice for construction of infectious JFH1-Con1 chimeras [**Fig. 26.2** and **ref.** *(6)*]. Moreover, as the C3 junction proved to be superior to the prototypic C6 junction in two of three further chimeric constructs, this site may be best applicable for construction of further inter- and intragenotypic chimeras (*see* **Note 1**). The techniques detailed below describe the experimental procedures for evaluation of efficiency of RNA replication, virus release, and infectivity of such chimeric constructs.

3.1. Cell Culture

1. Propagate Huh-7 (*see* **Note 2**) cells in DMEM complete and detach them by brief incubation with Trypsin/EDTA buffer. We routinely passage these cells twice weekly at a split ratio of 1:4 to 1:7.
2. Avoid seeding cells too sparsely or growing them to overconfluency. Usually, cells should be split when they have just reached confluence. Split ratios given above generally avoid too-sparse seeding.

3.2. In Vitro Transcription

The methods and reagents used in our laboratory to prepare cRNA for transfection experiments are described in **Chapter 11**, "HCV Replicons: Overview and Basic Protocols," **Section 3.1**, "*In vitro transcription.*" Note that the JFH1-Con1 constructs we used encode a ribozyme of the hepatitis delta virus at the 3′ end of the genome, followed by a T7-RNA polymerase terminator sequence. As a result of this configuration, linearization of template plasmids before in vitro transcription is not necessary because the catalytic activity of the ribozyme creates the authentic 3′ end of the HCV RNA.

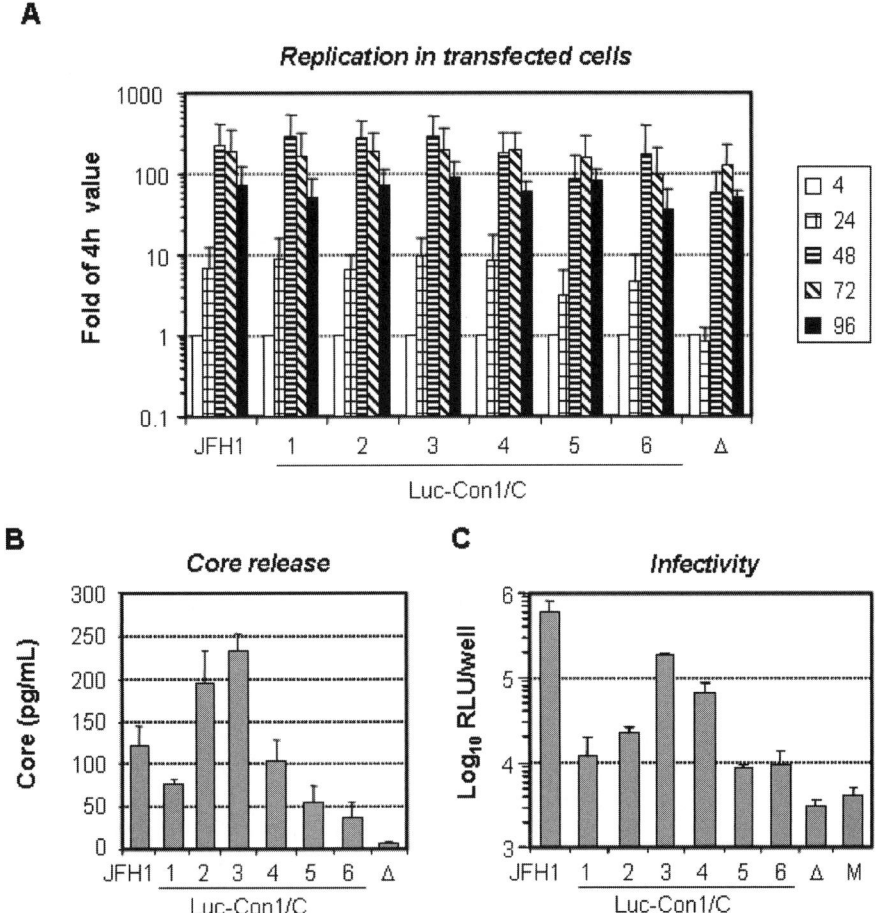

Fig. 26.2. Replication, virus production, and infectivity of chimeric Con1-JFH1 reporter virus genomes. **(A)** Huh7-Lunet cells transfected with the given constructs were harvested 4, 24, 48, 72, and 96 h after transfection, and luciferase activity was determined. To account for different transfection efficiency, we normalized absolute values for the luciferase activity detected 4 h after transfection, which was set to 1. **(B)** Release of Core protein from transfected Huh7-Lunet cells given in (A) 96 h after transfection, as determined by ELISA. Values were normalized for transfection efficiency and RNA replication by means of the relative light units (RLU) determined 96 h after transfection. **(C)** Determination of infectivity released from transfected cells (A). Huh7-Lunet cells were inoculated with cell-free supernatants harvested 96 h after transfection with genomes specified at the bottom. After 72 h, infected cells were harvested, and luciferase activity was determined. We normalized values to RLU in transfected cells to exclude variations caused by different transfection and replication efficiencies. Panels A–C show mean values of two independent experiments and the standard errors of the means. △, Luc-JFH1/△E1-E2; M, mock infected cells.

3.3. Electroporation of Huh-7 Cells

A detailed description of our electroporation protocol is given in **Chapter 11**, "HCV Replicons: Overview and Basic Protocols," **Section 3.2**, "*Electroporation of Huh-7 cells.*"

1. For transient replication and infection assays of chimeric reporter virus constructs, transfect 5 μg of in vitro transcribed RNA into 6×10^6 Huh-7.5 cells.

2. Transfer electroporated cells directly from the cuvette to 20 mL of DMEM complete and mix gently by inversion of the tube.

3. Subsequently, seed 2 mL portions of this suspension into parallel wells of a six-well plate. Under these conditions, transfected cells generally reach confluence 72–96 h after electroporation, and depending on the kinetic of virus release of the respective chimera, highest virus titers can be harvested between 48 and 96 h after transfection. Good transfection efficiency is a prerequisite for high yields of infectious HCVcc chimeras (*see* **Note 3**).

4. To obtain maximal virus titers from transient transfections, optimize transfection efficiency, monitor the kinetic of virus release of the given virus chimera, and seed transfected cells so that they reach confluence at the time of maximum virus production.

3.4. Quantification of HCV Replication and Infection by Firefly-Luciferase Assays

1. Collect culture fluid of transfected cells containing chimeric HCVcc and pass it through $0.45 \mu m$ pore-size filters. In parallel, wash cells once with 1 mL PBS, aspirate, add 1 mL of luciferase lysis buffer per well of a six-well plate, and freeze cells on the plate at $-20°$ or $-70°C$ (*see* **Note 4**).

2. Store cell-free filtered culture fluids at 4°C or $-70°C$ until use (*see* **Note 5**).

3. Seed Huh-7.5 target cells at a density of 6×10^4 cells per 12-well plate 24 h before inoculation with HCVcc. Inoculate cells with $500 \mu L$ cell-free virus preparation, incubate for 4 h at 37°C (*see* **Note 6**), and then terminate inoculation by aspiration of culture fluid and application of fresh medium.

4. Harvest infected cells 48–72 h after inoculation as described above. Generally we use $350 \mu L$ of lysis buffer per well of a 12-well plate.

5. Prepare two luminometer tubes per lysate and add $360 \mu L$ of luciferase assay buffer to each tube.

6. Freshly prepare luciferase substrate buffer and store it protected from light until use.

7. Thaw cell lysates at room temperature and store them on ice or at 4°C until use. Thoroughly resuspend lysate and add $100 \mu L$ to luminometer tube containing luciferase assay buffer, swirl gently, and put into luminometer for measurement.

8. The protocol is optimized for injection of $200 \mu L$ luciferase substrate buffer and a measuring time of 20 s.

3.5. Evaluation of Virus Release by Core ELISA

All reagents necessary for quantification of HCV Core in cell-culture fluids are components of the commercial Track-C ELISA and are used according to the instructions of the manufacturer.

1. To quantify cell-associated Core protein expression, wash cells once with 1 mL PBS, aspirate, and lyse them by resuspension in 1 mL (per well of a six-well plate) ice-cold lysis buffer.

2. Centrifuge cell lysates at 20,000g for 5 min at 4°C.

3. Transfer cleared lysates to a fresh tube and either use them directly for the ELISA or store them at −80°C until use (*see* **Note 7**).

3.6. Assessment of Transfection and Infection Efficiency by Fluorescence Microscopy

1. Wash transfected or infected cells once with PBS.

2. Fix cells by incubation with 4% PFA buffer for 10 min at room temperature.

3. Rinse the cells three times with PBS.

4. Permeabilize cells by incubation with PBS supplemented with 0.5% (v:v) Triton-X100 for 5 min at room temperature.

5. Rinse the cells three times with PBS.

6. Dilute primary HCV-specific antibodies as appropriate with staining buffer (*see* **Note 8**) and incubate the mixture with cells for 45 min at room temperature.

7. Rinse the cells three times with PBS.

8. Dilute secondary antibody (Anti-mouse IgG-Alexa 546 or 488; Invitrogen Molecular Probes, Karlsruhe, Germany) 1:1000 in staining buffer and incubate the mixture with cells for 30 min at room temperature. Keep protected from light.

9. Rinse once with PBS.

10. Incubate cells with DAPI diluted 1:3000 in PBS for 1 min and rinse three times with PBS.

11. Rinse cells once with H_2O and mount on a drop of Fluoromount-G.

12. For quantifying transfection efficiency, determine the percentage of transfected cells by comparing DNA- and HCV-specific staining in three independent microscopic views. For quantitative assessment of infectivity, determine the number of infected cell foci per inoculated well. When highly efficient HCV chimeras like Fl-J6/JFH1 *(1)* or Jc1 *(6)*, which produce high virus titers with rapid kinetics, are used, secondary rounds of infection may influence the virus titer measured by this method, so in this case, use of the limiting-dilution assay developed by Lindenbach and colleagues *(1)* is advisable.

4. Notes

1. The efficiencies of virus production sustained by various intra- and intergenotypic chimeras differ dramatically depending on the virus isolate that is connected with JFH1 [peak titers attained range from below 100 to more than 100,000 $TCID_{50}$/ml; *see* **ref.** *(6)*]. Determinants residing in the Core, E1, E2, and p7 NS2 region therefore modulate the

efficiency of virus production, resulting in largely divergent kinetics of virus release and peak titers. Ample evidence shows that chimeras producing low numbers of infectious virions can be improved by serial passage in cell culture. The interested reader is referred to **Chapter 27** for more details about cell-culture adaptation of infectious JFH1-based genomes. Please note that bicistronic firefly-luciferase reporter virus genomes are somewhat impaired in RNA replication and virion production compared to authentic genomes *(15)* Therefore, authentic genomes without insertion of reporter genes should be used if only low virus titers are obtained and if improvement of virus fitness by adaptation is desired. When passaging Huh-7.5 cells (*see* **Note 2**) infected with a Luc-Jc1 reporter virus genome *(15)*, we have noted loss of reporter activity and outgrowth of a virus variant that had eliminated the reporter gene within approximately two weeks of passaging (Artur Kaul, Eva Dazert, and Ralf Bartenschlager, unpublished observation).

2. Among all cell lines that have so far been tested for HCV replication, the Huh-7 hepatoma cell line and its derivatives sustain highest levels of HCV RNA replication and are therefore the natural choice for evaluation of infectious chimeric HCV gnomes. Huh-7-derived cell clones obtained by "curing" of Huh-7 replicon cell lines by means of inhibitor treatment, such as Huh7-Lunet cells or Huh-7.5 cells, were shown to be even more permissive for RNA replication than naïve Huh-7 cells *(16, 17)*. Because Huh-7.5 cells, in contrast to Huh7-Lunet *(18)*, express all known HCV receptors at high levels and are therefore highly permissive for HCVcc infection, they are the best cell line currently available for evaluation of infectious HCV chimeras.

3. Good transfection efficiency is of key importance for production of high-titer preparation of chimeric HCVcc viruses. Therefore, when problems are encountered with low virus production, transfection conditions should be optimized by monitoring of luciferase activity obtained 4 and 48 h after transfection (reflecting the amount of RNA delivered into the cells and the efficiency of replication, respectively) and by determination of the percentage of transfected cells by means of the immunofluorescence assay described in **Section 3.6** above. With the electroporation protocol described above, 40–80% of cells can generally be successfully transfected. The absolute numbers of relative light units (RLU) obtained after transfection of reporter virus genomes will differ widely with different luciferase assays and luminometers. Typically, when using reporter virus genomes, we obtain about 10^4–10^5 RLU per 100 µL of cell lysate 4 h after electroporation. The assay background is generally less than 10^3 RLU.

4. Freezing of cell lysates is not obligatory. Alternatively, cells can be lysed and used directly for luciferase assays, but given that cells are often harvested at different time points, and because luciferase activity is stable when samples are frozen, freezing is convenient and allows measurement of all samples side by side. The volume of lysis buffer can be reduced to increase sensitivity of the assay when desired.

5. HCVcc particles display a limited stability at 37°C (approximately half-life ca. 6 h; G. Koutsoudakis and T.P., unpublished observation) but are stable for at least 5–7 days when stored at 4°C. One cycle of freeze (−70°C) and thaw generally reduces infectivity twofold.

6. We have noted that, when JFH1 reporter viruses harvested after transient transfection are used, maximal infectivity is generally reached when Huh-7 target cells are inoculated for 4 h at 37°C. Although, depending on the titer and total volume of the inoculum, optimal duration of inoculation may vary to some extent, in most cases we terminate inoculation after 4 h of incubation at 37°C. Synchronized inoculation can be achieved by administration of cold HCVcc preparations to pre-cooled target cells for 1 h at 4°C, a temperature permissive for virus binding but generally accepted to be nonpermissive for subsequent steps of virus entry. Note, however, that use of the latter protocol reduces the efficiency of infection approximately 10-fold compared to the standard inoculation scheme. Inoculation volumes can be considerably reduced when inoculation is performed on a rocker.

7. Given the limited dynamic range of ELISA assays, finding a sample dilution appropriate for the assay is important. Although the quantity of intracellular Core expressed will vary depending on transfection efficiency and local conditions, the following guidelines may be helpful for initial establishment of the assay. For the transfection procedure detailed above, we dilute the cleared lysates harvested 4 h after transfection 1:10, 24 h lysates 1:100, and all subsequent time points 1:500 in PBS immediately before performing the assay. Please note that these dilutions are tailored for measuring bicistronic firefly-luciferase reporter constructs of the design described above. Authentic JFH1 genomes without reporter gene replicate considerably more efficiently, yielding high levels of intracellular Core protein only 24 h after electroporation.

8. The high replication level sustained by JFH1 in Huh-7 cells yields HCV protein expression levels easily detectable by immunofluorescence with different HCV-specific antibodies (1–3). In our hands, 9E10 hybridoma supernatant containing NS5A-specific antibodies (1) allows highly specific detection of HCV transfected and infected cells. With this antibody, cells infected with authentic JFH1-based chimeras can be identified

as early as 24 h after infection. When bicistronic reporter virus constructs are used, RNA replication occurs with less efficiency and delayed kinetics permitting robust detection of NS5A by immunofluorescence about 48 h after infection.

Acknowledgments

The author would like to thank George Koutsoudakis and Eike Steinmann for critical reading of the manuscript. This work was supported by an Emmy Noether fellowship from the Deutsche Forschungsgemeinschaft (PI 734/1-1), a grant from the Ministry of Science, Research and the Arts of Baden-Württemberg (Az. 23-7532.24-22-21-12/1), and by grants from the Helmholtz Association SO-024.

References

1. Lindenbach, B. D., Evans, M. J., Syder, A. J., Wolk, B., Tellinghuisen, T. L., Liu, C. C., et al. (2005) Complete replication of hepatitis C virus in cell culture. *Science* **309**, 623–626.

2. Wakita, T., Pietschmann, T., Kato, T., Date, T., Miyamoto, M., Zhao, Z., et al. (2005) Production of infectious hepatitis C virus in tissue culture from a cloned viral genome. *Nat. Med.* **11**, 791–796.

3. Zhong, J., Gastaminza, P., Cheng, G., Kapadia, S., Kato, T., Burton, D. R., et al. (2005) Robust hepatitis C virus infection *in vitro*. *Proc. Natl. Acad. Sci. USA* **102**, 9294–9299.

4. Yi, M., Villanueva, R. A., Thomas, D. L., Wakita, and Lemon, S. M. (2006) Production of infectious genotype 1a hepatitis C virus (Hutchinson strain) in cultured human hepatoma cells. *Proc. Natl. Acad. Sci. USA* **103**, 2310–2315.

5. Gottwein, J. M., T. K. Scheel, A. M. Hoegh, J. B. Lademann, J. Eugen-Olsen, G. Lisby, and J. Bukh. (2007). Robust hepatitis C genotype 3a cell culture releasing adapted intergenotypic 3a/2a (S52/JFH1) viruses. Gastroenterology **133**:1614–1626.

6. Pietschmann, T., Kaul, A., Koutsoudakis, G., Shavinskaya, A., Kallis, S., Steinmann, E., Abid, K., et al. (2006) Construction and characterization of infectious intragenotypic and intergenotypic hepatitis C virus chimeras. *Proc. Natl. Acad. Sci. USA* **103**, 7408–7413.

7. Scheel, T. K., J. M. Gottwein, T. B. Jensen, J. C. Prentoe, A. M. Hoegh, H. J. Alter, J. Eugen-Olsen, and J. Bukh. (2008). Development of JFH1-based cell culture systems for hepatitis C virus genotype 4a and evidence for cross-genotype neutralization. Proc. Natl. Acad. Sci. U. S. A**105**:997–1002..

8. Yi, M., Ma, Y., Yates, J., and Lemon, S. M. (2007) Compensatory mutations in E1, p7, NS2, and NS3 enhance yields of cell culture-infectious intergenotypic chimeric hepatitis C virus. *J. Virol.* 81, 629–638.

9. Griffin, S. D., Beales, L. P., Clarke, D. S., Worsfold, O., Evans, S. D., Jaeger, J., et al. (2003). The p7 protein of hepatitis C virus forms an ion channel that is blocked by the antiviral drug, Amantadine. *FEBS Lett.* **535**, 34–38.

10. Pavlovic, D., Neville, D. C., Argaud, O., Blumberg, B., Dwek, R. A., Fischer, W. B., et al. (2003) The hepatitis C virus p7 protein forms an ion channel that is inhibited by long-alkyl-chain iminosugar derivatives. *Proc. Natl. Acad. Sci. USA* **100**, 6104–6108.

11. Jones, C. T., C. L. Murray, D. K. Eastman, J. Tassello, and C. M. Rice. 2007. Hepatitis C virus p7 and NS2 proteins are essential for production of infectious virus. J. Virol. **81**:8374–8383.

12. Steinmann, E., F. Penin, S. Kallis, A. H. Patel, R. Bartenschlager, and T. Pietschmann. 2007. Hepatitis C Virus p7 Protein Is Crucial for Assembly and Release of Infectious Virions. PLoS. Pathog.3:e103.

13. Yamaga, A. K. and Ou, J. H. (2002) Membrane topology of the hepatitis C virus NS2 protein. *J. Biol. Chem.* **277**, 33228–33234.

14. Kalinina, O., Norder, H., Mukomolov, S., and Magnius, L. O. (2002) A natural intergenotypic recombinant of hepatitis C

virus identified in St. Petersburg. *J. Virol.* **76**, 4034–4043.

15. Koutsoudakis, G., Kaul, A., Steinmann, E., Kallis, S., Lohmann, V., Pietschmann, T., et al. (2006) Characterization of the early steps of hepatitis C virus infection by using luciferase reporter viruses. *J. Virol.* **80**, 5308–5320.

16. Blight, K. J., McKeating, J. A., and Rice, C. M. (2002) Highly permissive cell lines for subgenomic and genomic hepatitis C virus RNA replication. *J. Virol.* **76**, 13001–13014.

17. Friebe, P., Boudet, J., Simorre, J. P., and Bartenschlager, R. (2005) Kissing-loop interaction in the 3′ end of the hepatitis C virus genome essential for RNA replication. *J. Virol.* **79**, 380–392.

18. Koutsoudakis, G., Herrmann, E., Kallis, S., Bartenschlager, R., and Pietschmann, T. (2007) The level of CD81 cell surface expression is a key determinant for productive entry of hepatitis C virus into host cells. *J. Virol.* **81**, 588–598.

Chapter 27

Adaptation of the Hepatitis C Virus to Cell Culture

Artur Kaul, Ilka Wörz, and Ralf Bartenschlager

Abstract

A major breakthrough in the field of HCV research was the development of a system that supports the production of infectious virus particles. The key to this achievement was the molecular cloning of a genotype 2a genome, designated JFH1, which replicates to exceptionally high levels and at the same time supports virus particle assembly and release. A major drawback of this system was, however, the rather low yield of infectious particles obtained with the JFH1 genome as well as with most JFH1-derived virus chimeras. One approach to overcoming this hurdle is adaptation of the HCV genomes to cell culture. We found that both JFH1 and all chimeras, except one, can easily be adapted to cultured cells, increasing virus yields by up to three orders of magnitude. Surprisingly, adaptation is achieved by a multitude of mutations residing in both the structural and the nonstructural proteins. We therefore argue that a complex interaction between structural proteins and the HCV replicase takes place to allow efficient virus particle production.

Key words: Cell-culture adaptation, HCV adaptation, adaptive mutations, virus production, Huh-7 cells, nonstructural proteins.

1. Introduction

Production of infectious HCV in cell culture has become possible because of the unique properties of the JFH1 isolate, which was cloned from a Japanese patient with fulminant hepatitis (1). For reasons that are not yet understood, the JFH1 genome has an exceptional capacity to replicate in transfected or infected human hepatoma cells derived from the cell line Huh-7 (2), but despite efficient RNA replication, yields of infectious virus particles are rather low (3,4). Two alternative ways to increase virus yields have been devised: first, the construction of chimeric HCV genomes (5, 6) (see **Chapter 26**) and, second, selection for cell-culture-adapted JFH1 variants supporting high levels of virus production.

Hengli Tang (ed.), *Hepatitis C: Methods and Protocols, Second Edition, vol. 510*
© 2009 Humana Press, a part of Springer Science+Business Media
DOI 10.1007/978-1-59745-394-3_27 Springerprotocols.com

Cell-culture adaptation of JFH1 can be easily achieved by serial passage of JFH1-infected Huh-7 cells or by passage of virus-containing culture supernatants *(7–9)*. After only a few passages, virus variants emerge that support much higher levels of virus assembly and release than does the JFH1 wild type. On the average, virus titers of wild-type JFH1 are in the range of 10^3 tissue-culture-infectious doses the ($TCID_{50}$) per milliliter, and this titer can be increased to $10^5 - 10^6\, TCID_{50}/mL$ by cell-culture adaptation.

By the same approach, similar levels of cell-culture adaptation can be achieved with chimeric JFH1 genomes. Key to this adaptation, however, is the use of highly permissive Huh-7 cells. Most often, Huh-7.5 cells are used; these support high levels of HCV RNA replication, presumably because of a defect in the induction of double-strand RNA response *(10, 11)*. These cells also express very high levels of CD81 on the cell surface, which is a critical prerequisite for efficient infection *(12)*. Alternatively, Huh7-Lunet cells have been used, which are equally permissive for HCV RNA replication but express lower amounts of CD81 on the cell surface *(12)*.

Comparative studies suggest that cell-culture adaptation can be brought about by mutations in various viral proteins *(7–9)*. Interestingly, these reside not only in the structural-proteins core, E1 or E2, but also in the nonstructural proteins, such as the ion-channel protein p7, NS2, the NS3 protease/helicase, NS5A, and the NS5B polymerase. A complex interplay between the components forming virus particles and components of the replicase machinery therefore probably contributes to highly efficient virus production.

2. Materials

2.1. Cell Culture

1. Dulbecco's modified Eagle's medium (DMEM) (Invitrogen, Karlsruhe, Germany) supplemented with 2 mM L-glutamine, nonessential amino acids, 100 U/mL of penicillin, 100 µg/mL of streptomycin (Invitrogen, Karlsruhe, Germany), and 10% fetal bovine serum (FBS (PAA Laboratories, Cölbe, Germany)). Store at 4°C for no more than 6 weeks; reheat to 37°C before use.
2. Solution of trypsin (0.05%) and EDTA (0.02%) (Invitrogen, Karlsruhe, Germany).
3. Cell-culture plates and cell-culture dishes (BD Falcon, Heidelberg, Germany), 0.45 µm filters (Carl Roth, Karlsruhe, Germany), and 10 mL syringes (Braun, Melsungen, Germany).
4. Centricon Plus-70 Centrifugal Filter (Millipore, Schwalbach, Germany) for concentration of virus-containing cell-culture supernatants.

2.2. Vectors and Virus Genomes	1. pFK, the vector used to generate all replicon constructs, was derived from pBR322 by deletion of a *Sty*I–*Eco*RI fragment spanning the complete tetracycline-resistance gene and insertion of a multiple cloning site containing the recognition sequences for the restriction enzymes *Hin*dIII, *Cla*I, *Not*I, *Sal*I, *Eco*RI, and *Spe*I upstream of the promoter for the T3 RNA polymerase *(13)*. A modified T7 RNA polymerase promoter was fused upstream with the HCV genome and the hepatitis delta virus genomic ribozyme (dg), and the T7 terminator sequence was fused to the 3′ end of the HCV DNA insert to allow synthesis of run-off transcripts with the correct 5′ and 3′ termini.

2. JFH1 genome (DDBJ/EMBL/GenBank accession no AB047639). It corresponds to a consensus genome that replicates to very high levels in various Huh-7 clones.
3. Chimeric genomes J6/C3, Con1/C3, H77/C3, and 452/C6 *(6)* (**Fig. 27.1A**), derived from JFH1 by replacement of the region encoding for core to the C3 or C6 position of NS2 of JFH1, respectively, with the analogous region of the HCV isolates J6 (genotype 2a), Con1 (genotype 1b), H77 (genotype 1a), and 452 (genotype 3a) (*see* **Fig. 27.1A** and for further details *see* **Chapter 26**).

2.3. In Vitro Transcription and Electroporation

1. *See* **Chapter 11**, "HCV Replicons: Overview and Basic Protocols," **Section 2.2,** *In vitro transcription, electroporation, and northern hybridization*.

2.4. Immunofluorescence

1. Four percent paraformaldehyde, stored in aliquots at −20°C.
2. 4′,6′-diamidino-2-phenylindole dihydrochloride (DAPI) (Molecular Probes, Karlsruhe, Germany), stored in aliquots at 4°C in the dark.
3. Normal goat serum, to be diluted before use in PBS to a final concentration of 5%. Store undiluted serum in aliquots at −20°C.
4. Fluoromount G (Southern Biotechnology Associates, Birmingham, USA), stored at 4°C.

2.5. TCID$_{50}$ Assay

1. Saponin from Quillaja bark (Sigma-Aldrich, Steinheim, Germany).
2. Mouse anti-NS5A hybridoma supernatant 9E10 *(5)*.
3. Peroxidase-conjugated anti-mouse antibody (Sigma-Aldrich, Steinheim, Germany).
4. Vector NovaRED substrate kit (Linaris Biologische Produkte, Wertheim, Germany).

2.6. Long-Distance RT-PCR

1. Extraction of total RNA from infected cells with the Nucleo Spin RNAII Kit (Macherey-Nagel, Düren, Germany).

A

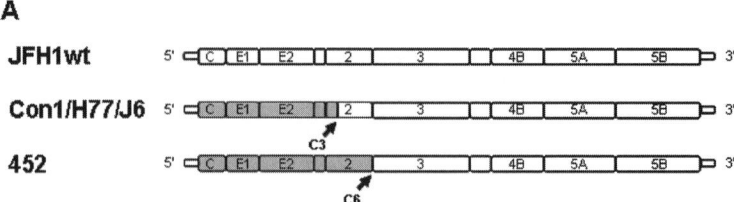

B

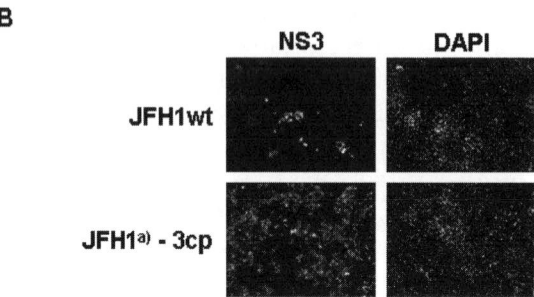

C

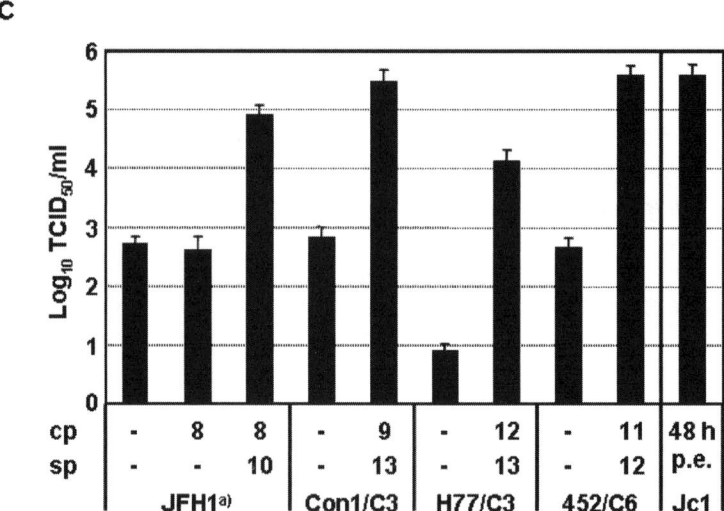

Fig. 27.1. Comparative analysis of nonadapted and adapted virus populations. (**A**) Schematic structure of the original JFH1 isolate and the virus chimeras used for cell-culture adaptation. Chimeras are composed of the core to the N-terminal (C3) NS2 region or the C-terminus of NS2 (C6) of different isolates and the remainder of the JFH1 isolate. The exact junction sites of the respective virus chimeras are indicated by *arrows (6)*. (**B**) Huh-7.5 cells were infected with cell-culture supernatant containing JFH1 wild-type (wt) viruses or adapted JFH1 viruses released after three cell passages (JFH1[a]: first adaptation experiment). Infected cells were visualized 3 days after infection, by NS3-specific immunofluorescence (*left side*). Nuclear DNA was stained with DAPI (right side). (**C**) Culture supernatants containing nonadapted viruses or supernatants collected after given numbers of cell passages (cp) and supernatant passages (sp) were collected, and $TCID_{50}$ titers were determined. For reference, the infectivity titer of the highly efficient Jc1 chimera (corresponding to J6/C3) 48 h after electroporation (p.e.) is shown on the *right*. Note that the titer of this chimera could not be increased further.

2. 100 mM deoxynucleoside triphosphate set (dNTP-Set) (Roche, Mannheim, Germany). Prepare a 10 mM solution of each deoxynucleoside triphosphate in H_2O and store in single aliquots at $-20°C$.

3. Expand reverse transcriptase and fivefold concentrated first-strand buffer (Roche, Mannheim, Germany).

4. Expand Long-Template PCR system consisting of enzyme and 10-fold concentrated buffer 3 (Roche, Mannheim, Germany).

2.7. Purification and Cloning of RT-PCR Products

1. NucleoSpin Extract Kit (Macherey-Nagel, Düren, Germany), useful for clean-up of PCR products and DNA extraction from agarose gel.

2. Agarose (Invitrogen, Karlsruhe, Germany).

3. TAE (40×): 193.82 g Tris-OH, 65.62 g NaAc, 29.78 g EDTA dissolved in water; adjust pH with approximately 50 mL acetic acid to 8.3 and add water to the final volume of 1 L.

4. Restriction endonucleases and corresponding 10-fold buffers (New England BioLabs, Frankfurt am Main, Germany).

5. T4 DNA ligase and corresponding 10-fold buffer (New England BioLabs, Frankfurt am Main, Germany).

6. LB-medium: 10 g trypton, 5 g yeast extract, 5 g NaCl per 1 L. Add for preparation of agar dishes 1.5% agar to LB-medium. Autoclave LB-medium and agar LB-medium, cool to 60°C, and add 100 μg/mL ampicillin.

3. Methods

A very critical issue in adaptation of HCV to cell culture is the choice of the proper host cell line, which must be highly permissive to allow adapted variants to spread rapidly in the culture. If poorly permissive cells are used, adapted viruses are eventually unable to spread sufficiently in the culture and will therefore be lost. In many cases, Huh-7.5 cells are used. An alternative possibility is Huh7-Lunet/CD81-high cells, which support HCV RNA replication to similar extents and also express high levels of CD81 on the cell surface (12).

3.1. In Vitro Transcription

1. See **Chapter 11**, "HCV Replicons: Overview and Basic Protocols," **Section 3.1,** *In vitro transcription.*

3.2. Transfection of Huh7-Lunet Cells with JFH1-Derived Genomes

1. See **Chapter 11**, "HCV Replicons: Overview and Basic Protocols," **Section 3.2,** *Electroporation of Huh-7 cells,* but make the following exceptions: use 5 μg of in vitro transcribed RNA (step 2), transfer the cells into 15–20 mL complete DMEM, and seed them into a 15 cm dish (step 7).

3.3. Infection of Huh-7.5 Cells and Adaptation of HCV Genomes by Cell Passage

1. Seed Huh-7.5 target cells into 12-well plates ($4.9\,cm^2$ per well) at a density of 4×10^4 cells per well and place the cells in the incubator overnight.

2. The producer cells, Huh7-Lunet transfected with authentic JFH1 or the virus chimeras, release moderate and construct-dependent virus titers 72 h and 96 h after transfection (see **Fig. 27.1C**; compare the low titer of H77/C3 before adaptation with all other nonadapted genomes and *see* **Note 1**). Filter medium containing virus through a $0.45\,\mu m$ filter. These virus preparations are stable at $4°C$ for approximately 1 week. In case of long-term storage at $-80°C$, note that virus titers are reduced approximately twofold by the freeze-thaw procedure.

3. For infection of Huh-7.5 target cells, replace the cell-culture medium with $500\,\mu L$ of filtered virus and incubate cells for 4 h in the incubator.

4. Replace the inoculum with $1.5\,mL$ of fresh medium and incubate infected cells until the cell monolayer reaches confluency.

5. Expand the confluent cells, typically 72–96 h after infection, into six-well plates (approximately $10\,cm^2$ per well). Remove culture medium from infected cells in 12-well plates, wash cells once with approximately $500\,\mu L$ PBS, add $500\,\mu L$ trypsin/EDTA to each well, incubate briefly at room temperature (RT), remove the trypsin/EDTA solution, and incubate for 5 min at $37°C$, until the cells completely detach from the plate. Harvest cells by rinsing each well with $500\,\mu L$ complete DMEM and transfer the cells to the well of a six-well plate prefilled with $1.5\,mL$ medium.

6. After an additional 72–96 h, or when cells become confluent, expand the cells into 6-cm-diameter dishes, each containing $4\,mL$ medium. Determine the status of adaptation by immunofluorescence analyses (see **Section 3.5** and **Fig. 27.1B**) and continue cell expansion into 10-cm-diameter dishes containing $8\,mL$ medium, if necessary (*see* **Notes 2, 3, 4, and 5**).

3.4. Adaptation of HCV Genomes by Passage of Cell-Culture Supernatants

1. Virus adaptation by cell-culture passage is the first and often major adaptation process. To increase virus titers to the highest levels and to avoid selection for cells that are not or only poorly permissive for JFH1 RNA replication, use supernatant passages in addition, for further virus adaptation (*see* **Fig. 27.1C**; JFH1 wt adaptation and *see* **Note 6**). To do so, follow **Section 3.3**, steps 1–4, and use the cell-culture supernatant collected 72–96 h after cell passage to infect new target cells in 12-well plates. Repeat these steps as often as necessary and determine the status of adaptation by immunofluorescence microscopy (*see* **Section 3.5** and **Fig. 27.1B**).

3.5. Determination of the Status of Adaptation by Immunofluorescence Microscopy

1. Seed target cells (Huh-7.5 or Huh7-Lunet/CD81-high) on glass coverslips in 24-well plates (approximately $1.8\,cm^2$ per well) at a density of 4–6×10^4 cells per well and incubate cells overnight in the incubator.

2. For infection, use $250\,\mu L$ of the filtered cell-culture medium collected 72–96 h after electroporation. Remove the inoculum after 4 h, add 1 mL fresh medium, and incubate cells for 72–96 h at 37°C.

3. Wash cells once with PBS before fixation with $500\,\mu L$ of 4% paraformaldehyde and incubate for 10 min at RT.

4. Wash cells three times with PBS, permeabilize cells by incubating them 5 min in $500\,\mu L$ of 0.5% Triton-X100 in PBS, and wash them three times with PBS before incubation with the primary antibody.

5. Perform immunostaining using a suitable HCV-specific antibody. As an example, we use a rabbit polyclonal serum against NS3, generated in our group *(14)*, at a dilution of 1:1000 in PBS containing 5% normal goat serum for 45 min.

6. After washing three times with PBS, detect bound primary antibodies using goat anti-rabbit antibodies conjugated to Alexa-Fluor 546 at a dilution of 1:1000 in PBS containing 5% normal goat serum for 30 min in the dark.

7. Add DAPI at a dilution of 1:3000 in PBS for 1 min to stain cell nuclei.

8. Finally, wash cells three times with PBS and once with water and mount them on glass slides with Fluoromount G.

9. The adaptation of virus is clearly visible in the form of an increased number of HCV antigen-containing cells (compare the wild-type and adapted virus in **Fig. 27.1B**).

3.6. Determination of Infectivity Titers by the $TCID_{50}$ Assay

The principle of the limiting-dilution assay for determining virus infectivity titers has already been reported *(5)*. We use this protocol with slight modifications.

1. Seed Huh-7.5 target cells at a density of 1.1×10^4 cells per well into a 96-well plate in a total volume of $200\,\mu L$ complete DMEM and incubate cells overnight in the incubator.

2. Generate a serial dilution of virus-containing supernatants in DMEM with 1:6 dilution steps. Infect eight wells of a 96-well plate per dilution and incubate cells for 72 h in an incubator. If necessary, pre-dilute the virus stock. For example, if an infectivity titer of about 10^5–$10^6\,TCID_{50}/mL$ is expected, pre-dilute the virus stock 1:60 for the first row and proceed with 1:6 dilutions for the remaining seven rows. For titers below $10^4\,TCID_{50}/mL$, start with a 1:6 dilution in the first row and proceed with 1:6 dilutions for another five rows.

3. Wash cells with PBS, fix them for 20 min with ice-cold methanol at −20°C, and wash again three times with PBS.

4. Permeabilize and block cells for 1 h with PBS containing 0.5% saponin, 1% bovine serum albumin, 0.2% dried skim milk, and 0.02% sodium azide.

5. Incubate cells with PBS containing 0.3% H_2O_2 (v:v) for 5 min at RT to block endogenous peroxidases.

6. After washing three times with PBS and once with PBS containing 0.5% saponin (PBS–saponin), perform immunostaining using a suitable HCV-specific antibody. As an example, we use an anti-NS5A antibody (hybridoma supernatant 9E10 (5)) at a 1:400 dilution in PBS–saponin for 1 h at RT or overnight at 4°C.

7. Wash cells again three times with PBS and once with PBS–saponin and detect bound 9E10 antibody by incubation with peroxidase-conjugated anti-mouse antibody diluted 1:200 in PBS–saponin for 1 h at RT.

8. Wash cells three times with PBS and once with PBS–saponin and detect peroxidase activity using the Vector NovaRED substrate kit.

9. Calculate virus titers (50% tissue culture infective dose, $TCID_{50}/mL$) by the method of Spearman and Kärber (15, 16).

3.7. Cloning of HCV Genomes by Long-Distance RT-PCR and Sequence Analysis

1. For isolation of total RNA from infected cells, infect a subconfluent Huh-7.5 cell layer seeded into a 6-cm-diameter culture dish with 800 μL of virus-containing cell-culture supernatant. After 4 h, replace the inoculum with 4 mL of fresh medium and incubate the cells for 96 h (see Note 7).

2. Wash cells once with 1 mL PBS and extract the total RNA from infected cells using the Nucleo Spin RNAII Kit as recommended by the manufacturer.

3. Amplify the complete open reading frame of the HCV genome in two overlapping fragments by reverse transcription-PCR (RT-PCR). First, for reverse transcription of viral RNA, mix 1 μg of total RNA with 50 pmol of primer A9482 (5′-GGA ACA GTT AGC TAT GGA GTG TAC C-3′) in a total volume of 10 μL.

4. Denature samples for 10 min at 65°C and subject them to reverse transcription with the Expand-RT system as recommended by the manufacturer in a total volume of 20 μL.

5. After 1 h at 43°C, use 2–4 μL of the reaction mixture for PCR with the Expand Long-Template PCR kit according to the manufacturer's instructions. Use these cycle conditions: an initial denaturation step for 3 min at 94°C, followed by 30 cycles, of 20 s at 94°C, 90 s at 52°C, and 8 min at 68°C. After 10 cycles, increase the extension time by 10 s for each additional cycle. Finally, incubate the reaction mixture for 12 min

at 68°C. To amplify the 5' half of the HCV genome, perform PCR with primers S59-*Eco*RI (5'-TGT CTT CAC GCA GAA AGC GCC TAG-3') and A4614 (5'-CTG AGC TGG TAT TAT GGA GAC GTC C-3'), and after restriction with *Eco*RI and *Spe*I, insert the PCR product into a suitable low-copy vector, for example, pFK-I$_{389}$Luc-EI/NS3-3'/JFH1-dg (Volker Lohmann, unpublished). Amplify the 3' half of the HCV genome with primers S3813 (5'-GGA CAA GCG GGG AGC ATT GCT CTC-3') and A9466-*Mlu*I (5'-AGC TAT GGA GTG TAC CTA GTG TGT GCC-3'), and after restriction with *Spe*I and *Mlu*I, insert the fragment into a suitable low-copy vector, for example, pFK-I$_{389}$neo/NS3 − 3'/Con1 *(13)*. Perform sequence analysis with a set of primers covering the complete HCV open reading frame. We routinely outsource the nucleotide sequence determination. Perform sequence analysis, for example, with the software Vector NTI (*see* **Note 8**).

---

4. Notes

1. The possibility of adapting HCV to cell culture and the extent of adaptation depend strongly on the isolate. The minimal requirements are efficient replication in the host cells and at least a minimum level of release of infectious virions into the culture medium (10^1–10^2 TCID$_{50}$/ml). On the basis of the H77/C3 chimera, a virus titer of about 10^1 TCID$_{50}$/mL is already sufficient for adaptation (**Fig. 27.1C** and **Note 4**), but this figure does not apply in all cases, as another JFH1-derived chimera replicating to the same level and also releasing virus titers of approximately 10^1 TCID$_{50}$/mL could not be adapted.

2. Infection of Huh-7.5 cells seeded into 12-well plates with 10^2–10^3 TCID$_{50}$ of nonadapted virus per well leads to less than 1% antigen-positive cells 4 days after infection (*see* **Fig. 27.1B**, left side). At this time, the cell monolayer is confluent and should be passaged. Depending on the initial virus titer, after three to six cell passages, growth of the cells is arrested by cytopathogenicity of JFH1, which by this time has infected most cells in the culture dish. Arrest is therefore a good indicator of efficient virus spread and, therefore, adaptation. The state of adaptation should then be determined by immunofluorescence microscopy (*see* **Fig. 27.1B**, right side).

3. Virus genomes releasing infectivity titers of more than about 10^5 TCID$_{50}$/mL probably cannot be adapted further (e.g., see Jc1 in **Fig. 27.1C**). We therefore assume that some cellular factor or conditions limit more efficient virus assembly

or release. In our hands, the highest virus titer achieved by transfection of Huh7-Lunet cells or Huh-7.5 cells is about 5×10^6 TCID$_{50}$/mL.

4. Genomes supporting a very low virus production (*see* **Fig. 27.1C**, H77/C3) can also be adapted. In this case we recommend starting with transfection of highly permissive cells and collecting culture medium at 48, 72, and 96 h after electroporation. Media can be pooled and concentrated up to about 100-fold by means of Centricon Plus-70 centrifugal filters. This concentrated virus stock can be used for infection of new target cells with which the passaging process can be started. However, this procedure is very susceptible for contaminations with the highly efficient Jc1 chimera. Because of its high efficiency to spread in cultured cells, Jc1 will overgrow the culture even after low dose contamination with virus, template DNA and/or *in vitro* RNA (**Sections 3.1 - 3.4**).

5 Yields of infectious virus particles obtained with nonadapted genomes are overall rather low and isolate dependent (*see* **Fig. 27.1C**, release of infectious particles before adaptation). Chimeric genomes that do not produce detectable amounts of infectious virus might be improved by introduction of highly adaptive mutations into the JFH1 sequence of the genome (C-terminal half of NS2 up to the C-terminus of NS5B). These "preadapted" chimeric genomes can be used to attempt adaptation.

6. The passage of transfected or infected cells is an applicable and convenient method for achieving adaptation, but passaged cells may release virus titers comparable to those of nonadapted parental virus (compare virus titer of JFH1 wild type and adapted JFH1 after eight cell passages in **Fig. 27.1C**). Including supernatant passages is therefore important (compare the JFH1 virus titer increase after eight cell passages with those after 10 additional supernatant passages in **Fig. 27.1C**). A major pitfall of cell passage is that JFH1 is cytopathogenic and induces growth arrest and eventually cell death. Poorly infected or transfected cells or cells that do not contain replicating HCV RNA therefore have a growth advantage. Also, cells that express, e.g., low amounts of CD81 or do not support HCV RNA replication efficiently will overgrow the culture. As a result, the cultured cells produce only small amounts of infectious HCV, and the virus may eventually be lost during continuous passage. For this reason, and to avoid selection for cells sustaining persistent JFH1 RNA replication, we recommend in addition supernatant passages.

7. For the isolation of sufficient amounts of viral RNA for RT-PCR, an almost completely (> 90%) infected cell monolayer grown on a 6-cm-diameter culture dish is required. This level should be achieved when cells are infected with

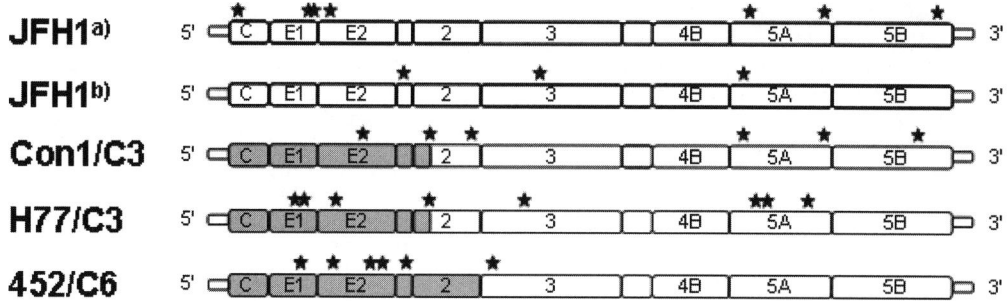

Fig. 27.2. Summary of conserved mutations found in cell-culture adaptations of the virus genomes listed. Conserved mutations identified in the different adapted isolates/chimeras are marked with asterisks (*see* **Note 9**). Silent mutations and nonconserved substitutions are not shown. JFH1[a]: first adaptation experiment; JFH1[b]: second adaptation experiment *(9)*.

a highly adapted virus population at a multiplicity of infection of $0.5\ TCID_{50}$/cell for 96 h. Alternatively, viral RNA can be directly isolated from culture supernatants by means of an appropriate RNA preparation kit.

8. In order to determine possible adaptive mutations, we routinely determine the nucleotide sequence of at least three complete HCV genomes. In all cases analyzed thus far, adaptive mutations were predominant in the sequence pool and at least two out of the three cDNA clones carried the mutation, but fixed mutations can also be found, which do not have an adaptive effect. They may be fixed on genomes containing other highly adaptive mutations.

9. Conserved mutations, that is, mutations that were present in at least two out of three cDNA clones, were identified in most viral genes. This was the case with various JFH1 chimeras but also with wild-type JFH1 adapted twice independently (*see* **Fig. 27.2**, JFH1[a] and JFH1[b]). In all cases analyzed thus far in our laboratory, mutations in the structural proteins made no or only a minor contribution to virus titers. High infectivity titers were rather mediated by mutations in nonstructural proteins.

Acknowledgments

This work was supported by grants of the German Research Council (SFB 638, Teilprojekt A5 and BA 1505/2-1).

References

1. Kato, T., Furusaka, A., Miyamoto, M., Date, T., Yasui, K., Hiramoto, J., et al. (2001) Sequence analysis of hepatitis C virus isolated from a fulminant hepatitis patient. *J. Med. Virol.* **64**, 334–339.

2. Kato, T., Date, T., Miyamoto, M., Furusaka, A., Tokushige, K., Mizokami, M., et al. (2003) Efficient replication of the genotype 2a hepatitis C virus subgenomic replicon. *Gastroenterology* **125**, 1808–1817.

3. Wakita, T., Pietschmann, T., Kato, T., Date, T., Miyamoto, M., Zhao, Z., et al. (2005) Production of infectious hepatitis C virus in tissue culture from a cloned viral genome. *Nat. Med.* **11,** 791–796.

4. Zhong, J., Gastaminza, P., Cheng, G., Kapadia, S., Kato, T., Burton, D. R., et al. (2005) Robust hepatitis C virus infection in vitro. *Proc. Natl. Acad. Sci. USA* **102,** 9294–9299.

5. Lindenbach, B. D., Evans, M. J., Syder, A. J., Wolk, B., Tellinghuisen, T. L., Liu, C. C., et al. (2005) Complete replication of hepatitis C virus in cell culture. *Science* **309,** 623–626.

6. Pietschmann, T., Kaul, A., Koutsoudakis, G., Shavinskaya, A., Kallis, S., Steinmann, E., et al. (2006) Construction and characterization of infectious intragenotypic and intergenotypic hepatitis C virus chimeras. *Proc. Natl. Acad. Sci. USA* **103,** 7408–7413.

7. Zhong, J., Gastaminza, P., Chung, J., Stamataki, Z., Isogawa, M., Cheng, G., et al. (2006) Persistent hepatitis C virus infection *in vitro*: coevolution of virus and host. *J. Virol.* **80,** 11082–11093.

8. Yi, M., Ma, Y., Yates, J., and Lemon, S. M. (2007) Compensatory mutations in E1, p7, NS2, and NS3 enhance yields of cell culture-infectious intergenotypic chimeric hepatitis C virus. *J. Virol.* **81,** 629–638.

9. Kaul, A., Woerz, I., Meuleman, P., Leroux-Roels, G., Bartenschlager, R. (2007) Cell culture adaptation of hepatitis C virus and in vivo viability of an adapted variant. *J. Virol.* **81,** 13168-79.

10. Blight, K. J., McKeating, J. A., and Rice, C. M. (2002) Highly permissive cell lines for subgenomic and genomic hepatitis C virus RNA replication. *J. Virol.* **76,** 13001–13014.

11. Sumpter, R., Jr., Loo, Y. M., Foy, E., Li, K., Yoneyama, M., Fujita, T., et al. (2005) Regulating intracellular antiviral defense and permissiveness to hepatitis C virus RNA replication through a cellular RNA helicase, RIG-I. *J. Virol.* **79,** 2689–2699.

12. Koutsoudakis, G., Herrmann, E., Kallis, S., Bartenschlager, R., and Pietschmann, T. The level of CD81 cell surface expression is a key determinant for productive entry of hepatitis C virus into host cells. *J. Virol.* **81,** 588–598.

13. Lohmann, V., Körner, F., Koch, J. O., Herian, U., Theilmann, L., and Bartenschlager, R. (1999) Replication of subgenomic hepatitis C virus RNAs in a hepatoma cell line. *Science* **285,** 110–113.

14. Bartenschlager, R., Ahlborn, L. L., Mous, J., and Jacobsen, H. (1993) Nonstructural protein 3 of the hepatitis C virus encodes a serine-type proteinase required for cleavage at the NS3/4 and NS4/5 junctions. *J. Virol.* **67,** 3835–3844.

15. Kärber, G. (1931) Beitrag zur kollektiven Behandlung pharmakologischer Reihenversuche. *Arch. Exp. Pathol. Pharmakol.* **162,** 480–487.

16. Spearman C. (1908) The method of "right and wrong cases" ("constant stimuli") without Gauss's formulae. *Brit. J. Psychol.* **2,** 227–242.

Chapter 28

Primary Human Hepatocyte Culture for HCV Study

Haizhen Zhu, John Elyar, Robin Foss, Alan Hemming, Eric Hall, Edward L. LeCluyse, and Chen Liu

Abstract

Studies of HCV pathogenesis and antiviral research have been hampered by the lack of adequate cell-culture and small-animal models. The culturing of human primary hepatocytes would greatly facilitate the model development in HCV research. The availability of robust infectious virus, JFH1 (i.e., genotype 2) strain, will further increase the interest in using primary hepatocyte cultures. This cell model system will significantly enhance research in the areas of antiviral research and host–virus interaction, but obtaining pure and viable human primary hepatocytes is not trivial. We have optimized a method of liver perfusion and primary hepatocyte isolation that allows us to establish robust and reliable human primary hepatocyte cultures. Moreover, we have demonstrated that these primary cultures are susceptible to authentic HCV infection in vitro.

Key words: Primary hepatocytes, infectious clone, JFH-1, apoptosis, real-time RT-PCR, liver perfusion.

1. Introduction

Hepatitis C virus (HCV) infects more than 170 million people in the world. It is a major etiology, causing severe chronic liver diseases, including liver failure, cirrhosis, and liver cancer *(1)*. Despite significant advances over the past two decades, no effective preventive vaccines are available, and the current interferon-based therapy is only partially effective. New specific and effective antiviral modalities are urgently needed.

Cell culture plays a critical role in virology research. Culturing HCV was a daunting task until the recent establishment of the JFH-1 HCV strain in cell culture *(2,3)*. Many studies were based on genetically modified viral proteins or HCV replicons, which

Hengli Tang (ed.), *Hepatitis C: Methods and Protocols, Second Edition, vol. 510*
© 2009 Humana Press, a part of Springer Science+Business Media
DOI 10.1007/978-1-59745-394-3_28 Springerprotocols.com

have obvious limitations. Because HCV is generally believed to infect predominantly human hepatocytes, all attention has been focused on culture of liver cells, including primary hepatocytes and liver cancer cell lines. Primary human hepatocyte culture has been extensively used in studies on liver physiology, toxicology, and pathology (4–6). Hepatocytes are highly differentiated cells, so stringent conditions are required to maintain their normal structure and function in culture dishes. In the literature, abundant information is available on culture of human hepatocytes for various study purposes. In HCV research, several publications have also discussed the use of human hepatocytes (7, 8), but the lower efficiency, lower reproducibility, and lack of standardized procedure prevent their wide use in HCV research.

Host cell culture is certainly critical for successful viral culture. Virus used to infect the cultured cells is also essential. Previous work with HCV infection of primary human hepatocyte culture depended on infectious human serum, the infectivity of which is variable and generally not efficient in infecting human cells in vitro. The newly available JFH-1 virus provides a uniform infectious viral source and may renew interest in using primary hepatocyte culture.

Our intention here is to share our experience in primary human hepatocyte isolation, culture, and infection with JFH-1 HCV. We have validated this system in our laboratory by using primary hepatocytes for the study of the interaction between host cell and HCV infection. These hepatocytes are susceptible for JFH-1 HCV infection. The protocol described below, isolation of human hepatocytes and analysis of HCV infection by real-time reverse transcription polymerase chain reaction (real-time RT-PCR) and immunofluorescence staining, may serve as a reference for use of this cell-culture system for HCV research.

2. Materials

2.1. Human Liver Tissue Collection

1. Two sources can be used for human hepatocyte isolation: surgical resection for metastatic colon cancer or unused liver tissue from brain-dead-but-beating-heart donors.
2. Perfuse resected liver tissue as soon as possible to obtain good-quality hepatocytes (see **Note 1**).
3. Donor liver tissue can be preserved in University of Wisconsin (UW) solution for up to 18 h. For convenience of perfusion, a piece of liver tissue with intact liver capsule covering the tissue is preferable, usually a wedge resection piece weighing approximately 40–100 g.

2.2. Supplies and Equipment

1. Tissue-culture biosafety cabinet, water bath, refrigerator-centrifuge and suitable centrifuge tubes, peristaltic pumps to ensure flow of appropriate buffers, heater unit to maintain

temperature of system at a constant 34–35°C, variable-sized tanks to accommodate liver tissue, and 6-well, 12-well, or 24-well tissue-culture plates (Falcon).

2. Suitable surgical instruments for dissecting liver tissue and other small items: pouch sterilizer, 1 pouch peel, 1 scalpel, 1 glove (6.5, 7.5, 8), connector 1/16″, 3/32″, 1/8″, forceps, 1 angio administration set, three 20-gauge needles, 10 plain sponges, two 1 mL syringes, one 5 mL syringe, 1 tray, 1 Wrap 36, 1 biohazard bag, 1 fluid shield mask, and chloroform pads.

2.3. Reagents

1. Type IV collagenase.
2. Two bottles of 500 mL HypoThermosol® Solution (HTS-FRS) (BioLife Solutions, Oswego, NY).
3. Two IV bags of 500 mL HypoThermosol® Solution (HTS-FRS) (BioLife Solutions, Oswego, NY).
4. Streptokinase (Sigma).
5. Heparin (American Pharmaceutical Partners).
6. Collagen type I (rat tail, BD Biosciences).
7. William's medium (Sigma).
8. Dulbecco's modified Eagle's medium (DMEM) (Gibco).
9. MEM NEAA (Gibco).
10. Fetal bovine serum (Gibco).
11. Penicillin-streptomycin (Cellgro).
12. L-glutamine (Cellgro).
13. Instant medical adhesive (Loctite 4013).
14. Percoll (Sigma, cat. no. P-4937) 90% isotonic solution, prepared fresh on each occasion (mix 45 mL of Percoll and 5 mL PBS (10×). Ensure solution is well mixed before use. Store at 4°C until use.
15. PBS, 10× (Gibco).
16. Washing medium: Ca^{2+}- and Mg^{2+}-free HBSS (Hanks balanced salt solution) containing 0.5 mM EDTA, 0.5% BSA, and 50 mg/L of ascorbic acid.
17. Digestion solution: DMEM medium containing 500 mg/L collagenase type IV and 0.5% BSA.
18. Suspension and attachment medium: Prepare 5% FBS, and penicillin–streptomycin (100 U/mL and 100 μg/mL, respectively) in DMEM. Filter-sterilize and store at 4°C. Complete medium by adding insulin (4 μg/mL, 1:1000 of stock) and 1 μM dexamethasone (1:10,000 of stock) just before use.
19. Rat-tail collagen (BD Biosciences, Palo Alto, CA) at 4 mg/mL.
20. Trypan blue (Sigma, cat. no. T-8154).
21. TRIzol reagent (Invitrogen) for RNA extraction, Lux primers and related reagents for HCV real-time RT-PCR analysis (Invitrogen), Huh-7.5 cells (gift of Charles Rice), and JFH-1 plasmid (gift of Wakita) for HCV particle production.

3. Methods

3.1. Safety Conditions

Informed consent must be obtained before liver tissue collection. Virological analysis of the patients from whom the liver samples have been obtained must be carried out before the surgery. The serology includes hepatitis A virus (HAV), hepatitis B virus (HBV), HCV, and human immunodeficiency virus (HIV). All steps of isolation and culture of hepatocytes must be carried out in a laminar vertical flow microbiology safety hood. Universal precaution is advised. All materials and liquid wastes must be decontaminated before discarding or autoclaved.

3.2. Liver Tissue Flushing

3.2.1. Tissue Procurement

1. Before the tissue procurement process, clean the pail with a chloroform pad.
2. Fill the pail with ice to the top line.
3. Return the yellow pail with ice to the laminar-flow hood.
4. Place the sterile drape.
5. Aseptically place the dissection instruments.
6. Assemble 1 mL syringes and 5 mL syringe with needles.
7. Remove a bag of 500 mL HypoThermosol® Solution from a 4°C refrigerator and place it in the hood.
8. Using a 5 mL syringe, remove 2 mL of HypoThermosol® Solution and slowly add it to the bottle of streptokinase β.
9. Gently roll the bottle to resuspend the streptokinase β. This process should take about 1 min.
10. Using a 1 mL syringe, remove all the heparin, and add it to the bag of 500 mL HypoThermosol® Solution.
11. Then, remove all the resuspended streptokinase β and add it to the bag.
12. Shake the bag to distribute heparin and streptokinase β evenly.
13. Place the bag in an IV pump bag. Before attaching the IV hose to the bag, close the IV hose.
14. Continue pumping the bag until the pressure reading is 150 mg at maximum. Shake the bag and open the IV hosing. Be certain to allow any air bubbles in the bag to flow through the hosing, as bubbles must not enter the tissue at any point.

3.2.2. Tissue Flushing

1. Depending on the size of the liver vasculature, choose the connectors to be used.
2. Gently place the IV hose connected to the appropriate connector into the vasculature of the tissue.
3. Open the IV hose flow regulator to begin perfusing blood from the tissue. Depending on the size of the tissue, the goal is to perfuse enough tissue with the HypoThermosol® Solution containing heparin and streptokinase β.

4. After the first HypoThermosol® Solution bag has emptied, wait for 2 min to allow the heparin and Streptokinase β to work.

5. After 2 min, repeat the procedure, as described above, but do not add streptokinase β to the HypoThermosol® Solution bag (*see* **Note 2**). Begin the perfusion again to remove the clots and any excess blood still remaining in the tissue. The goal is to remove all red blood cells from the tissue so that they will not lyse and release toxic products into the tissue, decreasing hepatocyte viability.

3.3. Liver Tissue Perfusion

1. Begin hepatocyte isolation, for which the flushed liver tissue is now ready. The procedure is based on a previously published protocol *(9, 10)*. Using a sterile gauze pad, dry the cut surface of the liver. Cannulate the chosen vessel(s) with a 200 µL pipette tip or a 16–22-gage Teflon cannula. Make a collar around the periphery of the cannula with medical adhesive at the point where it will join the tissue on the cut surface; then insert the cannula into vessel opening. Secure the cannula in place by adding more adhesive around the cannula–tissue interface. Seal all other openings on the cut surface using medical adhesive. Once again, dab dry the cut surface of the liver and cover with a thin layer of adhesive. Allow the medical adhesive to dry sufficiently before initiating the perfusion.

2. Once the adhesive has dried, place the liver in a container, connect perfusion tubing to the cannula(s), and slowly start the perfusion (5–10 mL/min initially). If no overt leaks are observed, make three small incisions along the encapsulated edge of the tissue and carefully place the liver tissue in the container containing washing buffer inside the perfusion unit, prewarmed to 34–35°C.

3. Slowly increase the flow rate for the washing buffer until residual blood and perfusate are observed flowing from the incisions along the encapsulated edge. The flow rate varies with the size of the tissue and how well it is sealed. The recommended flow rate is about 20–30 mL/min for resections weighing between 20 and 100 g.

4. While the washing medium is perfusing through the liver, prepare digestion solution with collagenase. For normal liver, use 50–80 mg collagenase per 100 mL of digestion solution; for cirrhotic or fatty liver, use 80–120 mg collagenase per 100 mL of digestion solution. Use about 100 mL of digestion solution per 10 g liver tissue.

5. After 10–15 min, stop the pump, carefully drain the tank as completely as possible of the washing medium, and then add a similar volume of prewarmed digestion solution.

6. Perfuse the liver tissue with digestion solution for about 15–25 min. Adjust the time according to the activity of the collagenase and the size of the liver tissue. Indications of complete digestion are softening and enlargement of the tissue. The perfusion time must not be overextend, as too long a digestion will lead to excessive cell damage and a significant loss of viability.

7. When the perfusion is complete, remove the liver from the tank, place it in a sterile beaker, and then proceed with hepatocyte isolation.

3.4. Isolation, Plating, and Culture of Primary Human Hepatocytes

1. Place enough ice-cold washing medium in the beaker containing the perfused liver tissue to cover the tissue completely. Chop the liver tissue with sterile scissors to release the hepatocytes into the medium. Gently shake the beaker to achieve maximum cell release and then filter out the undigested liver tissue through gauze-covered funnels.

2. Centrifuge the cell suspension at 100g for 5 min. Discard the supernatant and resuspend the cells in ice-cold suspension medium. Repeat this process three times.

3. Resuspend the cell pellet in suspension medium and 90% Percoll solution (3:1) and centrifuge at 100g for 5 min (*see* **Note 3**).

4. Gently remove the top layer without disturbing the cell pellet. Resuspend the cell pellet in suspension medium and centrifuge at 100g for 5 min.

5. Resuspend the cell pellet in suspension medium and do a cell count and viability count using trypan blue.

6. Coat the cell-culture plates or flasks with type I collagen. Other types of matrix, such as Matrigel, gelatin, or laminin can also be used, depending on the applications intended.

7. Adjust the cell density with hepatocyte culture medium, such as William's E with 10% FCS, to appropriate levels (*see* **Note 3**). The hepatocytes are round and uniform in cell suspension (**Fig. 28.1**). For the HCV infection study, we seeded the hepatocytes into six-well culture plates. The cell concentration should be 2.5×10^5/mL, and each well should contain 2 mL of hepatocyte suspension, for a final hepatocyte number of 5×10^5 per well.

8. Allow hepatocytes to attach, which takes 4–12 h. Change the medium daily. After 24 h of culture, the hepatocytes are usually ready for experimentation. The attached hepatocytes usually show morphology resembling that of normal hepatocytes (**Fig. 28.1**).

3.5. Infection of Primary Human Hepatocytes with HCV

1. Discard the culture medium in the hepatocyte culture and replace it with DMEM with 10% FCS in six-well cell-culture plates, and culture the cells in 5% CO_2 at 37°C for 24 h.

A   B

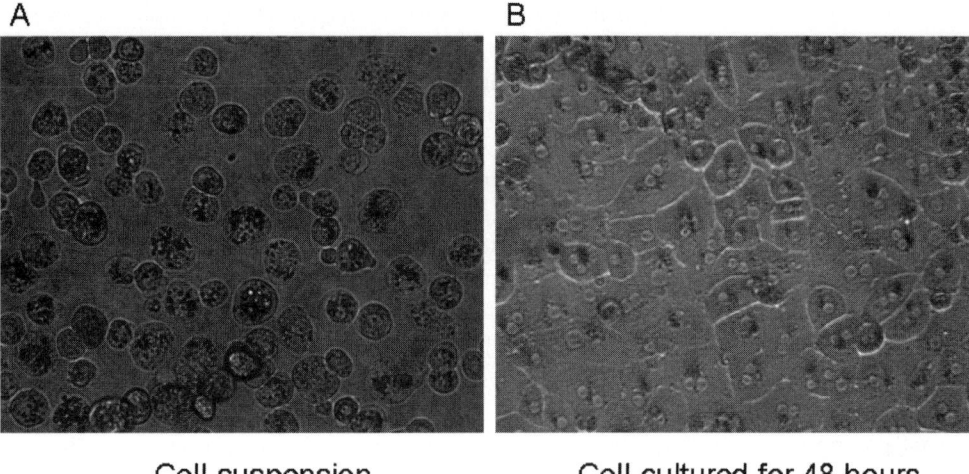

Cell suspension   Cell cultured for 48 hours

Fig. 28.1. Hepatocyte morphology. (**A**) Freshly isolated cells suspended in culture. (**B**) Hepatocytes on six-well tissue-culture plates after 48 h of culture. Original magnification 200×

2. To generate genomic JFH-1 and JFH-1/GND RNA, linearize the pJFH-1 and pJFH-1/GND plasmids (which contain HCV 2a full-length or mutant and which we obtained from Dr. Takaji Wakita, Tokyo Metropolitan Institute for Neuroscience, Toyko, Japan) at the $3'$ end of the HCV cDNA by *Xba*I digestion. Purify the linearized DNA and use it as a template for in vitro transcription according to the MEGAscript protocol (Ambion, Austin, TX). Deliver in vitro transcribed genomic JFH-1 or JFH-1/GND RNA to Huh-7.5 cells by electroporation (*see* **Note 4**). Briefly, wash trypsinized cells twice with PBS and then resuspended them in serum-free Opti-MEM (Invitrogen) at 1×10^7 cells/mL. Mix $10\,\mu$g of JFH-1 RNA with 0.4 mL of the cells in a 4 mm cuvette and use a Bio-Rad Gene Pulser system to deliver a single pulse at 0.26 kV, 100 ohms, and $960\,\mu$F. Plate the cells on two 100 mm dishes. Transfer the transfected cells to complete DMEM and culture them for the indicated period. Passage the cells every 3–5 days, and then filter and collect the supernatants at the indicated time points.

3. To generate viral stocks, dilute infectious supernatants in complete DMEM and use them to inoculate naïve 20% confluent Huh-7.5 cells. Trypsinize infected cells and replant them every 3–4 days. Harvest the supernatant from the infected cells 15 days after infection and store it at −80°C. Determine the titer of viral stock as described below.

4. Perform titration of infectious HCV as reported previously *(3)*. Briefly, serially dilute filtered cell supernatant 10-fold in the cell-culture medium and use it to infect 5×10^4 naïve Huh-7.5 cells per well in six-well plates. Incubate the

inoculum with cells for 24 h at 37°C and replace it with fresh medium. Determine the level of HCV infection 3 days after infection by immunofluorescence staining for HCV NS5A. Express the viral titer as focus-forming units per milliliter (ffu/mL), determined by the average number of NS5A-positive foci detected at the highest dilutions.

5. Add HCV to the hepatocytes for 8 h. Remove the tissue-culture medium containing HCV and replace it with fresh DMEM.

3.6. Detection of HCV in Primary Human Hepatocytes by Real-Time RT-PCR and Immunofluorescence Staining

1. Collect the cells every 24 h. Briefly, remove the cell-culture medium and wash three times with PBS. Lyse the cells with TRIzol reagent.

2. Isolate total RNA using the TRIzol reagent according to the manufacturer's protocol *(11)*.

3. Perform reverse transcription as described previously *(11)*. Synthesize first-strand cDNAs from total cellular RNA by reverse transcription (20 μL of reaction volume) using the SuperScript II (50 U reverse transcriptase per reaction) first-strand synthesis for RT-PCR kit (Invitrogen) primed with oligo $(dT)_{12-18}$ (Invitrogen) according to the manufacturer's instructions. Briefly, a single tube reaction (20 μL) should contain 400 ng RNA, DEPC-treated water, 0.0125 μg oligo(dT)12–18, 1× RT buffer, 5 mM $MgCl_2$, 0.5 mM dNTPs, 10 mM DTT, 2 units of RNaseOUT, and 2.5 units of SuperScript II reverse transcriptase. Incubate the tube at 25°C for 10 min. Use an RT profile of 42°C for 45 min, 70°C for 15 min, and then chill on ice. After the reverse transcription, perform real-time PCR as previously reported *(12)*. Each 50 μL PCR reaction should contain 10 μL of cDNA after reverse transcription, 200 nM FAM-labeled HCV-specific primer and JOE-labeled GAPDH-specific primer, 1× ROX reference dye, and 1× Platinum Quantitative PCR SuperMix-UDG (Invitrogen), including 200 μM each of dATP, dGTP, dCTP, and 400 μM dUTP, stabilizers, 1 U of UDG, 1.5 U platinum *Taq* DNA polymerase, 3 mM $MgCl_2$, 20 mM Tris–HCl, pH 8.4, and 50 mM KCl. Incubate reactions at 50°C for 2 min, 95°C for 2 min, and then cycle them 45 times through 95°C for 15 s, and 60°C for 30 s. Conduct reactions in a 96-well spectrofluoro-metric thermal cycler (ABI PRISM 7700 Sequence detector system, Applied Biosystems). Monitor fluorescence during every PCR cycle at the annealing step. Results for all experiments should represent triplicate determinations. Represent results as means ± SD. Fluorophore-labeled LUX primers and their unlabeled counterparts are obtained from Invitrogen. The primers for HCV are: 5′-CGCTCAATGCCTGG AGATTTG-3′ and 5′-GCACTCGCAAGCACCCTATC-3′;

for GADPH: 5′-TGCTGGCGCTGAGTACGTC-3′, 5′-GTGCAGGAGGCATTGCTGA-3′. Analyze the results with SDS 2.0 software from Applied Biosystems. Results for all experiments should represent triplicate determinations and be represented as means ± SD.

4. Fix primary human hepatocytes growing on coverslips in cold acetone for 5 min and wash them in PBS at room temperature. Block the slides in 1:50 diluted normal goat serum for 30 min at room temperature. Add the HCV NS5A-specific monoclonal antibody at a 1:100 dilution to the slides and incubate for 1 h at room temperature (*see* **Note 5**). After washing the slides with PBS three times for 5 min each, incubate them with FITC-labeled goat anti-mouse IgG for 45 min at room temperature. Wash the slides with PBS again. Stain cell nuclei with DAPI (blue) (Vector Laboratories, Burlingame, CA) and then examine them under a fluorescence microscope (Olympus) (*see* **Note 6**).

4. Notes

1. The source of the liver tissue is critical to obtaining adequate viable hepatocytes. Surgical resection tissues and spared donor liver tissues are the best sources. Liver tissues that have extensive fibrosis or fatty changes are not recommended. Flushing the liver tissue as soon as possible is extremely important. Addition of streptokinase to the flushing medium promotes breakage of blood clots, which increases hepatocyte viability. Streptokinase β should be kept at −20°C. The concentration of streptokinase β used in the tissue perfusion is 4 μg/mL. Heparin is stored at 4°C. The final concentration of heparin used in perfusion solution is 2 unit/mL.

2. Use of a Percoll gradient can significantly increase the cell uniformity. This centrifugation permits discarding of non-parenchymal cells and dead cells. Handling the Percoll gradient very carefully is critical to avoiding contamination.

3. Freezing primary hepatocytes is not advised because of the significant drop in cell attachment and viability after thawing, but for certain hepatocytes and special applications, the cells can be frozen in 10% DMSO in liquid nitrogen.

4. Good-quality DNA template for HCV full-length RNA is important for transcription. In vitro transcribed HCV RNA must also be purified by the phenol:chloroform extraction method.

5. Mouse anti-HCV NS5A monoclonal antibody was made at our institution. It can detect both genotype 1 and genotype 2 HCV NS5A protein. As we have reported, the antibody

is good for western blot analysis, immunofluorescence, and immunohistochemistry.

6. The typical efficiency of infection is approximately 10% and occasional apoptosis of the hepatocytes were observed in infected cells.

References

1. Liang, T. J., Rehermann, B., Seeff, L. B., and Hoofnagle, J. H. (2000) Pathogenesis, natural history, treatment, and prevention of hepatitis C. *Ann. Intern. Med.* **132**, 296–305.

2. Wakita, T., Pietschmann, T., Kato, T., Date, T., Miyamoto, M., Zhao, Z., et al. (2005) Production of infectious hepatitis C virus in tissue culture from a cloned viral genome. *Nat. Med.* **11**, 791–796.

3. Zhong, J., Gastaminza, P., Cheng, G., Kapadia, S., Kato, T., Burton, D. R., et al. (2005) Robust hepatitis C virus infection *in vitro*. *Proc. Natl. Acad. Sci. USA* **102**, 9294–9299.

4. Strom, S. C., Jirtle, R. L., Jones, R. S., Novicki, D. L., Rosenberg, M. R., Novotny, A., et al. (1982) Isolation, culture, and transplantation of human hepatocytes. *J. Natl. Cancer Inst.* **68**, 771–778.

5. Pichard, L., Raulet, E., Fabre, G., Ferrini, J. B., Ourlin, J. C., and Maurel, P. (2006) Human hepatocyte culture. *Methods Mol. Biol.* **320**, 283–293.

6. Gebhardt, R., Wegner, H,, and Alber, J. (1996) Perifusion of co-cultured hepatocytes: optimization of studies on drug metabolism and cytotoxicity *in vitro*. *Cell Biol. Toxicol.* **12**, 57–68.

7. Ito, T., Mukaigawa, J., Zuo, J., Hirabayashi, Y., Mitamura, K., and Yasui, K. (1996) Cultivation of hepatitis C virus in primary hepato-cyte culture from patients with chronic hepatitis C results in release of high titre infectious virus. *J. Gen. Virol.* **77**, 1043–1054.

8. Fournier, C., Sureau, C., Coste, J., Ducos, J., Pageaux, G., Larrey, D., et al. (1998) *In vitro* infection of adult normal human hepatocytes in primary culture by hepatitis C virus. *J. Gen. Virol.* **79**, 2367–2374.

9. LeCluyse, E. L. (2001) Human hepatocyte culture systems for the in vitro evaluation of cytochrome P450 expression and regulation. *Eur. J. Pharm. Sci.* **13**, 343–368.

10. LeCluyse, E. L., Alexandre, E., Hamilton, G. A., Viollon-Abadie, C., Coon, D. J., Jolley, S., et al. (2005) Isolation and culture of primary human hepatocytes. *Methods Mol. Biol.* **290**, 207–229.

11. Zhu, H. and Liu, C. (2003) Interleukin-1 inhibits hepatitis C virus subgenomic RNA replication by activation of extracellular regulated kinase pathway. *J. Virol.* **77**, 5493–5498.

12. Zhu, H., Nelson, D. R., Crawford, J. M., and Liu, C. (2005) Defective Jak-Stat activation in hepatoma cells is associated with hepatitis C viral IFN-α resistance. *J. Interferon Cytokine Res.* **25**, 528–539.

13. Zhu, H., Zhao, H., Collins, C. D., Eckenrode, S. E., Run, Q., McIndoe, R. A., et al. (2003) Gene expression associated with interferon alfa antiviral activity in an HCV replicon cell line. *Hepatology* **37**, 1180–1188.

Chapter 29

In Vivo Study of HCV in Mice with Chimeric Human Livers

Norman M. Kneteman and Christian Toso

Abstract

Estimates of hepatitis C virus infection include 170 million people worldwide, who face increased risk of development of cirrhosis, liver failure, and hepatocellular carcinoma. Standard of care therapy with pegylated interferon and ribavirin is effective in just half of patients, is challenged by substantial treatment-related morbidity, and is prohibitively expensive in most parts of the world. New therapeutics for treatment and prevention are clearly needed. Development of effective therapies has been significantly hampered by difficulties in establishing in vitro and in vivo models of viral replication. This chapter reviews development, validation, and early application of a mouse model with a chimeric human liver.

Key words: Chimeric liver, transgenic mice, liver transplantation, in vivo model, Scid mice, uPA mice.

1. Introduction

Estimates of hepatitis C virus infection include 170 million people worldwide (1), who face increased risk of development of cirrhosis, liver failure, and hepatocellular carcinoma (2, 3). Standard of care therapy with pegylated interferon and ribavirin is effective in just half of patients, is challenged by substantial treatment-related morbidity, and is prohibitively expensive in most parts of the world (4,5). New therapeutics for treatment and prevention are clearly needed. Development of effective therapies has been significantly hampered by difficulties in establishing in vitro and in vivo models of viral replication. This chapter reviews development, validation, and early application of a mouse model with a chimeric human liver.

Hengli Tang (ed.), *Hepatitis C: Methods and Protocols, Second Edition, vol. 510*
© 2009 Humana Press, a part of Springer Science+Business Media
DOI 10.1007/978-1-59745-394-3_29 Springerprotocols.com

2. Establishment of the Chimeric-Liver Mouse Model of HCV Infection

Attempts to establish cell culture or small animal models of HCV faced two formidable obstacles. First, HCV can survive only in hepatocytes of chimpanzee and man. Second, human hepatocytes in culture dedifferentiate rather rapidly and, in doing so, lose characteristics that are essential for support of infection by HCV. We attempted to surmount these obstacles by using an immunodeficient mouse supporting transplanted human hepatocytes as an in vivo culture dish able to support HCV infection and replication. Our initial attempts at human hepatocyte transplants in mice demonstrated low levels of engraftment that were lost within weeks, as evidenced by western blot of human albumin in mouse serum (**Fig. 29.1**, upper panel), and did not sustain HCV infection. Attempts to provide additional stimulation to hepatocyte proliferation by hepatectomy, irradiation, or administration of human hepatocyte growth factor increased and prolonged the expression of human albumin in mouse serum (unpublished data), but as in the efforts of other investigators, failed to achieve long-term or high-level human hepatocyte repopulation or HCV infection.

An intriguing mouse model of neonatal bleeding disorders was reported in 1990 by Heckel et al. *(6)*. The mice carried a urokinase-type plasminogen activator transgene controlled by an albumin promoter (Alb-uPA), which led to urokinase overproduction within the liver and a high risk of neonatal bleeding *(6)*. Although half of the animals died within a few days of birth, many of the initial survivors subsequently experienced liver failure, ultimately demonstrated to be due to the toxicity of the transgene *(7)*. Occasional individuals were able to survive both perinatal bleeding challenges and liver failure. This phenomenon could be explained by DNA rearrangement in the regenerating hepatocytes, which affected the transgene array tandem and prevented expression. This spontaneous transgene deletion caused urokinase-type plasminogen to return gradually to normal within the first 2 months of life *(7)*.

Subsequent studies demonstrated that these mice could also be rescued by transplantation of syngeneic *(8)* or xenogeneic (rat) hepatocytes *(9)*. Mice hemizygous for Alb-uPA were later developed on an immunodeficient RAG background and were shown to support human hepatocytes for at least 2 months after transplant *(10)*. Up to 15% human liver chimerism was reported, as was infection with human hepatitis B virus (HBV). Of interest, mice hemizygous for the uPA transgene have never been successfully demonstrated to support HCV infection.

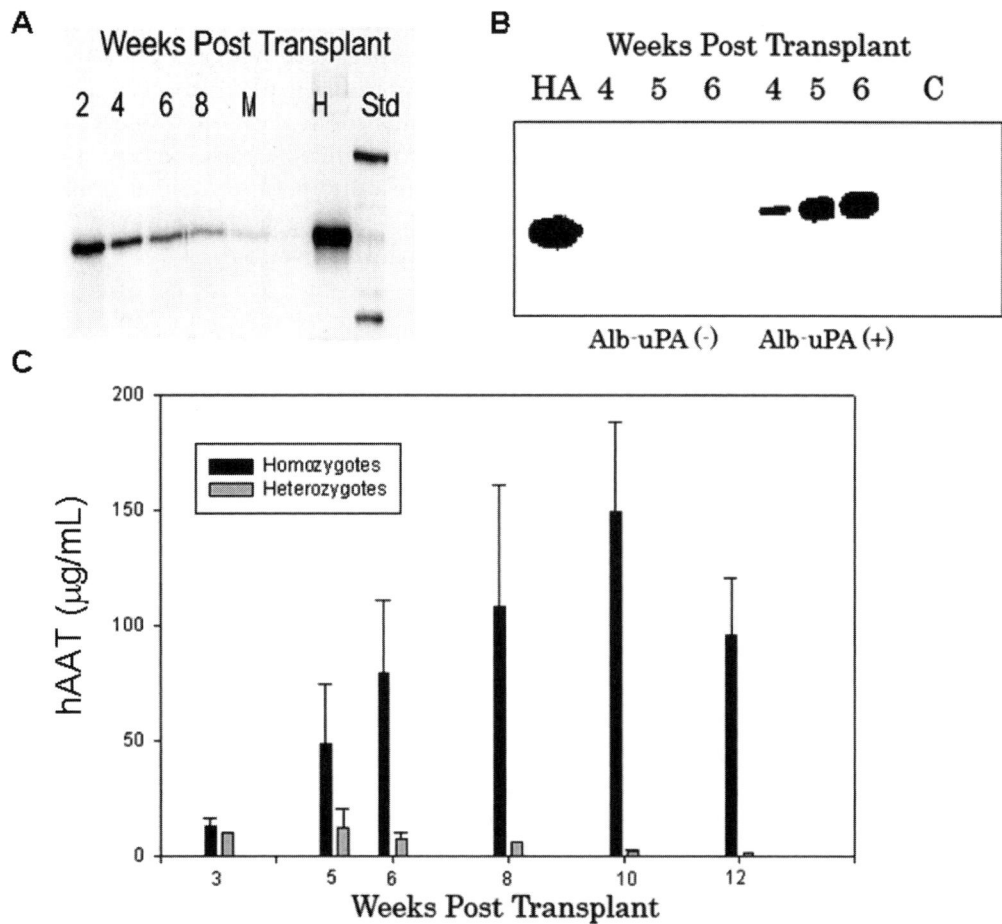

Fig. 29.1. **A,** Human hepatocytes transplanted into SCID mice displayed a limited duration, relatively low level, and steadily diminishing human albumin signal in mouse serum as reflected in this sequential western-blot analysis. These experiments demonstrated that an ongoing stimulus to human hepatocyte expansion was required. **B,** SCID-beige/Alb-uPA mice engrafted with human hepatocytes demonstrate increasing levels of human albumin in sera over the early post-transplant weeks that reach much higher levels than seen in nontransgenic SCID mice receiving fresh hepatocyte grafts. **C,** Hemizygous Alb-uPA mice demonstrate low levels of human protein including albumin or alpha-1 antitrypsin (hAAT), as illustrated here, that diminish within several weeks of transplant. Homozygous mice generate much higher levels of human protein in serum, reflecting much higher levels of human hepatocyte repopulation that can reach greater than 70%.

My laboratory began work on the establishment of a mouse model of HCV infection in 1994 *(11)*. Following our plan for development of an in vivo culture system for human hepatocytes, we first used human hepatocyte transplants in *scid-bg* mice. As noted above, this model showed modest early graft function and early loss of the human hepatocytes. We believed an ongoing stimulus to hepatocyte replication was required to produce high-level repopulation of the mouse liver with human hepatocytes. Our hypothesis was that an ongoing stimulus for hepatocyte replication could achieve high-level human chimerism that would support HCV infection and replication. With recognition that the subacute liver failure resulting from the uPA transgene might well provide the ongoing stimulus for replication, while simultaneously creating the "room" for human hepatocyte expansion that had been found important in other fields of cellular transplantation, we began to pursue this exciting approach to the development of a mouse with a humanized liver.

We bred transgenic Alb-uPA mice against a *scid-beige* background to generate *scid*-bg/Alb-uPA mice. Our concerns about the high mortality initially reported for uPA mice led us to accept the then general dogma in the field that homozygous uPA mice would not be a viable model because of excessive mortality in a breeding colony. We therefore began to explore the use of the uPA mouse for HCV using a breeding colony of hemizygous *scid*-bg/Alb-uPA mice. The first offspring revealed a different pattern of human albumin in sera over time, in which increasing protein levels (**Fig. 29.1**, center panel) contrasted with the diminishing levels seen in earlier work (**Fig. 29.1**, upper panel).

Evaluation of the initial series of human hepatocyte transplants to the offspring of these breeders generated what appeared to be distinct subpopulations, characterized by expression of human albumin in sera of the mice. Some mice never demonstrated significant levels of human albumin, others had good albumin expression initially but lost it within a few months of transplant, and a third group demonstrated high-level human albumin in sera and continued to express it for many months.

Attempts at inducing HCV infection provided additional insight. The first was the "Eureka!" moment in July of 2000 when we first confirmed HCV RNA by RT-PCR in the sera of a mouse. Next was the development of an understanding of why we had succeeded where others had failed. The only mice to demonstrate HCV infection were among those with high-level persistent human albumin expression in sera. The recognition that we could be witnessing the results of simple Mendelian distribution of the uPA transgene led to establishment of a Southern blot technique designed to clarify the zygosity of our mice. The mice with persistent high-level human albumin and alpha-1 antitrypsin (hAAT; **Fig. 29.1**, lower panel) expression (and by histological assessment high-level human chimerism) were uniformly homozygous. Mice hemizygous for uPA demonstrated very modest hepatocyte engraftment, but homozygous animals supported improved liver repopulation to levels well beyond 50% and sustained chimerism, with a median graft duration of >30.5 weeks *(12, 13)*. Despite repeated attempts across the spectrum of mice generated from the hemizygous breeding colony, the only offspring ever demonstrated to support infection and replication of HCV were the homozygous mice. We subsequently converted our colony to solely homozygous breeders. Careful attention to animal husbandry kept mortality to acceptable levels.

Later experiments with the model by Leroux-Roels' team more fully characterized the repopulation of the mouse liver and showed direct communications between biliary radicals of mouse and human origins on histological assessment. Regenerative nodules involved both murine and human hepatocytes. The human hepatocytes appear histologically quite normal, aside from glycogen deposition (14). Yoshizato's team carried out a series of functional assessments of human metabolic activity in homozygous uPA transgenic SCID mice with high-level hepatic repopulation with human hepatocytes (>70%). Substantial evidence was reported in a series of publications demonstrating the very "human" metabolic profile of the mice (15–18).

Others have explored use of the homozygous uPA mice with chimeric livers and have reported on the role of NK cell depletion and/or macrophage depletion as factors that can improve success rates of high-level human hepatocyte repopulation (19, 20). Our own studies are in support of this approach in *scid* mice that lack the beige mutation. We have also generated uPA mice on a Rag common gamma chain knock-out background with similar outcomes after transplant. Both freshly isolated hepatocytes and cryopreserved hepatocytes have been used for repopulation in the model with success, the common requirement for success being a strict need for high-quality, high-viability human liver cells (11, 12).

Animals can be infected with serum from patients carrying hepatitis C, as well as with infectious clones of HCV and with the JFH1 virus (21, 22). HCV titers have ranged from low levels, below quantitative assay range, to greater than 10^7 copies/mL, similar to the range of levels found in infected humans. Although infected mice differ considerably in viral titers, serum titers of individual animal usually remain quite stable for several weeks and often for many months (**Fig. 29.2**). HCV infection was attainable by inoculation with HCV-laden human serum in up to 90–100% of cohorts of homozygous animals with high-level human hepatocyte repopulation as reflected by high serum levels of hAAT. Initial increases in the total viral load of up to 1950-fold were demonstrated, as well as confirmation of replication by detection of negative-strand viral RNA in liver recovered from transplanted mice. HCV viral proteins were localized to human hepatocyte nodules, and infections were serially passaged through three generations of mice, confirming both synthesis and release of infectious viral particles (12). Successful infections have been established with viral genotypes 1a, 1b, 2a, 3a, 4a, and 6a (unpublished data).

Although challenging, this animal model with a chimeric human liver has been reproduced by several other groups,

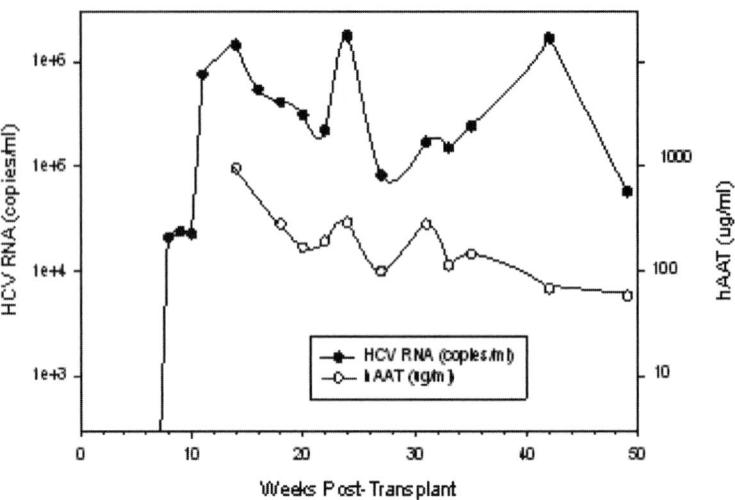

Apr18mLM - hAAT and HCV RNA

Fig. 29.2. Human hepatocyte chimerism and HCV infection are capable of long-term support in the uPA mouse model as demonstrated by maintenance of serum levels of human alpha-1 antitrypsin (Elisa) and HCV RNA (quantitative RT-PCR) beyond 1 y.

providing independent support for the utility of the chimeric mouse in study of human liver infections including HCV, HBV, and malaria (14, 19, 22). The model has also been demonstrated to be useful for in vivo evaluation of drug metabolism by human liver cells (15–18) and in vivo study of toxicity to human (liver) cells (23).

3. HCV Study in Mice with Chimeric Human Livers

This model demonstrates several important advantages in the study of HCV. HCV infection and replication occur in human hepatocytes, the virus' authentic environment, in contrast to other models such as the chimpanzee, the tamarin, and the tupaia (24–26). In addition, mice can be infected by inoculation with the serum of patients infected with HCV, demonstrating the susceptibility of the system to wild-type virus (without adaptive mutation) and to infection by the usual route. The model allows the study of the natural infection with easily measurable readouts, including serum measurement of HCV RNA by qualitative or quantitative PCR (27) and measurement of serum core antigen (28). Showing all stages of HCV infection from introduction into the host through bloodstream transport, cell entry, enzyme activation, RNA replication, viral particle production, and cellular egress, the model is well suited for studies of HCV biology and for generation of new therapeutic approaches. This model clearly represents a major step, as several compounds with in vitro activity against HCV enzymes had previously been advanced directly into clinical trials, with limited or no in vivo efficacy data and with broadly poor outcomes in clinical trial (Heptazyme, antisense, etc.)

Selection of animals for study should typically begin with measurement of human protein products in the sera of transplant-recipient mice, for confirmation of high-level human hepatocyte repopulation. Our initial studies used western blot and then semi-quantitative dot-blot assessment of human albumin. Other labs have reported use of ELISA for human albumin. At present, we use an ELISA for serum hAAT, in our experience a more reliably quantitative assay. Levels at 4–6 weeks after transplantation provide an initial screen for further assessment at 8 weeks. Probability of infection after HCV inoculation is correlated with the hAAT level, which itself reflects human hepatocyte engraftment. Mice in our colony with levels below $80 \mu g/L$ are rarely infectable, whereas those between 100 and $200 \mu g/L$ have a 20–30% chance of sustaining HCV infection; near-uniform infection results at levels higher than $500 \mu g/L$. We typically use mice with HCV RNA titers of at least 10^4 copies/mL for therapeutic studies, providing for evaluation of a minimum 2 log reduction (and up to 5 log reduction) in HCV titer with intervention (29).

4. Validation of the scid-Alb/uPA Chimeric Mouse Model of HCV Infection

Our initial stimulus for development of the chimeric mouse model was to facilitate evaluation of putative anti-HCV therapeutics for advancement to clinical study. Understanding of the fidelity of results in translation from the in vivo mouse model to clinical study is clearly a crucial measure of the value and utility of the model in the development of prophylactic or therapeutic products for use in those at risk or infected with HCV.

A range of developing and established therapeutics has been studied in the chimeric mouse that has also had evaluation in clinical application. These include interferon alpha 2b and molecules whose targets are validated HCV enzyme targets including protease inhibitors and polymerase inhibitors. Drugs aimed at other HCV targets as well as host targets important in HCV replication have also been studied, none of which has to date undergone clinical evaluation.

Several cohorts of HCV-infected mice were studied during and after treatment with interferon alpha 2b (Inferon, Schering) and outcomes compared to parallel control-vehicle-treated mice (29). Mice treated with 1350 IU/g/day interferon alpha 2b intramuscularly for 10–28 days demonstrated significantly lower viral titers than did control mice; dose responsiveness was demonstrated by lower impact with a 400 IU/g/day dose. As in clinical application, a two-phase decline in HCV titers has been routinely and reliably observed in HCV RNA titers. HCV titers typically decline by roughly 1 log within the first 4–7 days of therapy with intramuscular or subcutaneous interferon, and further decline by an average of 0.3 log per additional week of therapy out to 1 month (**Fig. 29.3**). Rates of viral titer decline suggest that at least 20 weeks of therapy would be required for viral clearance from the genotype 1a infection that is our standard inoculum. A very similar response to treatment with interferon has been reliably observed in over 10 cohorts of HCV-infected mice; as a result we now routinely use interferon treatment when a positive control treatment group is desired in addition to a standard vehicle-treated control group.

In contrast to the response of genotype 1a-infected mice, but in keeping with clinical experience, genotype 3a-infected mice have demonstrated much greater response upon treatment with interferon. HCV titers in 3a-infected mice have dropped 2–3 logs with 4–7 days therapy, and a substantial proportion have shown viral clearance (to below the detection limit of quantitative and

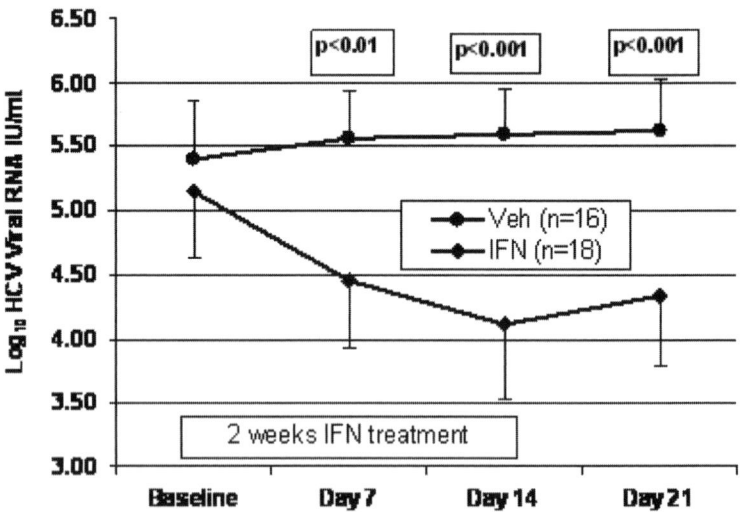

Fig. 29.3. Serum HCV RNA titers with interferon treatment. Chimeric SCID/uPA mice infected with genotype 1a were treated with 1350 IU/g body weight human interferon alpha 2b (IFN) by intramuscular injection once daily for 14 days. Viral titers dropped 0.8 log relative to control saline-treated (Veh) mice over the first week of treatment and showed a slower decline of 0.3 log after the second week of treatment. Subsequent mice treated for up to 4 weeks demonstrated very similar rates of viral decline of 0.3 log per week for the later weeks of treatment.

qualitative assays) and no evident rebound after an additional 2 weeks of follow-up *(29)*.

The first small-molecule anti-HCV therapeutic agent to demonstrate effect in clinical trial was the NS3/4A protease inhibitor BILN2061. This molecule produced dramatic 2- to 3-log decreases in HCV titer in patients infected with genotype 1a/1b after only 48 h of treatment in a q 12-h oral dosing regimen *(30)*. We developed a dosing regimen to achieve similar trough levels in the chimeric mouse using serum levels from a preliminary PK/toxicity study in infected mice. Administration of BILN2061 by oral gavage twice daily at a dose of 10 mg/kg resulted in viral titers declining by over 2 logs after 4 days of treatment in all three mice in the pilot series. Subsequent therapy for 14 days revealed a decrease of 1.4 logs at 7 days, and assessment at week 2 revealed a mean 1.1 log decrease from baseline *(29)*. Three of six mice demonstrated rebound of HCV titers during the second week of therapy. This outcome could be explained by either resistance development or deterioration in PK with ongoing treatment by such mechanisms as enzyme induction.

Sera and liver were recovered from two mice with rising viral titers despite ongoing treatment. HCV RNA was recovered and sequenced. Although the sequence alterations identified in previous replicon studies were not seen *(31, 32)*, a viral variant with Q to K mutation at amino acid 80 was identified. After cloning and insertion into a subgenomic replicon, studies of the Q80K mutant revealed a three- to fourfold lower sensitivity to BILN2061 than wild-type virus. Interestingly, the variant at amino acid 80 results in a sequence similar to that of HCV genotype 2, a genotype that has been demonstrated to have a reduced response to BILN2061 in clinical trials *(33, 34)*. Sufficient sera were not available from the treated mice for detailed PK analysis.

BILN has subsequently been halted in clinical development by Boehringer-Ingelheim, reportedly because of evidence of cardiotoxicity in primate studies. Subsequent directed evaluation in the uPA mouse model revealed evidence of cardiotoxicity from BILN2061 that had not been evident in our initial studies *(35)*. Although no deaths were noted during or after 14 days of treatment at 10 mg/kg bid during our own studies, treatment-related deaths were associated with histological evidence of cardiotoxicity with plasma levels of 840 and 4400 nM in two HCV-infected uPA-SCID animals in the subsequent mouse model evaluation by the Leroux-Roels group. These plasma levels were approximately 20-fold higher than those in the clinical trials and several fold higher than those in our own mouse studies.

A series of experiments in collaboration with Wyeth Research and Viropharma, Inc., provided further validation of the chimeric mouse as a useful predictor of clinical impact of anti-HCV treatment, while demonstrating the range of studies possible in a small-animal model that cannot be achieved in chimpanzee studies *(36)*. Initial proof-of-concept studies with the HCV NS5B RNA–dependent RNA polymerase inhibitor HCV796 administered every 8 h by oral gavage demonstrated a 2.02-log reduction in viral titers after 5 days of therapy and a rebound of 1 log within 1 week of the end of treatment. The outcomes were confirmed in a repeat study with a mean 1.78 ± 0.59 log decrease after 10 days of treatment with HCV 796, in comparison to a 1.11 ± 0.45 log decline in an interferon alpha positive control comparator group. Of interest, one of five mice in the 2-week study showed rebound of viral titers during the second week of single-agent therapy. A dose-response study compiling data from 10, 30, 50, 75, and 125 mg/kg groups suggested dose response peaked at above 50 mg/kg because of plateau in serum levels. Combination therapy of a suboptimal dosage of HCV796 of 30 mg/kg with interferon demonstrated at least additive impact, producing a decrease of 2.44 log after 10 days and a decline in all mice throughout treatment.

Subsequent clinical trials with single-agent HCV796 demonstrated 1.4-log reduction in HCV titers after 4 days, 1.3 log at 7 days, and maintenance of 0.7-log reduction at 2 weeks. Outcomes with combined treatment with interferon mirrored outcomes in the mouse experiments, producing substantially enhanced outcomes over single-agent therapy: 2.1- to 2.7-log reductions at day 7 and 3.3- to 3.5-log reduction at day 14. As in the mouse studies, no rebound was seen in any patient during combination therapy *(37)*. Results with these three successful anti-HCV therapies provide strong validation for the positive predictive value of the chimeric mouse model.

Support for the negative predictive value of the model is also available. The first lead molecule developed in the Wyeth/Viropharma collaboration on polymerase inhibitors for HCV therapy was the pyrranoindole series molecule HCV 371. This molecule demonstrated in vitro activity in the replicon model with EC50 at 5 µM *(38)*. In vivo studies in the chimeric mouse model failed to demonstrate significant decline in HCV titers at any point during 3 weeks of subcutaneous administration *(29)*. Subsequent phase 1b clinical trials failed to achieve efficacy targets, and the compound was dropped from further clinical evaluation *(39)*. At least 20 molecules have been evaluated for anti-HCV activity in the chimeric mouse model. Among them were eight small-molecule antivirals aimed at clinically validated viral or recipient targets and a further four exploring viral targets not yet confirmed by clinical experience. Four studies explored antibody preparations aimed at viral entry, one of several prevention strategies yet to demonstrate success. Interferon studies and evaluations of host targets round out the experience in our own laboratories and those of others.

Within this experience are two molecules that were ineffective in the chimeric mouse model but proceeded to clinical trial evaluation. Both failed. Five compounds were ineffective in the mouse model, and further development was halted. One drug could not be evaluated properly in the model, because of rapid metabolism that challenged achievement of adequate trough serum levels. This compound failed to reach clinical targets and was dropped from further development. We are aware of five drugs with demonstrated efficacy in reducing HCV titers in the mouse model. As noted above, four demonstrated efficacy in clinical evaluation, and one proved to have unacceptable pK in human studies. These results reflect the high positive and negative predictive value of the model in application for testing of antiviral drugs before clinical assessment.

The model has also been used to study atypical approaches to HCV therapy. We investigated a gene therapy directed against the NS3/NS4A serine protease, a key factor in replication of HCV. BH3 interacting death domain agonist (BID) is an apoptosis-inducing molecule that is part of the intrinsic caspase system involved in activation of the mitochondrial caspase cascade. The BID molecule was engineered to contain a cleavage site that specifically recognized the NS3/NS4A protease. Using delivery to hepatocytes by nonreplicating adenovirus in chimeric mice, we induced apoptosis in cells containing both the HCV NS3/NS4A protease and the modified BID. As a consequence, serum ALT increased within 1–2 days after adenovirus administration, reflecting hepatocyte death. HCV titers decreased by over 2 logs *(40)*. Subsequent studies with repeated doses of modified BID at 7-day intervals achieved viral clearance in five of seven mice (**Fig. 29.4**). hAAT levels initially dropped after administration, probably reflecting the loss of HCV-infected human hepatocytes. hAAT then rebounded to near baseline levels, as would be expected with human hepatocyte graft persistence and regeneration.

5. Other Applications of the Model

The development of a mouse model with a chimeric liver containing a high proportion of human hepatocytes provides the potential for a wide range of studies of human liver physiology, metabolism, toxicology, and infectious diseases in vivo in a small-animal model. Although to date the dominant interest in the chimeric mouse model has been in the study of HCV infection, it has numerous other potential applications *(11)*.

All pathogens that have an obligatory requirement to undergo parts of their life cycles in the liver could be studied

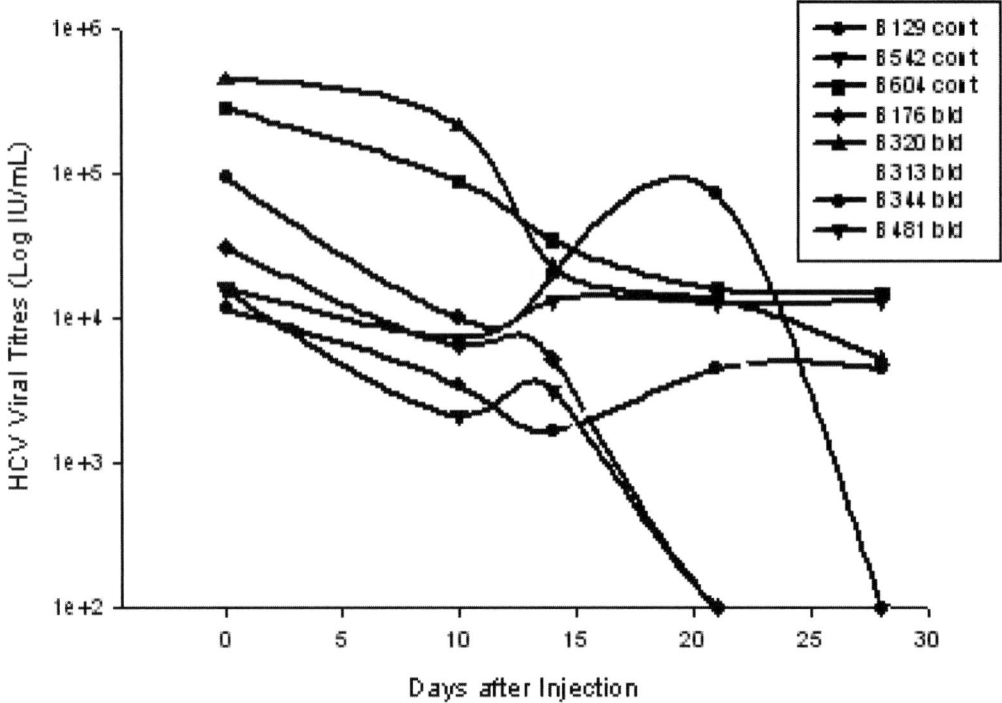

Fig. 29.4. Titers of HCV after administration of three doses of mBID. mBID was administered on days 0, 7, and 14, and samples were subsequently taken on days 10, 14, 21, and 28 after the initial injections of modified BID (bid) or control adenovirus (cont). Samples were assayed for HCV genomic RNA by RT-PCR. Although HCV titers remained unchanged in two of three control mice and declined by approximately 1 log in the third, four of five treated mice cleared virus after three treatments with modified BID.

with the model. For example, a murine model of malaria infection has been established. *Plasmodium falciparum* has such an obligate liver stage in its development. Chimeric mice have been infected with *P. falciparum* sporozoites and produced morphologically and antigenically mature liver-stage schizonts containing merozoites capable of invading human red blood cells. This model has a wide potential utility for assessing the biology of the parasite, potential antimalarial therapeutic agents, and vaccine design *(41)*. An effective vaccine against the liver stage of *P. falciparum* malaria could prevent maturation and progression to reinfection of red blood cells with merozoites and the subsequent development of the sexual stage of the disease, which is required for the transmission of the parasite to others. Study of the liver stage of infection has been particularly difficult because of limited access to such infected liver tissue (few indications lead to liver biopsy in malaria-infected patients). Isolation of liver-stage parasites could provide material for exploration of potentially unique liver-stage antigens to serve as targets for vaccine design.

A wider role for the chimeric mouse in study of the biology of hepatotropic infection is also possible. The model has been applied to evaluation of alterations in the genomics of human liver cells subsequent to infection with HCV with specific reference to alterations in the intrinsic immune system that might explain the high level of chronicity of HCV infection *(42)*. It may prove helpful in the study of cell entry by HCV and in the study of the persistence of HBV as a result of the generation of closed covalent circular DNA in vivo.

Mice with chimeric human livers may also prove extremely useful in the evaluation of drug toxicity to or metabolism within human liver. For example, TNF-related apoptosis-inducing ligand (TRAIL) is a newly identified member of the TNF family that induces tumor apoptosis. Unfortunately, simultaneous with early clinical development, TRAIL was reported to cause death of human hepatocytes in culture *(43)*. We reported that intravenous injection of TRAIL in mice with human chimeric livers and subcutaneous or intraperitoneal human cancer implants was associated with decreased tumor volumes without histological evidence of injury to human hepatocytes and with maintenance of baseline levels of serum levels of human proteins. Native TRAIL had no hepatotoxicity in the chimeric mouse, whereas injection of FasL caused apoptotic death of human hepatocytes and thus killed the mice. TRAIL could inhibit intraperitoneal and subcutaneous tumor growth without inducing apoptosis of human hepatocytes *(23)*. These results demonstrated the lack of in vivo toxicity of the nontagged TRAIL molecule and helped to facilitate its further clinical development.

The evaluation of metabolism of candidate drugs by human liver cells is a critical step in development of any new therapeutic. Lot-to-lot variability in commercially available human hepatocytes can challenge this process. The ability to generate cohorts of mice with stable high-level repopulation with human liver cells could provide a stable platform for such testing and thereby greatly facilitate drug development. Chimeric mice with human liver-cell repopulation to greater than 70% of hepatocyte mass have demonstrated cytochrome P450 subtype profiles similar to those of donor human livers and have retained normal pharmacological profile. As a consequence, mRNAs for hCYP3A4 and hCYP1A1/2 can be induced in the livers of mice treated with rifampicin and 3-methylcholanthrene *(19)*. The metabolism of debrisoquin, a CYP2D6 substrate, has further been studied. As expected, its metabolism was higher in animals with high human hepatocyte repopulation. Such metabolism could also be inhibited by quinidine, a CYP2D6 inhibitor. These tests confirm that the chimeric mice exhibit humanized profiles of drug metabolism *(15–18)*. A promising role for the chimeric mouse model in evaluation of drug metabolism and/or toxicity may well develop in the near future.

6. Limitations of the Model and Potential Future Improvements

This model is challenging as the animals are immunodeficient, are at high risk of neonatal bleeding from the uPA transgene, and suffer from a subacute liver failure, at least until the human hepatocyte xenograft is well established. To maximize the opportunity for human hepatocyte expansion before competition by murine liver cells that have spontaneously deleted the uPA transgene, we transplant between 5 and 21 days after birth by injection of 5×10^5–10^6 human hepatocytes into the inferior pole of the spleen. This procedure requires microsurgical techniques and should be done by an experienced operator to reduce mortality. The success of engraftment and the degree of repopulation with human hepatocytes can be monitored by assay of serum for levels of human proteins such as albumin or hAAT.

A key to success in study of HCV infection as well as in evaluation of human drug metabolism is the achievement of high hepatocyte repopulation, as reflected by the need to achieve high serum concentration of human albumin/hAAT to support HCV infection. The rate of liver repopulation is strongly related to the quality of the hepatocytes and the age of the hepatocyte donor, so access to large numbers of high-quality cells is mandatory. Ideally donors should be young and without liver disease or previous chemotherapy. Providing such tissue in sufficient quantity on a regular basis is challenging. The ability to use alternative sources for hepatocyte repopulation would without doubt improve access to the model by laboratories worldwide and facilitate a broadening of its use. Primary fetal hepatocytes *(44)*, human hepatocyte lines *(45)*, and monocyte-derived hepatocyte-like cells *(46)* are possible alternative candidates that need further evaluation.

Strategies that improve engraftment would improve the efficiency of the model and reduce the cost per mouse available for study. Inhibitions of NK cell and macrophage activity have both been reported to improve level of human hepatocyte repopulation. High-level engraftment has been reported to be associated with renal damage and mortality from human complement activity. Tateno et al. were able to sustain high-level engraftment without mortality by inhibiting the human complement activity resulting from human hepatocyte synthetic activity; repopulation rates up to 96% were thus achieved *(19)*. Strategies inhibiting cell apoptosis may be another means of facilitating improved engraftment *(47)*.

The chimeric human liver mouse model has multiple and expanding possible applications, including investigations of HBV, HCV, malaria, drug metabolism, and drug toxicity. At least 20 different anti-HCV drugs and drug candidates have been studied to date in the chimeric mouse, and the results have shown an excellent correlation with subsequent clinical outcome. This model should therefore improve the process of development of drugs for HCV infection and provide the potential to enter clinical trials with greater confidence of success and better guidance about PK and toxicity issues. Unfruitful clinical studies should be reduced, and time to clinical application shortened, so time should be saved, costs reduced, and human subjects spared unnecessary exposure to drugs with little likelihood of success in the clinic. Application of the chimeric mouse to the study of human hepatic metabolism of candidate molecules should further facilitate the improved and more cost-efficient development of novel therapeutics. This model should earn a wider role in the study of biology and therapeutics in future.

References

1. World Health Organization (1999) Global surveillance and control of hepatitis C. Report of a WHO Consultation organized in collaboration with the Viral Hepatitis Prevention Board, Antwerp, Belgium. *J. Viral Hepat.* **6**, 35–47.

2. Alter, M. J., Kruszon-Moran, D., Nainan, O. V., McQuillan, G. M., Gao, F., Moyer, L. A., et al. (1999) The prevalence of hepatitis C virus infection in the United States, 1988 through 1994. *N. Engl. J. Med.* **341**, 556–562.

3. Alter, M. J., Margolis, H. S., Krawczynski, K., Judson, F. N., Mares, A., Alexander, W. J., et al. (1992) The natural history of community-acquired hepatitis C in the United States. The Sentinel Counties Chronic non-A, non-B Hepatitis Study Team. *N. Engl. J. Med.* **327**, 1899–1905.

4. McHutchison, J. G., Gordon, S. C., Schiff, E. R., Shiffman, M. L., Lee, W. M., Rustgi, V. K., et al. (1998) Interferon alfa-2b alone or in combination with ribavirin as initial treatment for chronic hepatitis C. Hepatitis Interventional Therapy Group. *N. Engl. J. Med.* **339**, 1485–1492.

5. Moreno-Otero, R. (2005) Therapeutic modalities in hepatitis C: challenges and development. *J. Viral Hepat.* **12**, 10–19.

6. Heckel, J. L., Sandgren, E. P., Degen, J. L., Palmiter, R. D., and Brinster, R. L. (1990) Neonatal bleeding in transgenic mice expressing urokinase-type plasminogen activator. *Cell* **62**, 447–456.

7. Sandgren, E. P., Palmiter, R. D., Heckel, J. L., Daugherty, C. C., Brinster, R. L., and Degen, J. L. (1991) Complete hepatic regeneration after somatic deletion of an albumin-plasminogen activator transgene. *Cell* **66**, 245–256.

8. Rhim, J. A., Sandgren, E. P., Degen, J. L., Palmiter, R. D., and Brinster, R. L. (1994) Replacement of diseased mouse liver by hepatic cell transplantation. *Science* **263**, 1149–1152.

9. Rhim, J. A., Sandgren, E. P., Palmiter, R. D., and Brinster, R. L. (1995) Complete reconstitution of mouse liver with xenogeneic hepatocytes. *Proc. Natl. Acad. Sci. USA* **92**, 4942–4946.

10. Dandri, M., Burda, M. R., Torok, E., Pollok, J. M., Iwanska, A., Sommer, G., et al. (2001) Repopulation of mouse liver with human hepatocytes and in vivo infection with hepatitis B virus. *Hepatology* **33**, 981–988.

11. Kneteman, N. M. and Mercer, D. F. (2005) Mice with chimeric human livers: who says supermodels have to be tall? *Hepatology* **41**, 703–706.

12. Mercer, D. F., Schiller, D. E., Elliott, J. F., Douglas, D. N., Hao, C., Rinfret, A., et al. (2001) Hepatitis C virus replication in mice with chimeric human livers. *Nat. Med.* **7**, 927–933.

13. Meuleman, P., Vanlandschoot, P., and Leroux-Roels, G. (2003) A simple and rapid method to determine the zygosity of uPA-transgenic SCID mice. *Biochem. Biophys. Res. Commun.* **308**, 375–378.

14. Meuleman, P., Libbrecht, L., De Vos, R., de Hemptinne, B., Gevaert, K., Vandekerckhove, J., et al. (2005) Morphological and biochemical characterization of a human liver in a

uPA-SCID mouse chimera. *Hepatology* **41**, 847–856.

15. Katoh, M., Sawada, T., Soeno, Y., Nakajima, M., Tateno, C., Yoshizato, K., et al. (2007) In vivo drug metabolism model for human cytochrome P450 enzyme using chimeric mice with humanized liver. *J. Pharm. Sci.* **96**, 428–437.

16. Kakinuma, S., Asahina, K., Okamura, K., Teramoto, K., Tateno, C., Yoshizato, K., et al. (2007) Human cord blood cells transplanted into chronically damaged liver exhibit similar characteristics to functional hepatocytes. *Transplant. Proc.* **39**, 240–243.

17. Okumura, H., Katoh, M., Sawada, T., Nakajima, M., Soeno, Y., Yabuuchi, H., et al. (2007) Humanization of excretory pathway in chimeric mice with humanized liver. *Toxicol. Sci.* **97**, 533–538.

18. Yoshitsugu, H., Nishimura, M., Tateno, C., Kataoka, M., Takahashi, E., Soeno, Y., et al. (2006) Evaluation of human CYP1A2 and CYP3A4 mRNA expression in hepatocytes from chimeric mice with humanized liver. *Drug Metab. Pharmacokinet.* **21**, 465–474.

19. Tateno, C., Yoshizane, Y., Saito, N., Kataoka, M., Utoh, R., Yamasaki, C., et al. (2004) Near completely humanized liver in mice shows human-type metabolic responses to drugs. *Am. J. Pathol.* **165**, 901–912.

20. Morosan, S., Hez-Deroubaix, S., Lunel, F., Renia, L., Giannini, C., Van Rooijen, N., et al. (2006) Liver-stage development of *Plasmodium falciparum,* in a humanized mouse model. *J. Infect. Dis.* **193**, 996–1004.

21. Lindenbach, B. D., Evans, M. J., Syder, A. J., Wolk, B., Tellinghuisen, T. L., Liu, C. C., et al. (2005) Complete replication of hepatitis C virus in cell culture. *Science* **309**, 623–626.

22. Hiraga, N., Imamura, M., Tsuge, M., Noguchi, C., Takahashi, S., Iwao, E., et al. (2007) Infection of human hepatocyte chimeric mouse with genetically engineered hepatitis C virus and its susceptibility to interferon. *FEBS Lett.* **581**, 1983–1987.

23. Hao, C., Song, J. H., Hsi, B., Lewis, J., Song, D. K., Petruk, K. C., et al. (2004) TRAIL inhibits tumor growth but is nontoxic to human hepatocytes in chimeric mice. *Cancer Res.* **64**, 8502–8506.

24. Prince, A. M. and Brotman, B. (1998) Biological and immunological aspects of hepatitis C virus infection in chimpanzees, in *Hepatitis C Virus* (Reesink, H. W., ed.), Current Studies in Hematology and Blood Transfusion 62, Karger, Basel, New York, pp. 250–265.

25. Xie, Z. C., Riezu-Boj, J. I., Lasarte, J. J., Guillen, J., Su, J. H., Civeira, M. P., et al. (1998) Transmission of hepatitis C virus infection to tree shrews. *Virology* **244**, 513–520.

26. Beames, B., Chavez, D., Guerra, B., Notvall, L., Brasky, K. M., and Lanford, R. E. (2000) Development of a primary tamarin hepatocyte culture system for GB virus-B: a surrogate model for hepatitis C virus. *J. Virol.* **74**, 11764–11772.

27. Brass, V., Blum, H. E. and Moradpour, D. (2002) Of mice and men: a small animal model of hepatitis C virus replication. *Hepatology* **35**, 722–724.

28. Cagnon, L., Wagaman, P., Bartenschlager, R., Pietschmann, T., Gao, T., Kneteman, N. M., et al. (2004) Application of the trak-C HCV core assay for monitoring antiviral activity in HCV replication systems. *J. Virol. Methods* **118**, 23–31.

29. Kneteman, N. M., Weiner, A. J., O'Connell, J., Collett, M., Gao, T., Aukerman, L., et al. (2006) Anti-HCV therapies in chimeric scid-Alb/uPA mice parallel outcomes in human clinical application. *Hepatology* **43**, 1346–1353.

30. Lamarre, D., Anderson, P. C., Bailey, M., Beaulieu, P., Bolger, G., Bonneau, P., et al. (2003) An NS3 protease inhibitor with antiviral effects in humans infected with hepatitis C virus. *Nature* **426**, 186–189.

31. Lin, C., Lin, K., Luong, Y. P., Rao, B. G., Wei, Y. Y., Brennan, D. L., et al. (2004) In vitro resistance studies of hepatitis C virus serine protease inhibitors, VX-950 and BILN 2061: structural analysis indicates different resistance mechanisms. *J. Biol. Chem.* **279**, 17508–17514.

32. Lin, C., Gates, C. A., Rao, B. G., Brennan, D. L., Fulghum, J. R., Luong, Y. P., et al. (2005) In vitro studies of cross-resistance mutations against two hepatitis C virus serine protease inhibitors, VX-950 and BILN 2061. *J. Biol. Chem.* **280**, 36784–36791.

33. Reiser, M., Hinrichsen, H., Benhamou, Y., Reesink, H. W., Wedemeyer, H., Avendano, C., et al. (2005) Antiviral efficacy of NS3-serine protease inhibitor BILN-2061 in patients with chronic genotype 2 and 3 hepatitis C. *Hepatology* **41**, 832–835.

34. Thibeault, D., Bousquet, C., Gingras, R., Lagace, L., Maurice, R., White, P. W., et al. (2004) Sensitivity of NS3 serine proteases from hepatitis C virus genotypes 2 and 3 to the inhibitor BILN 2061. *J. Virol.* **78**, 7352–7359.

35. Vanwolleghem, T., Meuleman, P., Libbrecht, L., Roskams T., De Vos R., and Leroux-Roels, G. (2006) The human liver uPA-SCID mouse model allows for the rapid detection of antiviral and cardiotoxic effects of BILN2061.

13th International meeting on Hepatitis C Virus and Related Viruses, Cairns, Australia. Abstract 520.

36. Kneteman, N., Howe, A., Gao, T., Lewis, J., Collett, M., Pevear, D., et al. (2007) An NS5B polymerase inhibitor with potent anti-HCV impact in vitro, in mice with chimeric human livers in humans infected with the hepatitis C virus (HCV). *Hepatology,* submitted.

37. Viropharma Press Release (2006) HCV 796 Viropharma announces presentation of HCV 796 Phase Ib data at Digestive Disease Week. 21 May 2006. http://www.viropharma.com.

38. Gopalsamy, A., Aplasca, A., Ciszewski, G., Park, K., Ellingboe, J. W., Orlowski, M., et al. (2006) Design and synthesis of 3,4-dihydro-1H-[1]-benzothieno[2,3-c]pyran and 3,4-dihydro-1H-pyrano[3,4-b]benzofuran derivatives as non-nucleoside inhibitors of HCV NS5B RNA dependent RNA polymerase. *Bioorg. Med. Chem. Lett.* **16**, 457–460.

39. Viropharma Press Release (2003) HCV 371 Viropharma announces results of hepatitis C phase 1b Study. 15 July 2003. http://www.viropharma.com.

40. Hsu, E. C., Hsi, B., Hirota-Tsuchihara, M., Ruland, J., Iorio, C., Sarangi, F., et al. (2003) Modified apoptotic molecule (BID) reduces hepatitis C virus infection in mice with chimeric human livers. *Nat. Biotechnol.* **21**, 519–525.

41. Sacci, J. B., Jr., Alam, U., Douglas, D., Lewis, J., Tyrrell, D. L., Azad, A. F., et al. (2006) *Plasmodium falciparum* infection and exoery-throcytic development in mice with chimeric human livers. *Int. J. Parasitol.* **36**, 353–360.

42. Walters, K. A., Joyce, M. A., Thompson, J. C., Smith, M. W., Yeh, M. M., Proll, S., et al. (2006) Host-specific response to HCV infection in the chimeric SCID-beige/Alb-uPA mouse model: role of the innate antiviral immune response. *PLoS Pathog.* **2**, e59.

43. Jo, M., Kim, T. H., Seol, D. W., Esplen, J. E., Dorko, K., Billiar, T. R., et al. (2000) Apoptosis induced in normal human hepatocytes by tumor necrosis factor-related apoptosis-inducing ligand. *Nat. Med.* **6**, 564–567.

44. Mahieu-Caputo, D., Allain, J. E., Branger, J., Coulomb, A., Delgado, J. P., Andreoletti, M., et al. (2004) Repopulation of athymic mouse liver by cryopreserved early human fetal hepatoblasts. *Hum. Gene Ther.* **15**, 1219–1228.

45. Nguyen, T. H., Mai, G., Villiger, P., Oberholzer, J., Salmon, P., Morel, P., et al. (2005) Treatment of acetaminophen-induced acute liver failure in the mouse with conditionally immortalized human hepatocytes. *J. Hepatol.* **43**, 1031–1037.

46. Ruhnke, M., Ungefroren, H., Nussler, A., Martin, F., Brulport, M., Schormann, W., et al. (2005) Differentiation of in vitro-modified human peripheral blood monocytes into hepatocyte-like and pancreatic islet-like cells. *Gastroenterology* **128**, 1774–1786.

47. Fu, T., Blei, A. T., Takamura, N., Lin, T., Guo, D., Li, H., et al. (2004) Hypothermia inhibits Fas-mediated apoptosis of primary mouse hepatocytes in culture. *Cell Transplant* **13**, 667–676.

Part V

Immunity and Modeling

Chapter 30

Determination of HCV-Specific T-Cell Activity

Joo Chun Yoon and Barbara Rehermann

Abstract

The magnitude and breadth of T-cell responses against HCV are associated with the outcome of HCV infection. Parameters of HCV-specific T-cell responses that are frequently assessed in clinical immunological studies include proliferation of T cells in response to HCV antigens, frequency of HCV-specific T cells secreting cytokines, and changes in antigen specificity during the course of HCV infection. Common techniques for assessing these parameters such as ^{3}H-thymidine incorporation assay, cytokine ELISpot assay, and strategies for epitope mapping and identification of minimal optimal epitopes are outlined, and detailed protocols are described.

Key words: Hepatitis C virus, T cell, proliferation assay, ELISpot assay, epitope mapping.

---

1. Introduction

HCV is a noncytopathic, hepatotropic virus that elicits immune-mediated liver injury. Because the persistency rate of HCV infection is as high as 75–85% *(1)*, its impact on public health is considerable around the world *(2)*.

The outcome of HCV infection is determined by viral factors and by host factors that include innate immune response and adaptive immune response *(2–4)*. Spontaneously recovered patients usually mount vigorous multi-epitope-specific CD4$^+$ and CD8$^+$ T-cell responses that are readily detectable in the peripheral blood. In contrast, patients with persistent HCV infection develop late, transient, or narrowly focused T-cell responses *(5–8)* that select for viral escape mutations *(2, 4)*.

Studies on HCV-specific T-cell activity have been conducted with the aim to increase our understanding of the natural course

Hengli Tang (ed.), *Hepatitis C: Methods and Protocols, Second Edition, vol. 510*
© 2009 Humana Press, a part of Springer Science+Business Media
DOI 10.1007/978-1-59745-394-3_30 Springerprotocols.com

of HCV infection, the outcome of hepatitis C, and the effect of treatment. The analysis of HCV-specific T-cell responses includes the assessment (i) of T-cell proliferation in response to HCV antigens, (ii) of the frequency of cytokine-secreting HCV-specific T cells, and (iii) of changes in T-cell specificity and viral sequence during the infection. Here, we describe methods for assessing those parameters.

Proliferation assays are methods for measuring clonal proliferation of lymphocytes in peripheral blood mononuclear cells (PBMCs) in response to HCV antigens. The ^{3}H-thymidine incorporation assay has been the most frequently used assay, because of extensive published data available for comparison, the small number of cells required, and its high throughput (9). The amount of proliferation is detected after culture with antigens by measurement of incorporation of ^{3}H-thymidine into the DNA of proliferating lymphocytes.

The enzyme-linked immunospot (ELISpot) assay is a method for identification and enumeration of cytokine-secreting T cells. The assay employs the principle of the sandwich enzyme-linked immunosorbent assay (ELISA), with some modifications for investigation of cytokine-secreting cells rather than secreted cytokines in the supernatant. The ELISpot assay has the following advantages: high sensitivity, capacity for semiautomation, and good results even when cryopreserved and thawed PBMCs are used. Its sensitivity is 10- to 100-fold higher than that of intracellular cytokine staining (ICS) and flow cytometry (9). In the ELISpot assay, cytokine-secreting T cells are cultured for 24–40 h with individual peptides, pools of overlapping peptides, and recombinant proteins in culture plates coated with cytokine-specific antibodies, so that the number of cytokine spots formed by single cytokine-secreting T cell can be determined.

The matrix approach is an epitope screening method that employs peptide pools in a matrix array in ELISpot assays (10). The identity of individual, recognized peptides can be deduced from the pattern of recognized peptide pools. Minimal optimal epitopes can subsequently be determined from the peptides by performing IFN-γ ELISpot assays with individual truncated peptides. This process can be aided by computational prediction of potential epitopes by scanning HCV polypeptide sequences for specific HLA binding motifs.

2. Materials

2.1. ^{3}H-Thymidine Incorporation Assay with Recombinant HCV Proteins

1. Peripheral blood mononuclear cells (PBMCs): See **ref.** (11) for protocols regarding isolation of PBMCs. See **ref.** (9) for collection, shipment, freezing, and thawing of blood samples.

2. Hepatitis C proteins: Recombinant HCV core, NS3, helicase, NS4, NS5A, NS5B proteins, and protein buffer (Mikrogen, Neuried, Germany). Dilute with sterile $1\times$ phosphate-buffered saline (PBS; Mediatech, Herndon, VA) to $100\,\mu g/mL$.

3. Positive controls: Phytohemagglutinin (PHA; Sigma, St. Louis, MO) as mitogen and tetanus toxoid (Calbiochem/EMD, San Diego, CA) as recall antigen.

4. Complete medium: RPMI 1640 medium (Mediatech) supplemented with 10% fetal bovine serum (FBS; U.S. Bio-Technologies, Pottstown, PA), 2 mM l-glutamine (Mediatech), 100 IU/mL penicillin (Mediatech), and $100\,\mu g/mL$ streptomycin (Mediatech) (*see* **Note 1**).

5. 96-well round-bottom tissue culture plate (Nunc, Roskilde, Denmark).

6. Tritiated thymidine (^{3}H-thymidine; MP Biomedicals, Solon, OH): To prepare a working solution, dilute stock solution with RPMI 1640 or sterile $1\times$ PBS to $50\,\mu Ci/mL$.

7. Cell harvester (PerkinElmer, Wellesley, MA).

8. 96-well harvest plate (Millipore, Bedford, MA).

9. Harvest plate-sealing tapes, clear and opaque (Millipore).

10. Scintillation cocktail for microplate scintillation counters (MicroScint 40; PerkinElmer).

11. Microplate scintillation counter (TopCount; PerkinElmer).

12. Class II biological safety cabinet.

2.2. Ex Vivo IFN-γ ELISpot Assay with Overlapping HCV Peptides and Recombinant Proteins

1. Overlapping HCV peptides (OLPs): 15-mer peptides overlapping by 10 amino acids and covering the complete HCV genotype 1 polyprotein sequence (Mimotopes, Raleigh, NC) (*see* **Note 2**). Dissolve individual OLPs with sterile dimethylsulfoxide (DMSO; Fisher Scientific, Fair Lawn, NJ) at 20 mg/mL and dilute with sterile $1\times$ PBS to 1 mg/mL (1 mg OLP per 1 mL 5% DMSO/PBS). To prepare OLP pools, pool $50\,\mu L$ of each OLP solution (1 mg/mL) and adjust total volume with 5% DMSO/PBS to $2100\,\mu L$ (42-fold dilution). This concentration of each individual OLP in the pool is $24\,\mu g/mL$.

2. Hepatitis C proteins: Core, NS3, helicase, NS4, NS5A, NS5B, and protein buffer as negative control.

3. Positive controls: PHA (mitogen) and CMV control peptide pool, EBV control peptide pool, and influenza control peptide pool (CEF pools; recall antigens).

4. Coating antibody: anti-IFN-γ monoclonal antibody (M700A; Endogen/Pierce, Rockford, IL).

5. Detection antibody: biotinylated anti-IFN-γ monoclonal antibody (M701B; Endogen/Pierce) (*see* **Note 3**).

6. Alkaline phosphatase-conjugated streptavidin (streptavidin-AP; Dako, Carpinteria, CA) (*see* **Note 4**).

	A	B	C	D	E	F	G	H	I	J
AA	1	2	3	4	5	6	7	8	9	10
BB	11	12	13	14	15	16	17	18	19	20
CC	21	22	23	24	25	26	27	28	29	30
DD	31	32	33	34	35	36	37	38	39	40
EE	41	42	43	44	45	46	47	48	49	50
FF	51	52	53	54	55	56	57	58	59	60
GG	61	62	63	64	65	66	67	68	69	70
HH	71	72	73	74	75	76	77	78	79	80
II	81	82	83	84	85	86	87	88	89	90
JJ	91	92	93	94	95	96	97	98	99	100

OLP 23: hvdllvgaaa**fcsamyvgdl**
OLP 24: vgaaa**fcsamyvgdl**cgsvf
OLP 25: **fcsamyvgdl**cgsvfhdvga

Fig. 30.1. Matrix approach for epitope mapping. In this case, each matrix pool (A to J and AA to JJ) consists of five HCV overlapping peptides. A specific peptide is included in only one of the horizontal matrix pools (A to J) and only one of the vertical matrix pools (AA to JJ). Therefore, if pools D and CC yield positive responses, peptide 24 triggered the response. If pools C, D, E, and CC induce formation of a significant number of spots in the ELISpot assay, the responsible epitope probably consists of the shared sequence of peptides 23, 24 and 25 (**fcsamyvgdl**).

7. Alkaline phosphatase (AP) conjugates substrate kit (containing NBT/BCIP substrate; Bio-Rad, Hercules, CA).
8. Complete medium: RPMI 1640 medium supplemented with 10% human AB serum, 2 mM l-glutamine, 100 IU/mL penicillin, and 100 μg/mL streptomycin.
9. Sterile 1 × PBS.
10. PBS/Tween: Tween-20 (Sigma) in 1 × PBS (1:2000).
11. 1% BSA/PBS: 1% bovine serum albumin (BSA; MP Biomedicals) in 1× PBS, sterile filtered.
12. 1% BSA/PBS/Tween: 1% BSA in PBS/Tween.
13. ELISpot plates (MultiScreen-IP; Millipore).
14. Automated plate washer (Nunc).
15. Squirt bottles containing 1 × PBS or PBS/Tween.
16. ELISpot reader system (AID, Strassberg, Germany).
17. Class II biological safety cabinet.

2.3. Strategies for Epitope Mapping and Identification of Minimal Optimal Epitopes

1. Matrix pools of hepatitis C overlapping peptides: Pool OLPs according to a matrix map (e.g., **Fig. 30.1**), following step 1 in **Section 2.2**.
2. Materials for IFN-γ ELISpot assay in **Section 2.2**.

3. Methods

3.1. ³H-Thymidine Incorporation Assay with Recombinant HCV Proteins

3.1.1. Day 0: Assay Setup

Perform steps in a class II biological safety cabinet using aseptic techniques.

1. Predilute antigens to 2 µg/mL with complete medium in sterile 1.5 mL polypropylene tubes.
2. Resuspend PBMCs at 2×10^6 cells/mL with complete medium in a sterile 15 mL polypropylene tube.
3. Add 100 µL PBMC suspension to each well of a 96-well round-bottom tissue culture plate (2×10^5 cells/well) (see Note 5).
4. Add 100 µL of each antigen dilution at a final concentration of 1 µg/mL in quadruplicate to the culture plate.
5. Set up positive control wells using 1 µg/mL PHA and 1 µg/mL tetanus toxoid and negative control wells using protein buffer.
6. Incubate the plate at 37°C in a CO_2 incubator for 5 days.

3.1.2. Day 5: Pulsing ³H-thymidine

Follow standard radiation safety procedures when working with ³H-thymidine solution and ³H-thymidine-labeled cells. Perform steps in a class II biological safety cabinet using aseptic techniques.

1. Pulse the plate by adding 20 µL/well (1 µCi/well) ³H-thymidine.
2. Incubate the plate at 37°C in a CO_2 incubator for 18 h.

3.1.3. Day 6: Harvesting and Counting (See Note 6)

1. Harvest cells on a 96-well harvest plate with a cell harvester.
2. Dry the harvest plate in an oven at 56°C.
3. Seal the back of the plate with an opaque sealing tape, add 30 µL/well scintillation cocktail, and seal the front of the plate with a clear sealing tape.
4. Count the plate with a microplate scintillation counter.

3.1.4. Calculations and Reporting of Results

1. Data can be calculated and reported in two ways (see Note 7): (a) Δ counts per minute (Δ cpm) (mean cpm of experimental wells – mean cpm of background wells) and (b) stimulation index (SI) (mean cpm of experimental wells/mean cpm of background wells).
2. On the basis of the analysis of immune responses of healthy, HCV-uninfected control persons, a stimulation index of 3–5 is typically considered significant.

3.2. Ex Vivo IFN-γ ELISpot Assay with HCV Overlapping Peptides and Recombinant Proteins

Ex vivo ELISpot assays for cytokines other than IFN-γ and for other T-cell markers such as granzyme and TRAIL can be performed if appropriate coating and detection antibodies are used. The incubation time of the ELISpot cultured may need adjustment, depending on the cytokine to be analyzed.

3.2.1. Day -1: Coating

Perform steps in a class II biological safety cabinet using aseptic techniques.

1. Prepare coating antibody solution by adding 5 μL (5 μg) anti-IFN-γ monoclonal antibody to 10 mL sterile 1×PBS in a sterile 15 mL polypropylene tube (0.5 μg/mL final concentration) (*see* **Note 3**). Vortex.
2. Add 100 μL antibody solution to each well of an ELISpot plate (*see* **Note 8**).
3. Incubate the plate at 4°C overnight in a humidified box (*see* **Note 9**).

*3.2.2. Day 0: Assay Setup (See **Note 10**)*

Perform steps in a class II biological safety cabinet using aseptic techniques.

1. Wash the coated plate by adding 200 μL sterile 1 × PBS to each well using a multichannel pipette and flicking off PBS. Repeat the wash step three more times. Blot the plate on paper towel once after the last wash.
2. Immediately after blotting, block the plate with 200 μL/well 1% BSA/PBS for 1 h at room temperature in a class II biological safety cabinet. Do not let the plate dry before adding 1% BSA/PBS (*see* **Note 11**).
3. Predilute antigens at 2 μg/mL in sterile 1.5 mL polypropylene tubes with complete medium. For OLP pools, use 8 μL each pool per 200 μL complete medium. Each OLP in the pool will be diluted to 2 μg/mL.
4. Resuspend PBMCs at 3×10^6 cells/mL with complete medium in a sterile 15 mL polypropylene tube.
5. Wash the plate twice with PBS.
6. Block the plate for 30 min to 1 h with 200 μL/well complete medium at room temperature in a class II biological safety cabinet.
7. Discard medium and blot the plate once on a paper towel.
8. Add 100 μL PBMC suspension to each well (3×10^5 cells/well) (*see* **Note 12**).
9. Add 100 μL of each antigen dilution at least in duplicate, preferably in triplicate, to the plate. Final concentration of each antigen will become 1 μg/mL.
10. Set up positive control wells using 1 μg/mL PHA and 1 μg/mL CEF control peptide pools and negative control wells using 5% DMSO/PBS and protein buffer (*see* **Note 13**).
11. Incubate the plate at 37°C in a CO_2 incubator for 30 h (*see* **Note 14**).

3.2.3. Day 1: Detection Antibody

Aseptic techniques are no longer needed from this point.

1. Prepare detection antibody solution by adding 5 μL (2.5 μg) biotinylated anti-IFN-γ monoclonal antibody to 10 mL

1% BSA/PBS/Tween in a 15 mL polypropylene tube (0.25 µg/mL final concentration) (*see* **Note 3**). Vortex.

2. Wash the plate three times with PBS and four times with PBS/Tween using an automated plate washer (*see* **Note 15**).

3. Blot the plate once on a paper towel after the last wash.

4. Add 100 µL detection antibody solution to each well.

5. Incubate the plate at 4°C overnight in a humidified box (*see* **Note 16**).

3.2.4. Day 2: Development

1. Wash the plate four times with PBS/Tween using an automated plate washer.

2. Take the back cover off the plate, and wash the back of the plate membrane once with PBS/Tween using a squirt bottle. Blot dry. Reassemble the plate.

3. Prepare streptavidin-AP solution by adding 5 µL streptavidin-AP to 10 mL 1% BSA/PBS/Tween (1:2000 dilution).

4. Add 100 µL streptavidin-AP solution to each well.

5. Incubate the plate at room temperature for 1 h (see **Note 17**).

6. Prepare 1 × AP development buffer by adding 400 µL 25 × AP development buffer to 10 mL deionized water at room temperature. Vortex.

7. To prepare the AP development solution, add 100 µL color reagent A and 100 µL color reagent B to 10 mL 1 × AP development buffer. Vortex.

8. Allow the AP development solution to equilibrate to room temperature (*see* **Note 18**).

9. Wash the plate four times with PBS (no Tween-20, no BSA) using an automated plate washer (*see* **Note 19**).

10. Take the back cover off the plate. Rinse the back of the membrane extensively with PBS using a squirt bottle. Blot dry. Reassemble the plate.

11. Add 100 µL AP development solution to each well.

12. Wait for about 10–20 min until spots appear in positive control wells.

13. Rinse the plate under running tap water.

14. Remove the back cover. Blot dry. Dry the plate completely in inverted position in a laminar flow hood for 60–90 min or overnight in the dark (*see* **Note 20**).

15. Read the plate and count the spots using an automated ELISpot reader. Spots should be evaluated on the basis of size, shape, contrast, and density.

*3.2.5. Reporting of Results (See **Note 21**)*

1. Data can be reported as number of spot-forming cells (SFC) per 3×10^5 PBMCs.

2. Number of SFC against specific antigen can be calculated as follows: (mean SFC of experimental wells – mean SFC of background wells).

3.3. Strategies for Epitope Mapping and Identification of Minimal Optimal Epitopes

3.3.1. Screening of Epitopes by the Matrix Approach (IFN-γ ELISpot Assay with Matrix Peptide Pools)

1. Follow the standard IFN-γ ELISpot assay protocol in **Section 3.2**. Use matrix pools instead of individual OLPs or OLP pools.
2. Find wells showing significant numbers of spots (*see* **Note 21**).
3. Using the matrix map, locate specific peptides responsible for the putative responses (**Fig. 30.1**).
4. Repeat the IFN-γ ELISpot assay with the individual peptides that yielded positive responses in the matrix setting to confirm the putative response.
5. Compare the sequence of the confirmed peptides to those of previously described epitopes and identify known epitopes in the peptides. If no previously described epitopes appear among the confirmed peptides, determine the minimal optimal sequence by testing shorter, truncated peptides (**Fig. 30.1**).

3.3.2. Prediction of Minimal Optimal Epitopes with HLA Binding Motifs Databases

1. Scan the recognized peptides for HLA binding motifs using the Los Alamos National Laboratory's *HCV Databases* (http://hcv.lanl.gov/content/immuno/motif_scan/motif_scan) (*see* **Note 22**).
2. Find potential minimal optimal epitopes from the scan output.

3.3.3. Identification of Minimal Optimal Epitopes with Truncated Peptides

1. Synthesize truncated peptides of the confirmed peptides that contain potential epitopes.
2. Perform IFN-γ ELISpot assays using the truncated peptides and identify the shortest peptides that still induce T-cell responses. These are called minimal optimal epitopes (*see* **Note 23**).

4. Notes

1. Alternatively, use human AB serum instead of FBS.
2. Alternatively use 20-mer peptides. Both 15- and 20-mer peptides can be used to detect CD4$^+$ and CD8$^+$ T-cell responses, because breakdown products shorter than the original peptides are typically recognized by CD8$^+$ T cells, but 20-mer peptides are considered better for detection of CD4$^+$ T-cell responses, whereas much shorter peptides (less than 12-mer) are considered optimal for detection of CD8$^+$ T-cell responses *(9)*.
3. Titrate each batch of coating and detection antibody to find the optimal concentration. Use a monoclonal antibody for

coating to ensure specific binding to the cytokine of interest. For detection, however, a polyclonal antibody that binds to several epitopes of the cytokine may be used to amplify the signal.

4. Alternatively, horseradish peroxidase-conjugated streptavidin (streptavidin-HRP) and AEC substrate can be used for development, but red spots resulting from AEC fade faster than blue spots resulting from NBT/BCIP.

5. Alternatively, add as few as 5×10^4 cells/well.

6. Alternatively, harvest using glass-fiber filters and an appropriate cell harvester. After harvesting, dry the filters and punch them into scintillation vials. Add scintillation fluid to the vials and count with a scintillation counter.

7. Because different scintillation counters differ in the efficiency of counting, SI is better than Δ cpm when data from different laboratories are to be compared, but variations in the negative control can substantially affect the SI. Low cpm of the negative control may cause a false increase of SI, and high cpm of the negative control may cause a false decrease of SI. Therefore, we recommend indicating total cpm and/or the range of cpm of the negative controls in addition to the SI in publications (9).

8. If PVDF membrane plates, such as MultiScreen-IP, are used, prewetting membranes (wells) with 35% ethanol may be worthwhile. Prewetting enhances protein binding to the membrane. Before adding the coating antibody, prewet each well with 15 μL 35% ethanol for 1 min, decant ethanol, and rinse three times with 200 μL sterile 1 × PBS.

9. Alternatively, place a wet paper towel under the plate and wrap the plate with plastic wrap. Incubate at 4°C overnight. Generally, coated, wet plates can be stored at 4°C for about one week.

10. If necessary, preincubate PBMCs with antigens in a 96-well culture plate before setting up the ELISpot assay.

11. Alternatively, block wells with 200 μL/well complete medium for 2 h at 37°C in a CO_2 incubator.

12. Alternatively, add 200,000 cells/well. Using fewer than 200,000 cells/well is not recommended.

13. In addition to positive control and negative control, we recommend running other controls: background control (no cells, no antigens) and detection antibody control (PBS instead of detection antibody).

14. Do not disturb or move the plate during incubation. Doing so may cause streaks (cytokine trails caused by moving cells), irregular spots, or double spots. To ensure equal heat distribution, do not stack the plates during incubation. The incubation time may vary according to the cytokine to be detected. Perform a time kinetic and adjust incubation time accordingly.

15. If an automated plate washer is not available, add PBS/Tween using a multichannel pipette and completely flick off PBS/Tween.

16. Alternatively, incubate the plate at 37°C for 2 h or at room temperature for 4 h.

17. An incubation time of more than 1 h may result in increased background color. Dark background color prevents an automated ELISpot reader from recognizing spots.

18. Because alkaline phosphatase activity is reduced at low temperature, equilibration to room temperature prevents underdevelopment of the plate.

19. Washing the plates with PBS in the absence of Tween-20 is important because Tween-20 interferes with spot formation.

20. Spots may fade when they are exposed to light. The AP-developed, completely dried plate can be stored for at least six months in the dark. If AEC is used for development, the dried plate can be stored for up to five months.

21. The absolute number of spots does not always indicate a strong response, because of the possibility that the number of background spots is high. In addition, data from infected individuals should be compared with data from normal donors to reveal whether responses to specific antigens are significant. Therefore, setting a cut-off is necessary for reporting of the results. Internal cut-off values are defined as twofold higher than the background. External cut-off values are calculated as mean plus 2 or 3 standard deviations (SD) of number of SFC for normal donors (uninfected individuals). Ideally, both internal and external cut-off values should be used (9).

22. Alternatively, use the *SYFPEITHI Database of MHC Ligands, Peptide Motifs, and Epitope Prediction* (http://www.syfpeithi.de/).

23. If T-cell responses to minimal optimal epitopes are strong, direct ex vivo ICS/flow cytometry can be performed by stimulation of PBMCs with peptides containing the novel epitopes and gating on CD4+ and CD8+ T cells. If the T-cell responses are weak, ICS/flow cytometry can be performed with peptide-stimulated T-cell lines. HLA restriction may be determined by ICS/flow cytometry or cytotoxicity assay with partially HLA-matched and -mismatched heterologous B cell lines. For more details, see **ref.** (10).

Acknowledgments

This study was supported by the Intramural Research Program of the National Institute of Diabetes and Digestive and Kidney Diseases, National Institutes of Health.

References

1. Hoofnagle, J. H. (2002) Course and outcome of hepatitis C. *Hepatology* **36**, S21–S29.
2. Rehermann, B. and Nascimbeni, M. (2005) Immunology of hepatitis B virus and hepatitis C virus infection. *Nat. Rev. Immunol.* **5**, 215–229.
3. Chisari, F. V. (2005) Unscrambling hepatitis C virus-host interactions. *Nature* **436**, 930–932.
4. Shoukry, N. H., Cawthon, A. G., and Walker, C. M. (2004) Cell-mediated immunity and the outcome of hepatitis C virus infection. *Annu. Rev. Microbiol.* **58**, 391–424.
5. Diepolder, H. M., Zachoval, R., Hoffmann, R. M., Wierenga, E. A., Santantonio, T., Jung, M. C., et al. (1995) Possible mechanism involving T lymphocyte response to non-structural protein 3 in viral clearance in acute hepatitis C virus infection. *Lancet* **346**, 1006–1007.
6. Thimme, R., Oldach, D., Chang, K. M., Steiger, C., Ray, S. C., and Chisari, F. V., (2001) Determinants of viral clearance and persistence during acute hepatitis C virus infection. *J. Exp. Med.* **194**, 1395–1406.
7. Thimme, R., Bukh, J., Spangenberg, H. C., Wieland, S., Pemberton, J., Steiger, C., et al. (2002) Viral and immunological determinants of hepatitis C virus clearance, persistence and disease. *Proc. Natl. Acad. Sci. USA* **99**, 15661–15668.
8. Wedemeyer, H., He, X. S., Nascimbeni, M., Davis, A. R., Greenberg, H. B., Hoofnagle, J. H., et al. (2002) Impaired effector function of hepatitis C virus-specific CD8$^+$ T cells in chronic hepatitis C virus infection. *J. Immunol.* **169**, 3447–3458.
9. Rehermann, B. and Naoumov, N. V. (2007) Immunological techniques in viral hepatitis. *J. Hepatol.* **46**, 508–520.
10. Lauer, G. M., Barnes, E., Lucas, M., Tim, J., Ouchi, K., Kim, A. Y., et al. (2004) High resolution analysis of cellular immune responses in resolved and persistent hepatitis C virus infection. *Gastroenterology* **127**, 924–936.
11. Nascimbeni, M. and Rehermann, B. (2004) Determination of hepatitis B virus-specific CD8+ T-cell activity in the liver. *Methods Mol. Med.* **96**, 65–83.

Chapter 31

Analysis of HCV-Specific T Cells by Flow Cytometry

Masaaki Shiina and Barbara Rehermann

Abstract

Flow cytometry has become an essential research tool because of the increase in the number of its applications. The development of an increasing number of monoclonal antibodies (mAbs) and fluorochromes, and of instruments capable of multicolor detection, allows the acquisition of a large amount of phenotypic and functional information in a single assay. In addition, flow-cytometry techniques have overcome critical problems of conventional assays, such as the use of radioactive reagents to assess proliferation and cytotoxicity of virus-specific T cells. Here, we provide both an overview of available techniques as well as standard protocols that have proven valuable in the assessment of HCV-specific T-cell responses.

Key words: Flow cytometry, FACS, hepatitis C virus, proliferation, cytokines, cytotoxic T cells.

1. Introduction

Many techniques and protocols allow examination of antigen-specific T-cell responses (**Table 31.1**) *(1,2)*. Nevertheless, ex vivo analysis of HCV-specific T-cell proliferation, cytokine production, and direct tetramer staining are still considered difficult, because of the weakness of the HCV-specific T-cell response. Here, we discuss the major T-cell assays and provide selected protocols that have proven most valuable for HCV studies. Information regarding the combination and the number of monoclonal antibodies is kept basic so that it can be customized according to different instruments and skill levels.

Hengli Tang (ed.), *Hepatitis C: Methods and Protocols, Second Edition, vol. 510*
© 2009 Humana Press, a part of Springer Science+Business Media
DOI 10.1007/978-1-59745-394-3_31 Springerprotocols.com

Table 31.1
Major assays for examining the function of HCV-specific T cells

Protocol	Assay type	Possible combination(s)
CFSE dilution	Proliferation	Tetramers, ICS
BrdU incorporation	Proliferation	Tetramers, ICS, 7-AAD (cell cycle)
Intracellular cytokine staining (ICS)	Cytokine production	CFSE, BrdU, tetramers
Cytokine secretion assay (CSA)	Cytokine production	Tetramers
Cytokine bead array (CBA)	Cytokine production	
MHC tetramers	TCR detection	CFSE, BrdU, CSA
Live-count assay	Killing	
Annexin V, Caspase-3/8, TUNEL	Apoptosis	
CD107a/b	Degranulation	

1.1. Assays To Assess T-Cell Proliferation

1.1.1. Carboxyfluorescein Succinimidyl Ester Dilution

Carboxyfluorescein diacetate succinimidyl ester passively enters the cytoplasm of live cells, where it is converted to highly fluorescent CFSE. CFSE is equally divided among progeny cells, so division reduces the signal per cell by half *(3, 4)*. As a result of rapid proliferation, tumor cell lines or mitogen-stimulated cells typically show several CFSE peaks, each reflecting one cell division. Because of the typically weak HCV-specific T-cell response, such a pattern is rarely observed in HCV studies.

1.1.2. Bromodeoxyuridine (BrdU) Incorporation

BrdU, a DNA analog, is incorporated into newly synthesized DNA in the S phase and can be detected with an anti-BrdU antibody. Although this staining requires special treatment of the cells, it can be combined with other cytokine antibodies and with 7-amino-actinomycin D (7-AAD). Because of different mechanisms for the detection of proliferation, results of CFSE dilution experiments and BrdU incorporation may differ.

1.2. Assays to Assess Cytokine Production

1.2.1. Intracellular Cytokine Staining (ICS)

Newly synthesized cytokines are transferred from the ribosome to the Golgi apparatus and then secreted via the plasma membrane. The secretion and the transfer of cytokines can be blocked by inhibitors (e.g., brefeldin A and monensin), allowing the accumulation in the cell of cytokines and subsequent detection with anticytokine antibodies after fixation and permeabilization of the cells *(5, 6)*. The number of cytokines that can be detected in a single tube depends on both performance of the instrument and the combination of fluorochromes.

1.2.2. Cytokine Secretion Assay (CSA)

Secreted cytokines can be captured by means of chimeric antibodies that bind to the cytokine and to a surface molecule of the cell that secretes it. Captured cytokines are then detected with specific antibodies. Although the variety of cytokines detected by this method is limited by the small number of currently available chimeric antibodies, the advantage is that the cells remain alive throughout the analysis, allowing subsequent sorting by flow cytometry and/or magnetic beads and further culture.

1.2.3. Cytokine Bead Array (CBA)

The development of specific beads that are preconjugated with two fluorochromes (for accurate definition of specific bead populations in a flow-cytometry dot plot) and a large panel of anticytokine antibodies allows the quantification of multiple cytokines in serum or supernatant. Samples are first mixed with the capture beads and further incubated with PE-conjugated detection antibodies. The amount of cytokine in samples will be reflected by the intensity of PE signal. Unlike other protocols, the CBA is not a cell-based assay. In contrast to a conventional EIA, only a small sample volume is needed for an entire cytokine array.

1.3. Other T-Cell Assays

1.3.1. Staining with MHC Class I Tetramers

Tetramers consist of four molecules of peptide-refolded MHC and one avidin-fluorochrome conjugate (usually PE or APC) and have become a valuable reagent for visualizing epitope-specific T cells directly ex vivo *(7)*. The practical limitation of tetramers lies in the requirement that the sequence of the minimal optimal epitope and its specific HLA restriction must be known. Tetramers do not detect T-cell functions directly unless combined with other protocols (e.g., proliferation and IFN-γ production) *(8–10)*. MHC class II tetramers are also available, but the number of antigen-specific T cells detected by class II tetramers is typically 100- to 1000-fold lower than the number of cells detected by class I tetramers.

1.3.2. Cytotoxicity Assays

Specific killing is estimated by the reduction in the number of target cells cultured with proteins or peptides as compared to that of those cultured without the antigen. Target cells are labeled with a tracing dye (e.g., CFSE) in different concentrations (low and high separately), preincubated with or without antigens, and cocultured with effector cells *(11)*. This assay is also used to measure cytotoxic T-cell function in vivo.

1.3.3. Detection of Apoptosis

Apoptosis of target cells in the coculture reflects the killing activity of effector cells. Annexin V staining, activated-caspase 3/8 staining, and TUNEL (DNA fragmentation) can be applied.

1.3.4. CD107a/b Degranulation

CD107 is expressed on intracellular vesicles that usually contains perforin. If a cell degranulates, CD107 expression increases on the cell membrane. Because the release of perforin/granzyme induces cell death, the expression of CD107 is thought to reflect the cell's killing capacity *(12)*.

2. Materials

2.1. Common Materials

1. Cell culture medium: RPMI1640 (Mediatech) supplemented with 10% fetal bovine serum (U.S. Bio-Technologies), 100 IU/mL penicillin (Mediatech), 100 µg/mL streptomycin (Mediatech), and 2 mM L-glutamin (Mediatech).
2. Phosphate-buffered saline without Ca^{2+} and Mg^{2+} (PBS, Mediatech).
3. Lymphocyte separation medium (1.077–1.080 g/mL at 20°C, Mediatech). Protect lymphocyte separation medium from direct light and store at room temperature.
4. FACS staining buffer: PBS supplemented with 3% fetal bovine serum and 0.09% sodium azide (Cat. #S8032, Sigma-Aldrich).
5. FACS tubes: 5 mL polystyrene round-bottom tubes (e.g., Cat. #2058, BD Biosciences).
6. Culture plates: 96-well flat-bottom culture plates (Nunc or Costar).
7. HCV antigens: A single HCV peptide or protein or pools of HCV peptides or protein (1–10 µg/mL, final concentration) *(13)*.
8. Mitogens: Phytohemagglutinin (PHA, Cat. #10576-015, Gibco-Invitrogen) or staphylococcal enterotoxin B (SEB, Cat. #4881, Sigma-Aldrich).
9. Flow cytometer with 488-nm argon-ion laser or higher to allow the analysis of combinations of three or more fluorochromes (i.e., FITC, PE and PerCP or PE-Cy5).

2.2. Analysis of T-Cell Proliferation by CFSE Dilution

1. CellTrace CFSE Cell Proliferation Kit (Cat. # C34554, Molecular Probes).
2. Ice-cold 100% fetal bovine serum (FBS).
3. Ethidium monoazide (EMA, Cat. # E1374, Molecular Probes-Invitrogen). Working solution must be prepared for each experiment by dilution of the 5 mg/mL stock solution with PBS.
4. Fluorochrome-conjugated monoclonal antibodies (mAbs): PE-CD8 and PerCP-CD4. Optimal staining concentrations should be titrated in advance.
5. MHC class I tetramer: Optimal staining concentration should be determined in advance in a titration experiment.

2.3. Analysis of T-Cell Proliferation by BrdU Incorporation

1. FITC-BrdU Flow Kit (Cat. #559619, BD Biosciences). This kit contains BrdU, DNase, FITC-conjugated anti-BrdU antibody, Cytofix/Cytoperm Buffer, Cytoperm Plus Buffer, Perm/Wash Buffer, Perm Plus Buffer, and 7-AAD.

2. DNase working solution: Dilute DNase stock (provided with the kit) with PBS to prepare a 1 mg/mL solution of DNase I (Cat. #DN25, Sigma-Aldrich). Stock vials should be kept at −80°C. Prepare the working solution for each experiment; do not refreeze it.

3. FITC-BrdU staining solution: Dilute FITC-conjugated anti-BrdU antibody (provided with the kit) with Perm/Wash Buffer at 1:50 (v:v).

3. Methods

3.1. Analysis of T-Cell Proliferation by CFSE Dilution

3.1.1. Cell Culture

1. Prepare a single-cell suspension of peripheral blood mononuclear cells (PBMCs) by density-gradient centrifugation using lymphocyte separation medium (*see* **Note 1**).

2. Wash PBMCs (centrifuge for 5 min at $300 \times$ g and remove supernatant) with PBS twice and suspend at 5×10^6/mL in PBS (*see* **Note 2**).

3. Prepare a 50μM CFSE working solution immediately prior to the next step (*see* **Note 3**).

4. Add CFSE at 200–250 nM of the final concentration and vortex (*see* **Note 4**).

5. Incubate for 10 min at room temperature in the dark and briefly vortex every 3 min (*see* **Note 5**).

6. Add more than twofold volume of ice-cold FBS to the tube and mix.

7. Centrifuge ($300 \times$ g, 5 min) and remove supernatant completely.

8. Wash cells with culture medium.

9. Suspend cells at 2×10^6/mL in culture medium.

10. Prepare a 96-well culture plate with 250μL/well cell suspension.

11. Add HCV antigen(s) to wells according to the experiment design (*see* **Note 6**).

12. Leave more than two wells without antigen to allow for negative control and compensation.

13. Culture cells for 6 days at 37°C in a 5% CO_2 incubator (*see* **Note 7**).

14. On day 6, harvest cells into FACS tube by pipetting.

3.1.2. Staining and Analysis (for FACS with Three PMTs)

1. Wash cells and suspend in 100μL FACS staining buffer.

2. Add PE-CD8 mAb and PerCP-CD4 mAb, and then vortex briefly (*see* **Note 8**).

3. Incubate for 20 min on ice in the dark.

4. Wash cells and suspend in 500μL with FACS staining buffer (*see* **Note 9**).

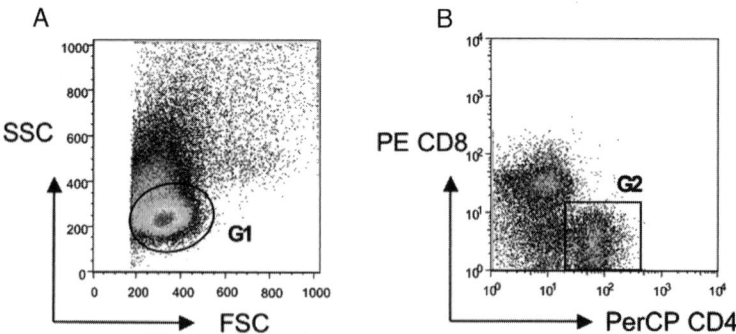

Fig. 31.1. Gating strategy using cultured cells. **A**. A plot of forward scatter (FSC) against side scatter (SSC) during sample acquisition. A threshold for data collection is set at FSC = 170 and excludes debris and dead cells. Gate G1 defines the main lymphocyte population. **B**. The population within the G1 gate is analyzed for CD4 and CD8 expression. The CD4 single positive population is clearly defined and gated as G2.

5. Confirm three-color compensation with CFSE-labeled cells, PE-stained cells, and PerCP-stained cells (*see* **Note 10**).
6. Run FACS with negative sample and define lymphocyte gating in the FSC/SSC plot (**Fig. 31.1A**).
7. Analyze the gated population for PE/PerCP expression and gate on the PerCP (i.e., CD4) single positive population (**Fig. 31.1B**).
8. Create a CFSE histogram or a CD4/CFSE plot to confirm the cut-off value (*see* **Note 11**).
9. Acquire more than 50,000 events in the gated population.
10. Continue acquisition with test samples using the same settings.
11. Cells with lower CFSE intensity than the control have proliferated (**Fig. 31.2**).

3.1.3. Staining and Analysis (for FACS with More Than Three PMTs)

3.1.3.1. Basic Protocol

1. Wash cells and suspend in 100 μL PBS.
2. Add 10 μL EMA working solution (5 μg/mL at final concentration) and mix (*see* **Note 12**).
3. Incubate for 10 min on ice under fluorescence light.
4. Wash cells with 2 ml PBS and suspend in 100 μL FACS staining buffer.
5. Add PE-CD8 mAb and APC-CD4 mAb, and then vortex briefly (*see* **Note 13**).
6. Incubate for 20 min on ice in the dark.
7. Wash cells and suspend in 500 μL with FACS staining buffer.
8. Run FACS with negative sample and define lymphocyte gating in FSC/SSC plot.
9. Analyze the gated population for FSC and EMA staining and gate on the EMA-negative population.

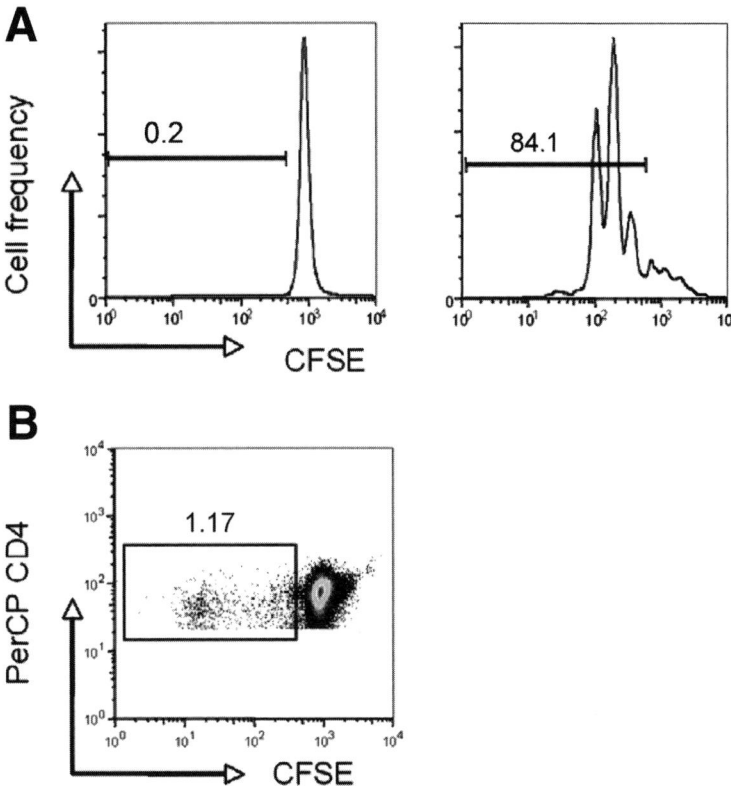

Fig. 31.2. Analysis of T-cell proliferation by means of carboxyfluorescein succinimidyl ester (CFSE) staining. **A**. CFSE histograms of a negative control sample (lymphocytes cultured without antigen, *left panel*) and a positive control (lymphocytes stimulated with PHA, *right panel*). Numbers indicate the percentage of proliferated cells among CD4-positive lymphocytes. **B**. A representative plot of stimulated lymphocytes from a subject that has recovered from HCV (HCV antibody positive and HCV RNA negative). Frozen and thawed peripheral blood mononuclear cells were cultured for 6 days with a mixture of overlapping peptides covering HCV core, E1 and E2 sequences.

10. Create a PE/APC plot and gate on the CD4 single positive population.
11. Continue analysis as described in **Section 3.1.2**, step 9.

3.1.3.2. Combination with MHC Class I Tetramer Staining (see Note 14)

1. Wash cells and suspend in 100 μL FACS staining buffer.
2. Add 10 μL EMA working solution (5 μg/mL at final concentration) and mix (*see* **Note 12**).
3. Incubate for 10 min on ice under fluorescence light.
4. Wash cells with PBS and suspend in 100 μL FACS staining buffer.
5. Add pretitrated volume of tetramer (consisting of APC-avidin) and mix (*see* **Note 15**).
6. Incubate for 1 h at 4°C (*see* **Note 16**).
7. Wash cells and suspend in 100 μL l FACS staining buffer.

8. Add PE-CD8 mAb and APC-CD4, and then vortex briefly (*see* **Note 13**).

9. Incubate for 20 min on ice in the dark.

10. Wash cells and suspend in 500 μL l FACS staining buffer (*see* **Note 9**)

11. Run FACS with negative sample and define lymphocyte gating in the FSC/SSC plot.

12. Analyze the gated population for FSC and EMA staining and gate on the EMA-negative population.

13. Gate on the CD8 single positive population and create a CFSE/APC plot to determine the proliferation of tetramer-positive and tetramer-negative cells.

3.2. Analysis of T-Cell Proliferation by BrdU Incorporation

3.2.1. Cell Culture

1. Prepare a single-cell suspension of PBMCs by density-gradient centrifugation using lymphocyte separation medium (*see* **Note 1**).

2. Wash PBMCs (centrifuge for 5 min at $300 \times$ g and remove supernatant) twice with PBS.

3. Suspend cells at 2×10^6/ml in culture medium.

4. Add 250 μL of the cell suspension to each well of a 96-well culture plate.

5. Add HCV antigen(s) into wells according to the experiment design (*see* **Note 6**).

6. Culture cells at 37°C in a 5% CO_2 incubator.

7. On day 3 after the stimulation, add 10 μM BrdU to each culture well (*see* **Note 17**).

8. Incubate under the same conditions for another 16 h (*see* **Note 18**).

9. Harvest cells into FACS tubes by pipetting.

3.2.2. Staining and Analysis

3.2.2.1. Basic Protocol

1. Wash cells and suspend in 100 μL FACS staining buffer.

2. Add PE-CD8 mAb and PerCP-CD4 mAb, and then vortex briefly (*see* **Note 8**).

3. Incubate for 20 min on ice in the dark.

4. Wash cells with FACS staining buffer and resuspend with 100 μL Cytofix/Cytoperm Buffer.

5. Incubate for 20 min on ice in the dark.

6. Wash cells with 1 mL Perm/Wash Buffer, centrifuge (5 min, 300g), and remove supernatant.

7. Resuspend cells with 100 μL Cytoperm Plus Buffer and incubate for 10 min on ice.

8. Wash cells with 1 mL Perm/Wash Buffer.

9. Refix cells with 100 μL Cytofix/Cytoperm Buffer for 5 min on ice.

10. Wash cells with 1 mL Perm/Wash Buffer.

11. Resuspend cells with 100 μL DNase working solution and incubate for 1 h at 37°C. Shake tubes gently every 20 min.

12. Wash cells with 1 mL Perm/Wash Buffer.

13. Resuspend cells with 50 μL FITC-BrdU staining solution and incubate for 20 min at room temperature in the dark.

14. Wash cells with 1 mL Perm/Wash Buffer.

15. Suspend cells with 500 μL FACS staining buffer (*see* **Note 9**).

16. Perform color compensation on the flow cytometer using either labeled beads or single color-stained cells.

17. Run a negative control on the flow cytometer, gate lymphocytes in the FSC/SSC plot, and define CD4 single positive lymphocytes.

18. Create an FITC(BrdU) and PerC(CD4) plot and acquire at least 50,000 events of the gated population (**Fig. 31.3**).

19. Continue acquisition. Set positive gate for BrdU using either DNase-untreated or FITC-isotype stained tube.

3.2.2.2. Combination with Intracellular IFN-γ Staining

1. Add respective HCV antigen(s) to each culture well 6 h before the harvest.

2. Incubate for 2 h at 37°C, add 10 nM brefeldin A, and then incubate for another 4 h.

3. Follow steps 1–12 of the basic protocol (**Section 3.2.2.1**).

4. Resuspend cells with 50 μL FITC-BrdU staining solution, add APC-IFN-γ mAb, mix, and incubate for 20 min at room temperature in the dark.

5. Follow steps 14–19 of the basic protocol (**Section 3.2.2.1**).

6. Define the IFN-γ-positive gate by using either unstained or APC isotype-matched control samples (*see* **Note 19**).

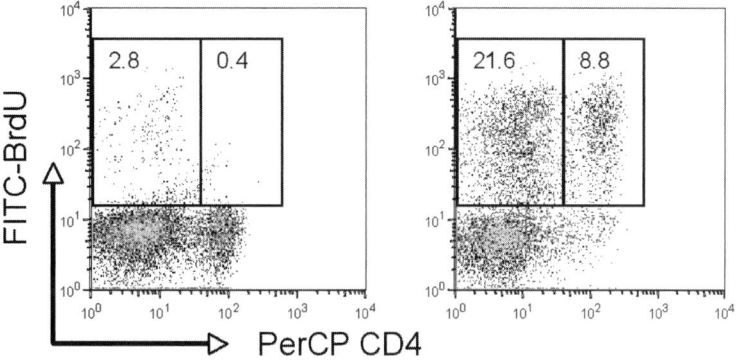

Fig. 31.3. Analysis of T-cell proliferation by BrdU staining: *dot plots* of CD4 against BrdU staining. Peripheral blood mononuclear cells were stimulated with either a mixture of overlapping peptides covering HCV core, E1, and E2 sequences (*left panel*) or with PHA (*right panel*). CD4-negative and -positive populations are clearly separated. Numbers indicate the percentage of gated cells within the lymphocyte population. On day 4 of culture, CD4-positive lymphocytes as well as CD4-negative lymphocytes (CD8 T cells and B cells) have proliferated.

4. Notes

1. Whole blood should not be used because cell density is strictly adjusted in this assay and the serum affects dye labeling. Lymphocyte separation medium should be used exactly according to manufacturer's instruction. Temperature affects the yield of PBMC separation. Frozen and thawed cells should be kept at 37°C for at least several hours before experiments.

2. Either 15 mL tubes or 50 mL tubes can be used for cell suspension according to sample volume.

3. Prepare fresh CFSE working solution for each experiment. Do not use previous stock because of the rapid decrease of signal intensity.

4. The optimal labeling concentration may vary according to the sample volume, cell density, and instrument. The labeling concentration should be changed if the CFSE signal is too high or too low. A very high CFSE concentration ($> 5\,\mu M$) affects cell viability during the culture.

5. Cells can be labeled at 37°C, but we prefer room temperature, as it prevents nonspecific activation of cells.

6. If the antigen volume exceeds $25\,\mu L$ (10% v:v of culture), the volume of cell suspension should be adjusted. To determine the sensitivity of the assay, add 1:100 (v:v) PHA or $5\,\mu g/mL$ SEB on day 2 as a positive control.

7. Shorter culture periods, such as 4 or 5 days, can be used if the antigen-specific proliferation is too strong after 6 days of culture.

8. CD8 can be replaced by CD3 for the detection of CD4 T-cell proliferation.

9. If particles (clumped cells) are visible, pass the cell suspension through a cell strainer or nylon mesh before sample acquisition. Stained samples can be fixed in 0.5–1% paraformaldehyde and stored in the refrigerator until the next day.

10. Compensation should be performed with CFSE-labeled cells because of the extreme spill-over of the CFSE signal into other channels.

11. Data may not be reliable if antigen nonspecific proliferation exceeds 1%.

12. EMA is stable in labeled cells and is recommended over propidium iodide and 7-AAD. Other antibody/fluorochrome combinations can be used, but the PerCP (or PE-Cy5) detection slot should be left for EMA. Alternatively, staining with the recently discovered amine reactive viability dyes may be considered for dead-cell exclusion *(14)*.

13. Sometimes, tetramer-positive cells are seen at a frequency of less than 0.02% (often used as a cut-off value on the basis

of data from healthy, uninfected control subjects) in direct ex vivo assays. Tetramer-positive cells may still be detectable after peptide-specific in vitro expansion, a protocol that can be combined with CFSE labeling. Dead-cell exclusion (e.g., EMA staining) is important in this situation because cell death increases during in vitro culture and tetramers stick to dead cells.

14. We recommend use of APC tetramers rather than PE tetramers because the strong spill-over from the FITC (CFSE) to PE wavelength reduces the brightness of the tetramer staining.

15. Optimal conditions for tetramer staining depend on the specific reagent. Usually, the incubation time can be shortened if staining is performed at 37°C.

16. To define the positive threshold for tetramer staining, test several control samples from HCV-negative and/or HLA-mismatched samples using the same staining procedure and instrument settings.

17. Do not stir, mix, or change medium. Doing so may stimulate cells.

18. Optimal length of BrdU exposure may vary. The incubation time can be either lengthened (e.g., to 24 h) or shortened (e.g., to 9 h) on the basis of the results.

19. To confirm the sensitivity of intracellular IFN-γ detection, inclusion of lymphocytes that have been stimulated for 6 h with SEB is advisable. Samples stained with all but one antibody, that is, stained with FITC-BrdU, PE-CD8, and PerCP-CD4 but not with APC-IFN-γ, represent the best negative control.

Acknowledgments

This study was supported by the Intramural Research Program of the National Institute of Diabetes and Digestive and Kidney Diseases, National Institutes of Health.

References

1. Rehermann, B., and Naoumov, N. V. (2007) Immunological techniques in viral hepatitis. *J. Hepatol.* **46,** 508–520.

2. Thiel, A., Scheffold, A., and Radbruch, A. (2004) Antigen-specific cytometry—new tools arrived! *Clin. Immunol.* **111,** 155–161.

3. Mannering, S. I., Morris, J. S., Jensen, K. P., Purcell, A. W., Honeyman, M. C., van Endert, P. M., et al. (2003) A sensitive method for detecting proliferation of rare autoantigen-specific human T cells. *J. Immunol. Methods* **283,** 173–183.

4. Lecoeur, H., Fevrier, M, Garcia, S, Riviere, Y., and Gougeon, M. L. (2001) A novel flow cytometric assay for quantitation and multiparametric characterization of cell-mediated cytotoxicity. *J. Immunol. Methods* **253,** 177–187.

5. Gauduin, M. C., Kaur, A., Ahmad, S, Yilma, T., Lifson, J. D., and Johnson, R. P. (2004) Optimization of intracellular cytokine staining for the quantitation of antigen-specific CD4+T cell responses in rhesus macaques. *J. Immunol. Methods* **288,** 61–79.

6. Nascimbeni, M., Shin, E. C., Chiriboga, L., Kleiner, D. E., and Rehermann, B. (2004) Peripheral CD4(+)CD8(+) T cells are differentiated effector memory cells with antiviral functions. *Blood* **104,** 478–486.

7. Altman, J. D. (2004) Flow cytometry applications of MHC tetramers. *Methods Cell Biol.* **75,** 433–452.

8. Lechner, F., Wong, D. K., Dunbar, P. R., Chapman, R., Chung, R. T., Dohrenwend, P., et al. (2000) Analysis of successful immune responses in persons infected with hepatitis C virus. *J. Exp. Med.* **191,** 1499–1512.

9. He, X. S., Rehermann, B., Boisvert, J., Mumm, J., Maecker, H. T., Roederer, M., et al. (2001) Direct functional analysis of epitope-specific CD8+T cells in the peripheral blood. *Viral Immunol.* **14,** 59–69.

10. He, X. S., Rehermann, B., Lopez-Labrador, F. X., Boisvert, J., Cheung, R., Mumm, J., et al. (1999) Quantitative analysis of hepatitis C virus-specific CD8(+) T cells in peripheral blood and liver using peptide-MHC tetramers. *Proc. Natl. Acad. Sci. USA* **96,** 5692–5697.

11. Devevre, E., Romero, P., and Mahnke, Y. D. (2006) LiveCount Assay: concomitant measurement of cytolytic activity and phenotypic characterisation of CD8(+) T-cells by flow cytometry. *J. Immunol. Methods* **311,** 31–46.

12. Betts, M. R., Brenchley, J. M., Price, D.A., De Rosa, S. C., Douek, D. C., Roederer, M., et al. (2003) Sensitive and viable identification of antigen-specific CD8+T cells by a flow cytometric assay for degranulation. *J. Immunol. Methods* **281,** 65–78.

13. Draenert, R., Altfeld, M., Brander, C., Basgoz, N., Corcoran, C., Wurcel, A. G., et al. (2003) Comparison of overlapping peptide sets for detection of antiviral CD8 and CD4 T cell responses. *J. Immunol. Methods* **275,** 19–29.

14. Perfetto, S. P., Chattopadhyay, P. K., Lamoreaux, L., Nguyen, R., Ambrozak, D., Koup, R. A., et al. (2006) Amine reactive dyes: an effective tool to discriminate live and dead cells in polychromatic flow cytometry. *J. Immunol. Methods* **313,** 199–208.

Chapter 32

Detection of Neutralizing Antibodies with HCV Pseudoparticles (HCVpp)

Marlène Dreux and François-Loïc Cosset

Abstract

Infectious HCV pseudoparticles (HCVpp) can be assembled by display of unmodified and functional HCV glycoproteins on retroviral and lentiviral core particles. HCVpp have been shown to mimic the early infection steps of parental HCV. The presence of a marker gene packaged within these HCV pseudoparticles allows reliable and fast determination of infectivity mediated by the HCV glycoproteins. With this highly flexible system, E1E2 from a broad range of HCV strains can be investigated, including autologous HCV strains from patients' virus, and it has allowed careful investigation of the humoral response to HCV.

Key words: Hepatitis C virus, glycoprotein, genotype, quasispecies, cell entry, neutralization, lipoprotein.

1. Introduction

1.1. The Humoral Immune Response to HCV

The efficiency of the humoral immune response to HCV has long been a controversial topic. Indeed, detection of neutralizing antibodies in patients' blood has been difficult so far, because of the lack of an efficient and reliable cell-culture system for HCV, but neutralizing antibodies have been identified by their ability to prevent both HCV replication in a lymphoid cell line and HCV infection in experimentally inoculated chimpanzees (1–4). They have been reported to emerge during the course of acute HCV infection (2) both in human patients and in experimentally infected chimpanzees (3, 4). Until recently, however, neutralizing responses remained difficult to detect and measure. Using a novel in vitro neutralization assay system based on infectious

Hengli Tang (ed.), *Hepatitis C: Methods and Protocols, Second Edition, vol. 510*
© 2009 Humana Press, a part of Springer Science+Business Media
DOI 10.1007/978-1-59745-394-3_32 Springerprotocols.com

retroviral pseudoparticles bearing HCV envelope glycoproteins (HCVpp) *(5, 6)*, we and others were able to confirm that HCV-infected patients' bloods neutralize in vitro infection. With this quantitative assay, high-titer neutralizing antibodies were detected in plasma from chronically infected chimpanzees and humans *(5–8)*. Moreover, using HCVpp infection assays, recent studies addressed the kinetics of humoral responses in cohorts of acute-phase, monoinfected patients *(9, 10)*. The emergence of a neutralizing response, of narrow specificity, was correlated with a decrease in high initial viremia, leading to control of viral replication.

1.2. HCV Genetic Diversity, Cross-neutralization, and Vaccine Design

HCV exhibits a high degree of genetic heterogeneity. The propensity for genetic change is associated primarily with the error-prone nature of its RNA-dependent RNA polymerase together with the high HCV replicative rate in vivo *(11, 12)*. Therefore, HCV can be classified into six genetically distinct genotypes and further subdivided into at least 100 subtypes, which differ by approximately 30% and 15% at the nucleotide level, respectively. Moreover, infected individuals harbor a diverse population of viral variants known as quasispecies, which arise through the high-level HCV replication with an error-prone viral RNA polymerase and evolve in response to a variety of selective pressures *(13, 14)*. This high genetic diversity renders the analysis of HCV humoral immune responses difficult and may explain why no efficient vaccine has been developed so far.

Induction of antibodies recognizing epitopes that are conserved in the majority of viral genotypes and subtypes is extremely relevant to vaccine design. This challenge is compounded because the E1E2 envelope proteins, the natural targets for the neutralizing response, are among the most variable HCV proteins. The flexibility of the HCVpp infection assay, allowing the identification of conserved neutralizing epitopes, offers significant hope for the development of successful HCV vaccines. Over 20 different HCV strains allowing production of infectious HCVpp, representing the most important genotypes and subtypes, have been developed *(15–17)* (**Table 32.1**).

1.3. Detection of Neutralizing Antibodies in Acute-Phase Patients by Means of Autologous HCV Particles

A role for antibodies in protection against natural HCV infection has been difficult to establish. Indeed, initial functional studies analyzing the neutralizing-antibody response during acute and chronic HCV infection suggested a lack of neutralizing antibodies in most patients with acute HCV infection *(8, 20–22)*. These studies were limited, however, because the surrogate viral particle against which the antibodies were tested was often derived from a different isolate than the virus present in the infected patient, thus precluding the detection of isolate-specific

Table 32.1
Infectivity of HCVpp harboring functional E1E2 proteins

Genotype	Subtype	Strain name	Accession number	Infectious titer[a]	Reference
1	a	H77	AF011752	+	(5, 6)
1	a	UKN1A14.8	–	+	(17)
1	a	UKN1A14.36	AY899303	++	(17)
1	b	UKN1B12.6	AY734975	++	(15, 17)
1	b	CG1b	AF333324	+	(15)
1	b	BK	M58335	+/–	(9)
1	b	CH35		–	(18)
1	b	Con1	AJ238799	+	(6)
1	b	HCJ4	AF054250	+/–	(8, 18)
1	b	OH8	AY54951	+	(18)
2	a	HCJ6	AF177036	+/–	(8)
2	a	UKN2A1.2	AY734977	+/–	(15, 17)
2	a	UKN2A2.4	AY734979	+	(15, 17)
2	a	J6CF	AF177036	+	(16)
2	a	JFH1	AB047639	+	(19)
2	b	UKN2B1.1	AY734982	++	(15, 17)
2	b	UKN2B2.8	AY734983	+	(15, 17)
3	a	UKN3A1.9	AY734985	–	(15)
3	a	UKN3A1.28	AY734984	–	(15)
3	a	UKN3A13.6	AY894683	–	(17)
3	a	S52(3a-11)	–	Nd	(16)
4	nd	UKN4.11.1	AY734986	+/–	(15, 17)
4	nd	UKN4.21.16	AY734987	+/–	(15, 17)
4	nd	UKN4.21.17	AY734988	++	(15, 17)
4	a	ED43(4a-1)	–	Nd	(16)
5	nd	UKN5.14.4	AY785283	+/–	(15, 17)
5	nd	UKN5.15.11	AY785283	–	(15, 17)
5	a	SA13(5a-12)	–	Nd	(16)

Table 32.1
(continued)

Genotype	Subtype	Strain name	Accession number	Infectious titer[a]	Reference
6	nd	UKN6.5.340	AY736194	+/–	(15, 17)
6	nd	C6a1	–	+/–	(18)
6	a	HK(6a-2.1)	–	Nd	(16)

[a]The infectivity of HCV pseudoparticles (HCVpp) harboring the E1E2 glycoproteins of the indicated HCV genotypes and subtypes was estimated relative to that of HCVpp with the H77 glycoproteins (H77pp), which have titers of ca. 10^5 i. u. /mL. (++), infectivity higher than H77pp; (+), infectivity similar to H77pp; (+/–), infectivity below H77pp; (–), infectivity at least 10-fold lower than H77pp.
nd: not determined.

antibodies. Functional studies using well-defined nosocomial or single-source HCV outbreaks with a defined inoculum and using the matching HCVpp model system resulted in very different findings. They demonstrated that neutralizing antibodies are induced in the early phase of infection and may play a role in control of viral infection or viral clearance (9, 10). A first study (9) addressed the kinetics of humoral immune responses in a cohort of acutely infected hemodialysis patients infected by a single viral strain during a nosocomial outbreak in a hemodialysis center, with various clinical and virological outcomes. This approach demonstrated for the first time (i) the presence of neutralizing antibodies at the acute phase of HCV infection, (ii) the inverse correlation of the emergence of neutralizing responses with HCV RNA kinetics, and (iii) the existence of human blood components that facilitate HCV infection. Thus, the results demonstrated for the first time the early emergence of a neutralizing response in patients who apparently evolved toward control of viral replication, as well as strong responses concomitant with steep HCV RNA decreases (over 4 logs) together with ALT normalization. The correlation of a relatively strong neutralizing response with a substantial loss of viremia was corroborated by the observation that, in a second group of patients, failure to reduce HCV RNA levels was associated with a lack of detection of a neutralizing response in blood. A second study (10) addressed the neutralizing-antibody responses in a cohort of women accidentally exposed to the same HCV strain of known sequence. In this single-source outbreak, viral clearance was associated with a rapid induction of neutralizing antibodies in the early phase of infection. Neutralizing antibodies decreased or disappeared after recovery from HCV infection. In contrast, chronic HCV infections were characterized by absent or low-titer neutralizing antibodies in the early phase of infection and the persistence of infection despite the induction of high-titer cross-neutralizing antibodies in the late phase of infection. Thus, these two studies from different cohorts of patients yielded very consistent results and highlighted the importance of detecting neutralizing antibodies from sera containing well-defined viral inocula by means of autologous HCV particles.

1.4. Escape Mechanisms from Neutralizing Antibodies

These studies indicated that humoral neutralizing-antibody responses were generated during acute hepatitis *(9, 10)*. Furthermore, several studies highlighted that high-titer cross-neutralizing antibodies are detected in plasma from chronically infected chimpanzees and humans *(5, 6, 8, 10, 21)*. Clearly, these responses are insufficient to achieve viral clearance in the majority of patients, suggesting that their effectiveness is limited in individuals who do not resolve the disease. The emergence of a spectrum of related but distinct sequences within infected individuals, known as quasispecies, may contribute to viral escape from the antibody-mediated immune response. Taking advantage of the meticulous follow-up of patient H since 1977, a recent study *(23)* demonstrated that HCV continuously escapes from neutralizing antibody.

Other studies indicate the presence of serum components that impair the detection of neutralizing antibodies *(9, 24–26)*, suggesting that HCV has evolved serum-dependent mechanisms by which to escape from or at least attenuate the humoral response. Indeed, high-density lipoprotein (HDL) is a serum component that both enhances HCV infection and attenuates neutralization by antibodies whose target is the critical cell entry step of interaction between E2 and CD81 *(25)* (**Fig. 32.1**). Surprisingly,

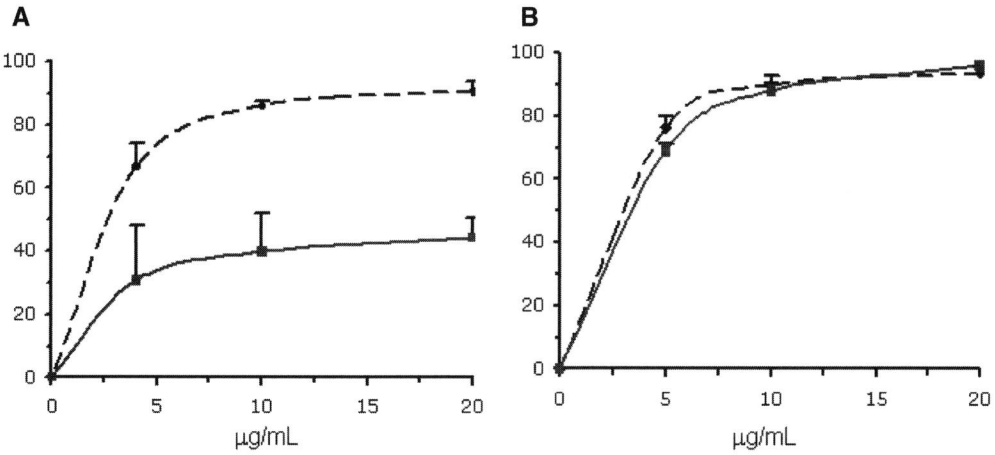

Fig. 32.1. Attenuation of HCVpp neutralization by human serum. Results of neutralization assays on Huh-7 cells using HCVpp of genotype 1a. HCVpp were produced in low-serum medium (0.1% FCS) to which 2.5% human serum (HS) and varied concentrations of the antibody were added during infection, as indicated. The infectious titers determined in the presence (plain) and the absence (dotted) of HS were used to draw the titration curves of the monoclonal antibody. For each of these conditions, the results are expressed as the mean percentages (mean ± SD; *n* = 6) of neutralization of the infectious titers relative to incubation with medium devoid of antibody. Similar results of attenuation were obtained when HDL rather than HS was used (not shown). No effect of HS or HDL could be detected on infectivity of control pseudoparticles harboring RD114 glycoproteins (not shown). (**A**) Results obtained with a monoclonal antibody that blocks the interaction between E2 and CD81. (**B**) Results obtained with E1 polyclonal antibodies that do not block the interaction between E2 and CD81.

HDL does not modulate HCV neutralizing efficiency by masking the neutralizing epitopes involved in this interaction. Rather, HDL-induced attenuation of neutralization operates through a novel mechanism involving the biological activity of the scavenger receptor BI (SR-BI), an HCV coreceptor *(27–30)*, through its capacity to mediate lipid transfer from HDL *(31)*. Remarkably, inhibitors of SR-BI-mediated lipid transfer fully restored the potency of neutralizing antibodies in infection assays conducted in the presence of HS/HDL *(25)*. Interestingly, some neutralizing antibodies that resist attenuation by HDL have been identified, for example, against HCV E1, and could be retrieved from patients or from immunized rodents *(25, 26)*. Such antibodies block cell entry steps that are different from the E2/CD81 interaction (**Fig. 32.1**). This result provides guidance for the screening of neutralizing monoclonal antibodies and for designing vaccines that induce efficient neutralizing responses in vivo.

2. Materials

2.1. Buffers and Solutions

1. 2× HEPES buffered saline (HBS) and 2 M $CaCl_2$: Calphos Mammalian Transfection Kit (Clontech, BD Biosciences).
2. Phosphate-buffered saline (PBS) without calcium and magnesium, without sodium bicarbonate, sterile (# 14190-169 Gibco/Invitrogen).
3. Trypsin-ethylenediaminetetraacetic acid (EDTA) 1× Hank's balanced salt solution without calcium and magnesium, sterile (# 25300-054 Gibco/Invitrogen).
4. Protein-G or A Sepharose (# 17-0618-02 Amersham Biosciences).
5. Binding buffer (100 mM phosphate, 150 mM NaCl, pH 5, #21011 Pierce).
6. ImmonoPure IgG Elution Buffer (ethanolamine 1 M, pH 2.8, # 21004 Pierce).

2.2. Media

1. Fetal calf serum (FCS) (#10270-106 from GIBCO/Invitrogen).
2. Dulbecco's modified Eagle's medium (DMEM) with 0.11 g/L sodium pyridoxine and pyridoxine (#41966-029 from GIBCO/Invitrogen). DMEM is supplemented with 10% FCS, 100 µg/L streptomycin, and 100 U/mL penicillin (stored at 4°C).
3. RPMI (#21875 from GIBCO/Invitrogen) medium supplemented with 10% FCS, 100 µg/L streptomycin, and 100 U/mL penicillin (stored at 4°C).

2.3. Cloning and Plasmids

1. Primer for isolation of patient HCV sequences (**Table 32.2**).
2. Envelope protein expression plasmids: (i) E1E2 expression constructs from several HCV strains have been described (**Table 32.1**). (ii) Control glycoproteins: a significant number of different viral glycoproteins have been described to incorporate well on pseudoparticles. The phCMV-RD expression vector, encoding glycoprotein of the cat endogenous virus RD114, is a most useful control in neutralization studies because it permits entry into liver-derived cell lines and is complement resistant.
3. A MLV-GFP plasmid, encoding a murine leukemia virus (MLV)–based transfer vector containing an internal transcriptional unit encoding the GFP marker gene *(32)*.
4. Retrovirus structural proteins (gagpol) expression plasmid: a CMV-Gag-Pol MLV packaging construct, encoding the MLV gag and pol genes *(32)*.

Table 32.2
Primers used for nested amplification of HCV glycoproteins, amino acids 170–746 (referenced to H77c)

Sense primers		
All genotypes	Outer	GGACGGGGTAAACTATGCAACAGG
	Inner	CACCCATGGGTTGCTCTTTTTCTATC
Antisense primers		
Genotype 1	Inner	AAAGTTTCTAGATTASGCCTCAGCYGTGGMTA
	Outer(1a)	GGGATGCTGCATTGAGTA
	Outer(1b)	CCGGCCACGGACGCCGCATTG
Genotype 2	Inner	AAACTTTCTAGATTACGCTTCGGCTTGGCCCA
	Outer	RGACCATTGGMRCTAGCAGC
Genotype 3	Inner	AAGATAAGCTTATGCTTCCGCCTGWGAWATC
	Outer	TGCGCTGAGGGCGTTCAG
Genotype 4	Inner	GACAGTTACGCCTGAACTTGACTTACCATAAACATC
	Outer	CACCAGCGGGTGAAGCAGCATTGA
Genotype 5	Inner	TTATGCTTCGGCCTGACAAACCAAG
	Outer	GCCAAGCGAAGCAAATAACGAGCGAACCCCAGAAAA
Genotype 6	Inner	TTATGCCTCTACCTGGCCGATGATCAACATGA
	Outer	GCAGGGCCAGGATTAGCAGGAGGAGCGGCCA

2.4. Cells

1. Producer cell lines: 293T human embryo kidney cells (CRL-1573) are the most suitable.
2. Target cells: Huh-7 human hepatocellular carcinoma (33); Hep3B human hepatocellular carcinoma (HB-8064); HepG2 human hepatocellular carcinoma (HB-8065) overexpressing CD81; PLC/PRF/5 human hepatoma (CRL-8024); and other liver-derived cell lines.

2.5. Consumables

1. Ø10 cm sterile petri dishes (#430167) from Corning; 24- or 96-well Costar plates with flat bottom (Corning).
2. 10 or 20 mL syringes and 0.45 μm filters (Millipore).

3. Methods

3.1. Isolation of E1 and E2 Glycoproteins from Patients and Expression Constructs

1. Isolate viral RNA from 100 £gl of serum using a total RNA extraction kit according to the manufacturer's instructions (Fluka). Genotype-defined primers have been described (**Table 32.2**) (15).
2. Generate virus-specific cDNA with a first-strand cDNA Kit (Amersham). PCR products represent amino acid residues 170–746 (referenced to strain H77) of the HCV polyprotein.
3. Generate products by nested PCR with genotype-specific primers, using Expand High-Fidelity Polymerase (Roche).
4. Cone PCR products into an expression vector.

3.2. Preparation of Sera and Immunoglobulin Purification

1. For preparation of sera, incubate blood on ice for 2 h and centrifuge it at 3400g for 20 min, harvest the supernatants, and store them in aliquots at 80°C.
2. Purify antibodies from sera or hybridoma supernatant using protein-G or A Sepharose (25). Mix 100–500 μL of sera or hybridoma supernatant with equal volumes of binding solution and incubate them with 200 μL of the protein-G or A Sepharose solution (according to species) for 1 h at room temperature.
3. Wash beads three times by adding 1 mL of binding solution and centrifuging them at 3800g.
4. Elute the antibody by incubating the beads with 400 μL of elution buffer for 10 min. Neutralize the acidic pH of the IgG-containing supernatant by adding 40 μL of 1 M Tris-HCl solution, pH 8. Finally, obtain the concentration of purified antibody by Bradford staining.

3.3. Production of HCV and Control Pseudoparticles

1. Day 0: Seed 1.8×10^6 293T cells the day before transfection in Ø10 cm plates in a final volume of 10 mL DMEM

2. Day 1: Cotransfect MLV packaging construct (8 µg) with the MLV-GFP plasmid (8 µg) and (a) the HCV E1E2 expression constructs of various genotypes (4 µg) or (b) phCMV-RD encoding the cat endogenous virus RD114 glycoprotein (4 µg) using the Clontech calcium-phosphate transfection system according to the manufacturer's instructions.

3. Day 2: Replace the medium, 15 h after transfection, with 7.5 mL of fresh DMEM (*see* **Note 1**).

4. Day 3: 48 h after transfection, harvest the supernatant of the producer cells, filter it through 0.45 µm pore-size membrane, and use it directly for infection reactions or store it at 4°C; it can be stored for 1 week without change in infectious titer. Of note, the transfection efficiency can be verified at this stage, as discussed in detail in **Chapter 21**.

3.4. Neutralization of HCVpp

1. Seed target cells the day before infection. In the case of Huh-7 cells, we usually seed 5.5×10^3 per well for 96-well plates or 4×10^4 per well for 24-well plates. Usually viral supernatant containing HCVpp is diluted two- to threefold in medium to produce a transduction between 5% and 10% of target cells (*see* **Note 1**). Because the infectious titers of the control pseudoparticles harboring RD114 glycoproteins (RD114pp) are higher than those of HCVpp, they are diluted 100-fold in medium to produce viral preparations containing equivalent input of infectious particles.

2. Mix the supernatant of the producer cells containing HCVpp or the medium-diluted control pseudoparticle with appropriate amounts of purified antibody or sera from HCV-immunized animals or HCV-positive patients and incubate them for 1 h at 37°C (*see* **Note 2**). Purified antibodies are usually tested in the concentration range of 0.1–100 µg/mL, and sera are analyzed in the range of dilution of 20- to 1000-fold. Several HCV polyclonal and monoclonal antibodies have been described that efficiently cross-neutralize HCVpp infection and can serve as positive controls for neutralization *(17, 24, 25, 34)*. Of note, monoclonal antibodies against RD114pp readily neutralize RD114pp infection and can serve as controls for the specificity of neutralization *(24)*.

3. Then add the mixture to Huh-7 cells and incubate the cells at 37°C for 4 h. After removal of the inoculum, replace the cells' media with fresh medium and incubate them at 37°C for 4 days.

3.5. Analysis

1. Seventy-two hours after inoculation, trypsinize cells (trypsin diluted at 1:5e in PBS) and transfer them to fluorescence-activated cell sorting (FACS) tubes or, alternatively, keep trypsinized cells in the 96-well plates for high-throughput (HTS) analysis (BD pharmingen).

2. Determine the proportion of infected cells by measurement of GFP by fluorescence-activated cell sorting.

3. Determine the neutralizing titers as the percentage of inhibition of the infectious titers relative to incubation with corresponding medium devoid of antibody. On the basis of serial purified antibody concentration or serial serum dilution, the neutralizing activity can be expressed as the IC_{50} or ID_{50}, respectively, defined as the concentration of antibody or the dilution of serum required to produce 50% inhibition of infection.

4. Notes

1. For analysis of the effect of human serum or HDL on the antibody neutralizing efficiency, pseudoparticles are produced in low-serum medium (0.1% FCS) or in lipoprotein-free medium (2% fetal calf lipoprotein-depleted serum, LPDS) *(24–26, 35)*. Before infection, dilute viral supernatants in low-serum medium or in lipoprotein-free medium to adjust input of infectious particles and to obtain viral supernatant that transduces 5%–10% of target cells.

2. For analysis of the effect of human serum or HDL on the antibody neutralizing efficiency, human serum or HDL is added to a final concentration of 2.5% or 6 μg/mL cholesterol-HDL to the mixture of pseudoparticles preincubated with purified antibody *(24–26)*. The neutralization attenuation of the purified monoclonal antibody or polyclonal antibody purified from sera of infected patients is equivalent as human serum or HDL is added to purified HCVpp/antibody immune complexes (purification by centrifugation upon sucrose cushion) or as human serum or HDL is added simultaneously with purified antibody to pseudoparticles *(25)* (Dreux and Cosset, unpublished data).

Acknowledgments

We are grateful to our coworkers and colleagues for encouragement and advice. Our work is supported by INSERM, ENS Lyon, and CNRS and by grants from the Ligue Nationale Contre le Cancer, the European Union (LSHB-CT-2004-005246), and the Agence Nationale de Recherches sur le SIDA et les Hépatites Virales (ANRS).

References

1. Shimizu, Y. K., Igarashi, H., Kiyohara, T., Cabezon, T., Farci, P., Purcell, R. H., et al. (1996) A hyperimmune serum against a synthetic peptide corresponding to the hypervariable region 1 of hepatitis C virus can prevent viral infection in cell cultures. *Virology* **223**, 409–412.

2. Shimizu, Y. K., Hijikata, M., Iwamoto, A., Alter, H. J., Purcell, R. H., and Yoshikura, H. (1994) Neutralizing antibodies against hepatitis-C virus and the emergence of neutralization escape mutant viruses. *J. Virol.* **68**, 1494–1500.

3. Farci, P., Alter, H. J., Wong, D. C., Miller, R. H., Govindarajan, S., Engle, R., et al. (1994) Prevention of hepatitis-C virus infection in chimpanzees after antibody-mediated *in vitro* neutralization. *Proc. Natl. Acad. Sci. USA* **91**, 7792–7796.

4. Farci, P., Shimoda, A., Wong, D., Cabezon, T., Gioannis, D. D., Strazzera, A., et al. (1996) Prevention of hepatitis C virus infection in chimpanzees by hyperimmune serum against the hypervariable region 1 of the envelope 2 protein. *Proc. Natl. Acad. Sci. USA* **93**, 15394–15399.

5. Bartosch, B., Dubuisson, J., and Cosset, F. L. (2003) Infectious hepatitis C virus pseudoparticles containing functional E1-E2 envelope protein complexes. *J. Exp. Med.* **197**, 633–642.

6. Hsu, M., Zhang, J., Flint, M., Logvinoff, C., Cheng-Mayer, C., Rice, C. M., et al. (2003) Hepatitis C virus glycoproteins mediate pH-dependent cell entry of pseudotyped retroviral particles. *Proc. Natl. Acad. Sci. USA* **100**, 7271–7276.

7. Bartosch, B., Bukh, J., Meunier, J. C., Granier, C., Engle, R. E., Blackwelder, W. C., et al. (2003) *In vitro* assay for neutralizing antibody to hepatitis C virus: Evidence for broadly conserved neutralization epitopes. *Proc. Natl. Acad. Sci. USA* **100**, 14199–14204.

8. Logvinoff, C., Major, M. E., Oldach, D., Heyward, S., Talal, A., Balfe, P., et al. (2004) Neutralizing antibody response during acute and chronic hepatitis C virus infection. *Proc. Natl. Acad. Sci. USA* **101**, 10149–10154.

9. Lavillette, D., Morice, Y., Germanidis, G., Donot, P., Soulier, A., Pagkalos, E., et al. (2005) Human serum facilitates hepatitis C virus infection, and neutralizing responses inversely correlate with viral replication kinetics at the acute phase of hepatitis C virus infection. *J. Virol.* **79**, 6023–6034.

10. Pestka, J. M., Zeisel, M. B., Blaser, E., Schurmann, P., Bartosch, B., Cosset, F. L., et al. (2007) Rapid induction of virus-neutralizing antibodies and viral clearance in a single-source outbreak of hepatitis C. *Proc. Natl. Acad. Sci. USA* **104**, 6025–6030.

11. Fukumoto, T., Berg, T., Ku, Y., Bechstein, W. O., Knoop, M., Lemmens, H. P., et al. (1996) Viral dynamics of hepatitis C early after orthotopic liver transplantation: evidence for rapid turnover of serum virions. *Hepatology* **24**, 1351–1354.

12. Neumann, A. U., Lam, N. P., Dahari, H., Gretch, D. R., Wiley, T. E., Layden, T. J., et al. (1998) Hepatitis C viral dynamics *in vivo* and the antiviral efficacy of interferon-alpha therapy. *Science* **282**, 103–107.

13. Gomez, J., Martell, M., Quer, J., Cabot, B., and Esteban, J. I. (1999) Hepatitis C viral quasispecies. *J. Viral. Hepat.* **6**, 3–16.

14. Pawlotsky, J. M. (2003) Hepatitis C virus genetic variability: pathogenic and clinical implications. *Clin. Liver Dis.* **7**, 45–66.

15. Lavillette, D., Tarr, A. W., Voisset, C., Donot, P., Bartosch, B., Bain, C., et al. (2005) Characterization of host-range and cell entry properties of the major genotypes and subtypes of hepatitis C virus. *Hepatology* **41**, 265–274.

16. Meunier, J. C., Engle, R. E., Faulk, K., Zhao, M., Bartosch, B., Alter, H., et al. (2005) Evidence for cross-genotype neutralization of hepatitis C virus pseudo-particles and enhancement of infectivity by apolipoprotein C1. *Proc. Natl. Acad. Sci. USA* **102**, 4560–4565.

17. Owsianka, A., Tarr, A. W., Juttla, V. S., Lavillette, D., Bartosch, B., Cosset, F. L., et al. (2005) Monoclonal antibody AP33 defines a broadly neutralizing epitope on the hepatitis C virus E2 envelope glycoprotein. *J. Virol.* **79**, 11095–11104.

18. McKeating, J. A., Zhang, L. Q., Logvinoff, C., Flint, M., Zhang, J., Yu, J., et al. (2004) Diverse hepatitis C virus glycoproteins mediate viral infection in a CD81-dependent manner. *J. Virol.* **78**, 8496–8505.

19. Keck, Z. Y., Xia, J., Cai, Z., Li, T. K., Owsianka, A. M., Patel, A. H., et al. (2007) Immunogenic and functional organization of hepatitis C virus (HCV) glycoprotein E2 on infectious HCV virions. *J. Virol.* **81**, 1043–1047.

20. Steinmann, D., Barth, H., Gissler, B., Schurmann, P., Adah, M. I., Gerlach, J. T., et al. (2004) Inhibition of hepatitis C virus-like particle binding to target cells by antiviral antibodies in acute and chronic hepatitis C. *J. Virol.* **78**, 9030–9040.

21. Bartosch, B., Bukh, J., Meunier, J. C., Granier, C., Engle, R. E., Blackwelder, W. C., et al. (2003) In vitro assay for neutralizing antibody to hepatitis C virus: evidence for broadly conserved neutralization epitopes. *Proc. Natl. Acad. Sci. USA* **100**, 14199–14204.

22. Netski, D. M., Mosbruger, T., Depla, E., Maertens, G., Ray, S. C., Hamilton, R. G., et al. (2005) Humoral immune response in acute hepatitis C virus infection. *Clin. Infect. Dis.* **41**, 667–675.

23. von Hahn, T., Yoon, J. C., Alter, H., Rice, C. M., Rehermann, B., Balfe, P., et al. (2007) Hepatitis C virus continuously escapes from neutralizing antibody and T-cell responses during chronic infection *in vivo*. *Gastroenterology* **132**, 667–678.

24. Bartosch, B., Verney, G., Dreux, M., Donot, P., Morice, Y., Penin, F., et al. (2005) An interplay between hypervariable region 1 of the hepatitis C virus E2 glycoprotein, the scavenger receptor BI, and high-density lipoprotein promotes both enhancement of infection and protection against neutralizing antibodies. *J. Virol.* **79**, 8217–8229.

25. Dreux, M., Pietschmann, T., Granier, C., Voisset, C., Ricard-Blum, S., Mangeot, P. E., et al. (2006) High density lipoprotein inhibits hepatitis C virus-neutralizing antibodies by stimulating cell entry via activation of the scavenger receptor BI. *J. Biol. Chem.* **281**, 18285–18295.

26. Voisset, C., Op de Beeck, A., Horellou, P., Dreux, M., Gustot, T., Duverlie, G., et al. (2006) High-density lipoproteins reduce the neutralizing effect of hepatitis C virus (HCV)-infected patient antibodies by promoting HCV entry. *J. Gen. Virol.* **87**, 2577–2581.

27. Bartosch, B., Vitelli, A., Granier, C., Goujon, C., Dubuisson, J., Pascale, S., et al. (2003) Cell entry of hepatitis C virus requires a set of co-receptors that include the CD81 tetraspanin and the SR-B1 scavenger receptor. *J. Biol. Chem.* **278**, 41624–41630.

28. Maillard, P., Huby, T., Andreo, U., Moreau, M., Chapman, J., and Budkowska, A. (2006) The interaction of natural hepatitis C virus with human scavenger receptor SR-BI/Cla1 is mediated by ApoB-containing lipoproteins. *FASEB J.* **20**, 735–737.

29. Scarselli, E., Ansuini, H., Cerino, R., Roccasecca, R., Acali, S., Filocamo, G., et al. (2002) The human scavenger receptor class B type I is a novel candidate receptor for the hepatitis C virus. *EMBO J.* **21**, 5017–5025.

30. Kapadia, S. B., Barth, H., Baumert, T., McKeating, J. A., and Chisari, F. V. (2006) Initiation of hepatitis C virus infection is dependent on cholesterol and cooperativity between CD81 and scavenger receptor B type I. *J. Virol.* **81**, 374–383.

31. Krieger, M. (2001) Scavenger receptor class B type I is a multiligand HDL receptor that influences diverse physiologic systems. *J. Clin. Invest.* **108**, 793–797.

32. Sandrin, V., and Cosset, F. L. (2006) Intracellular versus cell surface assembly of retroviral pseudotypes is determined by the cellular localization of the viral glycoprotein, its capacity to interact with Gag, and the expression of the Nef protein. *J. Biol. Chem.* **281**, 528–542.

33. Nakabayashi, H., Taketa, K., Miyano, K., Yamane, T., and Sato, J. (1982) Growth of human hepatoma-cell lines with differentiated functions in chemically defined medium. *Cancer Res.* **42**, 3858–3863.

34. Schofield, D. J., Bartosch, B., Shimizu, Y. K., Allander, T., Alter, A. J., Emerson, S. U., et al. (2005) Human monoclonal antibodies that react with the E2 glycoprotein of hepatitis C virus and possess neutralizing activity. *Hepatology* **42**, 1055–1062.

35. Voisset, C., Callens, N., Blanchard, E., Op De Beeck, A., Dubuisson, J., and Vu-Dac, N. (2005) High density lipoproteins facilitate hepatitis C virus entry through the scavenger receptor class B type I. *J. Biol. Chem.* **280**, 7793–7799.

Chapter 33

Mathematical Modeling of HCV Infection and Treatment

Harel Dahari, Emi Shudo, Ruy M. Ribeiro, and Alan S. Perelson

Abstract

In the last decade, viral kinetic modeling has played an important role in the analysis of HCV RNA decay after the initiation of antiviral therapy. Models have provided a means of evaluating the antiviral effectiveness of therapy and of estimating parameters, such as the rate of virion clearance and the rate of loss of HCV-infected cells, and they have suggested mechanisms of action for both interferon-α and ribavirin. The inclusion of homeostatic proliferation of infected and uninfected hepatocytes in existing viral kinetic models has allowed prediction of most observed HCV RNA profiles under treatment, for example, biphasic and triphasic viral decay and viral rebound to baseline values after the cessation of therapy. In addition, new kinetic models have taken into consideration the different pharmacokinetics of standard and pegylated forms of interferon and have incorporated alanine aminotransferase kinetics and aspects of immune responses to provide a more comprehensive picture of the biology underlying changes in HCV RNA during therapy. Here, we describe our current understanding of the kinetics of HCV infection and treatment.

Key words: Viral kinetics, mathematical modeling, drug effectiveness, drug efficacy.

1. Introduction

1.1. HCV RNA Kinetic Profiles

During chronic HCV infection, the level of serum HCV RNA does not vary significantly (< 0.5 log) on time scales of weeks to months (1), but when patients chronically infected with HCV are placed on antiviral therapy with interferon-α (IFN) or IFN plus ribavirin (RBV), HCV RNA generally declines in a biphasic manner. After a 7- to 10-h delay, a rapid first phase lasts for approximately 1–2 days during which HCV RNA may fall, on average, 1–2 logs in genotype 1–infected patients and as much as 3–4 logs in genotype 2–infected patients (2). Subsequently, a slower second phase of HCV RNA decline ensues. Triphasic viral

Hengli Tang (ed.), *Hepatitis C: Methods and Protocols, Second Edition, vol. 510*
© 2009 Humana Press, a part of Springer Science+Business Media
DOI 10.1007/978-1-59745-394-3_33 Springerprotocols.com

declines have also been observed in some patients where the frequency of measurement is sufficiently high. A triphasic decline consists of a first phase (1–2 days) in which virus load declines rapidly, a "shoulder phase" (4–28 days) in which virus load decays slowly or remains constant, and a third phase of renewed rapid viral decay. Hermann et al. *(3)* noted that approximately 50% of patients ($N = 34$) treated with pegylated interferon-α (PEG-IFN) and RBV, PEG-IFN alone, or IFN + RBV showed triphasic viral decays. Other investigators *(4–6)* have also observed triphasic viral decays. In particular, Bergmann et al. *(4)* observed triphasic viral declines in a large fraction (30–40%) of IFN-treated patients ($N = 200$). In some patients, however, a more complex kinetic pattern has been observed in which an initial decline in HCV RNA is followed by a transient increase or even a rebound *(7)*. In nonresponders, the first- or second-phase decline may be absent (null response), or the first phase may be followed by little or no second-phase decline (flat partial response). In **Fig. 33.1**, we show typical examples of these viral-load profiles under therapy.

1.2. Sustained Viral Response

The standard of care involving therapy with pegylated IFN and RBV has produced sustained viral response (SVR) rates of 50%, and no effective alternative treatment is available for nonresponders *(8, 9)*. Cases of nonresponders may be characterized by a

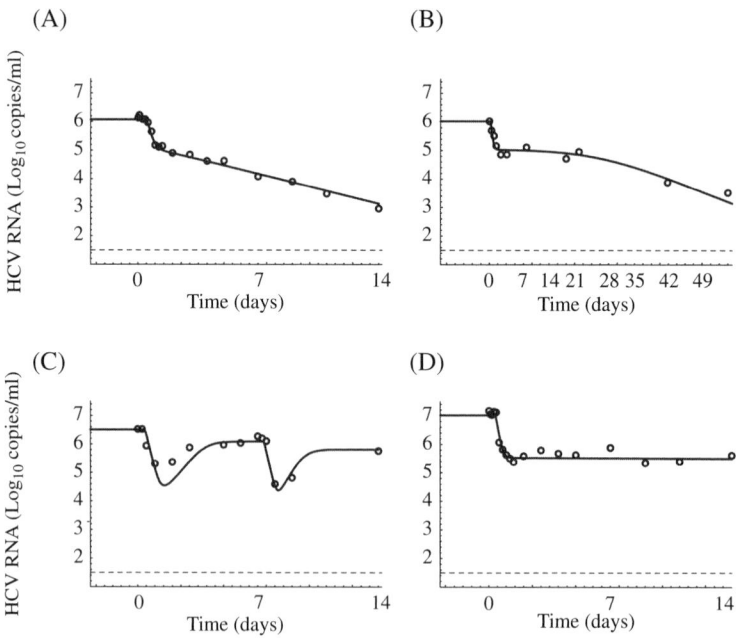

Fig. 33.1. Examples of different HCV RNA decay profiles under therapy. HCV RNA data (*circles*) and best fits of various dynamic models (*lines*) to the data. (**A**) biphasic decline, (**B**) triphasic decline, (**C**) viral rebound due to drug pharmacokinetics, and (**D**) flat partial response. All patients were treated with IFN-α alone or in combination with ribavirin (in the case of patients in (B) and (C), PEG-IFN-α). *Dashed lines* indicate the limit of detection for HCV RNA.

number of factors, both host related (e.g., liver fibrosis, ethnicity) and virus related (e.g., infection with genotype 1, high viral load), that limit the effectiveness of current treatments [reviewed in **ref.** *(10)*]. The reason these factors are correlated with therapy failure is not known.

2. Modeling HCV RNA Kinetics

The basic principle underpinning modeling of HCV kinetics under treatment is that a patient in the chronic stage of the disease, before therapy, has a constant viral load, that is, a steady-state or set-point level of HCV RNA that remains constant over days, weeks, and even months *(1)*. To maintain an approximately constant viral level, production must be in balance with viral clearance. When a patient is treated with an antiviral agent, viral production presumably decreases, but viral clearance remains unchanged leading to a decrease in HCV RNA. Modeling of the kinetics of viral-load decrease can be used to elucidate the biological processes that control viral production and clearance. This approach was first used to study HIV *(11, 12)* and later extended to the study of HCV *(2, 3, 13–18)*.

A model of HIV infection was adapted by Neumann et al. *(13)* for study of the kinetics of the HCV RNA response to high-dose daily IFN. The model used by Neumann et al. *(13)*, shown in **Fig. 33.2**, incorporates a population of cells susceptible to

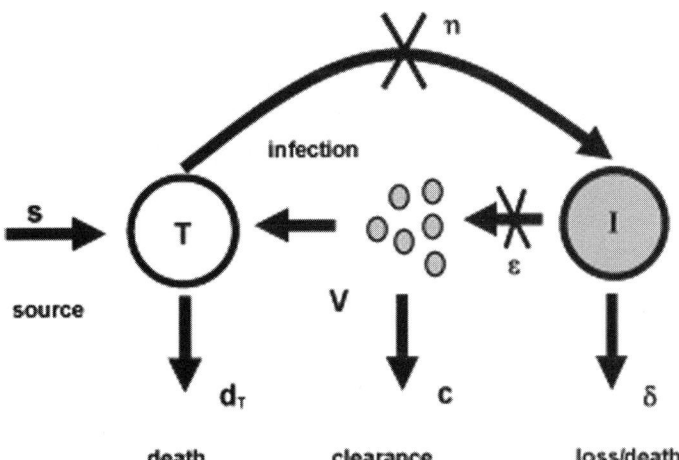

Fig. 33.2. Schematic illustration of the model of HCV infection and treatment from Equation [33.1].A population of cells, *T*, susceptible to infection by virus (*V*) is assumed to be produced at a constant rate, *s*, and to die at rate d_T per cell. Productively infected cells (*I*) are assumed to be lost at rate δ per cell and virus to be cleared at a rate *c*. Therapy is assumed to act by partially blocking virion production, with effectiveness ε and by reducing the rate of infection with effectiveness η.

infection, T, which are assumed to be produced at a constant rate, s, and to die at rate d_T per cell. Productive infection occurs when virus (V) interacts with target cells at a rate proportional to the product of their densities, that is, at rate βVT, where β is the infection rate constant. Productively infected cells (I) are assumed to be lost, by either death or loss of the infected state, at rate δ per cell. Virus is produced from productively infected cells at rate p per cell and is assumed either to infect new cells or to be cleared. In this model, loss of virus by cell infection is included in the clearance process, and virus is assumed to be cleared by all mechanisms at rate c per virion. Last, therapy is assumed to act by partially blocking virion production, with effectiveness ε, and by reducing the rate of infection, with effectiveness η, where ε and η vary between 0 and 1, and 1 means 100% effectiveness. The mathematical equations that describe this model are as follows:

$$\frac{dT}{dt} = s - d_T T - (1 - \eta)\beta VT$$

$$\frac{dI}{dt} = (1 - \eta)\beta VT - \delta I \qquad (33.1)$$

$$\frac{dV}{dt} = (1 - \varepsilon)pI - cV$$

2.1. Interferon's Mode of Action Against HCV

Neumann et al. *(13)* observed that the decline in HCV RNA levels during the first 14 days of daily IFN treatment with doses of 5, 10, and 15 million-IU IFN-α2b is biphasic and dose dependent; higher doses tended to produce greater declines. They solved Equation [33.1], assuming that early in treatment the number of target cells (T) remains constant. If IFN is assumed to block production of new virions with 100% efficacy ($\varepsilon = 1$), then virus decays exponentially at the virion clearance rate (c), and eventually virus is eradicated, but the experimental HCV RNA data reveal a two-phase exponential decay, that is, a biphasic linear decay when the log of viral load is plotted against time. If IFN treatment is assumed not to be 100% effective (i.e., if $0 < \varepsilon < 1$), and therefore some virus production is assumed to continue, then a biphasic viral decay is predicted by the model (**Fig. 33.3**).

On the other hand, if IFN is assumed to block viral infection with only efficacy $\eta > 0$, then the rate of decay of the virus reflects the death rate of productively infected cells (δ), and a single-phase viral decay is predicted *(13)*. Inclusion of the effect of IFN in blocking viral production again leads to the observed biphasic viral decay (13). Therefore, the conclusion drawn from the model is that the major mechanism of IFN action in HCV infection is to slow viral production ($\varepsilon > 0$). Indeed, recent studies using cell-culture models have shown that IFN blocks HCV replication *(19, 20)*.

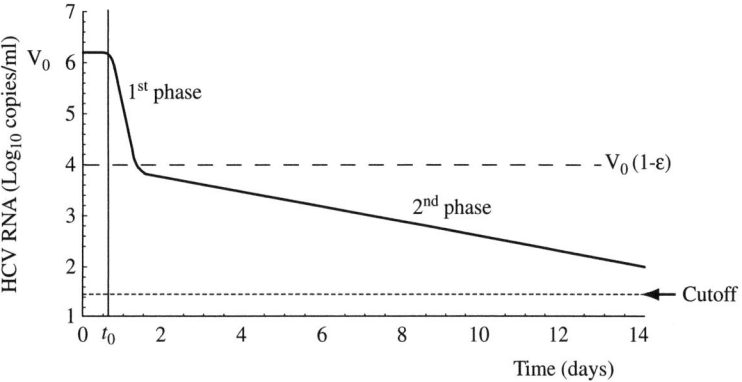

Fig. 33.3. Biphasic decline induced by high-dose daily IFN. After administration at time $t = 0$, the drug takes effect after a short delay, t_0. This delay is due both to the pharmacokinetics of the drug and to the need for IFN to signal through cell-surface receptors and cause induction of various IFN-stimulated genes. After t_0, the model, given by Equation [33.1], predicts that the HCV RNA will decline rapidly in a first phase from its initial level, V_0, by a factor $(1-\varepsilon)$, and then decline more slowly during a second phase.

The model therefore suggests that the first phase of HCV RNA decline is due to the partial blocking of HCV RNA production from cells infected before the beginning of therapy and unfettered viral clearance. As viral levels decline, the rate of de novo infection declines, and as infected cells are lost, they are not efficiently replaced. A slow net loss of infected cells occurs, therefore, and in the model, this net loss accounts for the second, slower phase of viral decline (**Fig. 33.3**).

2.2. Estimating the Effectiveness of IFN (ε) and the HCV RNA Clearance Rate (c)

If only the first few days of therapy are examined and on this time scale the infected cell number is assumed to remain close to its level before therapy, then the model (Equation [33.1]) predicts that viral load will decline from its baseline value, V_0, according to the following equation:

$$V(t) = V_0 \left[1 - \varepsilon + \varepsilon\, e^{-ct}\right] \tag{33.2}$$

This equation for the first phase of viral decline predicts that, after 24–48 h (i.e., at times long compared with $1/c$, the average free virion lifetime in serum), the viral load will decline to the value $(1 - \varepsilon)V_0$. Neumann et al. *(13)* fitted this equation to the experimental data collected over the first 2 days of treatment in each patient, and the parameters c and ε were estimated. For the 5 million–IU IFN dose, an average drug effectiveness, ε, of 81% was estimated, whereas the 10 and 15 million–IU IFN doses were much more effective and reduced viral production by approximately 95%. The viral half-life, that is, the time it takes for half the virus to be cleared from circulation, was estimated to be 2.7 h.

2.3. Viral and Infected-Cell Turnover

We noted that in most patients the viral infection system was at a steady state before treatment (i.e., target cells (T_0), infected cells (I_0), and virus (V_0) were at equilibrium), so Equation [33.1] implies that at baseline, viral production (pI_0) is in balance with viral clearance (cV_0). Because the viral clearance rate, c, and baseline viral load, V_0, measured in HCV RNA copies/ml, one can estimate the total amount of virus cleared per milliliter of serum per day by multiplying c by V_0. This number multiplied by total fluid volume of the patient gives the total amount of virus being cleared and, hence, produced daily. The calculated value is approximately 10^{12} virions per day *(13)*.

When the effects of therapy over periods longer than 2 days are considered, the assumption that the number of infected cells is constant must be relaxed. Solving Equation [33.1], assuming that target cells (T) are at a steady state during the first 14 days of therapy, and fitting the solution to the HCV RNA data produces an estimate of the death rate of infected cells *(13)*. Careful analyses *(3, 13, 18)* indicate that the slope of the second phase of viral decay is determined by the loss rate of infected cells (δ) and the effectiveness of the drug (ε). In particular, when total drug efficacy is close to 1, the slope of the final phase of viral decline is approximately the death rate of infected cells (δ), but for efficacies < 1 infection of new cells continues and a slower second-phase decline occurs, and the second-phase slope is only a minimal estimate of δ. Nonetheless, by fitting the model in Equation [33.1] or its various generalizations (see below) to the HCV RNA data, one can estimate δ.

There is no direct evidence that the parameter δ estimated from the second phase of HCV RNA decline is the loss rate of infected cells, but Neumann et al. *(13)* noted that δ was correlated with baseline alanine aminotransferase (ALT), a surrogate marker of liver-cell necrosis; the higher the initial ALT, the faster the turnover of infected liver cells and the larger the δ. This observation supports the conclusion that δ represents a preexisting cytotoxic cellular immune response, which had been suggested by others as necessary for successful IFN treatment *(21)*. An association between pretreatment ALT and δ has also been reported by Zeuzem et al. *(22)* in studies using pegylated IFN. In addition, an inverse correlation was found between baseline HCV RNA and δ. These findings support the notion that a cell-mediated immune response may be responsible for the loss of productively infected cells and that patients with high δ values have low viral load and high ALT levels. Further, the value of δ, estimated over the first 2 weeks of therapy as described above, was predictive of viral negativity at 12 weeks of therapy *(13)*. No patient that exhibited a $\delta < 0.1 \, \text{day}^{-1}$ achieved viral negativity (< 100 copies HCV RNA/ml) 12 weeks into treatment, but in many of these patients, δ could not be estimated because the second-phase viral decline was flat or viral rebound occurred.

2.4. Ribavirin's Mode of Action Against HCV

RBV, when given as monotherapy, has little or no antiviral activity against HCV *(23, 24)*, although in some patients a transient decline in viral load can be observed *(25)*. In combination with IFN, however, RBV produced substantially greater end-of-treatment response and SVR rates *(9, 26, 27)*. The antiviral mechanisms of RBV used in combination with IFN are still unknown. Both immunologic and direct antiviral effects, for example, lethal mutagenesis, have been suggested *(28–34)*. To determine how RBV can increase response rate, Dixit et al. *(14)* developed a viral dynamic model in which they assumed that, in a dose-dependent manner, RBV (alone or in combination with interferon) causes a fraction ρ of newly produced virions to be noninfectious, presumably through mutagenic action *(30, 32–35)*. Because the concentration of RBV increases during the first month of therapy, they also assumed that ρ increases with time in therapy, reaching saturation at 28 days *(36)*.

Analyses of this model showed that, when IFN effectiveness (ε) is high, RBV addition has a negligible influence on viral load, whereas when ε is low, RBV increases viral-load decay *(14)*. This effect is predicted to occur entirely in the second phase of decay of HCV RNA. These predictions are in agreement with recent clinical studies *(3, 17, 25)*, such as those observing that with 10 million–IU daily dosing of IFN, RBV addition does not affect HCV RNA decay in patients, such as Caucasians, with high values of ε but does increase second-phase decay in patients, such as African Americans, with low values of ε *(17)*. Because of the pharmacokinetics of PEG-IFN, its effectiveness is not high during the full 7-day dosing interval. The model predictions are therefore also in agreement with the observation by Herrmann et al. *(3)* of a faster final phase of viral decay in patients treated with PEG-IFN-α2a in combination with RBV than in those treated with PEG-IFN-α2a alone. Thus, the model reconciles the seemingly conflicting observations that RBV addition increases viral-load decline in some cases but not in others.

2.5. Pharmacokinetic and Pharmacodynamic Modeling of Interferon During Therapy

Some patients treated with PEG-IFN show a transient viral increase after an initial rapid HCV RNA decline. This increase tends to occur when serum levels of PEG-IFN are low, toward the end of the 7-day dosing interval *(37–41)*. A transient loss of IFN efficacy may therefore occur during phases of low IFN serum concentration. Indeed, Lam et al. *(42)* showed that, after a single dose of IFN, viral loads were decreased at 24 h but rebounded by 48 h. Because the pharmacokinetics of three-time-per-week standard IFN and once-weekly PEG-IFN are such that an assumption of constant IFN effectiveness may not be realistic, more recent models allow ε to vary. In a study where patients were treated with high-dose IFN induction followed by dose reduction, a model, in which ε was changed starting at the time of dose reduction from a level ε_1 to a lower level ε_2, provided a good fit to the viral-load data *(6)*. Others *(40,41)* allowed ε to depend on the varying IFN serum concentration, C, according to the pharmacodynamic model *(43)*.

$$\varepsilon(t) = \frac{C^n(t-\tau)}{EC_{50}^n + C^n(t-\tau)}, \qquad (33.3)$$

where the constant EC_{50} is the PEG-IFN concentration at which the drug's effectiveness in blocking viral production is half its maximum, n is a parameter called the Hill coefficient, and the time delay τ takes into account that IFN binds cellular receptors and initiates a signaling cascade before having an intracellular effect. **Figure 33.4** illustrates the relationship between PEG-IFN concentration and the drug's effectiveness as predicted by this pharmacodynamic model. This model was used to analyze kinetic data from patients coinfected with HCV and HIV who were treated with PEG-IFN-α2b plus RBV. Both HCV RNA and PEG-IFN concentrations were measured frequently after the first two weekly doses. The serum PEG-IFN concentrations were fitted to a pharmacokinetic model, and $C(t)$ was estimated. The Neumann et al. *(13)* model with $\eta = 0$ and $\varepsilon(t)$ chosen as in Equation [33.3] was then able to fit the end-of-the-week rebounds in HCV RNA seen in the patient data *(41)* (**Fig. 33.1C**).

Kinetic analysis of these data from 21 HCV/HIV-coinfected patients showed that drug concentration per se was not associated with long-term response but that features such as the maximum drug effectiveness computed from Equation [33.3] and

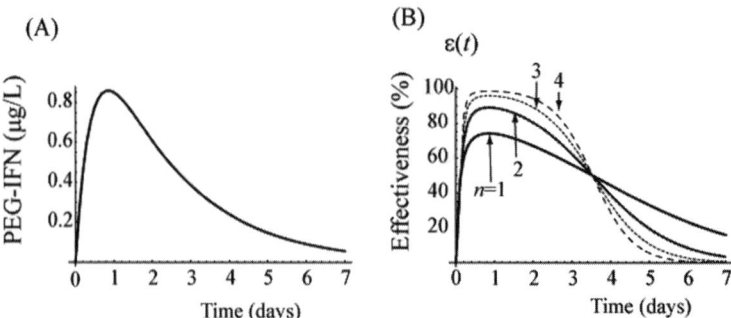

Fig. 33.4. **(A)** A typical profile of changes in PEG-IFN-α2b concentration with time after administration. **(B)** The corresponding drug effectiveness, given by Equation [33.3], with the effect of varying n, the Hill coefficient, illustrated. As n is increased, the effectiveness changes more abruptly.

the therapeutic quotient (the weekly average drug concentration divided by EC_{50}) were correlated with end-of-treatment response and SVR *(41)*. Rebounds have also been reported in patients chronically infected with HCV alone and treated with standard interferon, PEG-IFN-α2b or PEG- IFN-α2a, and may likewise be explained by the pharmacokinetics of the drug *(7, 37–39)*. We therefore believe the approach of incorporating pharmacokinetics and pharmacodynamics into viral dynamic models will also apply to PEG-IFN-α2a and possibly to other compounds.

The approach used by Talal et al. *(41)* is very useful as long as frequent simultaneous measurements are made of both PEG-IFN and HCV RNA, but commonly in clinical studies only HCV RNA is measured. Therefore, models that allow the effectiveness to vary but are independent of pharmacokinetic data, such as that of Bekkering et al. *(6)*, are needed. one such approach is to use the decreasing effectiveness model of shudo et al. *(51)*

2.6. The Critical Drug Efficacy: HCV Clearance or Flat Partial Response?

In the field of HIV therapy, the notion of a critical drug efficacy, ε_c, has been introduced *(44–46)*. If efficacy is not high enough, that is, is below a critical value, then theory predicts that HIV, rather than declining monotonically during therapy, will decline initially and then rebound, whereas if the efficacy is above a critical value, then viral levels will decay monotonically. Our model, Equation [33.1], in which target cell levels are allowed to vary with time, predicts similar behavior in the case of HCV infection. The critical efficacy is as follows:

$$\varepsilon_c = 1 - \frac{d_T c\delta}{sp\beta} = 1 - \frac{\bar{T}_0}{\bar{T}} \qquad (33.4)$$

where $\bar{T}_0/\bar{T}$ is the ratio of the number of uninfected hepatocytes in a chronically infected individual before treatment ($\bar{T}_0$) to the total number of hepatocytes in an uninfected individual ($\bar{T}$). We note that because $\bar{T} \approx \bar{T}_0 + \bar{I}_0$, the critical efficacy is also approximately equal to the fraction of infected hepatocytes, $\bar{I}_0/\bar{T}$. If a drug blocks both infection and viral production, then the total efficacy, defined as $\varepsilon_{tot} = 1 - (1 - \varepsilon)(1 - \eta)$, must be greater than the critical efficacy for continuous viral decay.

In the case of successful drug therapy $\varepsilon_{tot} > \varepsilon_c$, and the viral load will approach zero (**Fig. 33.1A and B**). Otherwise ($\varepsilon_{tot} < \varepsilon_c$), the system will converge to a new infected steady state with lower levels of virus and infected cells (**Fig. 33.1D**). The critical condition $\varepsilon_{tot} > \varepsilon_c$ is equivalent to the standard condition from epidemiology that the basic reproductive number R_0 is less than 1, where $R_0 = \frac{p\beta s}{c\delta d_T}$ *(44)*.

The notion of critical drug effectiveness explains why some patients do not clear virus. In particular, a flat partial response, that is, a rapid first phase of viral decay followed by a flat second phase, can result when drug effectiveness is too low *(13, 18)*.

2.7. Liver Cell Proliferation, Triphasic Viral Decay, and Ribavirin's Mode of Action

Hermann et al. *(3)* have suggested that the final phase of the triphasic viral decays observed in about 30–40% of patients treated with IFN or PEG-IFN alone or in combination with RBV *(3–6)* could be due to a delayed enhancement of an anti-HCV immune response by IFN or PEG-IFN. Further, they suggested that this immune enhancement is greater when RBV is included in the treatment regime, as the final-phase slope tended to be higher when RBV was present. Although this mechanism is certainly possible, the shoulder phase is difficult to explain in the Hermann et al. *(3)* and Neumann et al. *(13)* models, as it requires the assumption that the infected-cell loss rate, δ, is close to zero to generate a flat/shoulder phase. To address this difficulty, Dahari et al. *(18)* proposed an extension of the models of Neumann et al. *(13)* and Dixit et al. *(14)* that includes liver regeneration. In this model, any loss of hepatocytes due to infection is compensated for by the generation of new hepatocytes through proliferation *(47)*. In addition, because HCV infection is generally not cytopathic, the model allows both uninfected and infected hepatocytes to proliferate, albeit at different rates *(18)*.

Analyses of this extended model have shown that the flat second phase in flat partial responders (**Fig. 33.1D**) and in triphasic responses (**Fig. 33.1B**) may be a simple consequence of liver homeostasis, in which proliferation of hepatocytes compensates for the loss of infected cells *(18)*. In addition, the slope of the final-phase slope, that is, the second phase in biphasic declines and the third phase in triphasic declines, appears to be higher when ribavirin is included in the treatment regime *(3)*. This effect is accounted for in the model and is due to RBV's mutagenic effect *(14, 18)*. The authors noted that if, in the extended model, ribavirin is postulated to have a gradual effect in increasing either the loss rate of infected cells (δ) or the effectiveness of therapy (ε), then neither a shoulder phase nor a biphasic decline occurs; rather the second-phase slope slowly increases. This finding is of clinical relevance because as we move into the era of antivirals directed against HCV polymerase and protease *(48)*, the issue arises of whether ribavirin will still be needed, as it causes significant side effects, such as anemia *(49)*. If ribavirin does increase immune responses, then it might well be needed in future regimes, but if it plays a different role, say as a mutagen, as suggested by the Dixit et al. *(14)* and Dahari et al. *(18)* models, then it may not be needed in future regimes if they involve potent enough antivirals.

3. Fitting Models to HCV RNA Data

Many rigorous statistical approaches and numerical methods are available with which to fit models to data. Here we are concerned not with a full description of those but rather with explaining briefly the difficulties of fitting the models described above to data on HCV RNA decline under treatment. The main problem that usually arises is the relative sparseness of the data available. In fact, modeling and data fitting are generally only possible when the sampling protocol was designed from the start with such objectives in mind.

Appropriate sampling frequency depends on the overall goals of the research. The first tenet is that more data points are always better, but clinical and logistical constraints often dictate when and how much sampling is possible, and careful choices must be made. For example, if the main interest is to estimate the viral clearance rate constant, c, and the drug effectiveness in blocking viral production, ε, then multiple early samples are required. Often the patient is required to stay in the hospital at least for the first 48 h, and sampling could be done, as in Neumann et al. *(13)* at 0, 2, 4, 7, 10, 14, 19, and 24 h on the first day and at 5, 10, and 24 h on the second. Now that the fast turnover of HCV is well established, estimating c may be of less interest than estimating ε. In this case, less-intensive sampling during the first 2 days can be used, although enough samples are needed to discern the beginning and the end of the first phase. Therefore, if IFN were the drug being used, the samples at 2 and 4 h might be omitted, as we now know that a lag occurs before IFN induces an HCV RNA decline *(13)*, but if a new antiviral agent, such as protease inhibitor were being used, then these early samples should remain in the protocol. To yield information about the later phases of decline and to reveal whether the response is flat partial, biphasic, or triphasic, a longer schedule of HCV RNA measurements are needed, as in Herrmann et al. *(3)*, for example, days 0, 1, 2, 3, 7, 12, 15, 20, 25, 30, and 35. In a study employing PEG-IFN, and in particular PEG-IFN-α2b, HCV RNA levels have a substantial probability of rebounding toward the end of the weekly dosing interval, complicating the interpretation of HCV RNA kinetics. One approach is that suggested by Talal et al. *(41)*, in which the PEG-IFN concentration and viral load were simultaneously measured and appropriately modeled. The sampling schedule followed in that study was 0, 6, 12, and 24 h and on days 2, 3, 5, 6, and 7, and then after the second dose (on day 7) again at 6, 12, and 24 h and on days 9 and 14.

After the samples are acquired, HCV RNA should be measured for all samples at one time, so that assay variation is minimized. Software packages that perform nonlinear least-squares fitting are appropriate for fitting a viral kinetic model to the data from each individual patient. Another useful method is to use random effects models to fit all subjects simultaneously *(50)*. This methodology considers each patient to be a representative sample of the population of interest, and it estimates the average population parameters as well as the individual deviations from those *(2)*.

4. Conclusions

The application of rigorous quantitative analyses to experimental and clinical data on HCV infection has provided many insights into the dynamics of this infection. In particular, our understanding of viral turnover, viral-load decay under treatment, and the modes of action of different therapies have gained from application of the modeling techniques described here. As more information about HCV infection and the host response it generates becomes available, the mathematical models described here will need to be updated. The current models do not explicitly consider separate liver and blood compartments, nor do they attempt to describe the host immune response in detail, but the advances they have yielded are due to close collaborations between experimental or clinical researchers and modelers. We believe that this path, which has been so fruitful thus far, will continue to provide significant insights into HCV pathogenesis and treatment.

Acknowledgments

This research was performed under the auspices of the U.S. Department of Energy under contract DE-AC52-06NA25396 and supported by National Institutes of Health (NIH) grants RR06555 and AI28433 (to A.S.P). R.M.R. and H.D. were supported by grant P20-RR18754 from the National Center for Research Resources (NCRR), a component of the NIH.

References

1. Nguyen, T. T., Sedghi-Vaziri, A., Wilkes, L. B., Mondala, T., Pockros, P. J., Lindsay, K. L., et al. (1996) Fluctuations in viral load (HCV RNA) are relatively insignificant in untreated patients with chronic HCV infection. *J. Viral Hepat.* **3,** 75–78.

2. Neumann, A. U., Lam, N. P., Dahari, H., Davidian, M., Wiley, T. E., Mika, B. P., et al. (2000) Differences in viral dynamics between genotypes 1 and 2 of hepatitis C virus. *J. Infect. Dis.* **182,** 28–35.

3. Herrmann, E., Lee, J. H., Marinos, G., Modi, M., and Zeuzem, S. (2003) Effect of ribavirin on hepatitis C viral kinetics in patients treated with pegylated interferon. *Hepatology* **37,** 1351–1358.

4. Bergmann, C. C., Layden, J. E., Levy-Drummer, R. S., Layden, T. J., Haagmans, B. L., and Neumann, A. U. (2001) Clinical implications of a new tri-phasic model for hepatitis C viral kinetics during IFN-alpha therapy. *Hepatology* **34**, 345A.

5. Sentjens, R. E., Weegink, C. J., Beld, M. G., Cooreman, M. C., and Reesink, H. W. (2002) Viral kinetics of hepatitis C virus RNA in patients with chronic hepatitis C treated with 18 MU of interferon alpha daily. *Eur. J. Gastroenterol. Hepatol.* **14**, 833–840.

6. Bekkering, F. C., Neumann, A. U., Brouwer, J. T., Levi-Drummer, R. S., and Schalm, S. W. (2001) Changes in anti-viral effectiveness of interferon after dose reduction in chronic hepatitis C patients: a case control study. *BMC Gastroenterol.* **1**, 14.

7. Bekkering, F. C., Brouwer, J. T., Hansen, B. E., and Schalm, S. W. (2001) Hepatitis C viral kinetics in difficult to treat patients receiving high dose interferon and ribavirin. *J. Hepatol.* **34**, 435–440.

8. Fried, M. W., Shiffman, M. L., Reddy, K. R., Smith, C., Marinos, G., Goncales, F. L., Jr., et al. (2002) Peginterferon alfa-2a plus ribavirin for chronic hepatitis C virus infection. *N. Engl. J. Med.* **347**, 975–982.

9. Manns, M. P., McHutchison, J. G., Gordon, S. C., Rustgi, V. K., Shiffman, M., Reindollar, R., et al. (2001) Peginterferon alfa-2b plus ribavirin compared with interferon alfa-2b plus ribavirin for initial treatment of chronic hepatitis C: a randomised trial. *Lancet* **358**, 958–965.

10. Mihm, U., Herrmann, E., Sarrazin, C., and Zeuzem, S. (2006) Review article: predicting response in hepatitis C virus therapy. *Aliment. Pharmacol. Ther.* **23**, 1043–1054.

11. Wei, X., Ghosh, S. K., Taylor, M. E., Johnson, V. A., Emini, E. A., Deutsch, P., et al. (1995) Viral dynamics in human immunodeficiency virus type 1 infection. *Nature* **373**, 117–122.

12. Perelson, A. S., Neumann, A. U., Markowitz, M., Leonard, J. M., and Ho, D. D. (1996) HIV-1 dynamics in vivo: virion clearance rate, infected cell life-span, and viral generation time. *Science* **271**, 1582–1586.

13. Neumann, A. U., Lam, N. P., Dahari, H., Gretch, D. R., Wiley, T. E., Layden, T. J., et al. (1998) Hepatitis C viral dynamics in vivo and the antiviral efficacy of interferon-alpha therapy. *Science* **282**, 103–107.

14. Dixit, N. M., Layden-Almer, J. E., Layden, T. J., and Perelson, A. S. (2004) Modelling how ribavirin improves interferon response rates in hepatitis C virus infection. *Nature* **432**, 922–924.

15. Colombatto, P., Civitano, L., Oliveri, F., Coco, B., Ciccorossi, P., Flichman, D., et al. (2003) Sustained response to interferon-ribavirin combination therapy predicted by a model of hepatitis C virus dynamics using both HCV RNA and alanine aminotransferase. *Antivir. Ther.* **8**, 519–530.

16. Perelson, A. S., Herrmann, E., Micol, F., and Zeuzem, S. (2005) New kinetic models for the hepatitis C virus. *Hepatology* **42**, 749–754.

17. Layden-Almer, J. E., Ribeiro, R. M., Wiley, T., Perelson, A. S., and Layden, T. J. (2003) Viral dynamics and response differences in HCV-infected African American and white patients treated with IFN and ribavirin. *Hepatology* **37**, 1343–1350.

18. Dahari, H., Ribeiro, R. M., and Perelson, A. S. (2007) Triphasic decline of HCV RNA during antiviral therapy. *Hepatology* **46**, 16–21.

19. Guo, J. T., Bichko, V. V., and Seeger, C. (2001) Effect of alpha interferon on the hepatitis c virus replicon. *J. Virol.* **75**, 8516–8523.

20. Blight, K. J., Kolykhalov, A. A., and Rice, C. M. (2000) Efficient initiation of HCV RNA replication in cell culture. *Science* **290**, 1972–1974.

21. Nelson, D. R., Marousis, C. G., Davis, G. L., Rice, C. M., Wong, J., Houghton, M., and Lau, J. Y. (1997) The role of hepatitis C virus-specific cytotoxic T lymphocytes in chronic hepatitis C. *J. Immunol.* **158**, 1473–1481.

22. Zeuzem, S., Herrmann, E., Lee, J. H., Fricke, J., Neumann, A. U., Modi, M., et al. (2001) Viral kinetics in patients with chronic hepatitis C treated with standard or peginterferon alpha2a. *Gastroenterology* **120**, 1438–1447.

23. Di Bisceglie, A. M., Conjeevaram, H. S., Fried, M. W., Sallie, R., Park, Y., Yurdaydin, C., et al. (1995) Ribavirin as therapy for chronic hepatitis C. A randomized, double-blind, placebo-controlled trial. *Ann. Intern. Med.* **123**, 897–903.

24. Bodenheimer, H. C., Jr., Lindsay, K. L., Davis, G. L., Lewis, J. H., Thung, S. N., and Seeff, L. B. (1997) Tolerance and efficacy of oral ribavirin treatment of chronic hepatitis C: a multicenter trial. *Hepatology* **26**, 473–477.

25. Pawlotsky, J. M., Dahari, H., Neumann, A. U., Hezode, C., Germanidis, G., Lonjon, I., et al. (2004) Antiviral action of ribavirin in chronic hepatitis C. *Gastroenterology* **126**, 703–714.

26. Poynard, T., Marcellin, P., Lee, S. S., Niederau, C., Minuk, G. S., Ideo, G., et al. (1998) Randomised trial of interferon alpha2b plus ribavirin for 48 weeks or for 24 weeks versus interferon alpha2b plus placebo for 48 weeks for treatment of chronic infection with

hepatitis C virus. International Hepatitis Interventional Therapy Group (IHIT). *Lancet* **352,** 1426–1432.

27. McHutchison, J. G., Gordon, S. C., Schiff, E. R., Shiffman, M. L., Lee, W. M., Rustgi, V. K., et al. (1998) Interferon alfa-2b alone or in combination with ribavirin as initial treatment for chronic hepatitis C. Hepatitis Interventional Therapy Group. *N. Engl. J. Med.* **339,** 1485–1492.

28. Chung, R. T., He, W., Saquib, A., Contreras, A. M., Xavier, R. J., Chawla, A., et al. (2001) Hepatitis C virus replication is directly inhibited by IFN-alpha in a full-length binary expression system. *Proc. Natl. Acad. Sci. USA* **98,** 9847–9852.

29. Ning, Q., Brown, D., Parodo, J., Cattral, M., Gorczynski, R., Cole, E., et al. (1998) Ribavirin inhibits viral-induced macrophage production of TNF, IL-1, the procoagulant fgl2 prothrombinase and preserves Th1 cytokine production but inhibits Th2 cytokine response. *J. Immunol.* **160,** 3487–3493.

30. Crotty, S., Maag, D., Arnold, J. J., Zhong, W., Lau, J. Y., Hong, Z., Andino, R., and Cameron, C. E. (2000) The broad-spectrum antiviral ribonucleoside ribavirin is an RNA virus mutagen. *Nat. Med.* **6,** 1375–1379.

31. Young, K. C., Lindsay, K. L., Lee, K. J., Liu, W. C., He, J. W., Milstein, S. L., et al. (2003) Identification of a ribavirin-resistant NS5B mutation of hepatitis C virus during ribavirin monotherapy. *Hepatology* **38,** 869–878.

32. Contreras, A. M., Hiasa, Y., He, W., Terella, A., Schmidt, E. V., and Chung, R. T. (2002) Viral RNA mutations are region specific and increased by ribavirin in a full-length hepatitis C virus replication system. *J. Virol.* **76,** 8505–8517.

33. Lanford, R. E., Chavez, D., Guerra, B., Lau, J. Y., Hong, Z., Brasky, K. M., et al. (2001) Ribavirin induces error-prone replication of GB virus B in primary tamarin hepatocytes. *J. Virol.* **75,** 8074–8081.

34. Asahina, Y., Izumi, N., Enomoto, N., Uchihara, M., Kurosaki, M., Onuki, Y., et al. (2005) Mutagenic effects of ribavirin and response to interferon/ribavirin combination therapy in chronic hepatitis C. *J. Hepatol.* **43,** 623–629.

35. Dahari, H., Markatou, M., Zeremski, M., Haller, I., Ribeiro, R. M., Licholai, T., et al. (2007) Early ribavirin pharmacokinetics, HCV RNA and alanine aminotransferase kinetics in HIV/HCV co-infected patients during treatment with pegylated interferon and ribavirin. *J. Hepatol.* **47,** 23–30.

36. Glue, P. (1999) The clinical pharmacology of ribavirin. *Semin. Liver Dis.* **19**(Suppl. 1), 17–24.

37. Buti, M., Sanchez-Avila, F., Lurie, Y., Stalgis, C., Valdes, A., Martell, M., et al. (2002) Viral kinetics in genotype 1 chronic hepatitis C patients during therapy with 2 different doses of peginterferon alfa-2b plus ribavirin. *Hepatology* **35,** 930–936.

38. Formann, E., Jessner, W., Bennett, L., and Ferenci, P. (2003) Twice-weekly administration of peginterferon-alpha-2b improves viral kinetics in patients with chronic hepatitis C genotype 1. *J. Viral Hepat.* **10,** 271–276.

39. Levy-Drummer, R. S., Haagmans, B. L., Soulier, A., Germanidis, G., Lurie, Y., Herode, C., et al. (2004) Pharmacodynamic modeling of HCV kinetics during peg-interferon-alfa-2a (40KD) and ribavirin treatment of chronic hepatitis C genotype 1 patients in the DITTO-HCV study. *Hepatology* **40,** 390A.

40. Powers, K. A., Dixit, N. M., Ribeiro, R. M., Golia, P., Talal, A. H., and Perelson, A. S. (2003) Modeling viral and drug kinetics: hepatitis C virus treatment with pegylated interferon alfa-2b. *Semin. Liver Dis.* **23**(Suppl. 1), 13–18.

41. Talal, A. H., Ribeiro, R. M., Powers, K. A., Grace, M., Cullen, C., Hussain, M., et al. (2006) Pharmacodynamics of PEG-IFN alpha differentiate HIV/HCV coinfected sustained virological responders from nonresponders. *Hepatology* **43,** 943–953.

42. Lam, N. P., Neumann, A. U., Gretch, D. R., Wiley, T. E., Perelson, A. S., and Layden, T. J. (1997) Dose-dependent acute clearance of hepatitis C genotype 1 virus with interferon alfa. *Hepatology* **26,** 226–231.

43. Gabrielsson, J., and Weiner, D. (2000) *Pharmacokinetic and Pharmacodynamic Data Analysis: Concepts & Applications,* Swedish Pharmaceutical Society, Stockholm.

44. Callaway, D. S., and Perelson, A. S. (2002) HIV-1 infection and low steady state viral loads. *Bull. Math. Biol.* **64,** 29–64.

45. Wein, L. M., D'Amato, R. M., and Perelson, A. S. (1998) Mathematical analysis of antiretroviral therapy aimed at HIV-1 eradication or maintenance of low viral loads. *J. Theor. Biol.* **192,** 81–98.

46. Huang, Y., Rosenkranz, S. L., and Wu, H. (2003) Modeling HIV dynamics and antiviral response with consideration of time-varying drug exposures, adherence and phenotypic sensitivity. *Math. Biosci.* **184,** 165–186.

47. Dahari, H., Major, M., Zhang, X., Mihalik, K., Rice, C. M., Perelson, A. S., et al. (2005) Mathematical modeling of primary

hepatitis C infection: noncytolytic clearance and early blockage of virion production. *Gastroenterology* **128,** 1056–1066.

48. De Francesco, R., Tomei, L., Altamura, S., Summa, V., and Migliaccio, G. (2003) Approaching a new era for hepatitis C virus therapy: inhibitors of the NS3-4A serine protease and the NS5B RNA-dependent RNA polymerase. *Antiviral Res.* **58,** 1–16.

49. De Franceschi, L., Fattovich, G., Turrini, F., Ayi, K., Brugnara, C., Manzato, F., et al. (2000) Hemolytic anemia induced by ribavirin therapy in patients with chronic hepatitis C virus infection: role of membrane oxidative damage. *Hepatology* **31,** 997–1004.

50. Pinheiro, J. C., and Bates, D. M. (2002) *Mixed Effects Models in S and S-Plus,* Springer Verlag, New York.

51. Shudo, E., Riberio, R. M., Talal, A. H. and Perelson, A. S. (2008) A hepatitis C viral kinetic model that allows for time-varying drug effectiveness. Antiviral Ther. (in press).

INDEX

Printed in the United States of America